Hydroinformatics: Analysis and Modeling

Hydroinformatics: Analysis and Modeling

Editor: Roman Morris

Callisto Reference
www.callistoreference.com

Callisto Reference,
118-35 Queens Blvd., Suite 400,
Forest Hills, NY 11375, USA

Visit us on the World Wide Web at:
www.callistoreference.com

ISBN: 978-1-64116-609-6 (Hardback)

Cataloging-in-Publication Data

Hydroinformatics : analysis and modeling / edited by Roman Morris.
p. cm.
Includes bibliographical references and index.
ISBN 978-1-64116-609-6
1. Hydrology--Data processing. 2. Hydrology--Computer simulation. 3. Hydrologic models. I. Morris, Roman.
GB656.2.E43 H93 2022
551.480 285--dc23

Table of Contents

Preface VII

Chapter 1 **Attribution of satellite-observed vegetation trends in a hyper-arid region of the Heihe River basin, Western China** 1
Y. Wang, M. L. Roderick, Y. Shen and F. Sun

Chapter 2 **Multivariate bias adjustment of high-dimensional climate simulations: the Rank Resampling for Distributions and Dependences (R^2D^2) bias correction** 12
Mathieu Vrac

Chapter 3 **Attributing regional trends of evapotranspiration and gross primary productivity with remote sensing: a case study in the North China Plain** 33
Xingguo Mo, Xuejuan Chen, Shi Hu, Suxia Liu and Jun Xia

Chapter 4 **Retrospective forecasts of the upcoming winter season snow accumulation in the Inn headwaters (European Alps)** 49
Kristian Förster, Florian Hanzer, Elena Stoll, Adam A. Scaife, Craig MacLachlan, Johannes Schöber, Matthias Huttenlau, Stefan Achleitner and Ulrich Strasser

Chapter 5 **The development and evaluation of a hydrological seasonal forecast system prototype for predicting spring flood volumes in Swedish rivers** 65
Kean Foster, Cintia Bertacchi Uvo and Jonas Olsson

Chapter 6 **The WACMOS-ET project – Tower-scale evaluation of four remote-sensing-based evapotranspiration algorithms** 83
D. Michel, C. Jiménez, D. G. Miralles, M. Jung, M. Hirschi, A. Ershadi, B. Martens, M. F. McCabe, J. B. Fisher, Q. Mu, S. I. Seneviratne, E. F. Wood and D. Fernández-Prieto

Chapter 7 **Probabilistic inference of ecohydrological parameters using observations from point to satellite scales** 103
Maoya Bassiouni, Chad W. Higgins, Christopher J. Still and Stephen P. Good

Chapter 8 **Quantification of surface water volume changes in the Mackenzie Delta using satellite multi-mission data** 118
Cassandra Normandin, Frédéric Frappart, Bertrand Lubac, Simon Bélanger, Vincent Marieu, Fabien Blarel, Arthur Robinet and Léa Guiastrennec-Faugas

Chapter 9 **Evaluation of the ability of the Weather Research and Forecasting model to reproduce a sub-daily extreme rainfall event in Beijing, China using different domain configurations and spin-up times** 137
Qi Chu, Zongxue Xu, Yiheng Chen and Dawei Han

Chapter 10 **Use of reflected GNSS SNR data to retrieve either soil moisture or vegetation height from a wheat crop** 154
Sibo Zhang, Nicolas Roussel, Karen Boniface, Minh Cuong Ha, Frédéric Frappart, José Darrozes, Frédéric Baup and Jean-Christophe Calvet

Chapter 11 **High-resolution land surface modeling utilizing remote sensing parameters and the Noah UCM: a case study in the Los Angeles Basin** 172
P. Vahmani and T. S. Hogue

Chapter 12 **Geostatistical assessment of warm-season precipitation observations in Korea based on the composite precipitation and satellite water vapor data** 188
Sojung Park, Seon Ki Park, Jeung Whan Lee and Yunho Park

Chapter 13 **Assessment of the Weather Research and Forecasting (WRF) model for simulation of extreme rainfall events in the upper Ganga Basin** 206
Ila Chawla, Krishna K. Osuri, Pradeep P. Mujumdar and Dev Niyogi

Permissions

List of Contributors

Index

Preface

Every book is a source of knowledge and this one is no exception. The idea that led to the conceptualization of this book was the fact that the world is advancing rapidly; which makes it crucial to document the progress in every field. I am aware that a lot of data is already available, yet, there is a lot more to learn. Hence, I accepted the responsibility of editing this book and contributing my knowledge to the community.

The branch of informatics which deals with the application of information and communication technologies in addressing the problems related to the use of water is referred to as hydroinformatics. It primarily focuses on the problems of equitable and efficient use of water for various purposes. One of the closely related fields of hydroinformatics is computational hydraulics. Hydroinformatics primarily focuses on the numerical simulation of water flows and related processes. It makes use of the techniques of artificial intelligence such as artificial neural networks, support vector machines, etc. Hydroinformatics is a vast field which draws on various fields such as hydrology, hydraulics and environmental engineering. It is applied in urban drainage and water supply systems. This book consists of contributions made by international experts. The topics included herein on hydroinformatics are of utmost significance and bound to provide incredible insights to readers. This book will serve as a reference to a broad spectrum of readers.

While editing this book, I had multiple visions for it. Then I finally narrowed down to make every chapter a sole standing text explaining a particular topic, so that they can be used independently. However, the umbrella subject sinews them into a common theme. This makes the book a unique platform of knowledge.

I would like to give the major credit of this book to the experts from every corner of the world, who took the time to share their expertise with us. Also, I owe the completion of this book to the never-ending support of my family, who supported me throughout the project.

Editor

1

Attribution of satellite-observed vegetation trends in a hyper-arid region of the Heihe River basin, Western China

Y. Wang[1,2,3,4], M. L. Roderick[4,5,6], Y. Shen[1], and F. Sun[4,6]

[1]Key Laboratory of Agricultural Water Resources, Hebei Key Laboratory of Agricultural Water-Saving, Center for Agricultural Resources Research, Institute of Genetics and Developmental Biology, Chinese Academy of Sciences, Shijiazhuang, 050021, China
[2]University of Chinese Academy of Sciences, Beijing, 10049, China
[3]Research Center for Hebei Ecological Environmental Construction, Hebei Academy of Social Sciences, Shijiazhuang, 050051, China
[4]Research School of Biology, The Australian National University, Canberra, 0200, Australia
[5]Research School of Earth Science, The Australian National University, Canberra, 0200, Australia
[6]Australian Research Council Centre of Excellence for Climate System Science, Canberra, Australia

Correspondence to: Y. Shen (yjshen@sjziam.ac.cn)

Abstract. Terrestrial vegetation dynamics are closely influenced by both climate and by both climate and by land use and/or land cover change (LULCC) caused by human activities. Both can change over time in a monotonic way and it can be difficult to separate the effects of climate change from LULCC on vegetation. Here we attempt to attribute trends in the fractional green vegetation cover to climate variability and to human activity in Ejina Region, a hyper-arid landlocked region in northwest China. This region is dominated by extensive deserts with relatively small areas of irrigation located along the major water courses as is typical throughout much of Central Asia. Variations of fractional vegetation cover from 2000 to 2012 were determined using Moderate Resolution Imaging Spectroradiometer (MODIS) vegetation index data with 250 m spatial resolution over 16-day intervals. We found that the fractional vegetation cover in this hyper-arid region is very low but that the mean growing season vegetation cover has increased from 3.4 % in 2000 to 4.5 % in 2012. The largest contribution to the overall greening was due to changes in green vegetation cover of the extensive desert areas with a smaller contribution due to changes in the area of irrigated land. Comprehensive analysis with different precipitation data sources found that the greening of the desert was associated with increases in regional precipitation. We further report that the area of land irrigated each year can be predicted using the runoff gauged 1 year earlier. Taken together, water availability both from precipitation in the desert and runoff inflow for the irrigation agricultural lands can explain at least 52 % of the total variance in regional vegetation cover from 2000 to 2010. The results demonstrate that it is possible to separate the satellite-observed changes in green vegetation cover into components due to climate and human modifications. Such results inform management on the implications for water allocation between oases in the middle and lower reaches and for water management in the Ejina oasis.

1 Introduction

Terrestrial vegetation plays a key role in energy, water, and biogeochemical cycles, and changes in vegetation can also significantly influence atmospheric processes (Pielke et al., 1998; Gerten et al., 2004). Monitoring of terrestrial vegetation dynamics therefore underpins efforts to better understand the feedbacks between vegetation and the atmosphere (Bonan et al., 2003; Bounoua et al., 2010; Angelini et al., 2011). In particular, the Normalized Difference Vegetation Index (NDVI) derived from satellite observations of red and near-infrared reflectance has proven useful in assessing vegetation dynamics from regional to global scales (Tucker, 1979; Box et al., 1989; Fensholt et al., 2013).

Greening trends have been detected on global (Myneni et al., 1997; Nemani et al., 2003; Donohue et al., 2013; Fensholt et al., 2013) and regional (Fang et al., 2004; Herrmann et al., 2005; Donohue et al., 2009) scales, but attribution of those trends in terms of the underlying biophysical and socioeconomic causes remains a difficult task. The central challenge is that vegetation can change for a myriad of reasons including changes in the local climate (e.g., rainfall, radiation, temperature, humidity.) (Myneni et al., 1997; Goetz et al., 2005; Donohue et al., 2009), biogeochemistry (e.g., atmospheric CO_2, nutrient deposition.) (Lim et al., 2004; Bond et al., 2003; Donohue et al., 2013; Dirnböck et al., 2014), ecological processes (e.g. long-term successional recovery from disturbance, fire dynamics, disease.) (Thonicke et al., 2001; Bond et al., 2003), and via direct anthropogenic activity (e.g., land use change, irrigation, agriculture) (Hutchinson et al., 2000; Thonicke et al., 2001). One approach to handle this complexity is to use regional knowledge to constrain the problem.

In terms of the attribution of regional vegetation trends, the landscapes typical of Central Asia present a unique challenge. The region is hyper-arid with annual precipitation in many areas of often less than 50 mm, with hot summers and cold winters. What is of particular interest throughout Central Asia is the presence of many localized regions of irrigated agriculture that often support relatively large local populations. Those irrigation communities are usually located at oases that receive an annual input of water in the form of runoff from surrounding mountains. Much of the outflow from the mountains is recent precipitation (snowmelt) but a further complication of recent years is that glacier melt has augmented the snowmelt and increased inflow of water to many oases throughout Central Asia (Yao et al., 2004; Lioubimtseva and Henebry, 2009; Rahimov, 2009; Unger-Shayesteh et al., 2013). Time series of NDVI satellite imagery generally show greening trends over Central Asia (Fang et al., 2004; Mohammat et al., 2013). However, in terms of overall water resources management it is important to understand what caused the overall greening (or browning) trend. For example, did it arise because of an expansion (or contraction) of the area being irrigated? Or alternatively, the irrigated area might have remained more or less constant and any large-scale greening (or browning) trend in vegetation might be related to subtle yet detectable changes in vegetation cover in the extensive deserts of Central Asia. The management implications are quite different for those two scenarios and require a clear separation of these sources of variation.

In this paper, we investigate satellite-observed (MODIS) vegetation trends (NDVI) in a hyper-arid region of the Heihe River basin located in northwest China. The aim of this study is to test whether it is possible to separate the vegetation trends in a small relatively well-studied basin into components due to climate and due to changes in irrigation. We believe that the method may be widely applicable throughout Central Asia.

2 Data and methods

2.1 Study area

We examined part of the landlocked Heihe River basin in northwest China (40°20′–42°40′ N, 97°30′–101°45′ E) (Fig. 1). Our study area occupies the downstream (northern) part of the basin (Fig. 1b) and is serviced by the regional center, Ejina, which currently has a population of around 30 000 (http://www.geohive.com/cntry/cn-15.aspx). The hydroclimate of this predominantly desert environment is extreme. As an indication, at Ejina, the mean annual temperature is around 8 °C but daytime excursions in the summer reach 42 °C while nighttime temperatures drop to −36 °C during winter (Y. C. Zhang et al., 2011). The mean annual precipitation over the extensive flatlands is typically less than 50 mm (Fig. 1a) while the mean annual pan evaporation is typically around 3500 mm (Jin et al., 2010; Jia et al., 2011; Wang et al., 2013b). Agriculture is only possible via irrigation that is located immediately adjacent to the Heihe River (Fig. 1c). The study area ($\sim$ 80 000 km^2) is located within the broader Gobi Desert and also hosts the second largest area of *Populus euphratica* and *Haloxylon ammodendron* forests in China. The basin is generally considered to be the main eco-barrier in northern China (Fu et al., 2007; Qin et al., 2012).

The Heihe River (Fig. 1b) is the second longest inland river in China and is the sole river flowing through the Ejina Region (Guo et al., 2009). This river originates in the Qilian Mountains. After reaching the mountain outlet at the Yingluoxia hydrological gauge station (Fig. 1b), it flows through several oases (Zhangye, Gaotai, Dingxin, Ejina) before terminating at the East and West Juyan lakes. Zhengyixia station is located downstream of those main oases, where most water is consumed for agriculture. The discharge at Zhengyixia typically peaks around September each year while the growing season extends from April to October (Fig. 2a). Consequently, the irrigated crops in the northern parts of the basin use irrigation water that was discharged from the mountains some 6 months earlier.

The river discharge from the mountain regions has shown an increasing trend in past decades. Annual discharge observed at Yingluoxia site increased to 15.7×10^8 m^3 in the 1990s from around 14.4×10^8 m^3 in the 1960s (Fig. 2b). However, the discharge observed at Zhengyixia station located at the place after the river flowing through the oases decreased from around 10.5×10^8 m^3 in the 1960s to around 7.5×10^8 m^3 in the 1990s. The increasing water withdrawal in the upper and middle reaches since the 1960s was associated with increased irrigation (and associated industrial development and urbanization) that made significant reductions

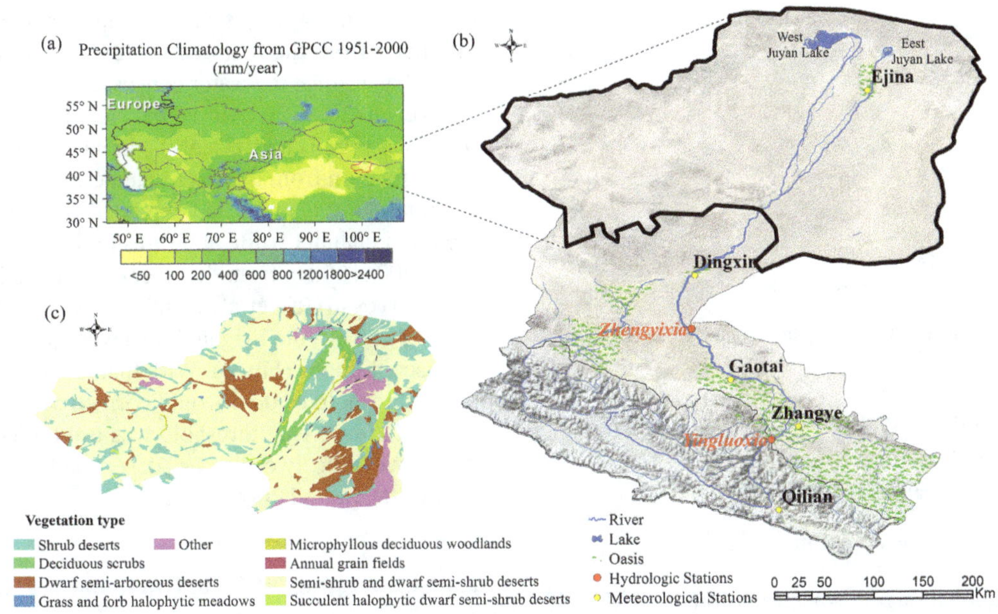

Figure 1. Details of the study area. Regional setting and the mean annual precipitation (1951–2000) **(a)**. Landscape of Heihe River basin with location of meteorological and hydrologic observation sites **(b)**. Vegetation map of lower Heihe River basin where the dashed line denotes the bounds of the possible irrigation area **(c)**.

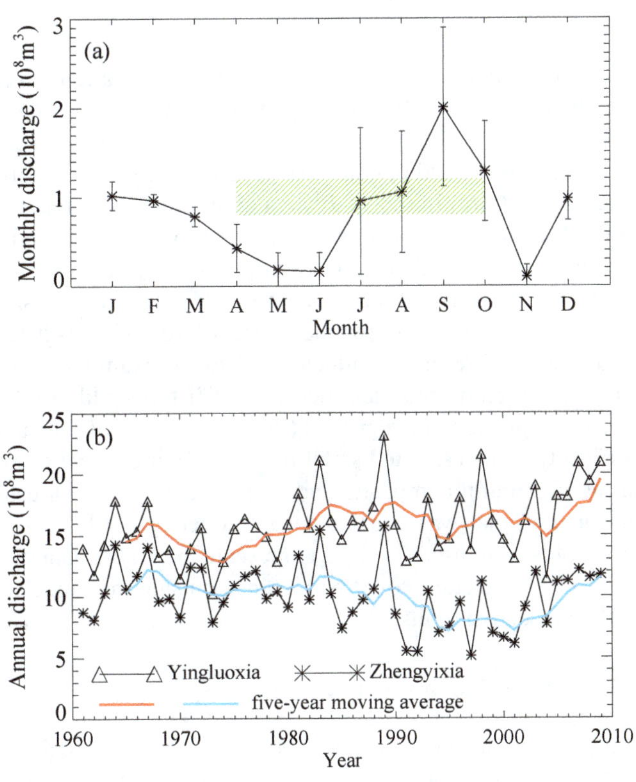

Figure 2. River flows in the Heihe River basin. Mean monthly river discharge (bars indicate standard deviation) of Heihe River (2000–2009) at Zhengyixia station in relation to the growing season (diagonal stripes) **(a)**. Annual discharge of Heihe River (1961–2009) at Yingluoxia and Zhengyixia stations **(b)**.

in river flows to the downstream oases and accelerated desertification in the northern parts of the basin (Guo et al., 2009; Jin et al., 2010). This phenomenon resulted in the drying up of East Juyan Lake in 1992 and the drying up of West Juyan Lake even earlier (Guo et al., 2009).

To restore the ecosystem of the downstream Heihe River basin the Ecological Water Conveyance Project (EWCP) was launched by the Chinese government. Water use has been regulated (reduced) since around the year 2000 in the middle parts of the basin, thereby delivering more water to the terminal lakes in the northern extremities of the basin (Y. C. Zhang et al., 2011). In the past decade (2000–2009) the average flow at Zhengyixia has increased to levels (about $10.5 \times 10^8 m^3$) not seen since the 1960s (Y. C. Zhang et al., 2011; Qin et al., 2012). One aim was to reduce degradation of the ecological environment in the northern extremities. Since 2000, an increase in native vegetation growth and species diversity has been attributed to increased groundwater recharge from the increased flows making their way into the northern parts of the basin (Jin et al., 2010; Jia et al., 2011).

In summary, the basin is a classic source–sink system with water sourced (via snowmelt and glacier melt) in the humid mountains in the south that subsequently flows northwards to terminal sinks at the East and West Juyan lakes. With this background, we note that many studies have reported trends in vegetation in particular subregions of the basin (Jin et al., 2010; Jia et al., 2011; Wang et al., 2011) but there has yet to be a comprehensive assessment of vegetation trends in the study area. A basin-wide assessment that is useful for hydrologic management requires separation of the overall

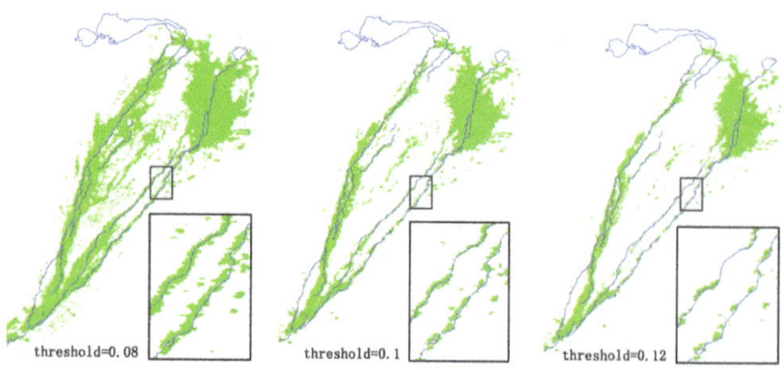

Figure 3. Irrigated areas derived using different NDVI thresholds.

vegetation trend into a component due to irrigation and a component due to changes in the desert vegetation. That is the aim of the current study.

2.2 MODIS satellite observations

Moderate Resolution Imaging Spectroradiometer (MODIS) Terra vegetation indices (MOD13Q1) data were acquired from the National Aeronautics and Space Administration (NASA) Earth Observing System Data and Information System (http://reverb.echo.nasa.gov) with spatial resolution of 250 m and temporal resolution of 16 days between April 2000 and December 2012. We used the Savitzky–Golay filter (Chen et al., 2004) to minimize noise in the NDVI series prior to further processing. Exploratory analysis highlighted anomalously low NDVI values during many of the winter months that coincided with snow/ice cover. To avoid those anomalous values we restricted the time series to cover the 7-month growing season (April–October).

2.3 Identifying the irrigated areas

The irrigation regions of interest are restricted to the immediate vicinity of the Heihe River (within the dashed line in Fig. 1c). Within that zone, irrigation is usually supplied by the extraction of groundwater and it is very difficult to distinguish agricultural vegetation from native vegetation that is drawing upon groundwater reserves in the satellite imagery. From the point of view of water resource management, both vegetation types use the same groundwater resources and we made no distinction between them. That enabled us to use a simple threshold approach to identify the vegetated areas of interest because they have much higher green vegetation cover during the April–October growing season.

To identify the irrigated area we created a composite image for each year (2000–2012) showing the maximum NDVI recorded during the April–October growing period. We first estimated the mean of the maximum NDVI over adjacent desert regions and found an NDVI of 0.0996 (±0.024). On that basis we initially defined the irrigated areas as being within the river zone (Fig. 1c) and having an annual maximum NDVI greater than 0.10. To test that threshold we used field surveys showing that irrigated vegetation (that includes native vegetation accessing groundwater) can exist up to a kilometer from the main east river channel in a central part of the basin since the water conveyance project was launched (Guo et al., 2009). We varied the NDVI threshold (0.08, 0.10, 0.12; see Fig. 3) and visually estimated the lateral extent of the vegetation from the river channel. At a threshold of 0.08, the implied irrigation area extended further than 1 km from the main channel while at a threshold of 0.10 the extent was some 200–1000 m from the river channel and consistent with field surveys (Guo et al., 2009). When the NDVI threshold was set at 0.12, the irrigated area was (incorrectly) shown to be discontinuous (Fig. 3). With that result we were confident that a threshold of 0.10 would correctly identify irrigated areas as defined. That threshold was used to classify the basin land cover into two classes, desert and irrigation, for each year of the period 2000–2012.

2.4 Converting NDVI to fractional vegetation cover

Fractional vegetation cover (f_V) was computed from NDVI (V) using a simple linear scaling (Carlson and Ripley, 1997):

$$f_V = (V - V_{min}) / (V_{max} - V_{min}), \tag{1}$$

where V_{min} and V_{max} represent zero green vegetation cover (i.e., bare soil, $f_V = 0$) and complete vegetation cover ($f_V = 1$), respectively. We assume that there are regions of bare soil (e.g., desert) and of complete vegetation cover (e.g., irrigated agriculture) in the study area of sufficient size relative to the MODIS spatial resolution (250 m) to define the limits of our scaling. To identify those limits we first composited the annual (April–October) maximum V image into a single maximum V image for the entire 13-year (2000–2012) study period. We then conducted a detailed examination of the desert regions and identified an NDVI threshold of 0.05 that was equated to bare ground ($f_V = 0$). To identify the upper limit, we investigated small regions of agricultural crops in the maximum composite and identified an NDVI threshold of 0.65 that was equated to full cover ($f_V = 1$). Those thresholds were used (in Eq. 1) to rescale the NDVI data into fractional vegetation cover with values outside the range set to the respective limits.

2.5 Attribution of vegetation changes

With the region split into two land cover types, the regional fractional vegetation cover (f_V) is determined by fractional vegetation coverage of the irrigated (f_I) and non-irrigated (f_D) areas and the respective areas (A_I, A_D) for each year,

$$f_V = \frac{A_I * f_I + A_D * f_D}{A_I + A_D}. \tag{2a}$$

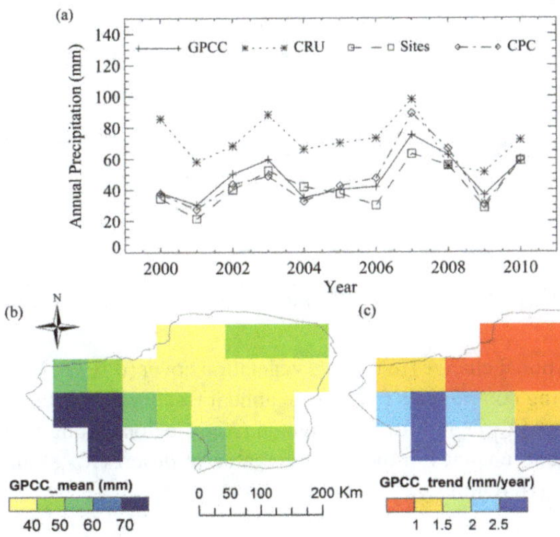

Figure 4. Spatiotemporal change of precipitation for the study area. **(a)** Annual precipitation for four different data sources (as indicated)**(a)**; Mean annual precipitation from GPCC **(b)**, and the trend from 2000 to 2010 **(c)**.

Defining the area fractions $A_{\mathrm{I}}^{*}\left(=\frac{A_{\mathrm{I}}}{A_{\mathrm{I}}+A_{\mathrm{D}}}\right)$ and $A_{\mathrm{D}}^{*}\left(=\frac{A_{\mathrm{D}}}{A_{\mathrm{I}}+A_{\mathrm{D}}}\right)$ with $A_{\mathrm{I}}^{*}+A_{\mathrm{D}}^{*}=1$, we rewrite Eq. (2a) as

$$f_{\mathrm{V}}=A_{\mathrm{I}}^{*} * f_{\mathrm{I}}+A_{\mathrm{D}}^{*} * f_{\mathrm{D}}. \tag{2b}$$

The total differential $\mathrm{d}f_{\mathrm{V}}$ is

$$\begin{aligned}\mathrm{d}f_{\mathrm{V}} &= \frac{\partial f_{\mathrm{V}}}{\partial f_{\mathrm{I}}}\mathrm{d}f_{\mathrm{I}}+\frac{\partial f_{\mathrm{V}}}{\partial A_{\mathrm{I}}^{*}}\mathrm{d}A_{\mathrm{I}}^{*}+\frac{\partial f_{\mathrm{V}}}{\partial f_{\mathrm{D}}}\mathrm{d}f_{\mathrm{D}}+\frac{\partial f_{\mathrm{V}}}{\partial A_{\mathrm{D}}^{*}}\mathrm{d}A_{\mathrm{D}}^{*},\\ &= A_{\mathrm{I}}^{*}\,\mathrm{d}f_{\mathrm{I}}+f_{\mathrm{I}}\,\mathrm{d}A_{\mathrm{I}}^{*}+A_{\mathrm{D}}^{*}\,\mathrm{d}f_{\mathrm{D}}+f_{\mathrm{D}}\,\mathrm{d}A_{\mathrm{D}}^{*}.\end{aligned} \tag{3}$$

The relative change in f_{V} is given by

$$\begin{aligned}\frac{\mathrm{d}f_{\mathrm{V}}}{f_{\mathrm{V}}} &= \frac{A_{\mathrm{I}}^{*}f_{\mathrm{I}}}{f_{\mathrm{V}}}\frac{\mathrm{d}f_{\mathrm{I}}}{f_{\mathrm{I}}}+\frac{f_{\mathrm{I}}}{f_{\mathrm{V}}}\mathrm{d}A_{\mathrm{I}}^{*}+\frac{A_{\mathrm{D}}^{*}f_{\mathrm{D}}}{f_{\mathrm{V}}}\frac{\mathrm{d}f_{\mathrm{D}}}{f_{\mathrm{D}}}+\frac{f_{\mathrm{D}}}{f_{\mathrm{V}}}\mathrm{d}A_{\mathrm{D}}^{*}\\ &= X_{f_{\mathrm{I}}}+X_{A_{\mathrm{I}}}+X_{f_{\mathrm{D}}}+X_{A_{\mathrm{D}}}.\end{aligned} \tag{4}$$

where the various X terms on the right-hand side denote the total change in f_{V} due to changes in the greenness ($X_{f_{\mathrm{I}}}, X_{f_{\mathrm{D}}}$) and fractional area ($X_{A_{\mathrm{I}}}, X_{A_{\mathrm{D}}}$).

2.6 Estimates of water availability

The vegetation trends are ultimately compared to estimates of trends in water availability over the desert and in the irrigation area. We used the monthly discharge gauged at Zhengyixia (Fig. 1b) as a measure of the inflow available for irrigation in Ejina. Over the desert parts of the region, precipitation represents the only input of water. To estimate water availability via precipitation we used three gridded databases (0.5° × 0.5°, monthly, 2000–2010) from the Global Precipitation Climatology Centre (GPCC) (Schneider et al., 2008), Climatic Research Unit (CRU) TS 3.10 (Harris et al., 2013), and the Climate Prediction Center (CPC) (Chen et al., 2002). We also averaged the data from two local meteorological sites (Ejina, Dingxin; see Fig. 1b) as a further check on the gridded databases. Initial analysis showed that precipitation (P) was generally a little higher in the CRU database (but with similar interannual variability) while the other two remaining databases gave similar spatial patterns, variability, and trends (Fig. 4 and Supplement Fig. S1). We were most familiar with the GPCC database following previous work (Sun et al., 2012) and subsequently adopted the GPCC database as the precipitation record for the study area (Fig. 4b, c). Note that final interpretations and our conclusions are not sensitive to the choice of precipitation database and we also present the complete analysis using the other spatial databases (CRU, sites, and CPC) in the Supplement.

3 Results

3.1 Vegetation trends

The fractional vegetation cover f_{V} in this hyper-arid region is very low, with mean growing season f_{V} of about 3–4 %. The oases systems are clearly distinguished by the much higher vegetation cover (Fig. 5a and Supplement Fig. S2). Over the 13-year period (2000–2012) the mean growing season fractional vegetation cover (f_{V}) showed an increase trend overall, especially in the irrigated oasis (Fig. 5b). The annual maximal fractional vegetation cover changed similarly (Fig. S2 in the Supplement). The mean annual fractional vegetation for the whole region increased steadily starting at about 3.2 % in 2000 and rising to around to 4.5 % in 2012 (Fig. 5c).

The mean growing season fractional vegetation cover in the extensive desert regions (f_{D}) more or less tracked the changes in the regional total (f_{V}). The area classified as irrigated only occupies around 3 % of total study area with a relatively high growing season average vegetation cover f_{I} of around 17%. Over the 2000–2012 period, the fractional irrigation area (A_{I}^{*}) showed a steady, increase (from 3 to 4 %) and the mean growing season fractional vegetation cover (f_{I}) also increased from around 16 to 18 %.

3.2 Sensitivity analysis and trend attribution

3.2.1 Sensitivity

After substituting the relevant numerical values ($f_{\mathrm{V}}=0.039$, $f_{\mathrm{I}}=0.174$, $A_{\mathrm{I}}^{*}=0.033$, $f_{\mathrm{D}}=0.035$, $A_{\mathrm{D}}^{*}=0.967$) derived from the mean growing season (2000–2012) into Eq. (4), the relative change in f_{V} is

$$\frac{\mathrm{d}f_{\mathrm{V}}}{f_{\mathrm{V}}}=0.15\frac{\mathrm{d}f_{\mathrm{I}}}{f_{\mathrm{I}}}+4.45\mathrm{d}A_{\mathrm{I}}^{*}+0.85\frac{\mathrm{d}f_{\mathrm{D}}}{f_{\mathrm{D}}}+0.88\mathrm{d}A_{\mathrm{D}}^{*}. \tag{5}$$

The coefficients in Eq. (5) denote the different sensitivities to change in the overall regional vegetation cover. From that

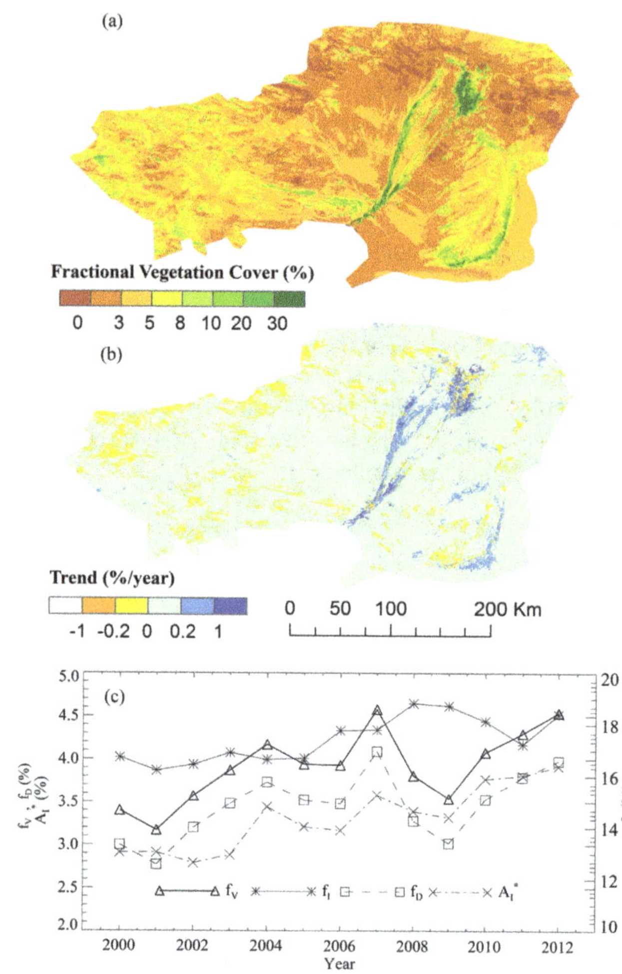

Figure 5. Spatiotemporal change of the growing season (April–October) mean annual fractional vegetation cover (2000–2012). Spatial pattern of growing season (April–October) mean annual fractional vegetation cover **(a)**; spatial trend of growing season (April–October) mean annual fractional vegetation cover (2000–2012) **(b)**; trends in the growing season (April–October) mean annual fractional vegetation for the whole region (f_V, left axis), desert (f_D, left axis), and irrigated (f_I, right axis) land cover classes, and for the fractional area of irrigation (A_I^*, left axis) **(c)**.

equation we infer that the relative change in fractional green vegetation is most sensitive to variations in the fractional irrigation area (A_I^*). Note that regional vegetation cover is also a factor of around 6 (0.85/0.15) times more sensitive to variations in greenness over the desert than over the irrigated regions because the desert land-cover type dominates the total area.

3.2.2 Trend attribution

From 2000 to 2012, the regional f_V increased by $\sim 25\,\%$. In terms of the underlying components, the causes of those changes varied from year to year (Fig. 6) but the largest contribution was generally due to changes in f_D (see X_{f_D}

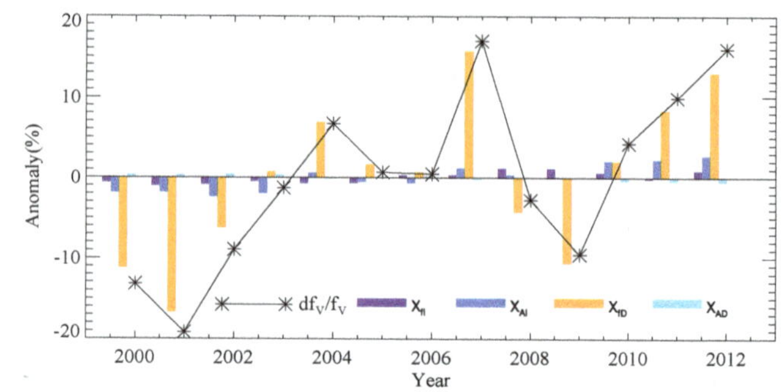

Figure 6. Annual changes in relative vegetation cover ($\mathrm{d}f_V/f_V$) and the underlying components from mean annual fractional vegetation for the desert (X_{f_D}) and for the irrigated (X_{f_I}) regions along with changes due to changes in the fractional area of desert (X_{A_D}) and irrigated (X_{A_I}), regions.

in Fig. 6) with a smaller contribution due to changes in A_I^* (X_{A_I} in Fig. 6). Variations in the remaining terms (X_{f_I}, X_{A_D}) had little impact on trends in the overall regional vegetation cover.

3.2.3 Uncertainty analysis

The major source of uncertainty in our results relates to the NDVI threshold used to distinguish the irrigated and desert regions (Sect. 2.3, Fig. 3). To evaluate the robustness of our results we also calculated the relative change in fractional vegetation cover using five different NDVI thresholds (0.08, 0.09, 0.10, 0.11, 0.12; Table 1). Those results show that while the numerical value of the sensitivity coefficients does change, the overall conclusion that regional vegetation cover is most sensitive to changes in irrigation area remains unchanged. Similarly, the attribution results for different thresholds show that most of the change in regional vegetation cover remained due to changes in desert greenness and in the area of irrigation.

3.3 Predicting regional vegetation cover based on water availability

The earlier results (Sects. 3.2.2, 3.2.3) show that the regional vegetation cover trend mainly depends on fractional vegetation cover over the extensive desert regions and the area of the irrigated lands. With that result, we approximate Eq. (4) as

$$\frac{\mathrm{d}f_V}{f_V} \sim \frac{f_I}{f_V}\mathrm{d}A_I^* + \frac{A_D^*}{f_V}\mathrm{d}f_D. \tag{6}$$

In this region, the area of land irrigated each year is dependent on the inflow at an earlier time while greenness in the desert areas is dependent on precipitation. To test this, we use the 12-month GPCC precipitation estimate to the end of the growing season (the previous November–October) to estimate the growing season desert vegetation cover. The results

Table 1. Sensitivity and attribution uncertainty with varied NDVI thresholds to define irrigated and non-irrigated regions (mean annual fractional vegetation for the whole region f_V, irrigated oasis f_I and desert f_D, fractional area of irrigation A_I^* and desert A_D^*, and their attribution $X_{f_I}, X_{A_I}, X_{f_D}, X_{A_D}$).

NDVI Threshold	Sensitivity	Attribution			
		X_{f_I}	X_{A_I}	X_{f_D}	X_{A_D}
0.08	$\frac{df_V}{f_V}=0.18\frac{df_I}{f_I}+2.97dA_I^*+0.82\frac{df_D}{f_D}+0.88dA_D^*$	14.0 %	16.8 %	74.1 %	−5.0 %
0.09	$\frac{df_V}{f_V}=0.16\frac{df_I}{f_I}+3.76dA_I^*+0.84\frac{df_D}{f_D}+0.88dA_D^*$	9.5%	19.4%	75.6%	−4.5%
0.1	$\frac{df_V}{f_V}=0.15\frac{df_I}{f_I}+4.45dA_I^*+0.85\frac{df_D}{f_D}+0.88dA_D^*$	7.8%	20.5%	75.8%	−4.1%
0.11	$\frac{df_V}{f_V}=0.14\frac{df_I}{f_I}+4.99dA_I^*+0.86\frac{df_D}{f_D}+0.89dA_D^*$	7.6%	20.3%	75.7%	−3.6%
0.12	$\frac{df_V}{f_V}=0.13\frac{df_I}{f_I}+5.42dA_I^*+0.87\frac{df_D}{f_D}+0.89dA_D^*$	7.3%	19.4%	76.4%	−3.2%

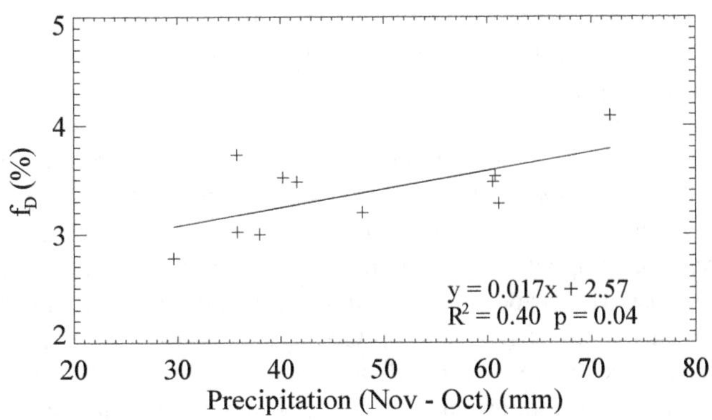

Figure 7. Relationship between growing season desert vegetation cover (f_D) and annual precipitation (per GPCC database 2000–2010).

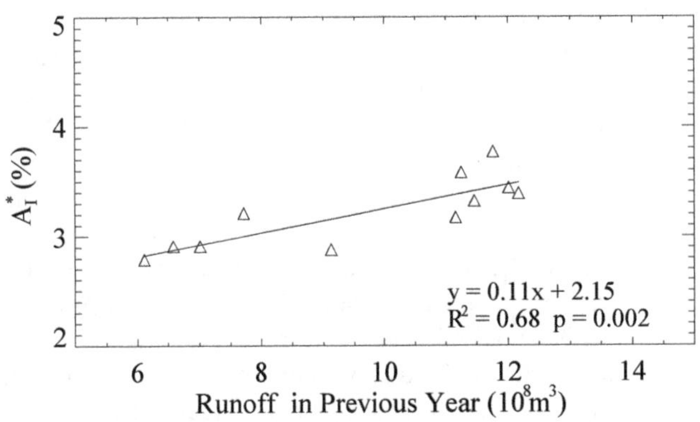

Figure 8. Relationship between fractional irrigated area A_I^* and runoff at Zhengyixia from the previous year 2000–2010.

show a positive relationship ($p < 0.05$, Fig. 7) and imply that the desert vegetation cover increases by 0.017 % for each additional mm of annual P. The analysis based on other precipitation databases (CRU, sites, and CPC) is included in the supporting materials (Fig. S3 and Table S1 both in the Supplement).

We sought a similar predictive relationship between the total runoff (R) in the previous calendar year at Zhengyixia and the fractional irrigated area (A_I^*) (Fig. 8). The results reveal a strong positive relationship ($p = 0.002$) where an increase in inflow at Zhengyixia of 1×10^8 m^3 will increase the fractional area of irrigation by around 0.1 %.

The results allow us to modify earlier expression by replacing dA_I^* with $\alpha\, dR$ (Fig. 8) and df_D with $\beta\, dP$ (Fig. 7), respectively,

$$\frac{df_V}{f_V} \sim \frac{f_I}{f_V}\alpha\, dR + \frac{A_D^*}{f_V}\beta\, dP. \tag{7}$$

Expressing that in a relative form we have

$$\frac{df_V}{f_V} \sim \frac{f_I}{f_V}\alpha R\frac{dR}{R} + \frac{A_D^*}{f_V}\beta P\frac{dP}{P}.$$

Taking the long-term mean annual values ($f_I = 0.17$, $f_V = 0.038$, $R = 9.67 \times 10^8$, $P = 47.5$, $A_D^* = 0.97$) and the empirical coefficients ($\alpha = 0.0011$, Fig. 8; $\beta = 0.00017$, Fig. 7) we have

$$\frac{df_V}{f_V} \sim 0.05\frac{dR}{R} + 0.21\frac{dP}{P} = X_R + X_P. \tag{8}$$

The empirically based equation predicts that a 1 % variation in runoff would lead to a 0.05 % variation in f_V while a 1 % variation in precipitation would increase f_V by 0.21 %. Finally, we use the runoff and precipitation data to estimate the relative changes in regional vegetation cover. The overall results show that the model developed here accounts for ca. 52 % of the total variance in regional vegetation cover (Fig. 9). It varied from 45 to 62 % depending on different precipitation databases (Fig. S4 in the Supplement.).

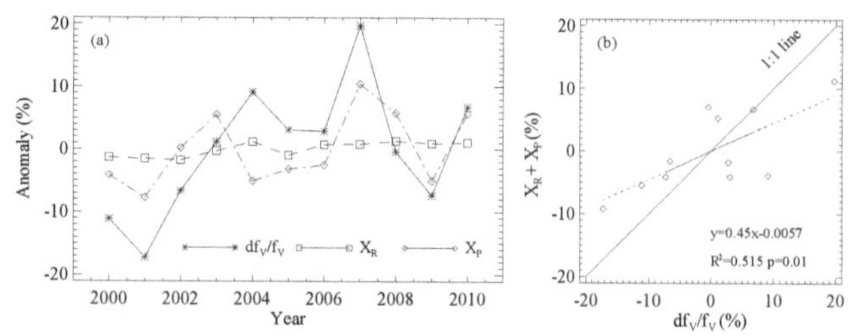

Figure 9. Relationship between regional vegetation cover and water availability. Relative variations **(a)**, and the observed annual changes in relative vegetation cover (df_V/f_V) versus predicted changes from water availability of runoff (X_R) and precipitation (X_P) **(b)**.

4 Discussion

In this study, we focused on a hyper-arid oasis–desert system where agricultural crops (artificial oasis) and groundwater-fed native vegetation (natural oasis) that occupy some 4 % of the entire region are concentrated along the Heihe River. As is well known, it is very hard to evaluate sparse desert vegetation cover in hyper-arid regions at a regional scale owing to coarse spatial and spectral resolution (Fensholt and Proud, 2012). However, the MODIS sensor has spectral bands that are specifically designed for vegetation monitoring and MODIS-based vegetation indices are known to perform well in discriminating vegetation differences in both sparsely and densely vegetated areas (Huete et al., 2002). Therefore, a vegetation index from MODIS with 250 m spatial resolution is likely to be appropriate for monitoring of fragmented landscapes of drylands, e.g. Central Asia (Dubovyk et al., 2013) and provides a potentially useful data source to evaluate regional vegetation change.

We found it necessary to restrict our analysis to the growing season to avoid noise apparently caused by snow and ice. With that pre-processing, our results showed that the mean growing season fractional vegetation cover (f_V) in Ejina showed a steady increase from ~3.2 % in 2000 and rising to ~4.5 % in 2012. The key question is what caused this change: the general climate variability or human-induced land use changes relating to irrigation?

We were able to identify the crops and green native vegetation along the river using the elevated NDVI signal during the April–October growing season, but we were unable to separate the crops from the native vegetation using the MODIS NDVI satellite data. Therefore, the entire region was split into two land cover types, denoted here as desert and irrigation oasis (that includes native vegetation along the river) for each year. Desert is distinguished from irrigated lands using a simple maximal NDVI threshold (Fig. 3).

Regional vegetation cover depends on both the desert and irrigated vegetation cover and their area fractions. Regional vegetation cover in the downstream of the Heihe River basin (2000–2012) is highly sensitive to variations in the area of irrigated land (Eq. 5). Over the whole period we found that the contributions to regional vegetation cover change due to changes in irrigated vegetation cover (f_I) and the area fraction (A_I^*) were 7.8 and 20.5 %, respectively, while changes in the non-irrigated vegetation cover (f_D) and area fraction (A_D^*) accounted for 75.8 %, and −4.1 %, respectively. Uncertainty analysis indicated that the fractional changes were not especially sensitive to the assumed NDVI threshold that was used to delineate desert from irrigated lands (Table 1). With that, our final result was that the relative vegetation change over the basin was most sensitive to changes in greenness of desert vegetation (~75 %, Table 1) and in the area of irrigation (~21 %, Table 1). The remaining terms (greenness of irrigated vegetation, area of desert) could be ignored.

To be able to prognostically estimate changes in relative vegetation cover, we sought an empirical relationship between desert greenness and precipitation (Fig. 7) and between the extent of irrigation and runoff in the previous year (Fig. 8). Gridded databases supplied a useful way to estimate the precipitation and its spatial distribution. Therefore, as expected, vegetation cover in the non-irrigated desert region (f_D) was strongly related to annual precipitation (Fig. 7) (Bhuiyan, 2008; D'Odorico and Porporato, 2006). The underlying basis of the aforementioned runoff–irrigation area relationship would be complex and would involve a lag because (i) farmers may anticipate future planting areas based on runoff from the previous years and (ii) the runoff recharges the local groundwater that is subsequently used by the local population (for irrigation) and by the native oasis vegetation. The lagged relationship between runoff and the area of irrigation may provide a useful empirical basis for forecasting and confirms the importance of managing human impact to achieve targeted improvements in the regional ecology.

It is difficult to discriminate between the effects of climate change and of human activities on regional vegetation change in arid regions (Zhou et al., 2013). In northwest China, previous work has suggested that precipitation is the most important factor (Ma and Frank, 2006) while other studies concluded that climate factors only played a small role, with the major cause of regional vegetation change being caused by agricultural activities (Kong et al., 2010; Zhou et al., 2013). Previous studies trying to separate the change into climate versus human causes have been limited to either qualitative distinctions (Dai et al., 2011), model estimation (C. X. Zhang et al., 2011; Zhou et al., 2013), and regression and residual approximation (Wang et al., 2012). To resolve those differences, we used a formal analytic framework to attribute the change of regional vegetation cover between climatic and human drivers.

The reason why we can use this method in this ecologically delicate and highly important area is that most human activities are focused around the irrigated oases that account for 3–5 % of the total area. This is typical of oasis–desert landscapes which dominate Central Asia, with widespread irrigation oases. In this system, the allocation of water resources is critical in achieving a balance among different oases as well as between human water appropriation for irrigation and ecological conservation. The overuse of water in the upper and middle reaches associated with increased irrigation made significant reductions in river flows to the downstream oases and accelerated desertification. A similar overuse of water for irrigation also happened in the Aral Sea. The water withdrawal for agricultural expansion (e.g. from about 4.5 Mha in 1960 to almost 7.9 Mha by 1999) led to a dramatic shrinkage of the Aral Sea that has attracted the attention of the international scientific community over the last few decades (Micklin, 1988; Whish-Wilson, 2002; Lioubimtseva et al., 2005). In the last few decades, runoff from mountains has shown a positive trend, with more precipitation and warmer climate (Unger-Shayesteh et al., 2013; Wang et al., 2013a). However, rational distribution and sustainable management of water resources is still a long-term and arduous task. Our results suggest that it is possible to use remotely sensed data to provide practical support in assessing the ecological status of irrigation regions that surround most Central Asian rivers.

5 Conclusions

We found that the regional fractional vegetation cover f_V in the downstream parts of the greater Heihe River basin increased by 25 % from 2000 to 2012. The largest contribution was due to a slight greening of the desert regions that was consistent with increased precipitation over the period. The other main contribution to the regional trend was an expansion of irrigated areas (including native vegetation dependent on groundwater) along the Heihe River that was found to be dependent on the runoff in the previous year. In conclusion, water availability both from precipitation and runoff can explain around 52 % of the total variance in regional vegetation cover over the period in this extremely arid environment. This study showed that it is feasible to separate the variations in regional vegetation cover that are due to changes in the climate from those due to changes in human activities given an appropriate regional context.

Acknowledgements. This study was supported by the National Program on Key Basic Research Project of China (2010CB951003) and the China Scholarship Council. The vegetation type data set is provided by Environmental and Ecological Science Data Center for West China, National Natural Science Foundation of China (http://westdc.westgis.ac.cn). We thank three anonymous reviewers for their insightful comments and suggestions that improved this manuscript.

Edited by: P. Saco

References

Angelini, I. M., Garstang, M., Davis, R. E., Hayden, B., Fitzjarrald, D. R., Legates, D. R., Greco, S., Macko, S., and Connors, V.: On the coupling between vegetation and the atmosphere, Theor. Appl. Climatol., 105, 243–261, doi:10.1007/s00704-010-0377-5, 2011.

Bhuiyan, C.: Desert vegetation during droughts: response and sensitivity, The International Archives of the Photogrammetry, Remote Sens. Spatial Inform. Sci., 37, 907–912, 2008.

Bonan, G. B., Levis, S., Sitch, S., Vertenstein, M., and Oleson, K. W.: A dynamic global vegetation model for use with climate models: concepts and description of simulated vegetation dynamics, Global Change Biol., 9, 1543–1566, doi:10.1046/j.1365-2486.2003.00681.x, 2003.

Bond, W., Midgley, G., and Woodward, F.: The importance of low atmospheric CO_2 and fire in promoting the spread of grasslands and savannas, Global Change Biol., 9, 973–982, doi:10.1046/j.1365-2486.2003.00577.x, 2003.

Bounoua, L., Hall, F. G., Sellers, P. J., Kumar, A., Collatz, G. J., Tucker, C. J., and Imhoff, M. L.: Quantifying the negative feedback of vegetation to greenhouse warming: A modeling approach, Geophys. Res. Lett., 37, L23701, doi:10.1029/2010gl045338, 2010.

Box, E. O., Holben, B. N., and Kalb, V.: Accuracy of the Avhrr Vegetation Index as a Predictor of Biomass, Primary Productivity and Net CO_2 Flux, Veget., 80, 71–89, doi:10.1007/Bf00048034, 1989.

Carlson, T. N. and Ripley, D. A.: On the relation between NDVI, fractional vegetation cover, and leaf area index, Remote Sens. Environ., 62, 241–252, doi:10.1016/S0034-4257(97)00104-1, 1997.

Chen, J., Jonsson, P., Tamura, M., Gu, Z. H., Matsushita, B., and Eklundh, L.: A simple method for reconstructing a high-quality NDVI time-series data set based on the Savitzky-Golay filter, Remote Sens. Environ., 91, 332–344, doi:10.1016/j.rse.2004.03.014, 2004.

Chen, M. Y., Xie, P. P., Janowiak, J. E., and Arkin, P. A.: Global land precipitation: A 50-yr monthly analysis based on gauge observations, J. Hydrometeorol., 3, 249–266, doi:10.1175/1525-7541(2002)003<0249:Glpaym>2.0.Co;2, 2002.

Dai, S. P., Zhang, B., Wang, H. J., Wang, Y. M., Guo, L. X., Wang, X. M., and Li, D.: Vegetation cover change and the driving factors over northwest China, J. Arid Land, 3, 25–33, doi:10.3724/sp.j.1227.2011.00025, 2011.

Dirnböck, T., Grandin, U., Bernhardt-Römermann, M., Beudert, B., Canullo, R., Forsius, M., Grabner, M. T., Holmberg, M., Kleemola, S., and Lundin, L.: Forest floor vegetation response to nitrogen deposition in Europe, Global Change Biol., 20, 429–440, doi:10.1111/gcb.12440, 2014.

D'Odorico, P. and Porporato, A.: Dryland ecohydrology, Springer, 2006.

Donohue, R. J., McVicar, T. R., and Roderick, M. L.: Climate-related trends in Australian vegetation cover as inferred from satellite observations, 1981-2006, Global Change Biol., 15, 1025–1039, doi:10.1111/j.1365-2486.2008.01746.x, 2009.

Donohue, R. J., Roderick, M. L., McVicar, T. R., and Farquhar, G. D.: Impact of CO_2 fertilization on maximum foliage cover across the globe's warm, arid environments, Geophys. Res. Lett., 40, 3031–3035, doi:10.1002/grl.50563, 2013.

Dubovyk, O., Menz, G., Conrad, C., Kan, E., Machwitz, M., and Khamzina, A.: Spatio-temporal analyses of cropland degradation in the irrigated lowlands of Uzbekistan using remote-sensing and logistic regression modeling, Environ. Monit. Assess., 185, 4775–4790, doi:10.1007/s10661-012-2904-6, 2013.

Fang, J. Y., Piao, S. L., He, J. S., and Ma, W. H.: Increasing terrestrial vegetation activity in China, 1982–1999, Sci. China Ser. C, 47, 229–240, doi:10.1360/03yc0068, 2004.

Fensholt, R. and Proud, S. R.: Evaluation of earth observation based global long term vegetation trends – Comparing GIMMS and MODIS global NDVI time series, Remote Sens. Environ., 119, 131–147, doi:10.1016/j.rse.2011.12.015, 2012.

Fensholt, R., Rasmussen, K., Kaspersen, P., Huber, S., Horion, S., and Swinnen, E.: Assessing Land Degradation/Recovery in the African Sahel from Long-Term Earth Observation Based Primary Productivity and Precipitation Relationships, Remote Sens., 5, 664–686, doi:10.3390/Rs5020664, 2013.

Fu, K., Chen, X. P., Liu, Q. G., and Li, C. H.: Land use and land cover changes based on remote sensing and GIS in Heihe River basin, China, IEEE International Symposium on Geoscience and Remote Sensing, 23–28 July, Barcelona, 1790–1793, 2007.

Gerten, D., Schaphoff, S., Haberlandt, U., Lucht, W., and Sitch, S.: Terrestrial vegetation and water balance – hydrological evaluation of a dynamic global vegetation model, J. Hydrol., 286, 249–270, doi:10.1016/j.jhydrol.2003.09.029, 2004.

Goetz, S. J., Bunn, A. G., Fiske, G. J., and Houghton, R.: Satellite-observed photosynthetic trends across boreal North America associated with climate and fire disturbance, Proc. Natl. Acad. Sci. USA, 102, 13521–13525, doi:10.1073/pnas.0506179102, 2005.

Guo, Q. L., Feng, Q., and Li, J. L.: Environmental changes after ecological water conveyance in the lower reaches of Heihe River, northwest China, Environ. Geol., 58, 1387–1396, doi:10.1007/s00254-008-1641-1, 2009.

Harris, I., Jones, P., Osborn, T., and Lister, D.: Updated high-resolution grids of monthly climatic observations–the CRU TS3. 10 Dataset, Int. J. Climatol., 34, 623–642, doi:10.1002/joc.3711, 2013.

Herrmann, S. M., Anyamba, A., and Tucker, C. J.: Recent trends in vegetation dynamics in the African Sahel and their relationship to climate, Global Environ. Chang., 15, 394–404, doi:10.1016/j.gloenvcha.2005.08.004, 2005.

Huete, A., Didan, K., Miura, T., Rodriguez, E. P., Gao, X., and Ferreira, L. G.: Overview of the radiometric and biophysical performance of the MODIS vegetation indices, Remote Sens. Environ., 83, 195–213, doi:10.1016/S0034-4257(02)00096-2, 2002.

Hutchinson, C. F., Unruh, J. D., and Bahre, C. J.: Land use vs. climate as causes of vegetation change: a study in SE Arizona, Global Environ. Chang., 10, 47–55, doi:10.1016/S0959-3780(00)00009-1, 2000.

Jia, L., Shang, H., Hu, G., and Menenti, M.: Phenological response of vegetation to upstream river flow in the Heihe Rive basin by time series analysis of MODIS data, Hydrol. Earth. Syst. Sci., 15, 1047–1064, doi:10.5194/hess-15-1047-2011, 2011.

Jin, X. M., Schaepman, M., Clevers, J., Su, Z. B., and Hu, G. C.: Correlation Between Annual Runoff in the Heihe River to the Vegetation Cover in the Ejina Oasis (China), Arid Land Res. Manag., 24, 31–41, doi:10.1080/15324980903439297, 2010.

Kong, W., Sun, O., Chen, Y., Yu, Y., and Tian, Z.: Patch-level based vegetation change and environmental drivers in Tarim River drainage area of West China, Landscape Ecol., 25, 1447–1455, doi:10.1007/s10980-010-9505-y, 2010.

Lim, C., Kafatos, M., and Megonigal, P.: Correlation between atmospheric CO_2 concentration and vegetation greenness in North America: CO_2 fertilization effect, Climate Res., 28, 11–22, doi:10.3354/cr028011, 2004.

Lioubimtseva, E. and Henebry, G. M.: Climate and environmental change in arid Central Asia: Impacts, vulnerability, and adaptations, J. Arid Environ., 73, 963–977, doi:10.1016/j.jaridenv.2009.04.022, 2009.

Lioubimtseva, E., Cole, R., Adams, J. M., and Kapustin, G.: Impacts of climate and land-cover changes in arid lands of Central Asia, J. Arid Environ., 62, 285–308, doi:10.1016/j.jaridenv.2004.11.005, 2005.

Ma, M. and Frank, V.: Interannual variability of vegetation cover in the Chinese Heihe River Basin and its relation to meteorological parameters, Int. J. Remote Sens., 27, 3473–3486, doi:10.1080/01431160600593031, 2006.

Micklin, P. P.: Desiccation of the Aral Sea – a Water Management Disaster in the Soviet-Union, Science, 241, 1170–1175, doi:10.1126/science.241.4870.1170, 1988.

Mohammat, A., Wang, X. H., Xu, X. T., Peng, L. Q., Yang, Y., Zhang, X. P., Myneni, R. B., and Piao, S. L.: Drought and spring cooling induced recent decrease in vegetation growth in Inner Asia, Agr. Forest. Meteorol., 178, 21–30, doi:10.1016/j.agrformet.2012.09.014, 2013.

Myneni, R. B., Keeling, C. D., Tucker, C. J., Asrar, G., and Nemani, R. R.: Increased plant growth in the northern high latitudes from 1981 to 1991, Nature, 386, 698–702, doi:10.1038/386698a0, 1997.

Nemani, R. R., Keeling, C. D., Hashimoto, H., Jolly, W. M., Piper, S. C., Tucker, C. J., Myneni, R. B., and Running, S. W.: Climate-driven increases in global terrestrial net primary production from 1982 to 1999, Science, 300, 1560–1563, doi:10.1126/science.1082750, 2003.

Pielke, R. A., Avissar, R., Raupach, M., Dolman, A. J., Zeng, X. B., and Denning, A. S.: Interactions between the atmosphere and terrestrial ecosystems: influence on weather and climate, Global Change Biol., 4, 461–475, doi:10.1046/j.1365-2486.1998.t01-1-00176.x, 1998.

Qin, D. J., Zhao, Z. F., Han, L. F., Qian, Y. P., Ou, L., Wu, Z. Q., and Wang, M. C.: Determination of groundwater recharge regime and flowpath in the Lower Heihe River basin in an arid area of Northwest China by using environmental tracers: Implications for vegetation degradation in the Ejina Oasis, Appl. Geochem., 27, 1133–1145, doi:10.1016/j.apgeochem.2012.02.031, 2012.

Rahimov, S.: Impacts of climate change on water resources in Central Asia, Documentos CIDOB. Asia, 33–56, 2009.

Schneider, U., Fuchs, T., Meyer-Christoffer, A., and Rudolf, B.: Global precipitation analysis products of the GPCC, Global Precipitation Climatology Centre (GPCC), DWD, Internet Publikation, 1–12, 2008.

Sun, F. B., Roderick, M. L., and Farquhar, G. D.: Changes in the variability of global land precipitation, Geophys Res. Lett., 39, L19402, doi:10.1029/2012gl053369, 2012.

Thonicke, K., Venevsky, S., Sitch, S., and Cramer, W.: The role of fire disturbance for global vegetation dynamics: coupling fire into a Dynamic Global Vegetation Model, Global Ecol. Biogeogr., 10, 661–677, 2001.

Tucker, C. J.: Red and Photographic Infrared Linear Combinations for Monitoring Vegetation, Remote Sens. Environ., 8, 127–150, doi:10.1016/0034-4257(79)90013-0, 1979.

Unger-Shayesteh, K., Vorogushyn, S., Farinotti, D., Gafurov, A., Duethmann, D., Mandychev, A., and Merz, B.: What do we know about past changes in the water cycle of Central Asian headwaters? A review, Global Planet. Change, Part A, 110, 4–25, doi:10.1016/j.gloplacha.2013.02.004, 2013.

Wang, P., Zhang, Y. C., Yu, J. J., Fu, G. B., and Ao, F.: Vegetation dynamics induced by groundwater fluctuations in the lower Heihe River Basin, northwestern China, J. Plant Ecol. UK, 4, 77–90, doi:10.1093/Jpe/Rtr002, 2011.

Wang, T., Sun, J. G., Han, H., and Yan, C. Z.: The relative role of climate change and human activities in the desertification process in Yulin region of northwest China, Environ. Monit. Assess., 184, 7165–7173, doi:10.1007/s10661-011-2488-6, 2012.

Wang, Y., Shen, Y., Chen, Y., and Guo, Y.: Vegetation dynamics and their response to hydroclimatic factors in the Tarim River Basin, China, Ecohydrology, 6, 927–936, doi:10.1002/eco.1255, 2013a.

Wang, Y., Feng, Q., Chen, L. J., and Yu, T. F.: Significance and Effect of Ecological Rehabilitation Project in Inland River Basins in Northwest China, Environ. Manage., 52, 209–220, doi:10.1007/s00267-013-0077-x, 2013b.

Whish-Wilson, P.: The Aral Sea environmental health crisis, Journal of Rural and Remote Environmental Health, 1, 29–34, 2002.

Yao, T. D., Wang, Y. Q., Liu, S. Y., Pu, J. C., Shen, Y. P., and Lu, A. X.: Recent glacial retreat in High Asia in China and its impact on water resource in Northwest China, Sci. China Ser. D, 47, 1065–1075, doi:10.1360/03yd0256, 2004.

Yu, J. and Wang, P.: Relationship between Water and Vegetation in the Ejina Delta, Bull. Chinese Acad. Sci., 26, 68–75, 2012.

Zhang, C. X., Wang, X. M., Li, J. C., and Hua, T.: Roles of climate changes and human interventions in land degradation: a case study by net primary productivity analysis in China's Shiyanghe Basin, Environ. Earth Sci., 64, 2183–2193, doi:10.1007/s12665-011-1046-4, 2011.

Zhang, Y. C., Yu, J. J., Wang, P., and Fu, G. B.: Vegetation responses to integrated water management in the Ejina basin, northwest China, Hydrol. Process., 25, 3448–3461, doi:10.1002/Hyp.8073, 2011.

Zhou, W., Sun, Z., Li, J., Gang, C., and Zhang, C.: Desertification dynamic and the relative roles of climate change and human activities in desertification in the Heihe River Basin based on NPP, J. Arid Land, 5, 465–479, doi:10.1007/s40333-013-0181-z, 2013.

2

Multivariate bias adjustment of high-dimensional climate simulations: the Rank Resampling for Distributions and Dependences (R^2D^2) bias correction

Mathieu Vrac

Laboratoire des Sciences du Climat et de l'Environnement (LSCE-IPSL, CNRS), Centre d'Etudes de Saclay, Orme des Merisiers, 91190 Gif-sur-Yvette, France

Correspondence: Mathieu Vrac (mathieu.vrac@lsce.ipsl.fr)

Abstract. Climate simulations often suffer from statistical biases with respect to observations or reanalyses. It is therefore common to correct (or adjust) those simulations before using them as inputs into impact models. However, most bias correction (BC) methods are univariate and so do not account for the statistical dependences linking the different locations and/or physical variables of interest. In addition, they are often deterministic, and stochasticity is frequently needed to investigate climate uncertainty and to add constrained randomness to climate simulations that do not possess a realistic variability. This study presents a multivariate method of rank resampling for distributions and dependences (R^2D^2) bias correction allowing one to adjust not only the univariate distributions but also their inter-variable and inter-site dependence structures. Moreover, the proposed R^2D^2 method provides some stochasticity since it can generate as many multivariate corrected outputs as the number of statistical dimensions (i.e., number of grid cell × number of climate variables) of the simulations to be corrected. It is based on an assumption of stability in time of the dependence structure – making it possible to deal with a high number of statistical dimensions – that lets the climate model drive the temporal properties and their changes in time. R^2D^2 is applied on temperature and precipitation reanalysis time series with respect to high-resolution reference data over the southeast of France (1506 grid cell). Bivariate, 1506-dimensional and 3012-dimensional versions of R^2D^2 are tested over a historical period and compared to a univariate BC. How the different BC methods behave in a climate change context is also illustrated with an application to regional climate simulations over the 2071–2100 period. The results indicate that the 1d-BC basically reproduces the climate model multivariate properties, 2d-R^2D^2 is only satisfying in the inter-variable context, 1506d-R^2D^2 strongly improves inter-site properties and 3012d-R^2D^2 is able to account for both. Applications of the proposed R^2D^2 method to various climate datasets are relevant for many impact studies. The perspectives of improvements are numerous, such as introducing stochasticity in the dependence itself, questioning its stability assumption, and accounting for temporal properties adjustment while including more physics in the adjustment procedures.

1 Introduction

Climate change impact studies aim to investigate and understand the consequences of the potential evolutions of the climate system. Impacts can be hydrological with changes in seasonal flows and water resources driven by precipitation changes (e.g., Schneider et al., 2013), agronomical with crop yields perturbed by heat stress and/or rainfall evolutions (e.g., Müller et al., 2010; Wheeler and von Braun, 2013) and ecological with plant and animal diversity (in terms of structures or spatial repartitions) modified by future climate conditions (e.g., Araújo and Rahbek, 2006; Tisseuil et al., 2012), among many others. The common point of those impact studies is that they use global (GCM) or regional climate model (RCM) simulations of different variables over future time periods according to some scenarios as inputs into impact models to project (e.g., hydrological, ecological) consequences of climate change. However, most of those climate simulations suffer from statistical biases with respect to observations – or more generally reference data. This means that some of their statistical properties, such as mean, variance, distribution or even temporal, spatial or inter-variable dependence structures may not be representative of what is observed in the reference dataset. Consequently, before employing climate simulations to feed an impact model, it is often mandatory to "bias correct" (or to "adjust") them in order to correct some of their statistical properties (e.g., Christensen et al., 2008; Muerth et al., 2013).

Over the last decade, most of the developed – and therefore applied – bias correction (BC) methods focused on the adjustment of the mean (e.g., Delta method, Xu, 1999), the variance (e.g., simple scaling adjustment, Berg et al., 2012) or more generally on the adjustment of the distribution (e.g., "quantile-mapping", Haddad and Rosenfeld, 1997). Bias adjustments of the whole distribution through quantile-mapping techniques have been quite popular since it allows for adjusting not only the mean and variance but also any quantile of the variable of interest. Hence, many variants have been proposed (e.g., Déqué, 2007; Michelangeli et al., 2009; Kallache et al., 2011; Tramblay et al., 2013; Vrac et al., 2016) and applied in different studies (e.g., Oettli et al., 2011; Colette et al., 2012; Tisseuil et al., 2012; Vigaud et al., 2013). Nevertheless, usually, those approaches only work in a univariate context, which means that they are designed to independently correct one variable at a time, for one location (e.g., grid cell) at a time. Therefore, if the marginal (i.e., univariate) distributions are generally improved, that is closer to the reference ones – even when the BC is used as a preliminary step to downscaling (e.g., Colette et al., 2012; Vrac and Vaittinada Ayar, 2017) – the inter-site and inter-variable dependence structures are usually conserved from the climate model simulations to be corrected. Indeed, 1d-BC methods preserving the ranks of the simulations – as it is the case for quantile-mapping approaches – will not correct the copula functions characterizing the dependencies between sites and/or between variables (e.g., Nelsen, 2006; Schölzel and Friederichs, 2008; Vrac et al., 2011; Bevacqua et al., 2017). Such a preservation of the model dependence can obviously cause some deficiencies in the subsequent impact studies that will use the 1-dimensional bias corrected simulations if the model copula function is far from that of the references. It is therefore crucial to adjust not only the marginal distributions of the climate simulations but also their multivariate dependence structures, which is the goal of the present study. A few multivariate methodologies have been proposed over the last few years (e.g., Bardossy and Pegram, 2012; Piani and Haerter, 2012; Mao et al., 2015; Vrac and Friederichs, 2015; Cannon, 2017; Dekens et al., 2017; Li et al., 2017). Most of these methods can be categorized into one of the two following approaches: the "marginal/dependence" correction approach and the "successive conditional" correction approach. The "marginal/dependence" BC methods (e.g., "matrix recorrelation" approach in Bardossy and Pegram, 2012; Vrac and Friederichs, 2015; Cannon, 2017; Li et al., 2017) separately correct the 1d-marginal distributions (e.g., one variable at one given location) and the dependence structure, usually under the form of the underlying copula function linking the different marginal distributions. Once those two components of the joint distribution have been corrected, they are reassembled to obtain adjusted data that respect both the univariate and multivariate dependencies. Although they also aim to adjust climate simulations in a multivariate fashion, the "successive conditional" BC methods (e.g., "sequential recorrelation" approach in Bardossy and Pegram, 2012; Piani and Haerter, 2012; Dekens et al., 2017) are based on a slightly different philosophy. They consist first of correcting one given variable (e.g., one variable at one specific location). Then, a second variable (e.g., another variable or another location) is corrected conditionally on the previously corrected variable. The procedure goes on successively for each dimension (variable/location), correcting variable n conditionally on previously corrected variables $(1, \ldots, n-1)$. However, this approach suffers from two main limitations. First, since at each step the correction is performed conditionally on previously corrected data, this reduces the number of data available for adjusting each simulation. Consequently, the higher the number of variables to correct, the fewer the number of data to perform the bias correction at each successive step, and therefore the less robust the correction. Second, the ordering of the variables in the successive corrections matters: different orderings generally produce different corrections with different qualitative results (e.g., in terms of multivariate properties; see Piani and Haerter, 2012; Vrac and Friederichs, 2015). For those reasons, the present study deals with the development of a multivariate BC method within the "marginal/dependence" approach. The proposed methodology relies on the "Empirical Copula – Bias Correction" (EC-BC) method (Vrac and Friederichs, 2015) and is intended to fill some of its weaknesses, mainly its lack of flexibility in terms of temporal properties as well as its deterministic aspect. Concerning the time-related weakness, it has to be noted that it is not possible to correct the multidimensional properties of the simulations without changing the rank sequence of the simulations. In other words, any multivariate BC method will necessarily modify the initial rank chronology of the simulated events. For example, the EC-BC method – belonging to the "marginal/dependence" correction family – allows for both the corrected 1d-distributions to evolve consistently with the modeled ones and to reproduce the dependence (copula) structure of the references. But the price for this reproduction is that the temporal sequence of the ranks of the corrected data is exactly that of the reference data over the calibration time period, even for an adjustment performed over a future time period (or more generally over a projection/correction time period different from the calibration one). Of course, there is no reason why the rank chronology should be the same. This also implies that this multivariate BC provides deterministic corrections, while some studies pointed out the need for stochastic corrections or at least the need for introducing some stochasticity and variability in the BC process (e.g., Wong et al., 2014; Mao et al., 2015; Volosciuk et al., 2017). Hence, the goals of this paper are

- to propose a multivariate BC (MBC) method for both multi-site and multi-variable simulations;
- to relax the temporal constraints of EC-BC on the corrected data ranks in order to let the climate model drive more temporal properties and their evolutions and therefore express its own temporal dynamics;
- to introduce some stochasticity in the MBC outputs, or at least to enable the proposed MBC method to provide multiple corrected scenarios.

The proposed method relies on a multivariate Rank Resampling for Distributions and Dependences bias correction and will here be referred to as R^2D^2. This paper is organized as follows: the reference and reanalysis datasets used in this study are presented in Sect. 2. The R^2D^2 method is then described in Sect. 3 after some reminders about the copula theory and the EC-BC approach. The design of experiments performed to evaluate R^2D^2 over a historical time period is presented in Sect. 4 and results are provided in Sect. 5. Section 6 displays an application of R^2D^2 to RCM simulations over a future time period. Conclusions, potential future developments and discussions are finally given in Sect. 7.

2 Reference and model data

To apply, investigate and evaluate the proposed R^2D^2 correction method, a reference dataset and a model dataset to be corrected is needed, as for any BC method. The reference data employed here are daily temperature and precipitation time series from the "Systeme d'Analyze Fournissant des Renseignements Atmosphériques à la Neige" (SAFRAN) reanalysis data (Quintana-Segui et al., 2008) over the southeast region of France ($2 - 7.5°$ E $\times$ $42 - 45°$ N) corresponding to 1506 continental grid cell with an approximate 8 km $\times$ 8 km spatial resolution. SAFRAN has been described, validated and employed in many studies (e.g., Quintana-Segui et al., 2008; Lavaysse et al., 2012; Vrac et al., 2012).

The ERA-Interim (hereafter ERA-I; Dee et al., 2011) daily reanalysis data with a 0.75° by 0.75° spatial resolution are used here as model data to be corrected. Temperature at 2 m (hereafter T2) and precipitation data (hereafter PR) have been extracted for the same spatial domain as for the reference data. The time period from 1 January 1980 to 31 December 2009 is retained for both reference and ERA-I data. Then, each ERA-I grid cell is first regridded by a simple nearest neighbor technique to the nearest SAFRAN grid cell center, in order to be associated with a unique reference SAFRAN grid cell. Next, each BC method to be tested is applied over two distinct periods of the year: one corresponding loosely to "winter" from 15 October to 14 April, the other to "summer" from 15 April to 14 October. For each "season", corrected ERA-I T2 and PR are obtained for the 1995–2009 "evaluation" time period based on BC models calibrated over 1980–1994. Note that, as they potentially still include some seasonality (especially in temperature), 6-month long seasons to condition the BC procedures are certainly not the most suited time intervals for practical use or applications. In the latter cases, regular 3-month seasons or even monthly conditioning could be preferred, provided that enough data are available for calibration and projection. This could nevertheless introduce some artificial "discontinuities" when going from one month or season to another, which may be detrimental to some specific applications. However, here, 6-month seasons correspond to a very convenient cutting for testing and illustrating how a newly developed BC method behaves, which is the main purpose of this article. Indeed, this cutting allows for (1) increasing the number of data points (e.g., with respect to a monthly cutting) and (2) restricting the number of figures and evaluations that would be multiplied by two in case of 3-month seasons, or by six with a monthly cutting.

Moreover, Sect. 6 of this article will present an illustration of how the suggested R^2D^2 method works when applied to RCMs, both in present and future climates. Nevertheless, the ERA-I reanalyses are primarily used as test data to be corrected because they ensure some consistency with the SAFRAN reference dataset. Indeed, when employing RCM data in a cross-validation context (or more generally when applying a BC method to RCM data over a projection time period different from the calibration time period), the changes in statistical properties (e.g., mean, variance) from the calibration to the projection time periods can be different for the reference and for the RCM data. Hence, when evaluating the results of a BC method over a projection period, it may be difficult to assess which remaining biases come from the disagreement between reference and RCM changes, and which come from the BC method itself. Using reanalysis data ensures more consistency with the reference and is therefore more appropriate for initial evaluation of a BC method.

3 Reconstructing multi-site and multi-variable dependence structures

3.1 A brief reminder on copulas

In many of the multivariate BC development papers, the notion of "copula functions" is used. Indeed, those functions characterize the rank dependence structure of most multivariate joint distributions (e.g., Nelsen, 2006; Schölzel and Friederichs, 2008) through the Sklar's theorem (Sklar, 1959). This theorem expresses that any multivariate cumulative distribution function (CDF) can be described by the univariate marginal CDFs of the multivariate random variable and a copula function. The latter is itself a multivariate CDF depicting the statistical dependence of the transformed random variables $U_j = F_{X_j}(X_j)$, where X_j is the j-th variable of the d-dimensional random variable $\boldsymbol{X} = (X_1, \ldots, X_d)^T$ and F_{X_j} the respective marginal CDF. Mathematically, Sklar's theorem states that any multivariate CDF $F_{\mathbf{X}}$ (e.g., temperature at several stations) can be written as

$$F_X = C_X\left(F_{X_1}, \ldots, F_{X_d}\right), \tag{1}$$

where C_X is the copula of $\boldsymbol{X}$. Therefore, any multivariate BC method will necessarily correct the copula of the simulations, explicitly (e.g., Piani and Haerter, 2012; Vrac and Friederichs, 2015) or implicitly (e.g., Cannon, 2017; Dekens et al., 2017).

3.2 A brief reminder of the "Empirical Copula – Bias Correction" (EC-BC) approach

The EC-BC approach (Vrac and Friederichs, 2015) takes advantage of the so-called "Schaake Shuffle", described by Clark et al. (2004) and employed in various studies to reconstruct multivariate dependence structures (e.g., Voisin et al., 2010; Verkade et al., 2013; Cannon, 2017, among others). The principle is the following: first, a 1d-BC is performed on each statistical dimension (i.e., for each variable at each location). Second, the univariate bias corrected data are reordered such that their rank time series is identical to that of the reference sample. This univariate shuffling performed separately on each variable allows us to reproduce both the temporal, inter-site and inter-variable dependencies of the reference data (see the synthetic example in Table 1 of Sect. 4c in Vrac and Friederichs, 2015), since it exactly reproduces the empirical copula function of the references. However, if the inter-site and inter-variable dependence structures can be assumed to be stable over time because they can be considered to be imposed by physical constraints over the region of interest, this is not the case for the temporal structures (or rank chronology of the climate events). For example, rain persistence can shorten or enlarge, or heat waves can increase and/or be more frequent, and seasonality of some specific (temperature or precipitation) events may change, depending on the geographical domain. It is therefore needed to relax the EC-BC temporal constraint to let the climate simulations express their temporal dynamics and evolutions through time. This is the goal of the proposed methodology.

3.3 The Rank Resampling for Distributions and Dependences (R^2D^2) bias correction approach

The R^2D^2 method is developed in the "marginal/dependence" context: The main idea of R^2D^2 is to take advantage of the Schaake Shuffle as in EC-BC but to relax the constraint of the reproduction of the temporal structure observed in the reference dataset. To do so, a historical time period is used as the calibration time period for which both climate simulations and reference datasets are at one's disposal. The correction is performed over a projection time period (e.g., a future time period) where only climate simulations are available. The R^2D^2 method proceeds as follows (please refer to Appendix A for a detailed mathematical description of the R^2D^2 algorithm):

1. As in EC-BC or any "marginal/dependence" approach, each dimension (variable/location) is first corrected independently from the others by a univariate BC method. In the present study, the CDF-t method is used (e.g., Vrac et al., 2012).

2. Then, a dimension is selected (i.e., one physical variable at one given location) to serve as a "reference dimension" for the shuffling. For this specific dimension, the time sequence of the ranks of the 1d-bias-corrected data is kept untouched. Note that this sequence is therefore the same as that of the ranks of the simulations to be corrected, at least with a BC method preserving the ranks as it is the case for CDF-t.

3. Next, for each time step t of the projection time period, R^2D^2 looks for the time step t^* in the calibration time period for which the rank of the reference dimension is the same as the current rank of the reference dimension, i.e., R^2D^2 searches t^* such that $\mathrm{rank_{dim}}^{\mathrm{1d-BC}}(t) = \mathrm{rank_{dim}}^{\mathrm{ref.data}}(t^*)$, where $\mathrm{rank_{dim}}^{A}(t)$ is the rank – in the dataset A – of the value taken by the reference dimension dim at time step t.

4. Once this time step t^* is found, the time series of the other dimensions (i.e., the other variables at the same location, and all variables at the other locations) are shuffled such that the inter-site and inter-variable rank structures of the reference dataset are reproduced. This means that the rank association found in the reference dataset for time t^* is reproduced for time t.

5. Steps 2 to 4 are then repeated successively until each dimension has served as the reference dimension.

Those different steps are expressed in more mathematical and algorithmic ways in the Appendix A.

An example is now given to illustrate the functioning of R^2D^2. Let's assume that the multivariate bias correction problem of interest concerns $P = 3$ statistical dimensions. Those can be one physical variable for three grid cell, or three physical variables for one grid cell. Each dimension is simulated and observed over $N = 4$ consecutive time steps (e.g., days). Of course, in practice, many more variables, grid cell and time steps can be treated. Let's say that the first step of independent univariate bias correction was performed and that the reference and 1d-bias-corrected data are those given in Table 1, where each second column indicates the ranks of the

Table 1. One example of 3-dimensional reference data and results from the 1d-bias correction of sample size 4 for illustration of the R^2D^2 method. $k()$ indicates the rank within the sample.

References						1d-BC					
$x_R^{(i)}$	$k(x_R^{(i)})$	$y_R^{(i)}$	$k(y_R^{(i)})$	$z_R^{(i)}$	$k(z_R^{(i)})$	$x_{1d}^{(i)}$	$k(x_{1d}^{(i)})$	$y_{1d}^{(i)}$	$k(y_{1d}^{(i)})$	$z_{1d}^{(i)}$	$k(z_{1d}^{(i)})$
0.3	1	1.1	1	2.1	2	0.7	3	1.3	2	1.9	1
0.5	2	1.7	3	1.8	1	0.5	2	1.8	4	2.9	4
0.9	4	1.2	2	3.0	4	0.2	1	1.1	1	2.0	2
0.8	3	1.9	4	2.7	3	0.9	4	1.4	3	2.6	3

Table 2. Results of the R^2D^2 correction method. As the initial data are 3-dimensional, three time series are provided by R^2D^2. The k indicates the rank within the sample. Within each 3-dimensional time series, the bold values and ranks indicate the dimension and rank sequence taken as reference.

3d-BC (1/3)						3d-BC (2/3)						3d-BC (3/3)					
$x^{(i)}$	k	$y^{(i)}$	k	$z^{(i)}$	k	$x^{(i)}$	k	$y^{(i)}$	k	$z^{(i)}$	k	$x^{(i)}$	k	$y^{(i)}$	k	$z^{(i)}$	k
0.7	**3**	1.8	4	2.6	3	0.9	4	**1.3**	**2**	2.9	4	0.5	2	1.4	3	**1.9**	**1**
0.5	**2**	1.4	3	1.9	1	0.7	3	**1.8**	**4**	2.6	3	0.9	4	1.3	2	**2.9**	**4**
0.2	**1**	1.1	1	2.0	2	0.2	1	**1.1**	**1**	2.0	2	0.2	1	1.1	1	**2.0**	**2**
0.9	**4**	1.3	2	2.9	4	0.5	2	**1.4**	**3**	1.9	1	0.7	3	1.8	4	**2.6**	**3**

values in the time series displayed in the previous column. The results provided by R^2D^2 are given in Table 2. First, a reference dimension is selected, starting with x in this illustration, and the 1d-BC time series of this dimension is preserved at this stage. We can note that the first column of "3d-BC (1/3)" in Table 2 is therefore the same as that given by the univariate BC of dimension x in Table 1. Then, for each time step (i.e., row in those two tables), the rank of the current 1d-BC value is calculated. The time step with the same rank is then searched into the reference data for this dimension (here, x, first and second columns in Table 1) and the ranks of the other dimensions y and z for this time step are taken to shuffle the 1d-corrections of those two dimensions. For example, for the first time step in Table 2, the value 0.7 of the reference dimension x has rank 3. Looking into Table 1, rank 3 is found at the last time step for x, and is associated with ranks 4 and 3 for y and z, respectively. Therefore, the $x = 0.7$ value is associated with values 1.8 and 2.6, which have ranks 4 and 3 for y and z, respectively, in the univariate bias correction (Table 1). This procedure is then repeated for each time step before changing the reference dimension and rank sequence. R^2D^2 then provides as many corrections as the total number dimensions – or at least as many as the number of reference dimensions employed. Indeed, for practical reasons, it may be necessary to apply this algorithm on a reduced number of reference dimensions, therefore reducing the number of corrected outputs. However, whatever the number of reference dimensions or correction scenarios selected, the multivariate corrected data should all have equivalent inter-site and inter-variable copula functions.

Moreover, step 4 assumes that these copula (dependence) functions are stable in time (i.e., stationary) and correspond to those from the reference data. This assumption makes it possible to apply the proposed R^2D^2 method in a high-dimensional context, e.g., more than 3000 statistical dimensions as will be illustrated in the following sections.

In the present study, the CDF-t univariate adjustment method (e.g., Vrac et al., 2012) is used to perform step 1 of the above algorithm. For the precipitation variable, CDF-t has been applied based on the relatively common "threshold adaptation" procedure. It consists of first defining a threshold th for which model data below th are set to zero (e.g., Schmidli et al., 2006; Lavaysse et al., 2012). This threshold is chosen such that the frequency of days with model precipitation greater than th is the same as the frequency of rainy days in the reference (observed) precipitation dataset. After this thresholding, only the positive values are corrected by CDF-t with respect to the strictly positive observed values. Other approaches are possible, such as applying a BC model directly on the whole time series including both dry days and rainy days, i.e., without separating the correction methodology into occurrence and intensity (e.g., Vrac et al., 2012; Vigaud et al., 2013; Vrac et al., 2016, among others). The latter approach has also been tested for preliminary tests and the results were not sensibly different from those presented in this article (not shown). Note also that other 1d-BC methods can of course be employed instead of CDF-t. For example, the regular quantile-mapping approach (e.g., Déqué, 2007) has also been tested within R^2D^2 and similar results were obtained (not shown).

4 Design of experiments

This section describes the comparisons that will be performed between different BC methods in the following for evaluating the proposed R^2D^2 bias correction methodology.

It is first reminded that, for each tested BC method applied to ERA-I reanalyses with SAFRAN data as reference, the calibration period is 1980–1994, while the correction/evaluation period is 1995–2009. Moreover, each calibration/evaluation is performed for daily temperature and precipitation time series on 1506 grid cell in the southeast of France over a 6-month "winter" (15 October to 14 April) and a 6-month "summer" (15 April to 14 October).

First of all, the 1-dimensional CDF-t bias correction (e.g., Michelangeli et al., 2009; Vrac et al., 2012) is performed. As it is also the 1d-BC method used in step 1 of the R^2D^2 algorithm, this will allow us to evaluate the contribution of the other steps in R^2D^2. Then, various configurations of the R^2D^2 method are applied and evaluated:

- a 2-dimensional R^2D^2 version, where each grid cell is corrected independently but temperature and precipitation are corrected jointly within each grid cell. As there are 1506 grid cell for the present dataset, 1506 applications of 2d-R^2D^2 are realized;
- a 1506-dimensional R^2D^2 version, where all 1506 grid cell time series are corrected jointly but separately in temperature and precipitation. Therefore, two 1506d-R^2D^2 are realized, one for temperature and one for precipitation;
- a 3012-dimensional version, where temperature and precipitation for all the 1506 grid cell are corrected jointly. Only one 3012d-BC is needed here.

Note that, as R^2D^2 can return as many datasets (or "scenarios") of correction as the number of statistical dimensions, the 2d-R^2D^2 versions return two corrected datasets for each grid cell. However, for the 1506d- and 3012d-R^2D^2 versions, a sub-sample of 10 reference dimensions has been selected for each version. Therefore, those versions provide 10 corrected datasets. The 10 reference dimensions have been chosen to uniformly cover the geographical domain for each physical variable.

In the following Sect. 5, the results of those four BC methods (1d, 2d, 1506d, 3012d) as well as the initial dataset to be corrected (ERA-I) are compared according to three different aspects evaluated on the 1995–2009 evaluation period. First, the inter-variable dependence properties are investigated in Sect. 5.1. Second, the inter-site dependence structures are compared in Sect. 5.2. Finally, although this aspect was not part of the correction design, the temporal properties are also evaluated in Sect. 5.3. Indeed, as any multivariate BC method will necessarily modify the initial rank chronology of the simulated events, it is interesting to understand – or at least to quantify – these modifications.

5 Results

In this section, all analyses are realized for the winter season but the main conclusions hold for the summer results that are displayed in the Supplement.

5.1 Inter-variable correlations

First, the BC results are compared in terms of inter-variable correlations. To do so, the spearman correlation between temperature and precipitation time series have been computed for each of the 1506 grid cell and the resulting maps are shown in Fig. 1. Note that the Pearson correlation maps have also been computed. The results were very similar (not shown) but, based on Pearson correlations, a larger gap between the ERA-I Pearson correlations Fig. 1b and the reference SAFRAN Pearson correlations Fig. 1a means that most maps only use a reduced number of colors, which is not convenient for visual evaluations. This is why only the Spearman correlation is used in Fig. 1. In this figure, it is clear that the ERA-I inter-variable correlation map Fig. 1b is very different from that of SAFRAN (a), with spearman rank correlations not only differing in their intensities but also in their structure. This strongly exemplifies the need for adjustment of this aspect. The 1d-BC CDF-t method Fig. 1c modifies the intensities of those correlations but does not change its structure, basically conserving that of ERA-I. However, the 2d- and 3012d-R^2D^2 methods (Fig. 1d and f, respectively) provide major improvements, since they allow us to approximate the temperature–precipitation correlations from SAFRAN (a). Correlation maps from those two R^2D^2 versions are strictly the same, indicating that the 3012d-version is a generalization of the 2d-configuration (at least in this inter-variable evaluation context). This is not the case for the 1506d-R^2D^2 configuration that basically destroys the inter-variable correlations. Indeed, as in this configuration, temperature and precipitation are adjusted separately and independently from each other, the obtained correlations are – by construction – close to 0. This version is designed to take care of the inter-site dependence but completely disregards and even ruins inter-variable correlations. Note that the 3012d-R^2D^2 configuration provides the same correlation map as presented in Fig. 1.f, whatever the reference dimension selected. This is also true for the 2d- and 1506d-versions where a different reference dimension still generates equivalent correlations.

5.2 Spatial correlations

The evaluation is now performed in terms of inter-site and spatial correlation. A principal component analysis (PCA) is first carried out on each physical variable (i.e., temperature and precipitation) separately but for the whole region of interest (i.e., 1506 grid cell). However, before applying the PCA, the daily areal mean has been removed from each

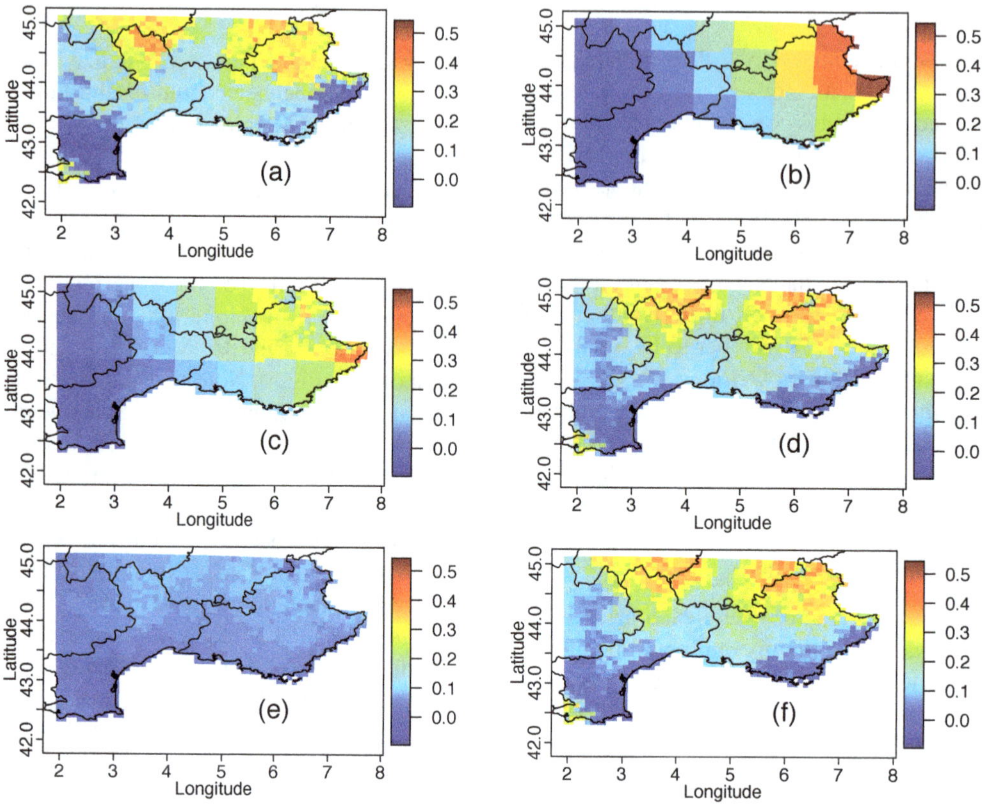

Figure 1. Inter-variable Spearman correlation maps in winter over the evaluation period from: **(a)** SAFRAN; **(b)** ERA-I; **(c)** 1d-BC (CDF-t); **(d)** 2d-R^2D^2; **(e)** 1506d-R^2D^2 on temperature and 1506d-R^2D^2 on precipitation; **(f)** 3012d-R^2D^2.

daily data. Indeed, the data present a high day-to-day variability within the region of interest. This strongly impacts the PCA that shows a predominant empirical orthogonal function (EOF) almost uniform over the region if the areal mean is not removed (not shown). Moreover, as precipitation presents a skewed distribution, all zero precipitation values are put to a non-zero but positive small value (3.3^{-4}) and the precipitation PCA is performed on the logarithm of the values (following, e.g., Vrac and Friederichs, 2015), where the areal mean has been removed. Although the log-precipitation values look more Gaussian than the initial ones, a PCA on those transformed data should still be interpreted with prudence. This is nevertheless a helpful means to describe spatial modes of variability. Figures 2 and 3 show the maps of the first EOFs obtained from PCAs applied to temperature or log-precipitation, respectively, from the different datasets. For both variables, ERA-I first EOF (Figs. 2b and 3b) maps are quite dissimilar from the SAFRAN EOF maps (Figs. 2a and 3a). The univariate BC (Figs. 2c and 3c) shows similar results as those from ERA-I, although less pronounced for precipitation (Fig. 3c). Concerning the results of the 2-dimensional version of R^2D^2 (Figs. 2d and 3d), for each grid cell, they are obtained based on selecting as reference dimension the "other" dimension. In other words, for precipitation the reference dimension is temperature, and for temperature the reference dimension is precipitation. Indeed, otherwise (i.e., if the reference dimension is the variable of interest), by construction, the spatial structures resulting from the 2d-R^2D^2 are exactly the same as those from the 1d-BC presented in Figs. 2c and 3c (not shown). In the present configuration of the 2d-R^2D^2 version, the spatial modes of variability (in Figs. 2d and 3d) are different from both the ERA-I and 1d-BC results. They visually look more similar to the SAFRAN results and seem to improve the inter-site dependence structure. But this is not the case for summer results (see Supplement) and they do present some major differences with respect to SAFRAN for both precipitation and temperature in the two seasons. However, the first EOF maps from the 1506-dimensional (Figs. 2e and 3e) and the 3012-dimensional versions (Figs. 2f and 3f) are very close to those from the reference SAFRAN dataset, indicating a satisfying modeling of the main modes of inter-site variability, both for temperature and (log-) precipitation. This is also confirmed by the eigenvalues and explained variance fractions of the leading EOF for temperature and log-precipitation given in Fig. 4, as well as by the correlograms (i.e., correlations in function of the distance) displayed in Fig. 5. In those figures, the results of the 1506d- and 3012d-R^2D^2 versions are

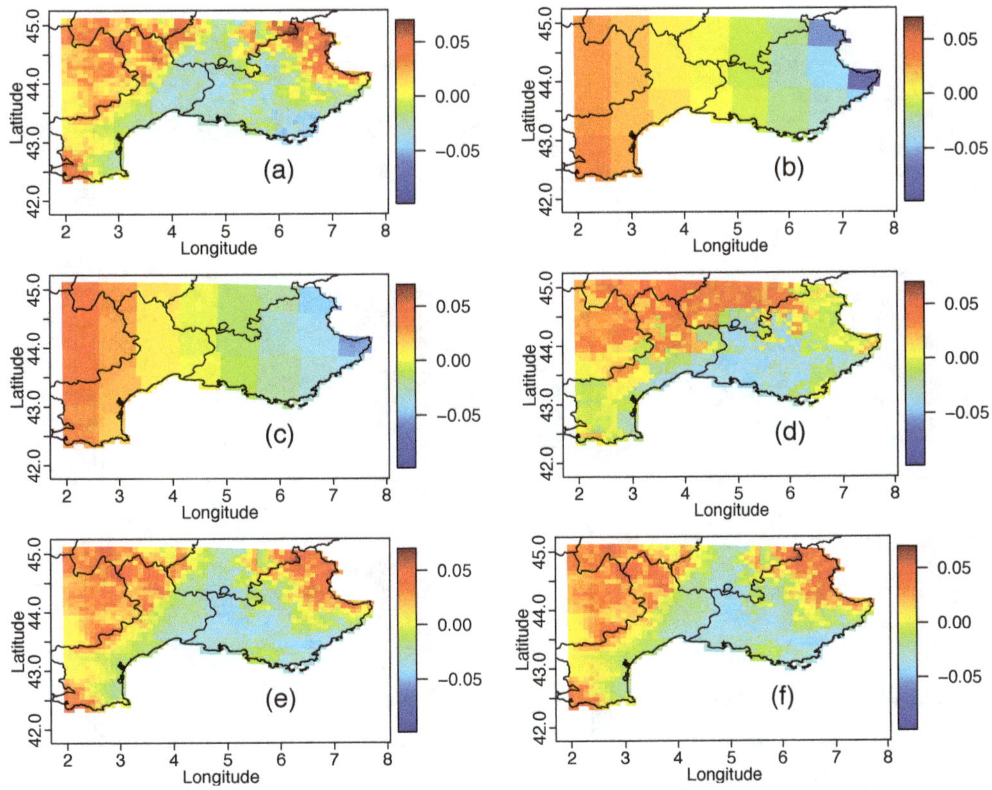

Figure 2. Maps of first temperature predominant empirical orthogonal functions (EOFs) in winter over the evaluation period for **(a)** SAFRAN; **(b)** ERA-I; **(c)** 1d-BC (CDF-t); **(d)** 2d-R^2D^2 (with PR as reference dimension); **(e)** 1506d-R^2D^2 (on T2 only); **(f)** 3012d-R^2D^2.

the same: they stick closely to the SAFRAN eigenvalues and explained variances (Fig. 4) and reproduce well its correlogram (Fig. 5), even at long distances. The other datasets show deviations from SAFRAN more or less pronounced and in agreement with previous Figs. 2 and 3: ERA-I results are relatively far away from SAFRAN, 1d-BC slightly modifies the spatial properties but stays comparable to ERA-I and 2d-R^2D^2 degrades the ERA-I spatial properties (at least in the present configuration). The same conclusions hold for summer (see the Supplement) where the 1506d- and 3012d-R^2D^2 versions follow SAFRAN spatial properties, although some differences appear between the correlograms at long distances (> 400 km) especially for temperature.

In order to get more quantification of those results, various Spearman and Pearson correlation matrices was computed for the different datasets (SAFRAN, ERA-I and the BC results) in the evaluation period over the 1506 locations:

- on temperature vs. temperature (resulting in a 1506 × 1506 spatial correlation matrix);
- on precipitation vs. precipitation (1506 × 1506 spatial correlation matrix);
- on temperature vs. precipitation (1506 × 1506 spatial correlation matrix across the two variables);
- on (temperature, precipitation) vs. (temperature, precipitation) (3012 × 3012 spatial and inter-variable correlation matrix).

The SAFRAN correlation matrix is then subtracted from the correlation matrix of each dataset (ERA-I and the BC results), therefore providing matrices that describe differences in correlations (hereafter referred to as $\mathrm{Diff}_{\mathrm{corr}}$). The absolute values of the elements of this matrix are then summed and the result – noted S_{corr} – gives a numerical indication of the global quality of the dataset dependence structure with respect to that of SAFRAN. The values of S_{corr} for each dataset and for the different types of correlations are given in Table 3. The results for the "T2 vs. T2" and "PR vs. PR" correlations are quite similar, showing the good behavior of the 1506d- and 3012d-BC methods, while it is clear that the 2d-R^2D^2 version deteriorates the ERA-I and 1d-BC correlations. For the "T2 vs. PR" correlations, the 2d-BC version is relatively equivalent to the ERA-I and 1d-BC but the 1506d-R^2D^2 slightly degrades those results, while the 3012-dimensional version is much better. Finally, for the "(T2,PR) vs. (T2,PR)" correlations, the 2d-BC version appears as not adapted, the 1506d-BC improves ERA-I and the 1d-BC but 3012d-R^2D^2 provides the best results.

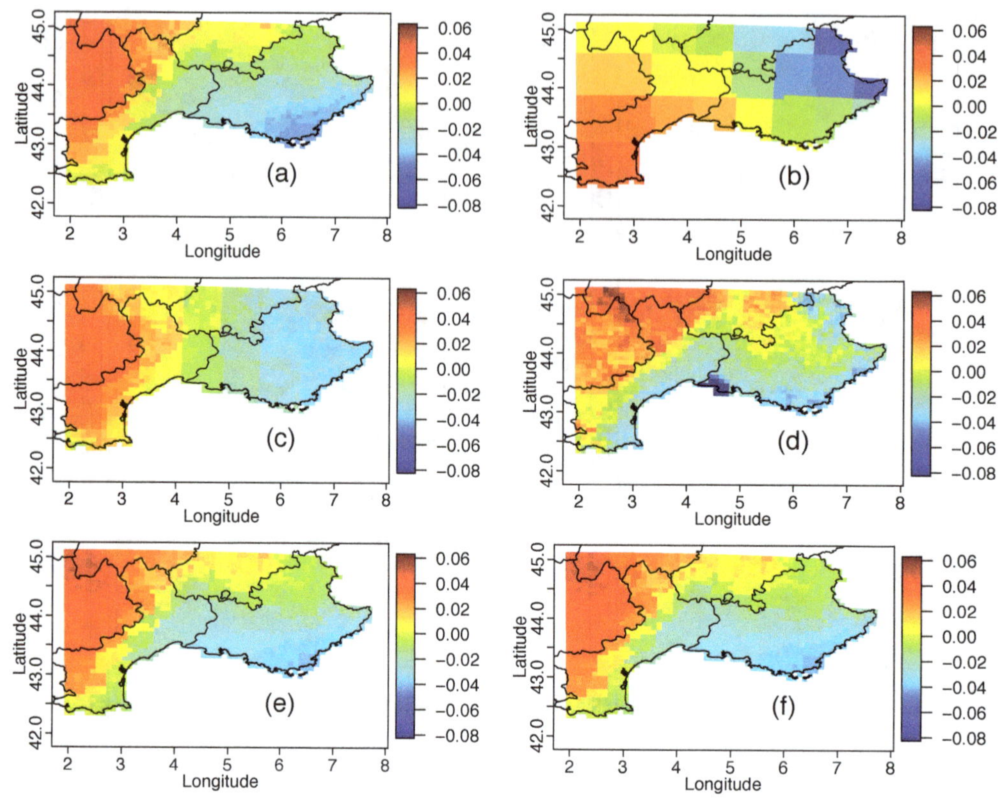

Figure 3. Same as Fig. 2 but for precipitation.

Table 3. Values of S_{corr}, corresponding to the sum of the absolute values of the elements of the difference correlation matrices for each dataset and for the different types of correlations (see text for details) in winter. Values have to be multiplied by 10^4. Some methods and applications provide different S_{corr} values depending on the reference dimension. For those cases, the mean S_{corr} value is indicated and the standard deviation is indicated between brackets. Values in bold font indicate the smallest values for the line.

	ERA-I	1d-BC	2d-R^2D^2	1506d-R^2D^2 (T2 or PR)	3012d-R^2D^2
Spearman (T)	20.1	20.1	96.4 (±107.9)	**5.4**	**5.4**
Pearson (T)	19	18.6	96.3 (±109.8)	**4.8**	**4.8**
Spearman (PR)	69	40.6	73.6 (±46.7)	**5.8**	**5.8**
Pearson (PR)	87.9	62.1	74.7 (±17.8)	**10.4**	**10.4**
Spearman (T2 vs. PR)	25.8	24.5	23.4 (±11)	30.4 (±5.8)	**8**
Pearson (T2 vs. PR)	16.3	13.3	14.8 (±6.7)	19.4 (±3.5)	**5.7**
Spearman (T and PR)	140.7	109.6	216.7 (±83.2)	71.9 (±11.6)	**27**
Pearson (T and PR)	139.5	107.3	200.6 (±105.5)	54 (±7.1)	**26.6**

Other analyses of the spatial properties derived for the different BC methods were also performed (e.g., quantile-quantile plots of the daily areal means) but are not provided here since their conclusions were the same as in the presented figures: 1d-BC approximately preserves ERA-I properties that are biased with respect to SAFRAN's; 2d-BC changes the ERA-I spatial statistics but does not necessarily improve them, while 1506d- and 3012d-BC via R^2D^2 provides satisfying spatial variability and dependence structures, close to those from SAFRAN.

5.3 Temporal correlations

The proposed R^2D^2 method is not designed to reproduce, correct or preserve the temporal structure of the simulations to be corrected. However, as any multivariate BC will necessarily modify their rank sequence, it is interesting to under-

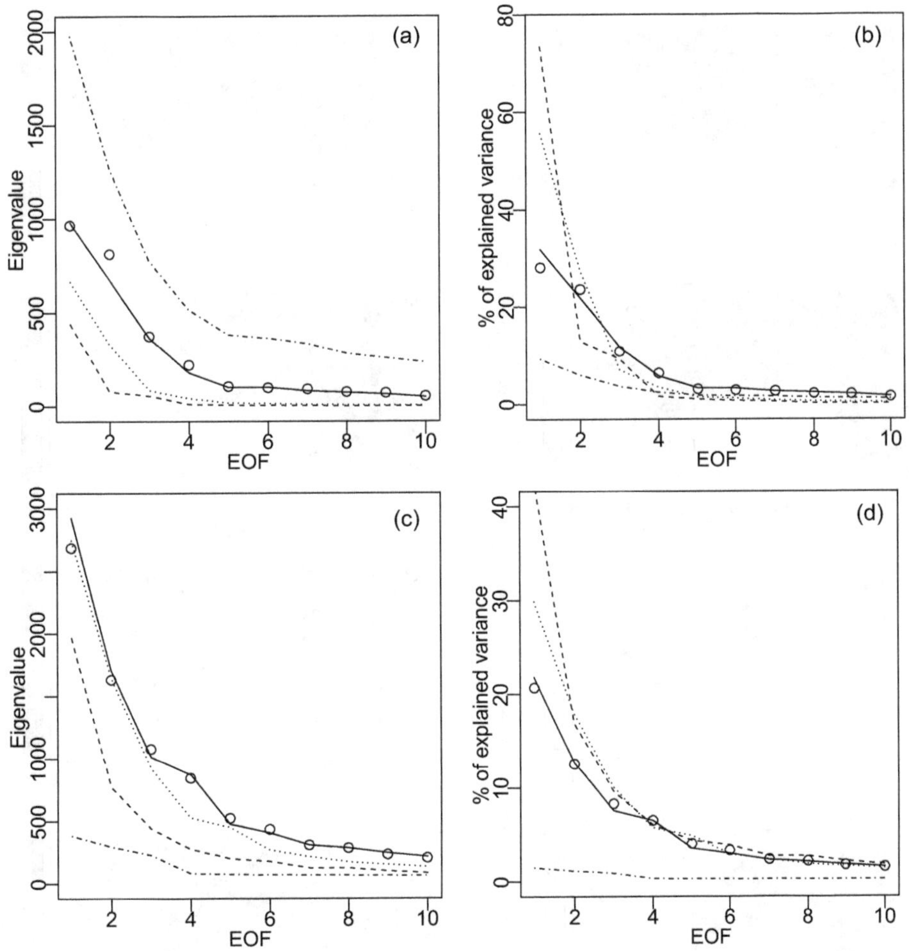

Figure 4. Eigenvalues **(a, c)** and percentage of explained variance **(b, d)** of temperature at 2 m **(a, b)** and precipitation **(c, d)** in winter over the evaluation period for: SAFRAN (circles); ERA-I (dashed); 1d-BC by CDF-t (dotted); 2d-R^{2}D^2 (dashed-dotted); 1506d-R^{2}D^2 (T2 or PR, long dashed); 3012d-R^{2}D^2 (solid line). Note that results of the 1506d- (long dashed) and 3012d-R^{2}D^2 (solid line) versions are the same and are therefore superimposed.

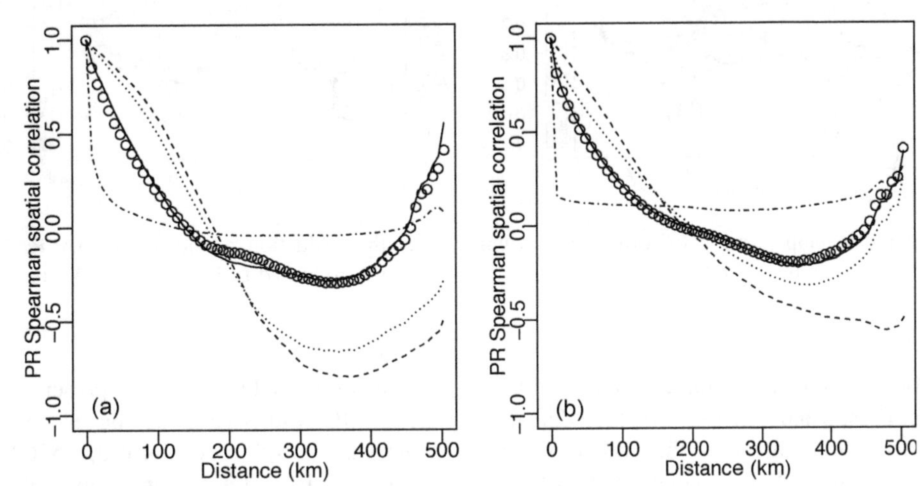

Figure 5. Correlograms in winter over the evaluation period for **(a)** temperature at 2 m; **(b)** precipitation, from SAFRAN (circles); ERA-I (dashed); 1d-BC (dotted); 2d-BC (dashed-dotted); 1506d-BC (T2 or PR, long dashed); 3012d-BC (solid line). Note that results of the 1506d- (long dashed) and 3012d-R^{2}D^2 (solid line) versions are the same and are therefore superimposed.

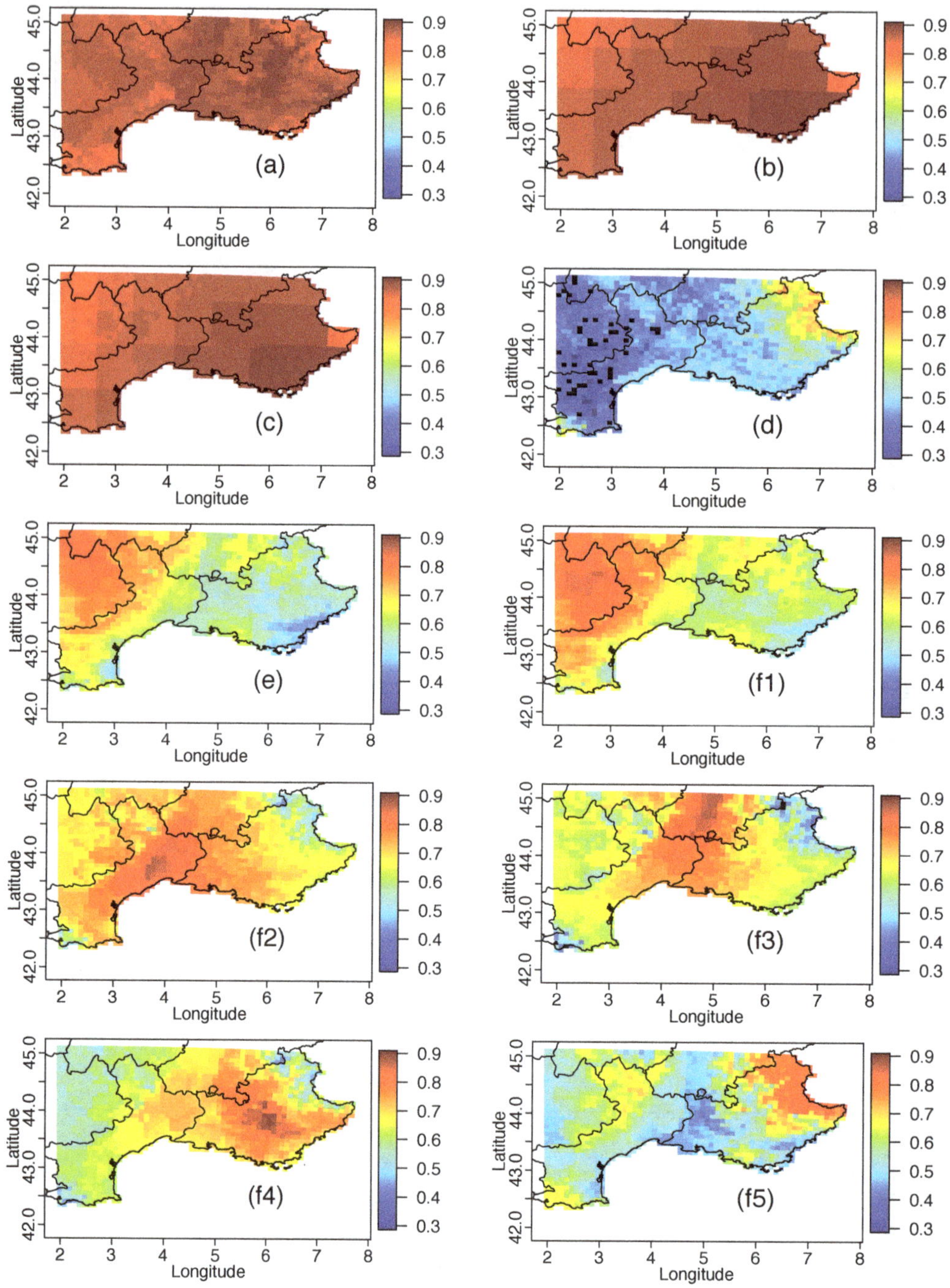

Figure 6. Maps of lag-1 day temperature auto-correlations in winter over the evaluation period for **(a)** SAFRAN; **(b)** ERA-I; **(c)** 1d-BC; **(d)** 2d-R^2D^2 (with PR as ref.dim. for each grid cell); **(e)** 1506d-R^2D^2; **(f1–5)** 3012d-R^2D^2 with five different reference temperature locations.

stand how strong those modifications are, depending on the R^2D^2 version. Hence, Figs. 6 and 7 display, for each dataset, the 1-day lag auto-correlation maps over the evaluation period for T2 and PR, respectively. For temperature, the ERA-I data (Fig. 6b) have high auto-correlation values between 0.8 and 0.9 in agreement with SAFRAN data (Fig. 6a), although the spatial structure is different (not highlighted here). Since the univariate CDF-t method preserves the rank sequence, the 1d-BC results (Fig. 6c) have similar auto-correlations. However, the other results (2d, 1506d and 3012d) deeply change the ERA-I auto-correlation values, with a strong reduction from the 2d-BC results (Fig. 6d). For the 3012d-R^2D^2 version, the auto-correlations depend on the statistical dimension serving as reference. Therefore, five illustrations

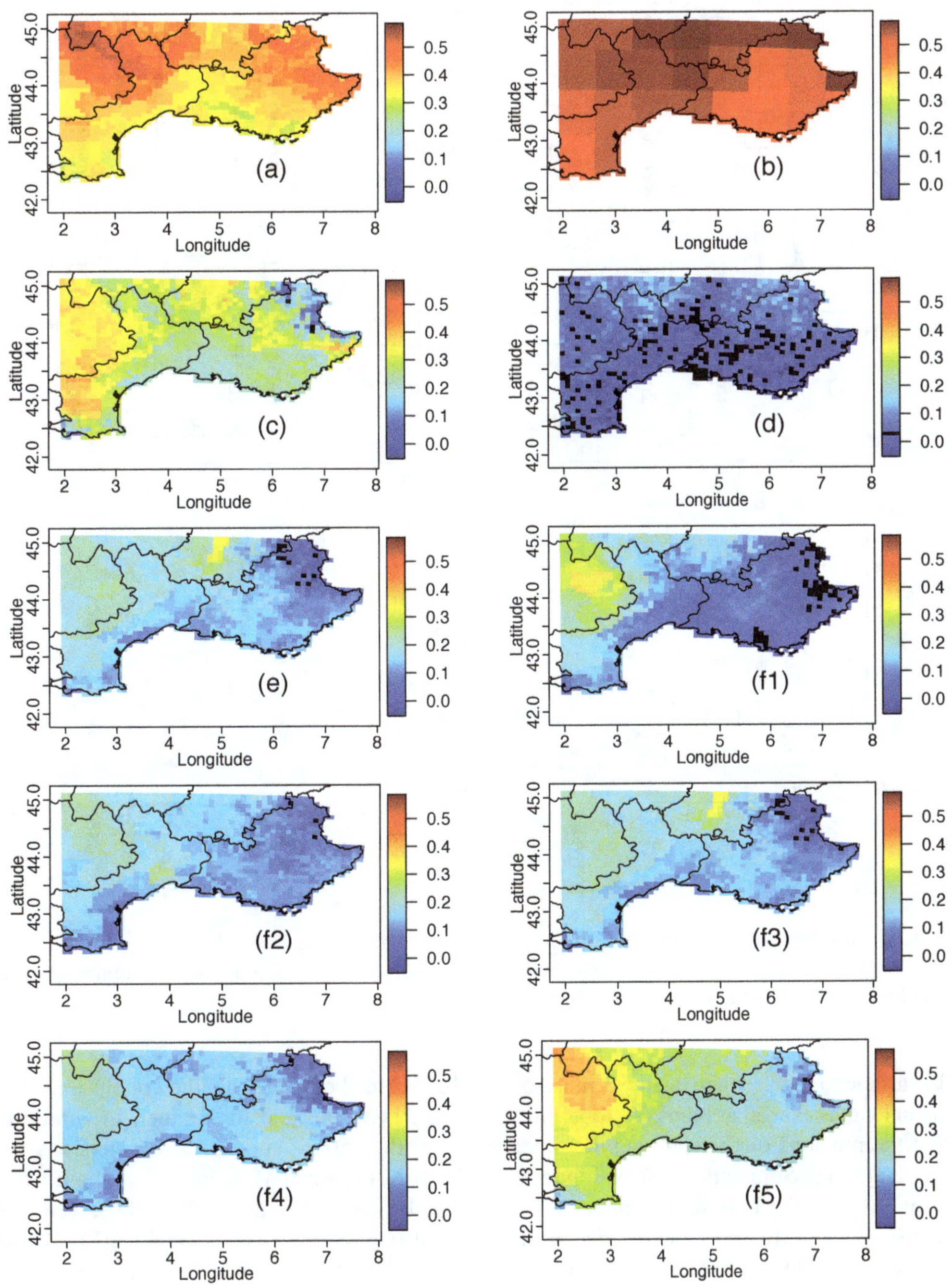

Figure 7. Same as Fig. 6 but for precipitation.

are provided in panels 6f1–f5 obtained from five reference dimensions, here corresponding to temperature at five locations. Interestingly, those five locations roughly correspond to the center of the red zones visible in panels Fig. 6f1–f5. Indeed, as the reference dimension preserves the rank sequence of the 1d-BC – and therefore of the model data to be corrected – the same auto-correlation values are found at this specific location. The obtained correlation is somehow also reproduced on a neighborhood more or less extended around this location, and rapidly decreases out of this neighborhood. For precipitation (Fig. 7), the same behavior is present although less pronounced. Moreover, the ERA-I auto-correlation results (Fig. 6b) are not in agreement with SAFRAN (Fig. 6a) anymore, and the 1d-BC results (Fig. 6c) appear quite different from ERA-I. The changes in behavior of the different BC results come from the precipitation occurrences that are modified both in frequency and in the structure of their sequence (e.g., spells). This is not shown

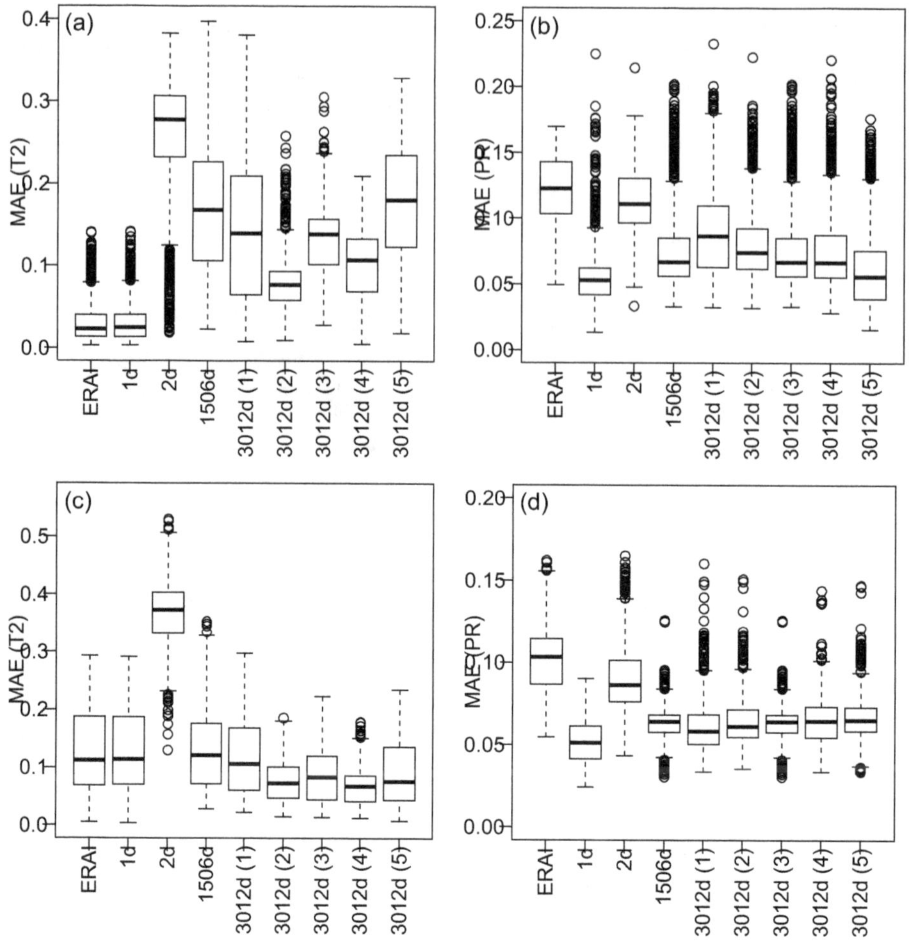

Figure 8. Box plots of the mean absolute error (MAE) values calculated on lag-1 to lag-7 day Pearson correlations for: ERA-I; 1d-BC; 2d-BC; 1506d-BC of T2 or PR (example for first reference variable); 3012d-BC with five different reference temperature locations. **(a, c)** for 2 m-temperature; **(b, d)** for precipitation; **(a, b)** winter; **(c, d)** summer.

here to constrain this article to a reasonable size but maps of wet and dry spell mean lengths as well as maps of probability of dry day given that the previous one is wet, and the other way around, are provided in the Supplement for both winter and summer. Nevertheless, in order to have a larger view on the temporal correlation of the different datasets, the mean absolute error (hereafter referred to as MAE) with respect to SAFRAN was computed over the evaluation period for each grid cell and physical variable, based on the first seven auto-correlation values:

$$\mathrm{MAE} = \sum_{n=1}^{7} |\rho_n(D) - \rho_n(\mathrm{SAFRAN})|, \quad (2)$$

where $\rho_n(D)$ is the n-day lag auto-correlation value of the dataset D. The resulting values are presented via box plots – summarizing the spatial variability of the MAE – in Fig. 8, and via maps as Supplement. For temperature (left panels of Fig. 8), except for the 2d-BC results that show a degradation of the MAE values compared to those from ERA-I or 1d-BC for both seasons, the conclusions are not exactly the same in winter and in summer. In winter, the MAE results from the 1506d- and 3012d-BC versions are of lower quality (i.e., higher MAE values) than those from ERA-I. This is not the case in summer where those versions present equivalent or even better (i.e., smaller) MAE values. For precipitation, however, winter and summer results are consistent: all tested BC methods generally improve the ERA-I MAEs – although only slightly for 2d-R^{2}D^2 – and the 1506d- and 3012d-R^{2}D^2 MAE are relatively close to those from the 1d-BC that presents the best (i.e., smallest) MAE values.

6 Bias correction of RCM simulations

6.1 GCM/RCM runs and scenario

For illustration purposes, in order to evaluate and compare the different BC methods when applied to regional climate simulations over a historical period and in a future climate change context, two RCMs driven by the same GCM are used to provide simulations to be corrected. Those RCMs are (i) the "Weather Research and Forecasting" (WRF) regional climate model (Skamarock et al., 2008) developed by the National Center for Atmospheric Research and (ii) the "Rossby Centre regional Atmospheric" model (RCA4; Samuelsson et al., 2011). Both RCMs provide daily simulations at a $0.11° \times 0.11°$ spatial resolution over the European domain of the Coordinated Regional Climate Downscaling Experiment (CORDEX; Giorgi et al., 2009; Jacob et al., 2014), and were forced by the "Institut Pierre Simon Laplace" (IPSL) global climate model (Marti et al., 2010; Dufresne et al., 2013) with a "historical" 1950–2005 run, as well as for the 2006–2100 time period under a scenario of representative concentration pathway associated with a radiative forcing of $+8.5\,\mathrm{W\,m^2}$ (RCP8.5) in the year 2100 with respect to the preindustrial period (IPCC, 2013). The calibration of the different BC methods is made over the SAFRAN 1980–2009 time period, and for the same winter and summer seasons as previously. The corrections of the WRF and RCA4 simulations are then performed over 1980–2009 and 2071–2100, and only with 1d-CDF-t, 2d-R^2D^2 (for T2 and PR together but for each of the 1506 grid cell separately) and 3012d-R^2D^2. The 1506d-BC version (either on T2 or PR) was not performed in this section since, in the previous one, it provided either equivalent or lower quality results than the 3012d-R^2D^2 version. In the following, results are given for the WRF model in winter but the WRF summer and RCA4 winter and summer results are provided as Supplement.

6.2 Historical evaluations and changes from historical to future climate simulations

This subsection contains a short evaluation of the BC methods applied to the RCM simulations over the 1980–2009 historical period, as well as an illustration of how the tested BC methods behave and differ from each other in a climate change context, both in terms of inter-variable and inter-site dependencies. As an objective of this sub-section is to evaluate the changes from the historical (1980–2009) to the future (2071–2100) time periods, in order to save space, the evaluations of the BC methods applied to the RCM simulations are performed directly over the whole historical period (1980–2009), without cross-validation. Nevertheless, when applying the same cross-validation exercise as was done with ERA-I in Sect. 5, the conclusions were exactly the same with RCM data (not shown). First, for each dataset, the inter-variable correlation between T2 and PR in winter is calculated for each grid cell for both the historical and future time period. The resulting maps are presented in the left panels of Fig. 9. Similarly to the BC of the ERA-I reanalyses, although the inter-variable correlations from WRF and its 1d-BC are quite distinct from the reference ones, the 2d- and 3012d-R^2D^2 versions (Fig. 9e and g, respectively) provide the same maps as that from SAFRAN (Fig. 9i), confirming their performance also on RCM simulations. However, the 2d-version does not do so well from the spatial perspective, as illustrated in Fig. 10 showing the temperature and precipitation correlograms. When driven by the "opposite" variable (i.e., T2 for PR correlograms and PR for T2 correlograms), the 2d-BC correlograms are away from both SAFRAN and RCM data, with a strong fall of correlation as soon as the very short distances (a few km) and a flat behavior after. As for the 3012d-BC of WRF, its correlogram nicely fits the empirical correlations calculated from SAFRAN for both variables. Regarding the RCM future climate simulations and their bias corrections, right panels of Fig. 9 show the changes (i.e., future – present) of the inter-variable correlations. The 1d-CDF-t method smoothes the RCM changes but preserves their structure, while, as expected, the 2d- and 3012d-BC versions do not present strong changes and therefore tend to provide an inter-variable correlation structure close to that of the SAFRAN data. For the changes in the temperature correlograms (Fig. 10a), the RCM simulations do not present much evolutions from the historical period to 2071–2100, and therefore the 1d-BC does the same. Moreover, neither the 2d-BC nor the 3012d-version show major changes and so the two versions are consistent with the raw simulations in terms of changes. For precipitation (Fig. 10b), the RCM simulations (solid and dashed black lines) do see some changes in the spatial dependence, and therefore, so does the 1d-BC (green lines). Interestingly, the 3012d-BC (red and orange, superimposed) also captures some changes, although slightly less pronounced. This means that 3012d-R^2D^2 allows a change (from historical to future) in the inter-site dependence structure that is consistent with the change provided by the RCM.

7 Conclusions and discussion

7.1 Conclusions

A new multivariate bias correction approach was proposed, allowing to correct not only the marginal (univariate) distributions of the climate variables of interest but also the statistical dependences between the variables, as well as the dependences between the different locations over a given geographical domain. This approach relies on the previously developed "Empirical Copula – Bias Correction" (EC-BC, Vrac and Friederichs, 2015) method, whose all dependence structures – inter-variable, inter-site and overall temporal - were taken from reference data and exactly reproduced by the EC-BC correction. The suggested BC approach is also based on a rank resampling to adjust the copula functions and therefore the dependences of the climate simulations, but this R^2D^2 method relaxes the EC-BC temporal constraint to let the climate model of interest express its temporal dynamics. Indeed, R^2D^2 is based on the assumption that the inter-site and inter-variable copula (dependence) functions

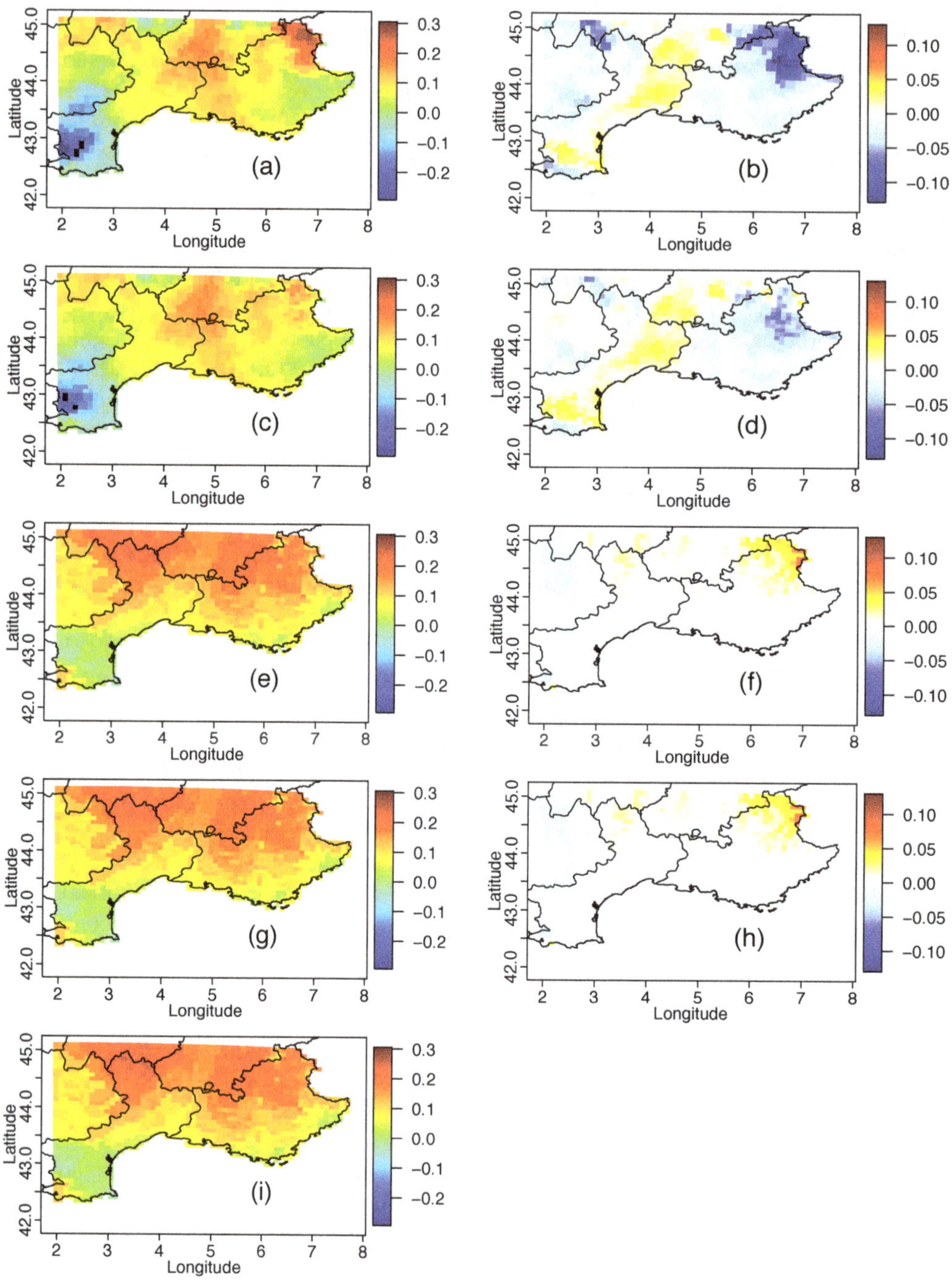

Figure 9. (left column) Inter-variable Pearson correlations between T2 and PR in winter for each grid cell and (right column) changes in inter-variable Pearson correlations from the historical period to the 2071-2100 period; **(a–b)** for the WRF RCM; **(c–d)** for its 1d- bias correction with CDFt; **(e–f)** for its 2d-R^2D^2 correction; **(g–h)** for its 3012d-R^2D^2 correction. Panel **(i)** corresponds to the correlations between T2 and PR for the SAFRAN reference data over the historical period.

are imposed by physical constraints over the region of interest and are therefore stable in time. Therefore, their dependence structures can be extracted and reconstructed from reference historical data. However, R^2D^2 is explicitly designed to partially respect the changes in the climate model (e.g., from historical to future periods) in terms of temporal (rank) properties. Since these evolutions can be distinct for different physical variables and/or grid cell, R^2D^2 generates multiple bias corrected scenarios, which can be considered as a stochasticity describing the possible variability in the different rank chronologies. As such, R^2D^2 can be seen as a method that is halfway between a multivariate correc-

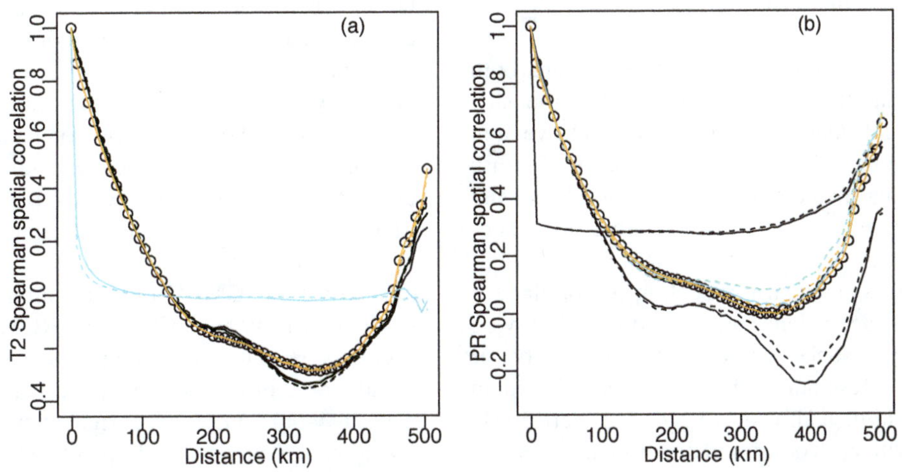

Figure 10. Spatial correlograms of temperature **(a)** and precipitation **(b)** in winter computed from daily areal mean-removed data. Correlations from SAFRAN are in circles; those from the WRF RCM are in black lines; 1d-BC in green; 2d-R^{2}D^2 driven by temperature in blue; 2d-R^{2}D^2 driven by precipitation in cyan; 3012d-R^{2}D^2 driven by temperature at a given location in red; 3012d-R^{2}D^2 driven by precipitation at the same location in orange. Solid lines indicate results for the historical period and dashed lines for the 2071–2100 period. Note that for temperature results **(a)**, green and blue lines are superimposed. For precipitation **(b)**, green and cyan are superimposed. Red and orange lines are always superimposed for both **(a)** and **(b)**.

tion method and a conditional multivariate stochastic weather generator. The assumption of stability of the copula function – which can hence be reproduced from the reference data – allows us to apply the multivariate bias correction in a high-dimensional context and at a reasonable computational cost. For example, the dataset generated by 3012d-R^{2}D^2 and analyzed in Sect. 5 (2734 winter days to be corrected for temperature and precipitation over 1506 grid cell) was obtained on a regular laptop computer with a 2.2 GHz Intel Core i7 processor and a 8 Go 1600 MHz DDR3 memory. On this computer, for each of the 1506 grid cell, the application of CDF-t (i.e., calibration and correction) takes about 0.05 s for temperature and 0.01 s for precipitation. Then, for one selected reference dimension, in the 3012-dimensional context, each application of the steps 2–4 of the R^{2}D^2 algorithm presented in Sect. 3.3 takes about 15 s. Consequently, the whole computation time of the 3012d-R^{2}D^2 version with 10 reference dimensions (and therefore 10 multivariate BC scenarios) took: $1506 \times (0.05+0.01) = 90.4$ s (for the univariate BC, step 1) + $10 \times 15 = 150$ s (for the 10 iterations of the steps 2–4), summing to about 240 s = 4 min.

R^{2}D^2 was first applied to adjust temperature and precipitation time series from ERA-Interim reanalyses (Dee et al., 2011) with respect to the SAFRAN dataset (Quintana-Segui et al., 2008) under a temporal cross-validation framework on 1506 locations. Different configurations of R^{2}D^2 were compared: a bivariate one (2d-R^{2}D^2) applied to jointly adjust temperature and precipitation but separately for each grid cell, a 1506-dimensional version (1506d-R^{2}D^2) applied jointly on the 1506 grid cell but separately for temperature and precipitation and a 3012-dimensional one (3012d-R^{2}D^2) where the two variables were jointly corrected over the 1506 grid cell. Those different versions were also compared to the univariate CDF-t bias correction method (e.g., Vrac et al., 2012) and to the raw ERA-I data. The results indicate that the 1d-BC by CDF-t generally reproduces the statistical dependence properties of the data to be corrected, from both the inter-variable, inter-site and temporal perspectives. Moreover, by construction, if 2d-R^{2}D^2 greatly improves the temperature–precipitation relationship, it does not do so well for inter-site dependences. This is the other way around for the 1506d-R^{2}D^2 that shows satisfying inter-site dependence reconstructions but disregards the inter-variable relationship. However, the 3012d-R^{2}D^2 performs well for both the inter-variable and inter-site property corrections. Regarding the temporal properties, except for the winter temperature with the tested datasets, most BC versions tend to provide autocorrelation getting slightly closer to SAFRAN's. However, it is worth keeping in mind that none of the multivariate BC versions were designed to adjust the temporal properties.

The different BC versions were then also tested and compared on climate simulations from the WRF and RCA4 regional climate models (RCMs) over the 1980–2009 historical period as well as the 2071–2100 future time period. The 2071–2100 bias corrections was not made to evaluate the methods (since no reference data are available for the future) but rather to illustrate how the different multivariate R^{2}D^2 versions behave in a climate change context. The evaluations over the historical period confirmed the results obtained on ERA-I, indicating a robustness of R^{2}D^2 to the dataset to be corrected.

7.2 Future work and discussion

The possible future developments of this work are both methodological and applied. First, as stated earlier, the variability/stochasticity introduced in the actual R^2D^2 version refers only to the timing of the events and does not perturb at all the corrected marginal distributions, neither the spatial dependence between sites and/or variables. More stochasticity could also be included into those properties. For example, the inference of a parametric modeling of copulas (or more generally of the dependence structures) would provide parameters generally associated with some uncertainty (or confidence intervals). Resampling those parameters based on this uncertainty would then allow us to generate "perturbed" copulas consistent with each other, and therefore multivariate corrections that are stochastic in their dependences.

Moreover, based on the results presented in this study, the assumption of conservation of the dependence structure sounds reasonable for the inter-site aspects (Fig. 10) but a bit more questionable for the inter-variable aspects, since the tested RCM shows some evolution of the inter-variable correlation in the future (Fig. 9). A generalization of this type of analysis to many more climate models is therefore needed to assess if the dependence preservation hypothesis is robust. This point can be reformulated as a practical question for multivariate BC developments: should the (inter-variable and/or inter-site) dependence structures evolve from calibration to projection periods? Due to the relative youth of the multivariate BC methods, this is still an open question in the literature that should be further investigated and debated.

Furthermore, the R^2D^2 method only partially preserves the temporal properties of the simulations to be corrected, and all multivariate BC methods necessarily modify the temporal structure and rank chronology of the simulations. If this is indirect for most of them (i.e., when accounting only for inter-site or inter-variable structures), some authors tried specifically to tackle the question of the temporal properties adjustment, such as Johnson and Sharma (2012) with a nesting 1d-BC model working across multiple timescales, Mehrotra and Sharma (2015) including inter-site dependence or Mehrotra and Sharma (2016) including multiple meteorological variables. However, no general comparison of the pros and cons of the two approaches has been performed and any BC method for both inter-site, inter-variable and temporal properties will necessarily consist of a trade-off between the temporal modifications brought by the multivariate adjustment and the correction of the temporal aspects, while respecting their changes from one time period to another.

More generally, there is not yet a complete intercomparison of the multivariate bias adjustment methods. As the need for such multivariate methods becomes crucial for many impact studies, intercomparison exercises are now essential to evaluate the various existing methodologies and to make distinctions, not only between "marginals/dependence" and "successive conditional" correction approaches for example but also between different methods and assumptions within each approach. If such an intercomparison study has to be performed first from the climate point of view (i.e., in terms of quality of the corrected climate variables and their various properties), it should also be conducted from the perspective of some specific impacts and impact models, trying to understand how the quality of the bias adjusted simulations transfer into the often non-linear impact model outputs. To do so, applying a high-dimensional R^2D^2 (and other methods) to various CMIP5 (and upcoming CMIP6) GCM simulations or to various CORDEX RCM runs would generate useful large datasets of multivariate bias corrected climate simulations. From the purely climatic point of view, those datasets would provide a corrected ensemble to conduct climate change studies, such as related to detection and attribution questions (e.g., Yiou et al., 2017), to the evolution in risks of compound events (e.g., Bevacqua et al., 2017) or more generally related to understanding of climate changes. From the societal and/or environmental point of view, those ensembles of multivariate corrected simulations would allow us to investigate how the correction of the dependence structures might modify the impacts of climate change. This question is quite large and concerns many domains, such as hydrology, agronomy, ecology, etc., and can have major consequences on adaptation and mitigation strategies.

Finally, the selection of an "optimal" reference dimension, or at least some preferential ones, is certainly a necessary future step. However, the notion of optimality here may depend on the context of the correction and on the subsequent use of the multivariate bias-corrected data. However, simple selection methods can be imagined. For example, a logical choice can be to select the dimension for which the temporal dynamics of the model to be corrected is the most similar to that of the observations over the calibration period. In such a case, that could correspond to the dimension for which the Spearman rank correlation (or an auto-correlation value) is the closest to that of the reference (observational) data. Of course, other selections are possible but this question is left for future work. In the same idea, we could also consider a "multivariate" reference vector. For example, instead of relying on a univariate reference dimension, the latter can be a couple (or more generally a n-dimensional vector) of dimensions. This would then ensure that the dependence structure linking those dimensions would be exactly that of the initial model and therefore "preserved" (i.e., not corrected). Another natural extension would also be to replace the univariate or multivariate "reference dimension" time series (used to condition the rank resampling in R^2D^2) by physical indices, such as NAO or ENSO indices, coming from the climate model to correct. Hence, by such an approach, R^2D^2 would be applied in a conditional process-oriented BC framework. Indeed, if the present study focused on the methodological aspects of the multivariate bias correction, it is worth keeping in mind that any application of a BC method should be performed with some physically based motivations: depending on their intrinsic skills to model specific features, some climate simulations cannot sensibly be corrected, especially in climate change context where artifacts of bias correction may appear while not visible in present climate evaluations (e.g., Maraun et al., 2017). So the development of BC methodologies allowing one to include some physics in the adjustment procedure is an important perspective of research, in order to have BC approaches not used as black boxes while they should be a support to increase the realism of the climate simulations based on physical knowledge.

Appendix A: Some more mathematical descriptions of the R^2D^2 bias correction method

The multivariate BC method is applied to P statistical dimensions. Those dimensions encompass several physical variables at several grid cell. For example, if there are V physical variables at each of the S grid cell, then $P = V \times S$. Each dimension is observed or simulated over N time steps, and in the following $V_p^A(t)$ is the value of the dimension p from the dataset A (reference, raw or corrected simulations) at time t.

The R^2D^2 method consists of the following steps:

1. Apply separate univariate BC to each dimension. We obtain P univariate time series of N values: $\{(V_1^{1\mathrm{dBC}}(1), \ldots, V_1^{1\mathrm{dBC}}(N)), \ldots, (V_P^{1\mathrm{dBC}}(1), \ldots, V_P^{1\mathrm{dBC}}(N))\}$;

2. Compute the **time series of ranks** for each 1d-bias corrected dimension p mong $(V_p^{1\mathrm{dBC}}(1), \ldots, V_p^{1\mathrm{dBC}}(N))$. For example, for dimension p, we compute $(\mathrm{rank}(V_p^{1\mathrm{dBC}}(1)), \ldots, \mathrm{rank}(V_p^{1\mathrm{dBC}}(N)))$ that will be denoted as $(r_p^{1\mathrm{dBC}}(1), \ldots, r_p^{1\mathrm{dBC}}(N))$. Therefore, for each time t, we have a P-dimensional vector of ranks: $\mathbf{R}^{1\mathrm{dBC}}(t) = (r_1^{1\mathrm{dBC}}(t), \ldots, r_P^{1\mathrm{dBC}}(t))$;

3. Compute the **time series of ranks** for each dimension p among $(V_p^{\mathrm{ref}}(1), \ldots, V_p^{\mathrm{ref}}(N))$ in the reference (calibration) dataset. For example, for dimension p, we compute $(\mathrm{rank}(V_p^{\mathrm{ref}}(1)), \ldots, \mathrm{rank}(V_p^{\mathrm{ref}}(N)))$ that will be denoted as $r_p^{\mathrm{ref}}(1), \ldots, r_p^{\mathrm{ref}}(N))$. Therefore, for each time t, we have a P-dimensional vector of ranks: $\mathbf{R}^{\mathrm{ref}}(t) = (r_1^{\mathrm{ref}}(t), \ldots, r_P^{\mathrm{ref}}(t))$;

4. Choose one dimension p (e.g., $p = 1$) and $(V_p^{1\mathrm{dBC}}(1), \ldots, V_p^{1\mathrm{dBC}}(N))$ as the reference dimension and sequence. Then, for each time t from 1 to N in the projection period:

   (a) Find t^* in the calibration period such that $r_p^{1\mathrm{dBC}}(t) = r_p^{\mathrm{ref}}(t^*)$ and therefore deduce $\mathbf{R}^{\mathrm{ref}}(t^*) = (r_1^{\mathrm{ref}}(t^*), \ldots, r_P^{\mathrm{ref}}(t^*))$;

   (b) For time t in the projection period, impose that the P-dimensional vector of ranks is $\mathbf{R}^{\mathrm{PdBC}}(t) = (r_1^{\mathrm{ref}}(t^*), \ldots, r_p^{1\mathrm{dBC}}(t), \ldots, r_P^{\mathrm{ref}}(t^*))$;

   (c) For all dimensions $d \neq p$, find the time step t_d such that $r_d^{\mathrm{ref}}(t^*) = V_d^{1\mathrm{dBC}}(t_d)$. Then, define the P-dimensional BC vector at time t as $\mathrm{MBC}(t) = (V_1^{1\mathrm{dBC}}(t_1), \ldots, V_p^{1\mathrm{dBC}}(t), \ldots, V_P^{1\mathrm{dBC}}(t_P))$. Thus, $\mathrm{MBC} = \{\mathrm{MBC}(t = 1), \ldots, \mathrm{MBC}(t = N)\}$ gathers the N P-dimensional vectors. In other words, MBC is a P-dimensional time series of length N and contains the multivariate bias corrected data via R^2D^2 with dimension p as reference dimension;

5. Repeat steps 4 (a–c) for all dimensions until P. This generates $\mathrm{MBC_{all}}$, which gathers P objects MBC (one per dimension as reference for the shuffling): $\mathrm{MBC_{all}} = (\mathrm{MBC}(\mathrm{ref.dim.} = 1), \ldots, \mathrm{MBC}(\mathrm{ref.dim.} = P))$.

Competing interests. The author declares that he has no conflict of interest.

Acknowledgements. This work has been partially supported by the ANR-project StaRMIP, the VW-project CE:LLO, the ERA4CS EUPHEME and CoCliServ projects, and the LABEX IPSL project. All computations were made in R. An R package containing functions for the R^2D^2 approach should soon be made available on the CRAN website (http://cran.r-project.org/) or upon request to the author.

Edited by: Florian Pappenberger

References

Araújo, M. and Rahbek, C.: How Does Climate Change Affect Biodiversity?, Science, 313, 1396–1397, https://doi.org/10.1126/science.1131758, 2006.

Bardossy, A. and Pegram, G.: Multiscale spatial recorrelation of RCM precipitation to produce unbiased climate change scenarios over large areas and small, Water Resour. Res., https://doi.org/10.1029/2011WR011524, 2012.

Berg, P., Feldmann, H., and Panitz, H.-J.: Bias correction of high resolution regional climate model data, J. Hydrol., 448–449, 80–92, https://doi.org/10.1016/j.jhydrol.2012.04.026, 2012.

Bevacqua, E., Maraun, D., Hobæk Haff, I., Widmann, M., and Vrac, M.: Multivariate statistical modelling of compound events via pair-copula constructions: analysis of floods in Ravenna (Italy), Hydrol. Earth Syst. Sci., 21, 2701–2723, https://doi.org/10.5194/hess-21-2701-2017, 2017.

Cannon, A.: Multivariate quantile mapping bias correction: An N-dimensional probability density function transform for climate model simulations of multiple variables, Clim. Dynam., 50, 31–49, https://doi.org/10.1007/s00382-017-3580-6, 2017.

Christensen, J., Boberg, F., Christensen, O., and Lucas-Picher, P.: On the need for bias correction of regional climate change projections of temperature and precipitation, Geophys. Res. Lett., 35, https://doi.org/10.1029/2008GL035694, 2008.

Clark, M., Gangopadhyay, S., Hay, L., Rajagopalan, B., and Wilby, R.: The Schaake shuffle: A method for reconstructing space-time variability in forecasted precipitation and temperature fields, J. Hydrometeor., 5, 243–262, 2004.

Colette, A., Vautard, R., and Vrac, M.: Regional climate downscaling with prior statistical correction of the global climate forcing, Geophys. Res. Lett., 39, L13707, https://doi.org/10.1029/2012GL052258, 2012.

Dee, D. P., Uppala, S. M., Simmons, A. J., Berrisford, P., Poli, P., Kobayashi, S., Andrae, U., Balmaseda, M. A., Balsamo, G., Bauer, P., Bechtold, P., Beljaars, A. C. M., van de Berg, L., Bidlot, J., Bormann, N., Delsol, C., Dragani, R., Fuentes, M., Geer, A. J., Haimberger, L., Healy, S. B., Hersbach, H., Hólm, E. V., Isaksen, L., Kållberg, P., Köhler, M., Matricardi, M., McNally, A. P., Monge-Sanz, B. M., Morcrette, J.-J., Park, B.-K., Peubey, C., de Rosnay, P., Tavolato, C., Thépaut, J.-N., and Vitart, F.: The ERA-Interim reanalysis: configuration and performance of the data assimilation system, Q. J. Roy. Meteor. Soc., 137, 553–597, https://doi.org/10.1002/qj.828, 2011 (data available at: https://www.ecmwf.int/en/forecasts/datasets/archive-datasets/reanalysis-datasets/era-interim/, last access: 1 June 2018).

Dekens, L., Parey, S., Grandjacques, M., and Dacunha-Castelle, D.: Multivariate distribution correction of climate model outputs: A generalization of quantile mapping approaches, Environmetrics, 28, e2454, https://doi.org/10.1002/env.2454, 2017.

Déqué, M.: Frequency of precipitation and temperature extremes over France in an anthropogenic scenario: Model results and statistical correction according to observed values, Global Planet. Change, 57, 16–26, 2007.

Dufresne, J.-L., Foujols, M.-A., Denvil, S., Caubel, A., Marti, O., Aumont, O., Balkanski, Y., Bekki, S., Bellenger, H., Benshila, R., Bony, S., Bopp, L., Braconnot, P., Brockmann, P., Cadule, P., Cheruy, F., Codron, F., Cozic, A., Cugnet, D., de Noblet, N., Duvel, J.-P., Ethé, C., Fairhead, L., Fichefet, T., Flavoni, S., Friedlingstein, P., Grandpeix, J.-Y., Guez, L., Guilyardi, E., Hauglustaine, D., Hourdin, F., Idelkadi, A., Ghattas, J., Joussaume, S., Kageyama, M., Krinner, G., Labetoulle, S., Lahellec, A., Lefebvre, M.-P., Lefevre, F., Levy, C., Li, Z. X., Lloyd, J., Lott, F., Madec, G., Mancip, M., Marchand, M., Masson, S., Meurdesoif, Y., Mignot, J., Musat, I., Parouty, S., Polcher, J., Rio, C., Schulz, M., Swingedouw, D., Szopa, S., Talandier, C., Terray, P., Viovy, N., and Vuichard, N.: Climate change projections using the IPSL-CM5 Earth System Model: from CMIP3 to CMIP5, Clim. Dynam., 40, 2123–2165, 2013.

Giorgi, F., Jones, C., and Asrar, G.: Addressing climate information needs at the regional level: the CORDEX framework, Bull. World Meteorol. Organ., 58, 175–183, 2009.

Haddad, Z. and Rosenfeld, D.: Optimality of empirical z-r relations, Q. J. Roy. Meteor. Soc., 123, 1283–1293, 1997.

IPCC: Climate Change 2013: The Physical Science Basis. Contribution of Working Group I to the Fifth Assessment Report of the Intergovern mental Panel on Climate Change, edited by: Stocker, T. F., Qin, D., Plattner, G.-K., Tignor, M., Allen, S. K., Boschung, J., Nauels, A., Xia, Y., Bex , V., and Midgley, P. M., Cambridge University Press, Cambridge, United Kingdom and New York, NY, USA, 2013.

Jacob, D., Petersen, J., Eggert, B., Alias, A., Christensen, O. B., Bouwer, L. M., Braun, A., Colette, A., Déqué, M., Georgievski, G., Georgopoulou, E., Gobiet, A., Nikulin, G., Haensler, A., Hempelmann, N., Jones, C., Keuler, K., Kovats, S., Kröner, N., Kotlarski, S., Kriegsmann, A., Martin, E., van Meijgaard, E., Moseley, C., Pfeifer, S., Preuschmann, S., Radtke, K., Rechid, D., Rounsevell, M., Samuelsson, P., Somot, S., Soussana, J.-F., Teichmann, C., Valentini, R., Vautard, R., and Weber, B.: EURO-CORDEX: New high-resolution climate change projections for European impact research, Reg. Environ. Change, 14, 563–578, 2014 (data available at: https://www.euro-cordex.net/060378/index.php.en, last access: 1 June 2018).

Johnson, F. and Sharma, A.: A nesting model for bias correction of variability at multiple time scales in general circulation model precipitation simulations, Water Resour. Res., 48, W01504, https://doi.org/10.1029/2011WR010464, 2012.

Kallache, M., Vrac, M., Naveau, P., and Michelangeli, P.-

A.: Non-stationary probabilistic downscaling of extreme precipitation, J. Geophys. Res.-Atmos., 116, D05113, https://doi.org/10.1029/2010JD014892, 2011.

Lavaysse, C., Vrac, M., Drobinski, P., Lengaigne, M., and Vischel, T.: Statistical downscaling of the French Mediterranean climate: assessment for present and projection in an anthropogenic scenario, Nat. Hazards Earth Syst. Sci., 12, 651–670, https://doi.org/10.5194/nhess-12-651-2012, 2012. .

Li, Z., Shi, X., and Li, J.: Multisite and multivariate GCM downscaling using a distribution-free shuffle procedure for correlation reconstruction, Climate Res., 72, 141–151, https://doi.org/10.3354/cr01460, 2017.

Mao, G., Vogl, S., Laux, P., Wagner, S., and Kunstmann, H.: Stochastic bias correction of dynamically downscaled precipitation fields for Germany through Copula-based integration of gridded observation data, Hydrol. Earth Syst. Sci., 19, 1787–1806, https://doi.org/10.5194/hess-19-1787-2015, 2015.

Maraun, D., Shepherd, T., Widmann, M., Zappa, G., Walton, D., Gutierrez, J., Hagemann, S., Richter, I., Soares, P., Hall, A., and Mearns, L.: Towards process informed bias correction of climate change simulations, Nat. Clim. Change, 7, 764–773, https://doi.org/10.1038/nclimate3418, 2017.

Marti, O., Braconnot, P., Dufresne, J., Bellier, J., R., B., Bony, S., Brockmann, P., Cadule, P., Caubel, A., Codron, F., de Noblet, N., Denvil, S., Fairhead, L., Fichefet, T., Foujols, M., Friedlingstein, P., Goosse, H., Grandpeix, J., Guilyardi, E., Hourdin, F., Idelkadi, A., Kageyama, M., Krinner, G., Lévy, C., Madec, G., Mignot, J., Musat, I., Swingedouw, D., and Talandier, C.: Key features of the IPSL ocean atmosphere model and its sensitivity to atmospheric resolution, Clim. Dynam., 34, 1–26, 2010.

Mehrotra, R. and Sharma, A.: Correcting for systematic biases in multiple raw GCM variables across a range of timescales, J. Hydrol., 520, 214–223, https://doi.org/10.1016/j.jhydrol.2014.11.037, 2015.

Mehrotra, R. and Sharma, A.: A Multivariate Quantile-Matching Bias Correction Approach with Auto- and Cross-Dependence across Multiple Time Scales: Implications for Downscaling, J. Climate, 29, 3519–3539, https://doi.org/10.1175/JCLI-D-15-0356.1, 2016.

Michelangeli, P., Vrac, M., and Loukos, H.: Probabilistic downscaling approaches: application to wind cumulative distribution functions, Geophys. Res. Lett., 36, L11708, https://doi.org/10.1029/2009GL038401, 2009.

Muerth, M. J., Gauvin St-Denis, B., Ricard, S., Velázquez, J. A., Schmid, J., Minville, M., Caya, D., Chaumont, D., Ludwig, R., and Turcotte, R.: On the need for bias correction in regional climate scenarios to assess climate change impacts on river runoff, Hydrol. Earth Syst. Sci., 17, 1189–1204, https://doi.org/10.5194/hess-17-1189-2013, 2013.

Müller, C., Bondeau, A., Popp, A., Waha, K., and Fader, M.: Climate Change Impacts on Agricultural Yields, Tech. rep., World Bank, Washington, DC, available at: https://openknowledge.worldbank.org/handle/10986/9065 (last access: 1 June 2018), 2010.

Nelsen, R. B.: An Introduction to Copulas, 2nd edition, Springer, New-York, 2006.

Oettli, P., Sultan, B., Baron, C., and Vrac, M.: Are regional climate models relevant for crop yield prediction in West Africa?, Environ. Res. Lett, 6, 014008, https://doi.org/10.1088/1748-9326/6/1/014008, 2011.

Piani, C. and Haerter, J. O.: Two dimensional bias correction of temperature and precipitation copulas in climate models, Geophys. Res. Lett., 39, L20401, https://doi.org/10.1029/2012GL053839, 2012.

Quintana-Segui, P., Le Moigne, P., Durand, Y., Martin, E., Habets, F., Baillon, M., Canellas, C., Franchistéguy, L., and Morel, S.: Analysis of near surface atmospheric variables: validation of the SAFRAN analysis over France, J. Appl. Meteorol. Clim., 47, 92–107, 2008 (data available at: https://donneespubliques.meteofrance.fr/?fond=produit&id_produit=230&id_rubrique=40, last access: 1 June 2018).

Samuelsson, P., Jones, C. G., Will'En, U., Ullerstig, A., Gollvik, S., Hansson, U., Jansson, E., Kjellström, C., Nikulin, G., and Wyser, K.: The Rossby centre regional climate model RCA3: model description and performance, Tellus, 63A, 4–23, https://doi.org/10.1111/j.1600-0870.2010.00478.x, 2011.

Schmidli, J., Frei, C., and Vidale, P.: Downscaling From Gcm Precipitation: A Benchmark For Dynamical And Statistical Downscaling Methods, J. Climatol., 26, 679–689, https://doi.org/10.1002/joc.1287, 2006.

Schneider, C., Laizé, C. L. R., Acreman, M. C., and Flörke, M.: How will climate change modify river flow regimes in Europe?, Hydrol. Earth Syst. Sci., 17, 325–339, https://doi.org/10.5194/hess-17-325-2013, 2013.

Schölzel, C. and Friederichs, P.: Multivariate non-normally distributed random variables in climate research – introduction to the copula approach, Nonlin. Processes Geophys., 15, 761–772, https://doi.org/10.5194/npg-15-761-2008, 2008.

Skamarock, W., Klemp, J., Dudhia, J., Gill, D., Barker, M., Duda, K., Huang, X., Wang, W., and Powers, J.: A description of the Advanced Research WRF Version 3, Tech. Note NCAR/TN-475+STR, NCAR, Natl. Cent. for Atmos. Res., Boulder, Colo, 2008.

Sklar, A.: Fonctions de répartition à n dimensions et leurs marges, Tech. Rep. 229–231, Publ. Inst. Statist. Univ. Paris 8, 1959.

Tisseuil, C., Vrac, M., Grenouillet, G., Gevrey, M., Oberdorff, T., Wade, A., and Lek, S.: Strengthening the link between hydro-climatic downscaling and species distribution modelling: Climate change impacts on freshwater biodiversity, Sci. Total Environ., 424, 193–201, https://doi.org/10.1016/j.scitotenv.2012.02.035, 2012.

Tramblay, Y., Ruelland, D., Somot, S., Bouaicha, R., and Servat, E.: High-resolution Med-CORDEX regional climate model simulations for hydrological impact studies: a first evaluation of the ALADIN-Climate model in Morocco, Hydrol. Earth Syst. Sci., 17, 3721–3739, https://doi.org/10.5194/hess-17-3721-2013, 2013.

Verkade, J., Brown, J., Weerts, A., and Reggiani, P. Weerts, A. H.: Post-processing ECMWF precipitation and temperature ensemble reforecasts for operational hydrologic forecasting at various spatial scales, J. Hydrol., 501, 73–91, https://doi.org/10.1016/j.jhydrol.2013.07.039, 2013.

Vigaud, N., Vrac, M., and Caballero, Y.: Probabilistic Downscaling of GCM scenarios over southern India, J. Climatol., 33, 1248–1263, https://doi.org/10.1002/joc.3509, 2013.

Voisin, N., Schaake, J. C., and Lettenmaier, D. P.: Calibration and downscaling methods for quantitative ensemble precipitation forecasts, Weather Forecast., 25, 1603–1627, 2010.

Volosciuk, C., Maraun, D., Vrac, M., and Widmann, M.: A combined statistical bias correction and stochastic downscaling method for precipitation, Hydrol. Earth Syst. Sci., 21, 1693–1719, https://doi.org/10.5194/hess-21-1693-2017, 2017.

Vrac, M. and Friederichs, P.: Multivariate – Intervariable, Spatial, and Temporal – Bias Correction, J. Climate, 28, 218–237, https://doi.org/10.1175/JCLI-D-14-00059.1, 2015.

Vrac, M. and Vaittinada Ayar, P.: Influence of Bias Correcting Predictors on Statistical Downscaling Models, J. Appl. Meteorol. Clim., 56, 5–26, https://doi.org/10.1175/JAMC-D-16-0079.1, 2017.

Vrac, M., Billard, L., Diday, E., and Chédin, A.: Copula Analysis of Mixture Models, Computation. Stat., 27, 427, https://doi.org/10.1007/s00180-011-0266-0, 2011.

Vrac, M., Drobinski, P., Merlo, A., Herrmann, M., Lavaysse, C., Li, L., and Somot, S.: Dynamical and statistical downscaling of the French Mediterranean climate: uncertainty assessment, Nat. Hazards Earth Syst. Sci., 12, 2769–2784, https://doi.org/10.5194/nhess-12-2769-2012, 2012.

Vrac, M., Noël, T., and Vautard, R.: Bias correction of precipitation through Singularity Stochastic Removal: Because occurrencesmatter, J. Geophys. Res.-Atmos., 121, 5237–5258, https://doi.org/10.1002/2015JD024511, 2016.

Wheeler, T. and von Braun, J.: Climate Change Impacts on Global Food Security, Science, 341, 508–513, https://doi.org/10.1126/science.1239402, 2013.

Wong, G., Maraun, D., Vrac, M., Widmann, M., Eden, J., and T. Kent, T.: Stochastic model output statistics for bias correcting and downscaling precipitation including extremes, J. Climate, 27, 6940–6959, https://doi.org/10.1175/JCLI-D-13-00604.1, 2014.

Xu, C.-Y.: From GCMs to river flow: a review of downscaling methods and hydrologic modelling approaches, Prog. Phys. Geog., 23, 229–249, https://doi.org/10.1177/030913339902300204, 1999.

Yiou, P., Jézéquel, A., Naveau, P., Otto, F., Vautard, R., and Vrac, M.: A statistical framework for conditional extreme event attribution, Adv. Stat. Clim. Meteorol. Oceanogr., 3, 17–31, https://doi.org/10.5194/ascmo-3-17-2017, 2017.

3

Attributing regional trends of evapotranspiration and gross primary productivity with remote sensing: a case study in the North China Plain

Xingguo Mo[1,2], Xuejuan Chen[1,2], Shi Hu[1], Suxia Liu[1,2], and Jun Xia[1]

[1]Key Laboratory of Water Cycle & Related Land Surface Processes, Institute of Geographic Sciences and Natural Resources Research, Chinese Academy of Sciences, Beijing 100101, China

[2]College of Resources and Environment, University of Chinese Academy of Sciences, Beijing, 100190, China

Correspondence to: Xingguo Mo (moxg@igsnrr.ac.cn)

Abstract. Attributing changes in evapotranspiration (ET) and gross primary productivity (GPP) is crucial for impact and adaptation assessment of the agro-ecosystems to climate change. Simulations with the VIP model revealed that annual ET and GPP slightly increased from 1981 to 2013 over the North China Plain. The tendencies of both ET and GPP were upward in the spring season, while they were weak and downward in the summer season. A complete factor analysis illustrated that the relative contributions of climatic change, CO_2 fertilization, and management to the ET (GPP) trend were 56 (−32) %, −28 (25) %, and 68 (108) %, respectively. The decline of global radiation resulted from deteriorated aerosol and air pollution was the principal cause of GPP decline in summer, while air warming intensified the water cycle and advanced the plant productivity in the spring season. Generally, agronomic improvements were the principal drivers of crop productivity enhancement.

1 Introduction

Terrestrial hydrological and carbon cycles are intimately coupled via transpiration and photosynthesis processes which are regulated by plant leaf stomata. Due to land use/cover changes, intensified agricultural management, and climatic change, terrestrial eco-hydrological processes have been noticeably shifted on multiple spatiotemporal scales (Tian et al., 2011; Douville et al., 2013); for example, prevailing irrigation and application of chemical fertilizers have raised soil moisture, evapotranspiration (ET), and crop productivity, etc. In some regions the effects of human activities are the same magnitude as, or even exceed, the impacts of global warming on the productions of agro-ecosystems (Haddeland et al., 2014). In the last decades, global consumptive water use and carbon fixation by terrestrial ecosystems have been demonstrated to slightly increase with more efficient water use, corresponding to changes in climatic factors and the fertilization effect of elevated atmospheric CO_2 concentration (Yan et al., 2013; Nayak et al., 2013). Consequently, spatiotemporal patterns of water and carbon fluxes on the regional scale are changing under global change (Zeng et al., 2014; Liu et al., 2012).

As ET is the major component of water budget in the water-limited basins, its long-term tendency has been taken as a critical indicator for diagnosing the intensification of the regional water cycle. The complementary relationship between actual and potential ET may reveal some clues to hydrological changes. Observations illustrated that potential evaporation rates (ET_p) (represented as pan evaporation) decreased in Europe, the US, China, India, and Australia in recent decades (Brutsaert, 2006; Katul et al., 2012), implying the decline of available energy and aerodynamics devoted to latent heat flux over the land surface. The climatic factors dominating ET_p change are usually diverse. For example, over the North China Plain (NCP) the changes in ET_p are mainly attributed to declines in global radiation and near-surface wind speed (Tang et al., 2011; Song et al., 2009). However, in southern Turkey a noticeable decline in ET_p was attributed to enhanced air humidity associated with the expansion of irrigation acreage and more water evaporated into the atmospheric boundary layer (Ozdogan and Salvucci, 2004). Burn and Hesch (2007) revealed that decreasing wind speed and raised water vapour deficit responded to the trend of ET_p in the Canadian Prairies. On a large scale, precipitation is usually the principal factor determining actual ET change; for example, Qian et al. (2007) showed that an increase in ET in the Mississippi River basin followed precipitation propensity, while the effects of solar radiation and air temperature changes were minor.

Terrestrial eco-hydrological processes are driven by climate and modulated by human activities. Generally climate warming enhances atmospheric evaporative demand, while CO_2 fertilization stimulates photosynthesis and inhibits leaf stomatal conductance, leading to more biomass accumulation and higher water productivity (Field et al., 1995; Buckley and Mott, 2013). Simultaneously, land use change and land management also noticeably affect the ecosystem production and hydrological fluxes (Shi et al., 2011). Separating the contributions of climatic change, CO_2 enrichment, and human activities to the long-term trends of water and carbon cycles is critical for assessment of ecosystem responses and resilience to environmental changes. Some researchers have explored the relative contributions of climate change and vegetation dynamics to changes in global land surface evapotranspiration and river runoff (Betts et al., 2007; Piao et al., 2007; Alkama et al., 2010; Liu et al., 2012; Chen et al., 2014, 2015; Banger et al., 2015), but the conclusions are still inconsistent. Climate change dominated the interannual variability of ET, while land use changes and agricultural practices and techniques exerted more discernable effects on the water cycle in the long term (Liu et al., 2012). However, Alkama et al. (2010) and Shi et al. (2011) demonstrate that climate change is the predominant driver of the global ET tendency in the 20th century. Commonly, the contributions of climate change to vegetation productivities on a large scale are quantified by the ecosystem models or statistical models. For example, Piao et al. (2015) documented that elevated atmospheric CO_2 and nitrogen deposition were the critical contributors to terrestrial greening over China; Baker et al. (2010) figured out that climate anomalies in springtime were the most frequent drivers of annual GPP variations in North America; Nayak et al. (2013) reported that climate change had a relatively small but significant control (15 %) on the trend of terrestrial net primary production (NPP) over India. In the crop ecosystems, contributions of climate change, cultivar renewal, and agronomic management to change in crop yield have been separated with the crop or statistical models (Yu et al., 2012; Song et al., 2014; Bai et al., 2016; Guo et al., 2014; Z. Wang et al., 2016). The impacts of climate change on crop yield may be positive or negative in different regions, depending on the tendencies of the dominant factors (Ewert et al., 2015).

As one of the granaries in China, the North China Plain (NCP) is confronted with challenges of agriculture sustainability due to global change and social and economic development. Thereby, it is crucial to understand the impacts of climate change on the productions of cropping systems. Over the plain, a winter wheat–summer maize double cropping system prevails, supported by irrigation, fertilizer, and agronomic techniques. In situ measurements, agricultural annals, and the regional remotely sensed vegetation index dataset all illustrate that both wheat and maize productivities have accrued remarkably since the 1970s (Mo et al., 2009; Yuan and Shen, 2013); correspondingly, the seasonal water consumption and water use efficiency are also slightly improved (Zhang et al., 2011). The achievement of long-term increasing grain production is related to the active adoption of new varieties for stabilization, extending the length of the crop growth period, as well as agronomic technique advancement (Liu et al., 2010; Sacks and Kucharik, 2011). Currently, water amount for food production consists of 65 % of total water consumption here. Further, along with the gradually augmented domestic and industrial water requirement, groundwater in some areas of the plain has been overexploited, and the environmental water requirement is generally under deficit conditions (e.g. MWR, 2010). Faced with the rapid deteriorating agricultural environment, some critical issues are still unclear, such as what mechanisms drive the evolutions of eco-hydrological processes over the plain? What are the impacts of climate change on the cropping systems?

In this study, the VIP eco-hydrological dynamic model integrated with the NOAA-AVHRR remotely sensed normalized difference of vegetation index (NDVI) is employed to assess the spatiotemporal evolutions of ET and vegetation GPP over the NCP from 1981 to 2013. By numerical experiments with the VIP model and the factor separation method, the contributions of climate change, fertilization of atmospheric CO_2 enrichment, agronomic practices, and technological advancement to crop water consumption and productivity are then analysed and the relevant mechanisms are discussed.

2 Method and materials

2.1 Study region

The NCP is one of the country's granaries, extending from latitude 32°00′ to 40°24′ N and longitude 112°48′ to 122°45′ E (Fig. 1a, b). It is located in the eastern part of China, with an area of 40×10^4 km^2, which is an alluvial plain developed by the intermittent flooding of the Huang, Huai, and Hai rivers, and 85 % is cultivated as farmland. The warm temperate climate varies gradually from subhumid in the southern to semi-arid in the northern parts. The annual precipitation ranges from 500 to 1000 mm, occurring irregularly among the seasons, and more than 70 % falls in summer. Soil moisture deficit happens widely during the spring and early summer period. Besides soybean/millet/sorghum/cotton, the double cropping system of winter wheat–summer maize prevails in the plain, where wheat and maize are the most common harvest crops in the summer and autumn seasons, respectively. Due to insufficient precipitation, the spring crops (such as wheat) usually need supplemental irrigation to form favourable production.

2.2 The VIP eco-hydrological model

The physical-process-based VIP (Vegetation Interface Processes) eco-hydrological model is designed to simulate the exchanges of energy, water, and carbon between the terrestrial ecosystem and the atmosphere (Mo et al., 2014). In the model, ET is termed the summation of canopy transpiration and evaporation from the canopy intercept and the soil surface, computed separately with the Penman–Monteith equation. Transpiration and photosynthesis processes are coupled through the Ball–Berry relationship between leaf stomatal conductance and net assimilation rate. Regarding the carbon cycle aspect, leaf carbon fixations on sunlit and shaded leaves are predicted with the biochemical schemes for C3 (Farquhar et al., 1980) and C4 plants (Collatz et al., 1992). In the radiation budget scheme, shortwave radiation transfer in the canopy distinguishes the leaf spectral properties of visible and near-infrared radiation, as well as the fraction of direct beam and diffusive irradiance in global radiation. Precipitation throughfall–runoff generation over the land surface is calculated with a curve-number (CN) type equation on a daily scale, using the daily net precipitation and the moisture deficit of the upper soil layer. Simulation of soil water movement in the root zone is carried out with a discrete Richards equation in three layers. The crop and natural vegetation growth modules are also embedded in the model to simulate the biomass accumulation and carbon cycle.

2.3 Data

The VIP model input data include land use/cover, soil physical properties, and atmospheric forcing variables. The GIMMS AVHRR 15-day normalized difference of vegetation index time series (NDVI3g; values are scaled by 1000) (https://nex.nasa.gov/nex/projects/1349/wiki/general_data_descriptionand_access/) (Pinzon and Tucker, 2014) from 1981 to 2013 is used to retrieve the vegetation leaf area index and other land surface characteristics. The land use classification originates from both Landsat TM images (www.resdc.cn) and MODIS remote sensing products, in which the farmland is classified as rice paddy and dryland. Soil textural data are available on a scale of 1 : 1 000 000 represented as fractions of sand, silt, and clay, by which the parameters of soil porosity (θ_{sat}) and saturated hydraulic conductivity (K_{ws}, mm s^{-1}) are estimated as in Bonan (1996). Daily climate variables (air temperature, water vapour pressure, wind speed, sunshine duration, and precipitation) recorded at 87 climatic stations (http://data.cma.cn)

Voisin, N., Schaake, J. C., and Lettenmaier, D. P.: Calibration and downscaling methods for quantitative ensemble precipitation

in and around the study area are available for generating the spatial atmospheric forces. The NDVI data are error-checked and the erroneous data are replaced by interpolation with the preceding and subsequent values according to the time series by the Savitzky–Golay (SG) filter (Savitzky and Golay, 1964), and then the daily values are derived with the Lagrange polynomial. Vegetation leaf area index (LAI) is retrieved from the NDVI with empirical relationships for different plant functional types.

The data used for model validation are field flux measurements with an eddy covariance technique at the Yucheng (116°38′ E, 36°57′ N), Daxing (116°25′ E, 39°37′ N), Miyun (117°19′ E, 40°38′ N), and Guantao (115°8′ E, 36°31′ N) sites over the plain. The cropping systems at the Yucheng, Daxing, and Guantao sites are all rotation of winter wheat–summer maize, while land cover is dwarf shrub at the Miyun site. The eddy covariance data are processed with general procedures (Liu and Xu, 2013; Liu et al., 2013). GPP data are available only at the Yucheng site. In addition to eddy covariance fluxes, grain yield records of wheat and maize in county statistics are also used to verify the GPP predictions at the regional scale. The method of crop yield converted to GPP follows Xin et al. (2015).

2.4 Model implementation and numerical experimental design

2.4.1 Simulation setup

The model simulations were conducted at 8 km spatial resolution and a half-hour time step. The cropland practices are classified into wheat–maize and wheat–rice rotations. Atmospheric driving forces are interpolated from daily meteorological variables recorded at the climatic stations to grid cells with a gradient inverse distance square method (GIDS), which accounts for the effects of elevation, latitude, and longitude (Nalder and Wein, 1998). Estimated with sunshine duration in a linear relationship, the global radiation is subdivided into direct visible and near-infrared parts, as well as direct beam and diffusive components with Weiss and Norman (1985). The daily air temperature is extended to hourly values with a sinusoidal function based on the daily maximum and minimum temperatures (Campbell and Norman, 1998). During the winter wheat growing period, irrigation water is supplied when water storage in the root zone is below 60 % of the field capacity. Summer maize is set to be irrigated not more than one time in its growth period. The simulation is conducted with prescribed daily LAI series retrieved from remotely sensed NDVI series for eco-hydrological prediction from 1980 to 2013, in which the first year is taken as warming up.

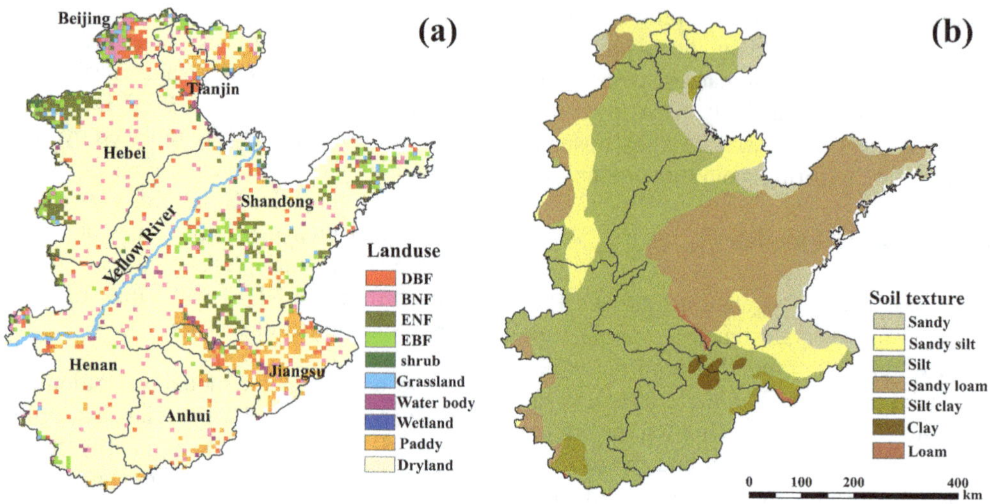

Figure 1. Land use/cover **(a)** and soil texture **(b)** of the North China Plain (NCP) (DBF: deciduous broadleaf forest; BNF: broadleaf and needleleaf mixed forest; ENF: evergreen needleleaf forest; EBF: evergreen broadleaf forest).

2.4.2 Separation of climate change and management effects

By using a general function, f, the scalar fluxes (water vapour and carbon) between land surface and the atmosphere are determined by climate factors (M), fertilization of elevated atmospheric CO_2 (C), and agronomic management and technological advancement (in this study we assume the long-term trend of leaf area index (LAI) may represent the effects of human activities on crop and natural ecosystems. Human activities (also referred to as management practices) in the agro-ecosystem include renewals of cultivars, irrigation facility improvement, fertilizer application, soil quality amelioration, etc.), namely,

$$f = F(M, \mathrm{C}, \mathrm{LAI}, \ldots). \tag{1}$$

The changes in f contributed by a single factor (expressed as f_i) and its interaction with another factor (expressed as f_{ij}) can be decomposed by the Taylor expansion as

$$f_i = \frac{\partial F}{\partial x_i}\Delta x_i + \frac{1}{2!}\frac{\partial^2 F}{\partial x_i^2}\Delta x_i^2 + \cdots + \frac{1}{n!}\frac{\partial^n F}{\partial x_i^n}\Delta x_i^n, \tag{2}$$

$$f_{ij} = f_i + f_j + \frac{\partial^2 F}{\partial x_i \partial x_j}\Delta x_i \Delta x_j + \cdots, \tag{3}$$

where x_i and x_j ($i \neq j$) represent M, C, and LAI, respectively. The factor separation methodology from Stein and Alpert (1993) and Alkama et al. (2010) is used to categorize the contributions of climate change, CO_2 fertilization, and LAI, as well as their interactions to long-term trends of ET and GPP. Similar to Alkama et al. (2010), the total effect, f_{123}, is expressed as

$$f_{123} = f_1 + f_2 + f_3 + f^{12} + f^{13} + f^{23} + f^{123}, \tag{4}$$

with

$$f^{12} = f_{12} - f_1 - f_2, \tag{5}$$

$$f^{13} = f_{13} - f_1 - f_3, \tag{6}$$

$$f^{23} = f_{23} - f_2 - f_3, \tag{7}$$

where f_1, f_2, and f_3 are the direct contributions of climate change, atmospheric CO_2 enrichment fertilization, and agronomic management, respectively; f^{12} is the contribution of interactions of climate change and CO_2 enrichment; f^{13} is the contribution of interactions between climate change and agronomic management; f^{23} is the contribution of interactions between CO_2 fertilization and agronomic management; f^{123} is the contribution of interactions between climate change, CO_2 fertilization, and agronomic management.

Seven numerical experiments designed to fully distinguish the contributions of climate change, CO_2 fertilization, and agronomic management are conducted by the VIP model over the NCP from 1980 to 2013. The experiments are as follows:

1. f_{123} ("all factors"): current climate, CO_2, and LAI spatiotemporal pattern;

2. f_1 ("climate change effect"): current climate, but the atmospheric CO_2 concentration is fixed at year 1981, and the LAI pattern is set as the multi-year average;

3. f_2 ("CO_2 fertilization effect"): climate and LAI are fixed at a specific year, but with the current CO_2 concentration;

4. f_3 ("management effect"): climate and CO_2 concentration are fixed at a specific year, but the current LAI pattern is used;

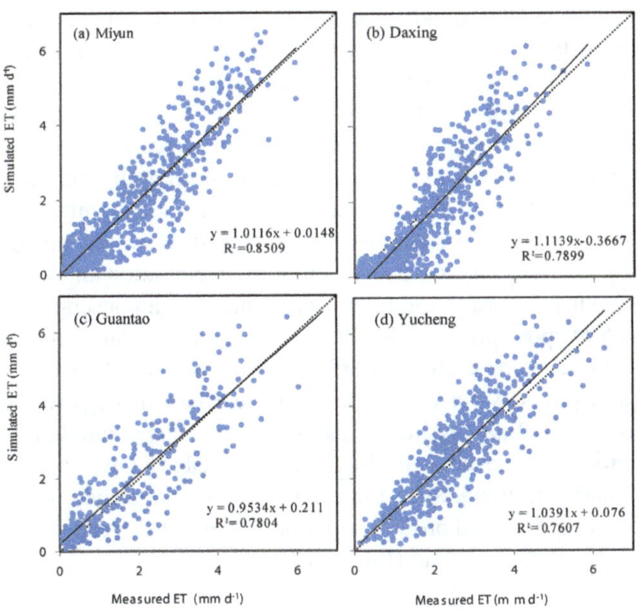

Figure 2. Comparisons of the simulated daily ET with eddy covariance measurements.

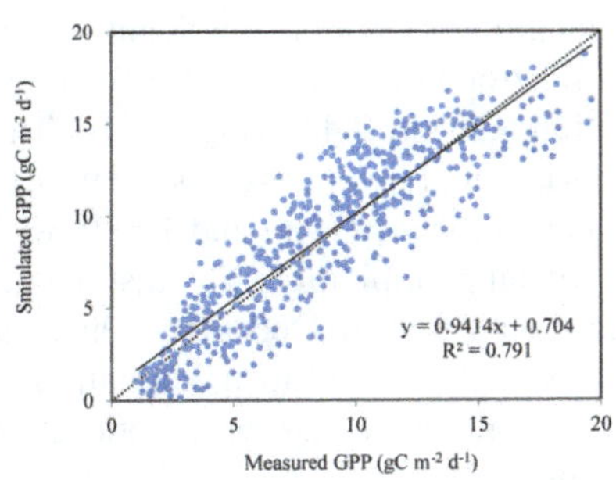

Figure 3. Comparison of the simulated GPP with eddy covariance measurements at the Yucheng site.

5. f_{12} ("climate change and CO_2 fertilization effects"): LAI pattern is fixed, but the current climate and CO_2 concentration are used;

6. f_{13} ("climate change and management effects"): CO_2 concentration is fixed, but the current climate and LAI pattern are used; and

7. f_{23} ("CO_2 fertilization and management effects"): climate is fixed at 1981, but current CO_2 and LAI are used.

The trends of annual ET and GPP in the above experiments are calculated, and then, according to Eqs. (4) to (7), the contributions of climate change and CO_2 fertilization and management practices to ET and GPP long-term trends are then separated.

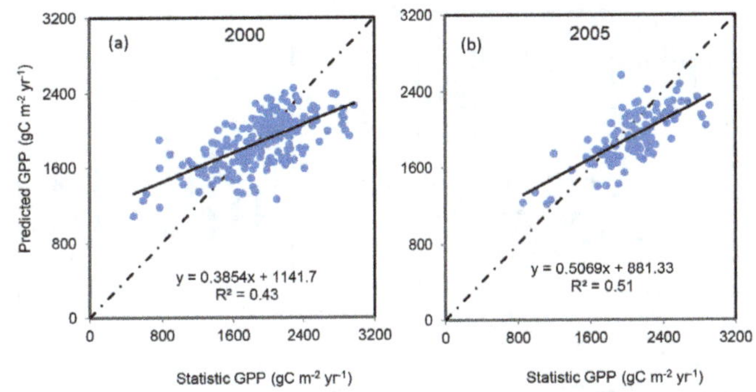

Figure 4. Comparisons of the predicted GPP and statistic yield derived GPP at county scale in 2000 and 2005.

3 Result analysis

3.1 Model verification

3.1.1 Validation with eddy covariance measurements

The VIP model is used to simulate the hydrological, energy partitioning, and crop growth processes at the four sites of eddy flux measurement. Here, eddy covariance measurements of daily ET and GPP are employed to verify the model predictions (ET is available at all the sites, but GPP is only available at one site). The land surface characteristics surrounding the flux mast are relatively homogeneous, ensuring the footprint for flux measurement. The meteorological information measured at each site is used to drive the VIP model. It is shown that the agreements are quite satisfactory for both ET and GPP (Figs. 2 and 3). In total, there are 9-year daily ET data and 3-year daily GPP data for comparison with the model simulations. The coefficients of determination (R^2) are above 0.76 for all the sites. On an annual scale, the relative absolute biases of predicted ET ranged from 1.5 to 12.6 % in the 9-year dataset, and biases of GPP are from 2.0 to 8.8 % in 3-year data. Therefore, the model performance is quite good and reliable for predictions of vegetation/crop productivity and water consumption. The biases may have stemmed from both measurements and model parameter uncertainty. Mo et al. (2012) showed that canopy leaf area index (LAI) and photosynthetic capacity (carboxylation rate for C3 crops and photon quantum use efficiency for C4 crops) were the parameters most sensitive to the model efficiency. Here, taking the Yucheng site as an example, annual ET and GPP may increase by 1.6 % (2.6 %) and 3.0 % (15.9 %), respectively, as LAI (photosynthesis capacity) is increased by 20 %.

3.1.2 Validation with the statistical yield records

The simulated GPP is also validated with the staple crop grain yield statistics at county level. The crop grain yield per hectare is converted to the equivalent GPP per square metre. As shown in Fig. 4 (years 2000 and 2005 are used), the

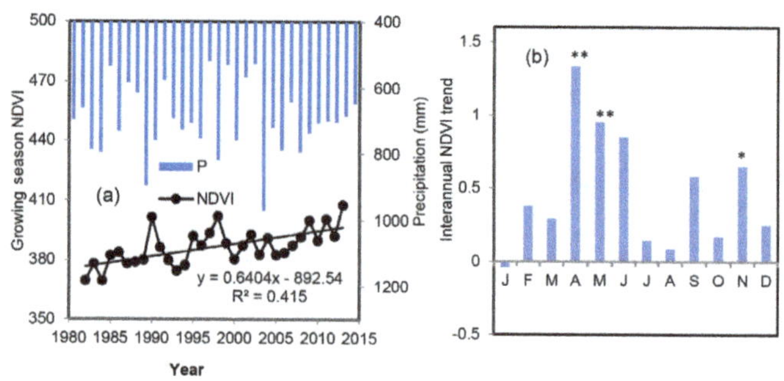

Figure 5. Spatial average trends of the growing season NDVI at annual **(a)** and monthly **(b)** scales (significance levels: * is $p<0.05$; ** is $p<0.01$; the NDVI is scaled by 1000).

agreement is satisfactory, with coefficients of determination (R^2) of 0.43 and 0.51 ($p<0.001$), respectively. It is noted that the model overestimated low-yield but underestimated high-yield zones. The errors may have resulted from two aspects, namely, biases in the LAI-NDVI empirical relationship and the harvest index in low- and high-yield fields. There are remarkable spatial variations in crop yields that resulted from diverse climates, soil physical texture, chemical composition, and management practices. In the simulation, it is found that the spatiotemporal evolution of greenness is a crucial factor regulating the yield patterns. Greenness represented by the vegetation index is an appropriate indicator of crop productivity under environmental stresses (Hu et al., 2014). In the areas with a high vegetation index and favourable irrigation facilities, the yield losses may be caused by heat waves or pest infections in the mature stage.

3.2 Trends of climate, crop productivity, and ET

Changes in climate variables and agro-ecosystem management are the dominant driving forces of the evolution of regional eco-hydrological processes. Intra-seasonal variations of climatic variables may exert different impacts on the crop water consumption and carbon assimilation. In the last 3 decades, air temperature has been rising, but sunshine duration and wind speed are decreasing significantly over the plain, associated with global climate change, aerosol, and air pollution exacerbation. Soil amelioration, genetic improvement, irrigation facility constructions, and application of chemical synthesis fertilizer are considered to be the principal drivers that have propelled the crop productivity close to the attainable level (Yu et al., 2012; Lobell and Burke, 2010).

3.2.1 Changes in climatic variables

Grid averages of the climatic variables were analysed over the North China Plain from 1980 to 2013. Nevertheless, inhomogeneous distributions of the climatic variables and the spatially averaged trends were clear (Table 1). On an annual scale, global radiation, air temperature (especially minimum temperature), and wind speed changed significantly ($p<0.01$). On a monthly scale, radiation declined in all the months except March, but only trends in June to September were significant ($p<0.01$); a significant increase in air temperature occurred in spring (February and March) and early summer (May to July); wind speed decreased significantly ($p<0.01$) in all the months except August. However, no significant trends were detected for both precipitation and water vapour pressure throughout each month. As a consequence, water vapour pressure deficit was exaggerated along with the rise in air temperature, which was expected to intensify the atmospheric water vapour demand and offset the negative effects of declining radiation and wind speed on potential evaporation. These changes in climatic variables have exerted remarkable impacts on the crop phenological stages, water consumption, and productivity during the last 3 decades over the North China Plain (Liu et al., 2010).

3.2.2 Changes in greenness and GPP

Here the remotely sensed NDVI is expressed as vegetation greenness. Averaged over the growth period (from March to October), vegetation greenness significantly increased from the 1980s, with a trend of $0.64\,yr^{-1}$ ($p<0.001$) and a coefficient of variation (CV) of 2.4 % (Fig. 5a, b). The maximum inter-annual variation of greenness was 5.4 % between 2 consecutive dry and wet years (1989 and 1990), with a 280 mm difference in annual precipitation. The distinct differences in greenness occurred in June to December. On an annual scale, greenness was weakly related to precipitation; however, in the growing season, greenness was noticeably correlated with monthly precipitation in April, May, June, September, and November ($r=0.29$–0.53, $p<0.1$). The reason is that the monthly rainfall is generally lower than the atmospheric evaporative demand in spring, and the water deficit to transpiration generally stresses the plant growth. Unlike precipitation, greenness anomalies are positively correlated with the detrended air temperature ($r^2=0.16$, $p<0.05$), implying that recent climate warming has stimulated vegetation growth by extending the growing stage and by pushing photosynthesis in water non-limited regions (Mao et al., 2012; Nemani et al., 2003).

Regional greenness trends were remarkably diverse (Fig. 6a, b). Except for climate change, human activities also exerted critical impacts on the terrestrial greenness variations. In the low plain of Hebei province, saline-alkali land amelioration and irrigation facility improvement contributed greatly to the vegetation density and greenness enhancement in the 1980s to the 1990s. In addition, atmospheric nitrogen deposition is also regarded as a positive driver of the vegetation improvement, since the nitrogen deposition increased on average by 25 % from the 1990s to the 2000s in North China (Jia et al., 2014; Piao et al., 2015). Spatially, greenness increased over 91.3 % of the NCP, in which the most distinc-

Table 1. Inter-annual trends of monthly sunshine duration (Sun), precipitation (P), air temperature (T), relative humidity (RH), and wind speed (U).

	Jan	Feb	Mar	Apr	May	Jun	Jul	Aug	Sept	Oct	Nov	Dec
Sun ($h\,yr^{-1}$)	−0.026	−0.036*	0.015	−0.012	−0.015	−0.039**	−0.040**	−0.052**	−0.060**	−0.018	−0.017	−0.020
P ($mm\,yr^{-1}$)	−0.129	0.212	−0.279	0.354	0.049	−0.472	0.608	0.456	0.694	−0.706	0.223	0.084
T ($°C\,yr^{-1}$)	0.022	0.058*	0.078**	0.030	0.042**	0.029*	0.030*	0.017	0.020	0.044*	0.025	0.018
RH ($\%\,yr^{-1}$)	−0.065	0.115	−0.260*	−0.079	−0.139	−0.074	−0.080	−0.074	0.057	−0.125	−0.094	−0.091
U ($m\,s^{-1}\,yr^{-1}$)	−0.016**	−0.012**	−0.012**	−0.020**	−0.021**	−0.019**	−0.015**	−0.006	−0.014**	−0.017**	−0.015**	−0.012**

Significance levels: * for $p < 0.05$; ** for $p < 0.01$.

tive grids were distributed in the southern parts and the belt along the Yellow River channel, where water supply was usually sufficient. In the northern part of the plain, however, the tendencies of greenness in a number of grids decreased significantly at the $p = 0.1$ level, which were assumed to have resulted from less irrigation supply to crops in springtime and rapid expansion of built-up occupations around cities and towns, such as the Beijing and Tianjin metropoles.

The spatially averaged GPP was $1913 \pm 584\,gC\,m^{-2}\,yr^{-1}$, with a CV of 6.8 % predicted by the VIP model from 1981 to 2013 showing noticeable spatial variability (Fig. 7). Low crop productivity resulted from fields with saline-alkali soil in the low lands near the coast of the Bohai Sea, where almost no favourable water was available for irrigation purposes in springtime. On average, the increasing trend of GPP was significant, with a slope of $8.2\,gC\,m^{-2}\,yr^{-2}$ ($r = 0.60$, $p < 0.01$). It was noticed that the average annual GPP increased steadily from the 1980s to the 2000s, compounded by decadal variations of climate and elevated atmospheric CO_2, as well as the improvement of agricultural practices and techniques. It was revealed that trends of annual GPP were positive over 87.9 % of the study region. As shown in Fig. 7, the obvious increasing trends were located in the mid and southern areas, while most of the decreasing trends occurred in the eastern and northern parts, where water for irrigation was considerably reduced in the spring season because of the competing demands of domestic and industrial water use.

On a monthly scale, GPP increased in all the months except for July, August, and September (Fig. 8). The positive trends were contributed principally by the summer harvest crops (wheat as the main crop), while the negative trends were mainly contributed by the autumn harvest crops (maize as the major crop). Regressive analysis showed that the downward trends of GPP in the summer season resulted from the significant declines in monthly sunshine duration and shortwave radiation ($r = 0.38$ to 0.57 from June to August, $p < 0.05$).

3.2.3 Changes in ET

Water loss from the vegetation surface as ET is directly regulated by atmospheric vapour demand and leaf stoma physiological functioning (Buckley and Mott, 2013). Inter-annual variation of ET is controlled by climate variability/change and agronomic managements. Generally potential ET (ET_p) is used to represent the available energy for water vaporization on land surface. As shown in Fig. 9a, ET_p was slightly decreasing, which resulted from offsetting between the negative effects of reduced global radiation and declining wind speed and the positive effect of increasing water vapour deficit (Song et al., 2009); simultaneously, actual ET was predicted to be slightly increasing ($p < 0.05$), consistent with the enhancement of greenness. It was noticed that the evolutions of potential and actual ET coincided with the hypothesis of complementary relationship. On a monthly scale, ET significantly increased from February to April, but it decreased in August (Fig. 9b). This implies that climate warming may be beneficial to spring crops by advanced recovering date from dormancy, whereas a decline of net radiation (R_n) (especially in August, significance level $p < 0.001$) may lead to the downward tendency of ET rates in summer.

Over the whole plain, spatially averaged actual ET and transpiration were $627 \pm 162\,mm\,yr^{-1}$ (about 92 % of annual precipitation) and $416 \pm 129\,mm\,yr^{-1}$ (about 67 % of ET), respectively. The trend of annual ET ($p < 0.1$) with CV of 0.05 was $0.88\,mm\,yr^{-2}$ from 1981 to 2013, which was less significant than that of the NDVI ($p < 0.01$). Decadal ET amounts in the 1980s, 1990s, and 2000s were 610, 626, and 640 mm, respectively, corresponding to the slightly rising trend of precipitation. It is found that GPP increased with a higher significance level than that of ET, implying the enhancement of water productivity in the plain. Spatially, the trends of ET were positive over 86.0 % of the study region, mainly occurring in the mid and southern parts, while negative trends were mainly found in the northern part (Fig. 10a, b), which was consistent with the pattern of GPP tendencies. By using a water balance model, Zeng et al. (2014) also presented an increasing trend of ET over the North China Plain from 1982 to 2009.

3.3 Contributions of climate change, atmospheric CO_2 fertilization, and agronomic management to changes in ET and GPP

3.3.1 Spatial patterns of the contributions

The contributions from climate change, atmosphere CO_2 enrichment fertilization, and agronomic management illus-

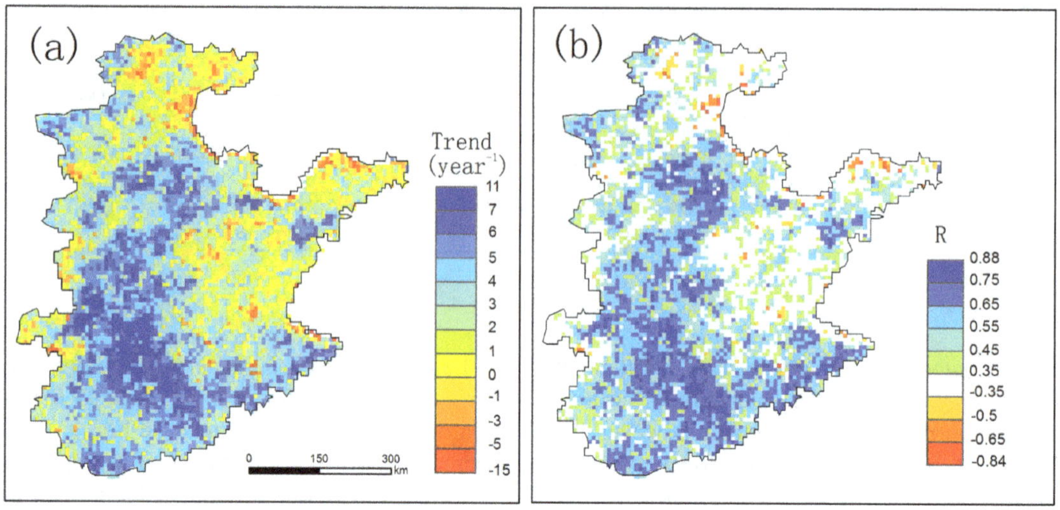

Figure 6. **(a)** Spatial distribution of the NDVI trends in the growing season and **(b)** the Pearson coefficients of the NDVI trends above the $p < 0.05$ significance level (the NDVI is scaled by 1000).

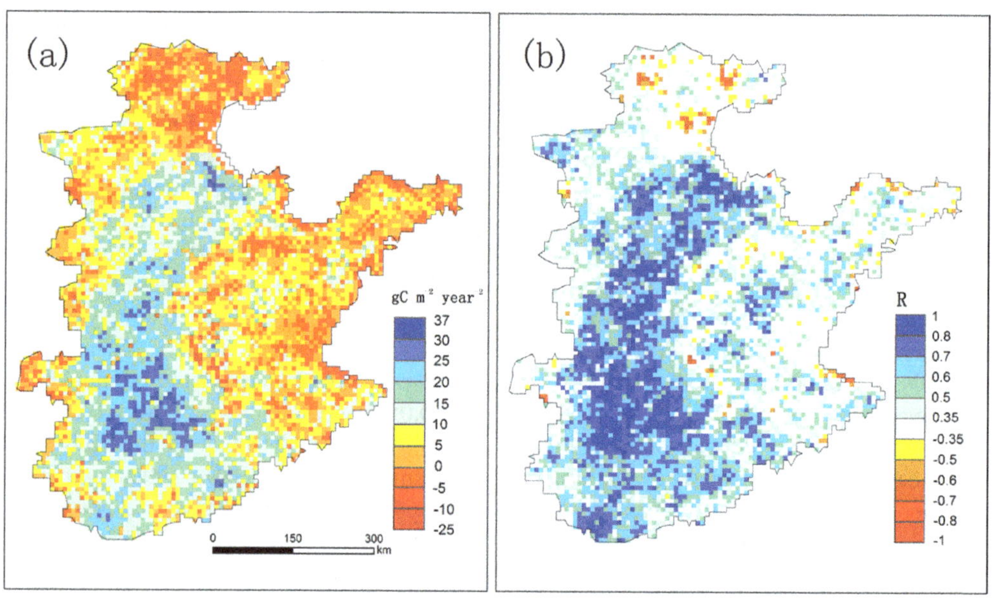

Figure 7. **(a)** Spatial distribution of GPP trends and **(b)** Pearson coefficients of trends at the $p < 0.05$ significance level.

trated considerable spatial heterogeneity for both ET and GPP (Fig. 11). Over the whole plain, climate change exerted a positive impact on water vapour exchange from the land surface to the atmosphere (f_1), especially in the eastern hilly part where precipitation increased slightly. As is generally known, air CO_2 enrichment stimulates the crop leaf stomatal closing and then reduces transpiration, but its fertilization effect enhances the photosynthetic rate and water use efficiency (Buckley and Mott, 2013). Descriptions of the separate effects are presented as follows.

The climate change has intensified the hydrological cycle, resulting in a 0 to 4 mm increment of ET per year. The effect of climate change was much stronger in the mid to eastern zones with high crop productivities, contributed mainly by air temperature increase. The contribution of CO_2 enrichment to ET is negative in most areas, ranging from 0 to -1 mm per year. The attributions of agronomic practices and technological advancement represented by LAI increase are somewhat complex, namely a remarkable increase ranging from 0 to 6 mm per year in the mid–western areas where irrigation facilities and soil conditions have been ameliorated greatly in the recent decades through land consolidation and de-salinization. Renewal of cultivars and improved agronomic practices also contributed to ET intensifying (Zhang et al., 2011). In contrast, ET in the grids with expansions of built-ups appeared to have negative trends.

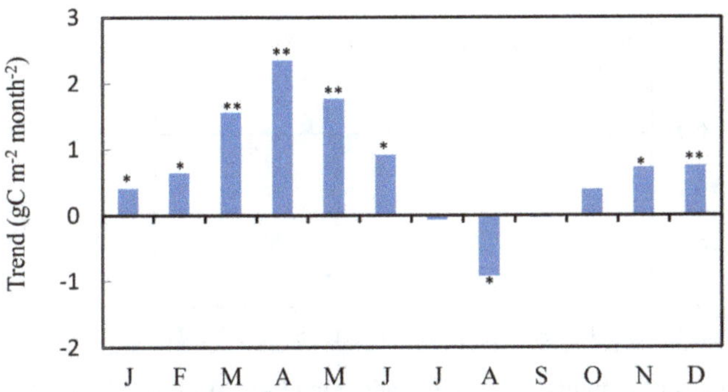

Figure 8. Monthly trend of GPP from 1981 to 2013 (significance levels: * is $p<0.05$; ** is $p<0.01$).

Annually, the contribution of climate change to GPP is negative, ranging from 0 to $-12\,\mathrm{gC\,m^{-2}}$ per year. Low rates occurred in the southwestern and northern parts. Air warming and heat waves, declines in precipitation, and global radiation are the main causes of crop production reduction (Lobell et al., 2011; Guo et al., 2014). In addition to the spatial variability, climate change effects are associated with the relevant land use/cover and cropping systems. In the hilly (western and mid) and eastern coast areas, negative effects were slight, where air warming and air pollution were relatively weak. CO_2 enrichment effects were positive over the whole plain, ranging from 0 to $6\,\mathrm{gC\,m^{-2}}$ per year. It was noticed that the higher effects were associated with higher cropland with favourable irrigation and high productivity, while the lower rate was related to low-productivity croplands and natural vegetation communities. Similarly, the effects of human activities on ET were positive in the mid to western areas, ranging from 0 to $35\,\mathrm{gC\,m^{-2}}$ per year, associated with croplands of high productivity. The negative effects mainly occurred in the eastern and northern parts, where there was remarkable expansion of urban and dwelling buildings in the study period.

3.3.2 Regionally averaged contributions

Regarding the aspect of regional average, some characteristics of the contributions to water and carbon assimilation are revealed. As shown in Fig. 12a, the contributions of climatic variable change (f_1), elevated atmospheric CO_2 concentration (f_2), and agronomic management (represented by leaf area index (LAI) increment) (f_3), and their interactions with the long-term trend of ET were positive, while the contribution of elevated atmospheric CO_2 was negative in the last 3 decades. It was shown that the contribution of climate change was less than that of agronomic improvement. The relative direct contributions of climatic change, CO_2 fertilization, and agronomic management and technological advancement to the long-term trend of evapotranspiration were 56, −28, and 68 %, respectively. Compared with the contributions of direct effects, the relative contributions by their interactions were low (the cumulative effect of f^{12}, f^{13}, f^{23}, and f^{123} was only about 4 %). Although the global radiation reaching the ground was diminished by a higher aerosol concentration and deteriorated pollution in the atmosphere (Che et al., 2005), its negative effect on terrestrial ET was offset by the positive effects of air warming and its related higher vapour pressure deficit (VPD) on ET on an annual scale. Reduction of transpiration by enriched atmospheric CO_2 caused by closure of plant leaf stomata at high CO_2 concentrations for both C3 and C4 plants may mediate the extra water demand by air warming. The dominant contribution to ET changes was from the renewal of cultivars and improvement of agricultural techniques and management. In the study period, agronomic management has greatly improved, including the establishment of irrigation facility, prevalent uses of chemical fertilizers, and pesticides. For example, irrigated area in the northern part (mainly the Hebei Plain) has increased by 2.5 times, and chemical synthetic fertilizer input has increased about 4 times; as a consequence, crop grain production enhanced about 2 times from the 1980s to the 2000s (Xu et al., 2005). Climate change and agronomic management improvement (irrigation practice, synthetic fertilizer supply, and new cultivar adoption) are the main contributors to ET intensifying over the plain.

As shown in Fig. 12b, the enriched atmospheric CO_2 fertilization and agronomic management improvement presented a positive contribution to the GPP trend during the study period. It was somewhat against expectations that the contribution of climate change to GPP was negative on an annual scale. The relative contributions of climate change, CO_2 fertilization, and management to the vegetation GPP enhancement were −32, 25, and 103 %, respectively. It was confirmed that the improvement of agricultural management was the dominant driver of GPP increase. The positive effects on GPP were associated with human activities and natural factors, such as input of synthetic fertilizers and atmospheric nitrogen deposition, irrigation, and other agronomic technology improvement, as well as fertilization of enriched atmospheric CO_2. The negative contribution by climate change mainly happened in summertime (Fig. 8). Since there was less benefit of CO_2 enrichment to summer maize (C4 type), the reduced maize productivity due to global radiation decline was not fully offset. Some studies, such as Piao et al. (2015), also reported that climate change exerted a negative impact on the vegetation greening phenology, and Liu et al. (2010) attributed the reduction of crop productivity to shortening of vegetative growth length under climate warming. As shown in Fig. 4b, it was illustrated by NDVI time series that greenness in the summer season was quite stable; however, it significantly increased in spring and autumn, indicating that climate warming was beneficial for crop growing in the cool seasons. Additionally, carbon assimilated by summer crops was larger than that by spring crops. Thereby, the sign of the annual GPP trend was determined by its trend in the summer season. As current air temperature increase

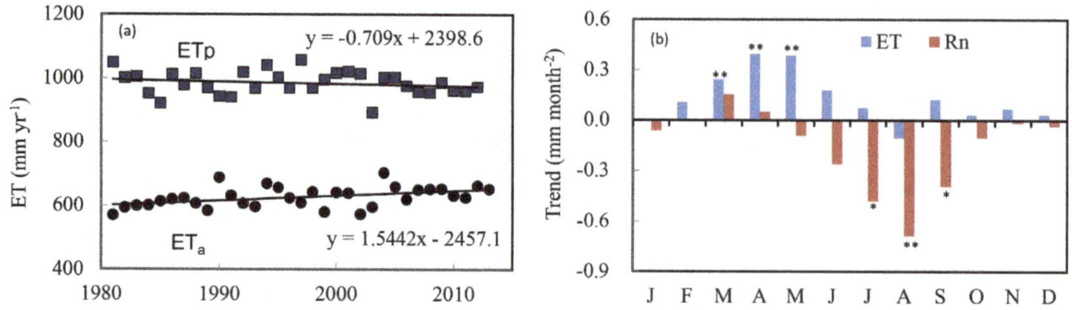

Figure 9. Inter-annual trends of potential and actual annual ET **(a)** and trends of monthly ET and net radiation (R_n) **(b)** from 1981 to 2013 (significance levels: * is $p<0.05$; ** is $p<0.01$).

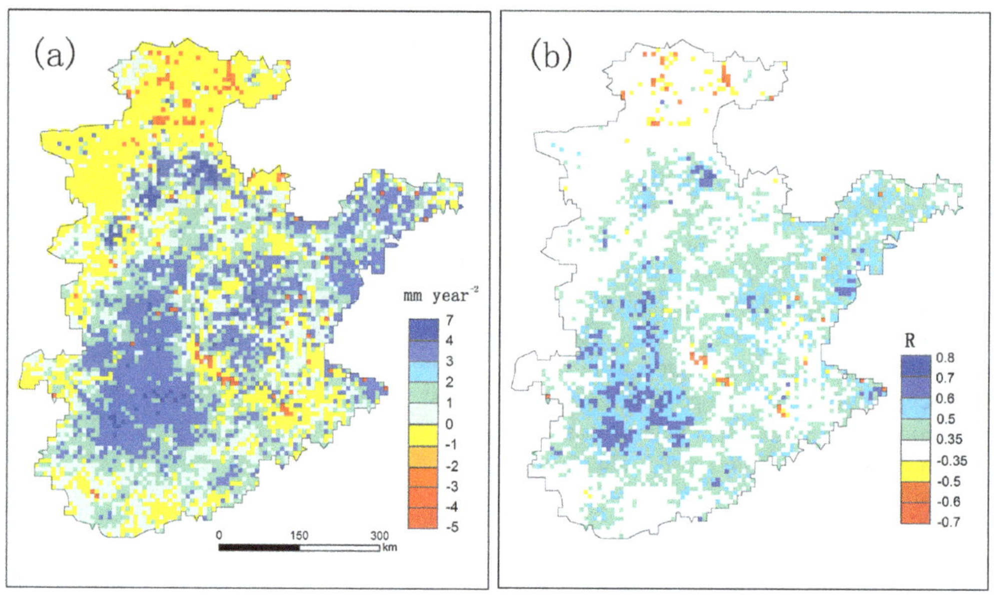

Figure 10. (a) Spatial distribution of ET trends, and **(b)** their temporal Pearson coefficients above the $p<0.05$ significance level from 1981 to 2013.

was not yet so detrimental to maize growth, the decline of downward shortwave radiation was considered to be responsible for GPP decline.

3.3.3 Effects of climatic variables on monthly ET and GPP trends

To attribute the responses of cropping systems to the trends of single climatic variables, the VIP model is used to diagnose the effects of climate change on ET and GPP at the Beijing meteorological observation site. The contributions of a single variable to the trends of ET and GPP are expressed by their differences simulated with the current and de-trended variables, respectively. Here only the climatic variables of radiation, air temperature, and wind speed were linearly de-trended on a monthly scale, since no significant trends of precipitation and specific humidity are detected. As shown in Table 2, while the global radiation was de-trended, the negative correlation coefficient of monthly GPP with time was reversed from negative to positive in springtime, and from significant ($r<-0.6$, $p<0.01$) to insignificant levels ($r<-0.15$, $p>0.01$) in July and August. It was affirmed that the decline in global radiation was the dominant factor for reduction of crop GPP in the summer period (June to August), but in the autumn season the changes in radiation, temperature, and wind speed were all responsible for GPP changes. From Table 2, it could be deduced that the effect of temperature rising on crop productivity was positive and significant in spring (March and April) and autumn (September and October), whereas its effect was relatively weak in summer (May to August). It was noticed that the effect of radiation change was quite weak in March, when no significant trend of shortwave radiation was detected. In the spring season, sunshine duration increased from the 1980s to the 2010s (Wang and Yang, 2014). In June, except for global radiation, changes in temperature and precipitation have contributed to GPP increasing. Additionally, the fertilizing effect

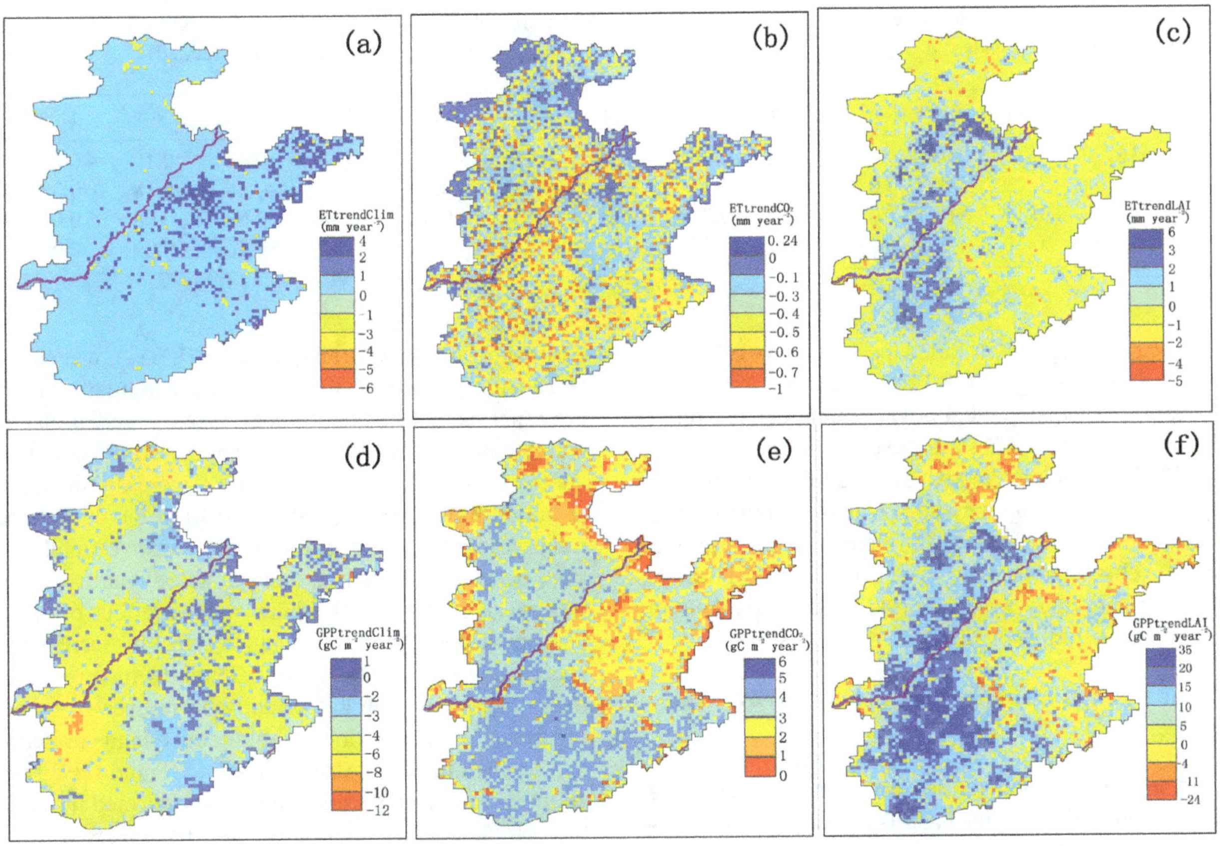

Figure 11. Contributions of climate changes, CO_2 enrichment, and management to ET (**a, b, c**) and GPP (**d, e, f**), respectively (**a** and **d** are for climate, **b** and **e** for CO_2, and **c** and **f** for management).

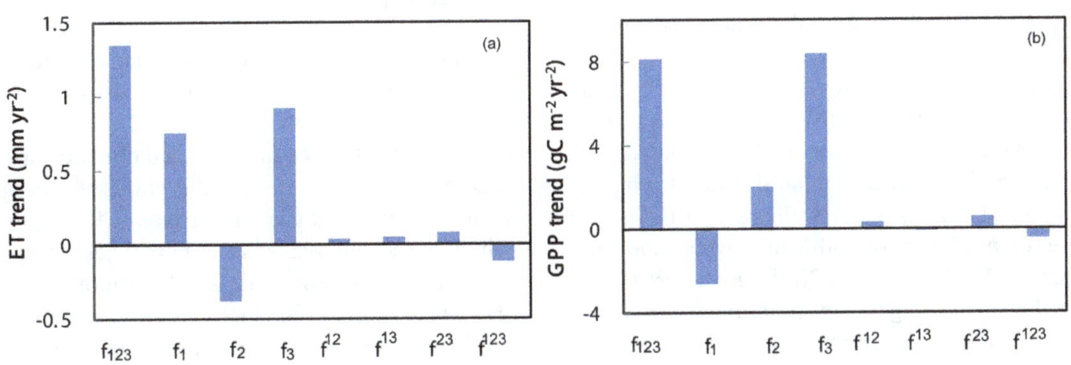

Figure 12. **(a)** Contributions of climate change, atmospheric CO_2 enrichment, and agronomic management to ET and **(b)** contributions to GPP trends.

of enriched atmospheric CO_2 on C3 crops is a non-ignorable factor in GPP increasing. However, maize as the C4 crop does not benefit much from atmospheric CO_2 enrichment. It is suggested that new cultivars with higher light use efficiency should be adopted to sustain the maize productivity under declined global radiation condition resulting from exacerbating aerosol concentration and air pollution.

Comparatively, the effect of climatic change on ET was less significant than that of vegetation GPP. The model simulations showed that ET enhanced by air temperature rising mainly occurred in August to October, while the negative effect of solar radiation decreasing was detected from June to September in the maize growing period.

4 Discussion

Our simulations suggested that annual ET and vegetation GPP increased over the North China Plain from 1981 to

Table 2. Monthly Pearson correlation coefficients of GPP trends ($r_$ALL: all variables are not de-trended; $r_$R: radiation is de-trended; r_T: air temperature is de-trended; $r\_R\_T\_U$: radiation (R), temperature (T), and wind speed (U) are all de-trended).

	Jan	Feb	Mar	Apr	May	Jun	Jul	Aug	Sep	Oct	Nov	Dec
r_ALL	0.17	0.14	0.36	0.31	0.25	−0.16	−0.56	−0.37	−0.75	−0.48	−0.19	−0.03
r_R	0.29	0.15	0.38	0.40	0.23	0.49	−0.15	−0.07	−0.67	−0.46	−0.18	−0.03
r_T	0.11	−0.06	0.02	0.09	0.30	0.05	−0.64	−0.30	−0.48	0.08	0.08	0.05
$r\_R\_T\_U$	0.10	−0.05	0.00	0.11	0.33	0.35	−0.20	−0.08	−0.13	0.10	0.09	0.06

2013. Climate change contributed positively to ET intensification, but negatively to GPP enhancement. Agronomic management and technological advancement are the dominant factors in promoting GPP increase. The use of remote sensing NDVI series has greatly improved the reliability of vegetation water consumption and productivity prediction at spatial and temporal scales, even if there were uncertainties in vegetation characteristics retrieved from the NDVI dataset. The results in this study were supported and consistent with the most relevant studies on field and regional scales.

4.1 Is the trend of ET upward or downward over the NCP?

Although the crop productivities are steadily increasing, whether the actual ET over the NCP is increasing or decreasing during the last 3 decades is controversial from the reported literatures. By using the complementary relationship models (Brutsaert and Stricker, 1979), actual ET and potential ET both decreased (Cao et al., 2014; Gao et al., 2012). However, ET increased, predicted by the process-based VIP model from 1981 to 2013, which was consistent with the increasing trend of terrestrial greenness (S. Wang et al., 2016). Yuan and Shen (2013) found that in the northern part of the NCP (Hebei Province), ET was positively correlated with recorded crop grain yield, and agricultural water use has increased. Field measurements under well-watered fields also showed that seasonal ET rates of both winter wheat and summer maize increased (Zhang et al., 2011). Brutsaert (2006) acknowledged that decreasing pan evaporation was evidence of increasing terrestrial evaporation. As is generally known, the sign of ET change should be the same as that of vegetation greenness. Over the NCP a positive trend of ET was more believable, in the light of a significantly increasing NDVI over the growing season, especially in spring. The positive effects of warming with higher water vapour deficit on ET might be offset by the negative effect of declining solar radiation and wind speed on potential evaporation. In view of the complementary relationship hypothesis (Hobbins et al., 2001), alteration of available energy partitioned into latent heat flux (or ET) is dominated by the atmospheric water vapour deficit, namely, while more vapour is evaporated into the atmosphere boundary layer, its water vapour deficit is then accordingly relaxed, resulting in a lower rate of ET_p. However, while declining global radiation is the dominant factor in the ET trend, actual ET (ET_a), wet surface ET (ET_w), and potential ET (ET_p) trace the trend of available energy (net radiation). In this study case, net radiation declined at a rate of -5.58 MJ yr^{-1} ($r = 0.56$, $p < 0.01$) over the NCP. As the trends of both ET_p and ET_w were dominated by the radiation trend, the ET_a estimated from the complementary relationship definitely followed the negative trend of radiation, because the positive trend of aerodynamic evaporation was weak as a tradeoff of the positive effect of rising water vapour deficit and the negative effect of decreasing wind speed. However, the declining ET trend which resulted from the reduced radiation has actually been reversed by increasing green leaf area, which would reduce land surface albedo and temperature, etc. Consequently, ET and GPP showed a slight increase. This study case also confirmed limitations of the complementary relationship for diagnosing the evaporation trend under the condition of declining radiation.

4.2 Effects of climate change and CO_2 on the cropping systems

Without adaptation measures, climate change is shown to exert negative effects on the productivity of cropping systems in the NCP during the last 3 decades (Mo et al., 2013; Liu et al, 2010). Changes in individual climate variables affected specific cropping systems differently, which was associated with crop type and growing season. Climate warming in the winter and spring seasons was beneficial to vegetative growth of winter wheat (Mo et al., 2013). Although air warming has shortened the growing length, the autonomously adopted cultivars with higher thermal requirements usually maintained the crop growth length and accumulated more photosynthesis products, which may outweigh the extra respiration consumption under warmer climates (Wang et al., 2012). So far, the effects of global warming on wheat production have been inhomogeneous in the North China Plain, which were positive in the northern part but negative in the southern part (Zhang et al., 2013). The reasons are that during the wheat growth period air temperature is still below the favourable conditions in the high-latitude part of the plain; therefore, recent global warming is benign to the wheat growth, whereas air warming may be detrimental in the southern part, especially for rainfed wheat (Xiao and Tao, 2014).

However, dominated by summer monsoon in the North China Plain, the climate is hot in the summer maize growth period. Due to maize being a tropically originated species with a high thermal requirement, it can tolerate relatively high air temperature. Our study showed that maize has not suffered noticeably from air warming in the recent decades, confirmed also by Xiong et al. (2012). However, Guo et al. (2014) reported that effect of air warming on maize was adverse with an Agro-Ecological Zones model, and the decreased daily temperature range (DTR) may be detrimental to crop yields. Nevertheless, adaptation measures may boost the production, such as harvest time delay (Wang et al., 2012) or planting date advancement (Sacks and Kucharick, 2011).

As atmospheric aerosols from industrial production and combustion have increased, global radiation has declined in many parts of the world, in which direct components decreased but diffuse components increased, so-called "global dimming" (Liepert, 2002; Ren et al., 2013). The decline of global radiation has resulted in less pan evaporation and carbon assimilation in crop and natural vegetation communities (Xiong et al., 2012; Xiao and Tao, 2014); nevertheless, plant canopies can use the diffuse radiation with higher efficiency than direct beams (Gu et al., 2002). In springtime, while the atmospheric circulation is shifting from continental to ocean monsoon in East Asia, the wind speed is relative high, and consequently air pollution and aerosols are usually low; therefore, global radiation is not reduced obviously in the wheat growth period (Table 1), illustrating that air warming and precipitation variability other than radiation decrease are the principal climate factors contributing to the tendency of wheat production. In contrast, global radiation declines significantly in the summer season in the North China Plain (Table 1); as a result, productivities of autumn harvest crops such as maize are mainly affected. For example, Guo et al. (2014) reported that maize potential productivity was reduced by 20 kg hm^{-2} due to global radiation decline in the last decades over China.

During the study period, atmospheric CO_2 concentration increased from 340 to 396 ppm, which contributed to enhancement of crop productivity. However, most studies with statistical analysis models neglected the contribution of CO_2 fertilization (e.g. Lobell and Burke, 2010; Song et al., 2014). As confirmed by FACE experiments, elevated atmospheric CO_2 concentration is accelerating plant photosynthesis and reducing transpiration, whose fertilizer effect is 0.065 % per ppm increment for C3 plants (Field et al., 1995; Long et al., 2006; Ainsworth et al., 2008). In this study, the contribution of CO_2 to GPP is positive, but negative for ET. The positive CO_2 effect on GPP almost compensates for the negative effects of climatic variable changes. However, it should be borne in mind, as Ainsworth et al. (2008) pointed out, that the CO_2 fertilizer effect may be overestimated by the process-based crop/ecosystem models.

4.3 Effects of agronomic practice, technique advancement and other factors

The remotely sensed NDVI is an excellent indicator of long-term changes of vegetation covers. Here we assumed that climate change did not modify the tendencies of vegetation covers, but dominated the inter-annual variation of vegetation density. Renewal of crop cultivars, applications of synthetic fertilizer and irrigation, as well as conservancy tillage and nitrogen deposition all contribute to the crop and/or natural productivity improvements (Yu et al., 2012; Bai et al., 2016; Piao et al., 2015). National statistical records of grain yields on county scale showed a rapid increase from 1980 to 1990 and a moderate increase in the 2000s. Enhancement of crop yields mainly stemmed from more biomass accumulation and a higher harvest index than previous varieties (Zhang et al., 2013). In our simulations the upward trend of GPP was more significant than that of ET, which was consistent with the increasing trend of the cumulative NDVI. Agronomic practices and technology advancement contributed 103 % of GPP changes in our study. By using crop models, Yu et al. (2012) and Song et al. (2014) reported, respectively, the relative contributions of 92 and 62 % by agronomic management and renewal of cultivars for rice, and Guo et al. (2014) showed that 99.6 to 141.6 % of maize yield increases was contributed by technological advancement in China since the 1980s. The previous studies showed that, if no adoption measures were taken, climate change generally contributed negatively to crop productivities in the mid-latitude areas, but the negative effects were usually compensated for by genetic improvements, applications of fertilizer and irrigation, pest and weed control, as well as CO_2 and nitrogen deposition effects (Liu et al., 2010; Lobell et al., 2011; Guo et al., 2014; Bai et al., 2015). Under warming climate conditions, it is expected that water requirement by crops and natural plants will increase, but the intensified ET may be limited by insufficient soil moisture availability. Therefore, the sustainability of crop productions greatly depends on the improvement of agronomic management and technological advancement on variety breeding.

5 Summary and conclusions

Climate change and human activities have greatly altered the hydrological regime and crop productivity in the North China Plain, with warm temperate climate during the recent 3 decades. The VIP ecological model integrated with the NOAA-AVHRR NDVI data series predicted that spatial average annual actual ET weakly increased, while vegetation primary productivity (GPP) significantly increased ($p < 0.01$) from 1981 to 2013, consistent with the remotely sensed NDVI trend. The increases in actual ET and GPP mainly occurred in spring, while ET and GPP were obviously decreasing in August owing to global radiation diminishing.

Climate change, elevated atmospheric CO_2 fertilization, and agronomic management (new cultivars, irrigation, and chemical fertilizer) all contribute to the inter-annual trends of ET and crop GPP. The relative direct contributions of climatic change, CO_2 fertilization, and agronomic management to ET increase are 56, −28, and 68 %, while the contributions to GPP tendency are −32, 25, and 103 %, respectively. Air warming intensifies the crop water requirement and enhances the production of crops harvested in summer. The decline of global radiation resulting from exaggerated aerosol concentration and air pollution is considered to be the main cause of GPP reduction in August. The study confirms the necessary for imminent control of air pollution and aerosol to sustain the productivity of the agricultural system.

Acknowledgement. This work was supported by the Natural Science Foundation of China grants (41471026 and 31171451) and the State's Key Project of Research and Development Plan (no. 2016YFC0401402). We thank the China Meteorological Administration (CMA) for providing the meteorological data and on-site soil moisture data used in this paper.

Edited by: A. Wei

References

Ainsworth, E. A., Leakey, A. D. B., Ort, D. R., and Long, S. P.: FACE-ing the facts: inconsistencies and interdependence among field, chamber and modeling studies of elevated [CO_2] impacts on crop yield and food supply, New Phytol., 179, 5–9, 2008.

Alkama, R., Kageyama, M., and Ramstein, G.: Relative contributions of climate change, stomatal closure, and leaf area index changes to 20th and 21st century runoff change: A modelling approach using the ORCHIDEE land surface model, J. Geophys. Res., 115, D17112, doi:10.1029/2009JD013408, 2010.

Bai, H. Z., Tao, F. L., Xiao, D. P., Liu, F. S., and Zhang, H.: Attribution of yield change for rice-wheat rotation system in China to climate change, cultivars and agronomic management in the past three decades, Climatic Change, 135, 539–553, 2016.

Baker, I. T., Denning, A. S., and Stockli, R.: North American gross primary productivity: regional characterization and interannual variability, Tellus B, 62, 533–549, 2010.

Banger, K., Tian, H., Tao, B., Ren, W., Pan, S., Dangal, S., and Yang, J.: Terrestrial net primary productivity in India during 1901–2010: contributions from multiple environmental changes, Climatic Change, 132, 575–588, 2015.

Betts, R. A., Boucher, O., Collins, M., Cox, P. M., Falloon, P. D., Gedney, N., Hemming, D. L., Huntingford, C., Jones, C. D., Sexton, D. M. H., and Webb, M. J.: Projected increase in continental runoff due to plant responses to increasing carbon dioxide, Nature, 448, 11037–1041, 2007.

Bonan, G. B.: Sensitivity of a GCM simulation to subgrid infiltration and surface runoff, Clim. Dynam., 12, 279–285, 1996.

Brutsaert, W. and Stricker, H.: An advection-aridity approach to estimate actual regional evapotranspiration, Water Resour. Res., 15, 443–450, 1979.

Brutsaert, W. B.: Indications of increasing land surface evaporation during the second half of the 20th century, Geophys. Res. Lett., 33, L20403, doi:10.1029/2006GL027532, 2006.

Buckley, T. N. and Mott, K.: Modelling stomatal conductance in response to environmental factors, Plant Cell Environ., 36, 1691–1699, 2013.

Burn, D. H. and Hesch, N. M.: Trends in evaporation for the Canadian Prairies, J. Hydrol., 336, 61–73, 2007.

Campbell, G. S. and Norman, J. M.: An introduction to environmental biophysics, Springer-Verlag, New York, 1998.

Cao, G., Han, D., and Song, X.: Evaluating actual evapotranspiration and impacts of groundwater storage change in the North China Plain, Hydrol. Process., 28, 1797–1808, 2014.

Che, H. Z., Shi, G. Y., Zhang, X. Y., Arimoto, R., Zhao, J. Q., Xu, L., Wang, B., and Chen, Z. H.: Analysis of 40 years of solar radiation data from China, 1961–2000, Geophys. Res. Lett., 32, L06803, doi:10.1029/2004GL022322, 2005.

Chen, L., Zhang, L., Wang Y., Yu, Q., Eamus, D., and O'Grady, A.: Impacts of elevated CO_2, climate change and their interactions on water budgets in four different catchments in Australia, J. Hydrol., 519, 1350–1361, 2014.

Chen, X., Liu, X., Zhou, G., Han, L., Liu, W., and Liao, J.: 50-year evapotranspiration declining and potential causations in subtropical Guangdong province, southern China, Catena, 128, 185–194, 2015.

Collatz, G. J., Ribas-Carbo, M., and Berry, J. A.: Coupled photosynthesis-stomatal conductance model for leaves of C4 plants, Aust. J. Plant Physiol., 19, 519–538, 1992.

Douville, H., Ribes, A., Decharme, B., Alkama, R., and Sheffield, J.: Anthropogenic influence on multidecadal changes in reconstructed global evapotranspiration, Nature Climate Change, 3, 59–62, 2013.

Ewert, F., Rotter, R. P., Bindi, M., Webber, H., Trnka, M., Kersebaum, K. C., Olesen, J. E., van Ittersum, M. K., Janssen, S., Rivington, M., Semenov, M. A., Wallach, D., Porter, J. R., Stewart, D., Verhagen, J., Gaiser, T., Palosuo, T., Tao, F., Nendel, C., Roggero, P. P., Bartosova, L., and Asseng, S.: Crop modelling for integrated assessment of risk to food production from climate change, Environ. Modell. Softw., 72, 287–303, 2015.

Farquhar, G. D., von Caemmerer, S., and Berry, J. A.: A biochemical model of photosynthetic CO_2 assimilation in leaves of C3 species, Planta, 149, 78–90, 1980.

Field, C., Jackson, R., and Mooney, H.: Stomatal responses to increased CO_2: Implications from the plant to the global scale, Plant Cell Environ., 18, 1214–1255, 1995.

Gao, G., Xu, C., Chen, D., and Singh, V. P.: Spatial and temporal characteristics of actual evapotranspiration over Haihe River basin in China, Stoch Env. Res. Risk A., 26, 655–669, 2012.

Gu, L. H., Baldocchi, D., Verma, S. B., Black, T. A., Vesala, T., Falge, E. M., and Dowty, P. R.: Advantages of diffuse radiation for terrestrial ecosystem productivity, J. Geophys. Res.-Atmos., 107, 4050, doi:10.1029/2001JD001242, 2002.

Guo, J., Zhao, J., Wu, D., Mu, J., and Xu, Y.: Attribution of Maize Yield Increase in China to Climate Change and Technological Advancement Between 1980 and 2010, J. Meteorol. Res., 28, 1168-1181, doi:10.1007/s13351-014-4002-x, 2014.

Haddeland, I., Heinke, J., Biemans, H., Eisner, S., Florke, M., Hanasaki, N., Konzmann, M., Ludwig, F., Masaki, Y., Schewe, J., Stacke, T., Tessler, Z. D., Wada, Y., and Wisser, D.: Global water resources affected by human interventions and climate change, P. Natl. Acad. Sci. USA, 111, 3251–3256, 2014.

Hobbins, M. T., Ramırez, J. A., Brown, T. C., and Claessens, L. H. J. M.: The complementary relationship in estimation of regional evapotranspiration: The Complementary Relationship Areal Evapotranspiration and Advection-Aridity models, Water Resour. Res., 37, 1367–1387, 2001.

Hu, S., Mo, X., and Lin, Z.: Optimizing the photosynthetic parameter Vcmax by assimilating MODIS-fPAR and MODIS-NDVI with a process-based ecosystem model, Agr. Forest Meteorol., 198–199, 320–334, 2014.

Jia, Y., Yu, G., He, N., Zhan, X., Fang, H., Sheng, W., Zuo, Y., Zhang, D., and Wang, Q.: Spatial and decadal variations in inorganic nitrogen wet deposition in China induced by human activity, Sci. Rep., 4, 3763, doi:10.1038/srep03763, 2014.

Katul, G. G., Oren, R., Manzoni, S., Higgins, C., and Parlange, M. B.: Evapotranspiration: A process driving mass transport and energy exchange in the soil-plant-atmosphere-climate system, Rev. Geophys., 50, RG3002, doi:10.1029/2011RG000366, 2012.

Liepert, B. G.: Observed reductions of surface solar radiation at sites in the United States and worldwide from 1961 to 1990, Geophys. Res. Lett., 29, 61-1–61-4, 2002.

Liu, M., Tian, H., Lu, C., Xu, X., Chen, G., and Ren, W.: Effects of multiple environment stresses on evapotranspiration and runoff over eastern China, J. Hydrol., 426–427, 39–54, 2012.

Liu, S. and Xu, Z.: Multi-scale surface flux and meteorological elements observation dataset in the Hai River Basin (Miyun site-automatic weather station), Cold and Arid Regions Science Data Center at Lanzhou, doi:10.3972/haihe.001.2013.db, 2013.

Liu, S. M., Xu, Z. W., Zhu, Z. L., Jia, Z. Z., and Zhu, M. J.: Measurements of evapotranspiration from eddy-covariance systems and large aperture scintillometers in the Hai River Basin, China, J. Hydrol., 487, 24–38, 2013.

Liu, Y., Wang, E., and Yang, X.: Contributions of climatic and crop varietal changes to crop production in the North China Plain, since 1980s, Glob. Change Biol., 16, 2287–2299, 2010.

Lobell, D. B. and Burke, M. B.: On the use of statistical models to predict crop yield responses to climate change, Agr. Forest Meteorol., 150, 1443–1452, 2010.

Lobell, D. B., Schlenker, W., and Costa-Roberts, J.: Climate trends and global crop production since 1980, Science, 333, 616–620, 2011.

Long, S. P., Ainsworth, E. A., Leakey, A. D. B., Nosberger, J., and Ort, D.: Food for thought: lower-than-expected crop yield stimulation with rising CO_2 concentrations, Science, 312, 1918–1921, 2006.

Mao, J., Shi, X., Thornton, P. E., Piao, S., and Wang, X.: Causes of spring vegetation growth trends in the northern mid–high latitudes from 1982 to 2004, Environ. Res. Lett., 7, 014010, doi:10.1088/1748-9326/7/1/014010, 2012.

Ministry of Water Resources of China (MWR): China water resources bulletin, Ministry of Water Res. of China, Beijing, available at: www.mwr.gov.cn/zwzc/hygb/szygb/qgszygb/201204/t20120426_319624.html (last access: 10 January 2017), 2010.

Mo, X., Liu, S., and Lin, Z.: Evaluation of an ecosystem model for wheat-maize double cropping system over the North China Plain, Environ. Modell. Softw., 32, 61–73, 2012.

Mo, X., Guo, R., Liu, S., Lin, Z., and Hu, S.: Impacts of climate change on crop evapotranspiration with ensemble GCM projections in the North China Plain, Climatic Change, 120, 299–312, doi:10.1007/s10584-013-0823-3, 2013.

Mo, X., Liu, S., and Meng, D.: Climate variability impacts on evapotranspiration and primary productivity by assimilating remotely sensed data with a process-based model over Songhua River Basin, Int. J. Climatol., 34, 1945–1963, doi:10.1002/joc.3813, 2014.

Mo, X. G., Liu, S. X., Lin, Z. H., and Guo, R. P.: Regional crop yield, water consumption and water use efficiency and their responses to climate change in the North China Plain, Agr. Ecosyst. Environ., 134, 67–78, 2009.

Nalder, I. A. and Wein, R. W.: Spatial interpolation of climatic Normals: test of a new method in the Canadian boreal forest, Agr. Forest Meteorol., 92, 211–225, 1998.

Nayak, R. K., Patel, N. R., and Dadhwal, V. K.: Inter-annual variability and climate control of terrestrial net primary productivity over India, Int. J. Climatol., 33, 132–142, 2013.

Nemani, R. R., Keeling, C. D., Hashimoto, H., Jolly, W. M., Piper, S. C., Tucker, C. J., Myneni, R. B., and Running, S. W.: Climate-driven increases in global terrestrial net primary production from 1982 to 1999, Science, 300, 1560–1563, 2003.

Ozdogan, M. and Salvucci, G. D.: Irrigation-induced changes in potential evapotranspiration in southeastern Turkey: Test and application of Bouchet's complementary hypothesis, Water Resour. Res., 40, W04301, doi:10.1029/2003WR002822, 2004.

Piao, S., Friedlingstein, P., Ciais, P., de Noblet-Ducoudré, N., Labat, D., and Zaehle, S.: Changes in climate and land use have a larger direct impact than rising CO_2 on global river runoff trends, P. Natl. Acad. Sci. USA, 104, 15242–15247, 2007.

Piao, S. L., Yin, G. D., Tan, J. G., Cheng, L., Huang, M. T., Li, Y., Liu, R. G., Mao, J. F., Myneni, R. B., Peng, S. S., Poulter, B., Shi, X. Y., Xiao, Z. Q., Zeng, N., Zeng, Z. Z., and Wang, Y. P.: Detection and attribution of vegetation greening trend in China over the last 30 years, Glob. Change Biol., 21, 1601–1609, doi:10.1111/gcb.12795, 2015.

Pinzon, J. E. and Tucker, C. J.: A Non-Stationary 1981–2012 AVHRR NDVI3g Time Series, Remote Sens., 6, 6929–6960, 2014.

Qian, T., Dai, A., and Trenberth, K. E.: Hydroclimatic trends in the Mississippi River from 1948 to 2004, J. Climate, 20, 4599–4614, 2007.

Ren, X. L., He, H. L., Zhang, L., Zhou, L., Yu, G. R., and Fan, J. W.: Spatiotemporal variability analysis of diffuse radiation in China during 1981–2010, Ann. Geophys., 31, 277–289, doi:10.5194/angeo-31-277-2013, 2013.

Sacks, W. J. and Kucharik, C. J.: Crop management and phenology trends in the U.S. Corn Belt: Impacts on yields, evapotranspiration and energy balance, Agr. Forest Meteorol., 151, 882–894, 2011.

Savitzky, A. and Golay, M. J. E.: Smoothing and differentiation of data by simplified least squares procedures, Anal. Chem., 36, 1627–1639, 1964.

Shi, X., Mao, J., Thornton, P. E., Hoffman, F. M., and Post, W. M.: The impact of climate, CO_2, nitrogen deposition and land use change on simulated contemporary global river flow, Geophys. Res. Lett., 38, L08704, doi:10.1029/2011GL046773, 2011.

Song, Y., Wang, C., and Ren, G.: The relative contribution of climate and cultivar renewal to shaping rice yields in China since 1981, Theor. Appl. Climatol., 120, 1–9, doi:10.1007/s00704-014-1089-z, 2014.

Song, Z. W., Zhang, H. L., Snyder, R. L., Anderson, F. E., and Chen, F.: Distribution and Trends in Reference Evapotranspiration in the North China Plain, J. Irrig. Drain. E-ASCE, 136, 240–247, 2009.

Stein, U. and Alpert, P.: Factor separation in numerical simulations, J. Atmos. Sci., 50, 2105–2115, 1993.

Tang, B., Tong, L., Kang, S., and Zhang, L.: Impacts of climate variability on reference evapotranspiration over 58 years in the Haihe river basin of north China, Agr. Water Manage., 98, 1660–1670, 2011.

Tian, H., Lu, C., Chen, G., Xu, X., Liu, M., Ren, W., Tao, B., Sun, G., Pan, S., and Liu, J.: Climate and land use controls over terrestrial water use efficiency in monsoon Asia, Ecohydrology, 4, 322–340, 2011.

Wang, J., Wang, E., Yang, X., Zhang, F., and Yin, H.: Increased yield potential of wheat-maize cropping system in the North China Plain by climate change adaptation, Climatic Change, 113, 825–840, 2012.

Wang, S., Mo, X., Liu, S., Lin, Z., and Hu, S.: Validation and trend analysis of ECV soil moisture data on cropland in North China Plain during 1981–2010, Int. J. Appl. Earth Obs., 48, 110–121, 2016.

Wang, Y. W. and Yang, Y. H.: China's dimming and brightening: evidence, causes and hydrological implications, Ann. Geophys., 32, 41–55, doi:10.5194/angeo-32-41-2014, 2014.

Wang, Z., Ye, T., Wang, J., Cheng, Z., and Shi, P. J.: Contribution of climatic and technological factors to crop yield: empirical evidence from late paddy rice in Hunan Province, China, Stoch. Environ. Res. Risk Assess., 30, 2019–2030, doi:10.1007/s00477-016-1215-9, 2016.

Weiss, A. and Norman, J. M.: Partitioning Solar-Radiation into Direct and Diffuse, Visible and near-Infrared Components, Agr. Forest Meteorol., 34, 205–213, 1985.

Xiao, D. and Tao, F.: Contributions of cultivars, management and climate change to winter wheat yield in the North China Plain in the past three decades, Eur. J. Agron., 52, 112–122, 2014.

Xin, Q., Broich, M., Suyker, A. E., Yu, L., and Gong, P.: Multi-scale evaluation of light use efficiency in MODIS gross primaryproductivity for croplands in the Midwestern United States, Agr. Forest Meteorol., 201, 111–119, 2015.

Xiong, W., Holman, I., Lin, E., Conway, D., Li, Y., and Wu, W. B.: Untangling relative contributions of recent climate and CO_2 trends to national cereal production in China, Environ. Res. Lett., 7, 044014, doi:10.1088/1748-9326/7/4/044014, 2012.

Xu, Y., Mo, X., Cai, Y., and Li, X.: Analysis on groundwater table drawdown by land use and the quest for sustainable water use in the Hebei Plain in China, Agr. Water Manage., 75, 38–53, 2005.

Yan, H., Yu, Q., Zhu, Z.-C., Myneni, R. B., Yan, H.-M., Wang, S. Q., and Shugart, H. H.: Diagnostic analysis of interannual variation of global land evapotranspiration over 1982–2011: Assessing the impact of ENSO, J. Geophys. Res.-Atmos., 118, 8969–8983, doi:10.1002/jgrd.50693, 2013.

Yu, Y., Huang, Y., and Zhang, W.: Changes in rice yields in China since 1980 associated with cultivar improvement, climate and crop management. Changes in rice yields in China since 1980 associated with cultivar improvement, climate and crop management, Field Crop. Res., 136, 65–75, 2012.

Yuan, Z. and Shen, Y.: Estimationn of agricultural water consumption from meteorological and yield data: A case study of Hebei, North China, PloS ONE, 8, e558685, doi:10.1371/journal.pone.0058685, 2013.

Zeng, Z. Z., Wang, T., Zhou, F., Ciais, P., Mao, J. F., Shi, X. Y., and Piao, S. L.: A worldwide analysis of spatiotemporal changes in water balance-based evapotranspiration from 1982 to 2009, J. Geophys. Res.-Atmos., 119, 1186–1202, doi:10.1002/2013JD020941, 2014.

Zhang, X., Chen, S., Sun, H., Shao, L., and Wang, Y.: Changes in evapotranspiration over irrigated winter wheat and maize in North China Plain over three decades, Agr. Water Manage., 98, 1097–1104, 2011.

Zhang, X., Wang, S., Sun, H., Chen, S., Shao, L., and Liu, X.: Contribution of cultivar, fertilizer and weather to yield variation of winter wheat over three decades: A case study in the North China Plain, Eur. J. Agron., 50, 52–59, 2013.

4

Retrospective forecasts of the upcoming winter season snow accumulation in the Inn headwaters (European Alps)

Kristian Förster[1,2,3], Florian Hanzer[3,4], Elena Stoll[2], Adam A. Scaife[5,6], Craig MacLachlan[5], Johannes Schöber[7], Matthias Huttenlau[2], Stefan Achleitner[8], and Ulrich Strasser[3]

[1]Leibniz Universität Hannover, Institute of Hydrology and Water Resources Management, Hanover, Germany
[2]alpS – Centre for Climate Change Adaptation, Innsbruck, Austria
[3]Institute of Geography, University of Innsbruck, Innsbruck, Austria
[4]Wegener Center for Climate and Global Change, University of Graz, Graz, Austria
[5]Met Office Hadley Centre, Exeter, Devon, UK
[6]College of Engineering, Mathematics and Physical Sciences, University of Exeter, Exeter, UK
[7]TIWAG, Tiroler Wasserkraft AG, Innsbruck, Austria
[8]Unit of Hydraulic Engineering, Institute of Infrastructure, University of Innsbruck, Innsbruck, Austria

Correspondence: Kristian Förster (foerster@iww.uni-hannover.de)

Abstract. This article presents analyses of retrospective seasonal forecasts of snow accumulation. Re-forecasts with 4 months' lead time from two coupled atmosphere–ocean general circulation models (NCEP CFSv2 and MetOffice GloSea5) drive the Alpine Water balance and Runoff Estimation model (AWARE) in order to predict mid-winter snow accumulation in the Inn headwaters. As snowpack is hydrological storage that evolves during the winter season, it is strongly dependent on precipitation totals of the previous months. Climate model (CM) predictions of precipitation totals integrated from November to February (NDJF) compare reasonably well with observations. Even though predictions for precipitation may not be significantly more skilful than for temperature, the predictive skill achieved for precipitation is retained in subsequent water balance simulations when snow water equivalent (SWE) in February is considered. Given the AWARE simulations driven by observed meteorological fields as a benchmark for SWE analyses, the correlation achieved using GloSea5-AWARE SWE predictions is $r = 0.57$. The tendency of SWE anomalies (i.e. the sign of anomalies) is correctly predicted in 11 of 13 years. For CFSv2-AWARE, the corresponding values are $r = 0.28$ and 7 of 13 years. The results suggest that some seasonal prediction of hydrological model storage tendencies in parts of Europe is possible.

1 Introduction

Seasonal prediction based on climate models (CMs) is an emerging field in hydrology (e.g. Yuan et al., 2015; Svensson et al., 2015; Mackay et al., 2015) complementing current progress in predicting long-term developments in changing hydrological conditions as a consequence of anthropogenic greenhouse gas emissions. In contrast to climate change projections, seasonal predictions focus on hydrological states of the upcoming months from their dependence on initial states (Warner, 2011). These can provide "climate services": a set of tools, products, and information serving decision makers and practitioners and bringing all types of information on climate research into practice at all levels of society (Vaughan and Dessai, 2014). This makes them relevant for detecting anticipated short-term changes in hydrological systems as requested by international research programmes such as the World Climate Research Programme (WCRP, e.g. Kirtman and Pirani, 2009), Future Earth (Greenslade and Berkhout, 2014), and more specialized programmes such as the International Network for Alpine Research Catchment Hydrology (INARCH; Pomeroy et al., 2015), which is part of the Global Energy and Water Cycle Exchanges Project (GEWEX; Chahine, 1992). In this context, seasonal predictions contribute both to coping with the WCRP "Grand Challenges" (Reid et al., 2010) and to detecting short-term changes in coupled hydrological–societal systems. The latter consideration of water and humans (Sivapalan et al., 2012) seeks to better understand interactions between society and hydrological systems for which seasonal predictions can also be seen as relevant. The goal of the current scientific decade, Panta Rhei – Everything Flows, of the International Association of Hydrological Sciences (IAHS) is to better understand these interactions over different timescales (Montanari et al., 2013).

Seasonal outlooks of hydrological variables have been prepared for decades. Antecedent hydrological and meteorological data have been used to predict monthly to seasonal streamflow using statistical methods (e.g. regression models) in various hydrological regimes (Pagano et al., 2004; Robertson and Wang, 2012; Schick et al., 2015). Another common way to predict future hydrological states is to run a process-based hydrological model based on known initial states and force it with ensembles of meteorological data observed in the past. This methodology is well known and referred to as ensemble streamflow prediction (ESP; Wood and Lettenmaier, 2008). The development of this method goes back to the 1970s and 1980s (Twedt et al., 1977; Day, 1985) and framed the development of statistical seasonal hydrological forecasting. ESP is a very useful method for studying the influence of meteorological boundary conditions (obtained from observed long-term records) on the results of hydrological forecasting models. In contrast, the reversed ESP experiment is based on actual meteorological forcing but involves an ensemble of initial states, which makes it an appropriate method to study the influence of initial conditions on forecast results. The combination of both methods is also the subject of recent research on predictability of hydrological systems (e.g. the VESPA approach; Wood et al., 2016). In the last decades, coupled atmosphere–ocean general circulation models (AOGCMs) have become a viable method for seasonal predictions (Svensson et al., 2015). These climate model-based forecasts provide future meteorological/climatological conditions for the following weeks (sub-seasonal forecasts), months (seasonal forecasts), or decades (decadal forecasts) on a physical basis rather than based on statistics. An overview of the state of the art of CM-based seasonal predictions is provided by Doblas-Reyes et al. (2013) and Yuan et al. (2015). Like numerical weather prediction, seasonal forecasts are an initial-state problem since predictions of the atmospheric states of the upcoming months strongly depend on the initial states of the atmosphere, oceans, land, and sea ice. In contrast to weather predictions however, the need for considering ocean and sea ice dynamics is crucial since these components of the climate system affect atmospheric phenomena on timescales beyond typical weather predictions. Another important difference from numerical weather predictions is the dependence of seasonal predictions on boundary conditions. Like long-term climate predictions, which are based on anthropogenic greenhouse gas emissions, CM-based seasonal predictions require adequate definitions of boundary conditions (Doblas-Reyes et al., 2013).

The skill of CM-based seasonal predictions is not distributed equally in space and time (Smith et al., 2012; Kim et al., 2012; Kirtman et al., 2014). For instance, the skill in Europe is much lower than in the tropics, where phenomena like El Niño–Southern Oscillation (ENSO) are predictable with higher accuracy (Yuan et al., 2015). Current progress on improving predictability has been recently reported by Scaife et al. (2014), who demonstrated skilful prediction of the North Atlantic Oscillation (NAO), a feature that is relevant for seasonal predictions in Europe. Bruno Soares and Dessai (2015) found that there is a mismatch in supply and demand regarding seasonal forecast products which is limited by skill levels in some regions although the authors also detected additional non-scientific reasons for this mismatch like, e.g. insufficient communication of forecasts to the users.

In general, hydrological forecast models are quite sensitive to initial hydrological conditions such as antecedent rainfall, soil moisture, and snow water equivalent (SWE). Uncertainties in the data of antecedent meteorological conditions influence the quality of process-based hydro-meteorological models at hourly resolution, e.g. in the case of 2-day flood forecasts (Achleitner et al., 2012) or 1-month sub-seasonal streamflow drought forecasts (Fundel et al., 2013). Statistical seasonal streamflow forecast models can be improved when initial conditions with respect to soil moisture and groundwater flow (Robertson et al., 2013) or snow water equivalent (Pagano et al., 2004) are considered. Discharge from alpine catchments is known to be related to snow and ice melt (Viviroli et al., 2003; Kaser et al., 2010). For hydropower generation it is interesting to know if a winter season is above or below average regarding the accumulation of snow. For water management demands such as efficient hydropower production, large efforts have been made to measure SWE in catchments of reservoirs (Painter et al., 2016; Krajči et al., 2017; Schattan et al., 2017), to simulate distributed SWE in basins of reservoirs and water intakes (Schöber et al., 2012; Hanzer et al., 2016), to improve flood forecasts with distributed SWE data (Schöber et al., 2014), and to model future runoff under climate change conditions in snow- and ice-melt-dominated catchments (Barnett et al., 2005; Schaefli et al., 2007; Finger et al., 2012; Hanzer et al., 2017). Gridded SWE data used for initialization of a process-based hydrological model improved predictions of SWE with lead times up to 1 month (Jörg-Hess et al., 2015). Seasonal streamflow and reservoir inflow predictions in snow-dominated basins were quite skilful during the snowmelt season and showed larger uncertainties during the rest of the year (Schick et al., 2015; Anghileri et al., 2016).

Besides hydropower, seasonal prediction of the accumulation of snow may be relevant to estimate the future evolution of snow depth on skiing slopes for the winter tourism business (Abegg et al., 2013; Marke et al., 2015). Well-focussed, sustainable operation of artificial snow production could re-

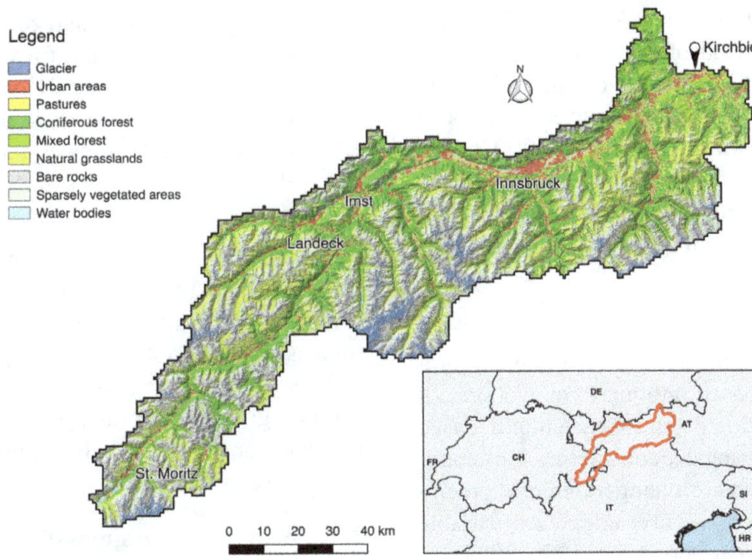

Figure 1. Map of the Inn headwaters upstream of the Kirchbichl gauging station. Major land use classes are shown along with the topography.

sult in significant savings with respect to energy costs and water use (Hanzer et al., 2014; Olefs et al., 2010).

In the present study, we focus on a way to make seasonal hydrological predictions more exploitable in the context of water resource planning in the Alps. We present a systematic evaluation of predicting above- and below-average snow accumulation, which is expected to significantly influence runoff in spring and early summer. To achieve this goal, CM-based seasonal forecasts are employed as input data to a water balance model that predicts snow water equivalent (SWE) and runoff in the Inn headwaters. A new aspect of this work is the focus on hydrological storage instead of instantaneous hydrological fluxes and the seasonal prediction of SWE in general. It is expected that the focus on integrated storage (e.g. mid-winter snow accumulation) is more robust than considering instantaneous fluxes (e.g. precipitation, runoff) in seasonal predictions.

Moreover, we focus on the winter season as extratropical seasonal forecasts appear to have the highest skill in this season (e.g. Riddle et al., 2013; Scaife et al., 2014; Kang et al., 2014). There are a number of reasons for this, including winter being the season when the stratosphere is active, which is known to affect predictions (e.g. Domeisen et al., 2015; Scaife et al., 2016; Butler et al., 2016). The winter season also shows much stronger dynamical connections to the tropics, allowing high predictability of tropical rainfall (Kumar et al., 2013) to be transmitted into the extratropics (Greatbatch et al., 2012; Molteni et al., 2015; Scaife et al., 2017).

Based on this, the research question remains as follows: can we detect above- or below-average snow conditions based on CM-based seasonal predictions in the Alps? To answer this, CM-based and hydrological modelling is applied in an Alpine case study. In Sect. 2 the relevant information about the study area, the climate data, the CM-based seasonal predictions, the water balance model, and the methodology for detecting the predictability of snow accumulation are described. In Sect. 3, the results are presented, compiled, and discussed. Finally, Sect. 4 provides concluding remarks and an outlook for future work.

2 Material and methods

2.1 Study area

The Inn headwaters catchment upstream of the Kirchbichl gauging station covers an area of $9310\,km^2$ and is located in Switzerland and Austria (see Fig. 1). The Inn river is the main tributary to the upper Danube. Elevations in the catchment range between 486 and 4049 m a.s.l., with a mean elevation of approximately 2000 m a.s.l. About 3 % of the catchment area is covered by glaciers. During the winter season runoff is lowest since a major fraction of precipitation is accumulated as snow cover. In spring snowmelt causes an increase in runoff reaching its maximum in August, when glacier melt is highest. For the period 1985–2009, the average areal precipitation and runoff amount to 1225 and $1000\,mm\,yr^{-1}$ respectively. In the second half of the 20th century, several reservoirs were built in the study area. Their total capacity is $638 \times 10^6\,m^3$.

2.2 Climate data and seasonal predictions

2.2.1 Climate data

The climate data provided by the HISTALP project (Historical Instrumental Climatological Surface Time Series of the Greater Alpine Region; Auer et al., 2007) constitute a suitable data set for studying climatology and long-term changes of temperature and precipitation in the Alps. The data have been compiled for a long period of time (1800–2010) and include a dense observational station network from different countries in the greater Alpine region. Moreover, it has been quality-checked and homogenized (Auer et al., 2007; Chimani et al., 2013). Mean temperature and precipitation depth are provided on a grid with a temporal resolution of 1 month and a spatial resolution of 5 arcmin (approx. 6 km).

2.2.2 Climatological forecasts

In the framework of this study, the term "climatological forecasts" refers to simulations based on long-term averages of air temperature and precipitation depth for each month based on the HISTALP data. For instance, considering a climatological forecast for January, mean air temperature and precipitation depth are computed through averaging each variable over all Januaries in a multi-year period (i.e. 1996–2009).

2.2.3 Climate model-based seasonal predictions

In this study, two different AOGCMs are utilized as input data for further analyses of seasonal predictions. As outlined earlier, the requirements of CM-based seasonal predictions exceed the extent of numerical weather predictions with respect to the forecast horizon and the number of subsystems of the climate system that need to be considered. Due to the extended forecast horizon, oceans and sea ice need to be incorporated in the models as well (see, e.g., Smith et al., 2012; Doblas-Reyes et al., 2013; Yuan et al., 2015). In this study, two different AOGCMs are applied independently:

- The NCEP (National Centers for Environmental Prediction) Coupled Forecast System model version 2 (CFSv2; Saha et al., 2014) is an operational seasonal prediction system. Forecasts are initialized 4 times a day. The horizontal resolution is 0.5° (approx. 40 km). In order to derive monthly forecasts, runs between the 8th day of the previous month and the 7th day of the current month are utilized in order to generate a lagged ensemble. This methodology was proposed by Yuan et al. (2013), who applied this method to re-forecasts. Since re-forecasts are only available for every 5th day, a typical ensemble of CFSv2 re-forecasts comprises 24 members per month. The archive of re-forecasts includes data from 1985 to 2009. The maximum lead time is 9 months.

- MetOffice Global Seasonal forecast system version 5 (GloSea5) is a seasonal prediction system that runs operationally at the MetOffice (MacLachlan et al., 2015; Scaife et al., 2014). Compared to CFSv2, it has a higher ocean horizontal resolution (0.25°, approx. 20 km). The data applied in this study were provided by the SPECS project ("Seasonal-to-decadal climate Prediction for the improvement of European Climate Services", http://www.specs-fp7.eu/) and cover the period between 1996 and 2010. Re-forecasts for winter were used with initial start dates: 25 October, 1 November, and 9 November. For each date, three runs are available which gives a lagged ensemble of nine members per winter. This subset of hindcasts has a lead time of 4 months for each run.

Systematic analyses are performed for 1996–2009 (the period in which both models are available). Only those re-forecasts that start in November are considered. The lead time is limited to 4 months to predict snow conditions in February. Monthly grids of the climate models with their original grid spacing (as specified above) are used as forcing data for the water balance model which is described in the next section.

2.3 Water balance simulations using AWARE

The Alpine Water balance and Runoff Estimation model (AWARE; Förster et al., 2016) is a deterministic hydrological model operating on a regular grid at 1-month time steps. The model has been designed to estimate anomalies in hydrological variables at the catchment scale from anomalies in meteorological fields predicted by climate models. The coarse temporal resolution allows one to carry out seasonal predictions considering a large number of individual runs at a minimum of computational costs which justifies the coarse time step. As the study's focus is on anomalies in seasonal characteristics, using a monthly scale water balance model is feasible (Kling et al., 2012; Bock et al., 2016), and these models are also applied for seasonal hydrological predictions (Bell et al., 2017).

Required meteorological forcing data include both mean monthly air temperature and monthly precipitation totals provided as grids or station data, which makes the model parsimonious with respect to data requirements. Altitudinal gradients are applied in order to realistically redistribute temperature and precipitation on the model grid. In general, this feature results in a decrease in temperature with increasing elevation and an increase in precipitation on the mountains. For each grid cell the relative contributions of rainfall and snowfall are computed taking into account two threshold temperature values. If the air temperature falls below the lower threshold temperature, the monthly precipitation depth is assumed to be snowfall only. Likewise, air temperatures exceeding the upper threshold indicate rainfall only. In order to enable the occurrence of both snow and rain, a transition range between both thresholds is defined. Based on air temperature, the fraction of rain and snow is linearly interpolated between these two thresholds. Even though the model is also capable of reading shortwave radiation fields (Förster et al., 2016) in order to improve ice-melt prediction, only a simplified snow- and ice-melt simulation using air temperature is possible. This simplification considers the fact that air temperature and precipitation are readily available and more predictable compared to some other meteorological fields. In order to perform simulations with this minimal input of data, the Thornthwaite (1948) evapotranspiration approach is applied. The soil water balance is computed following the approach of McCabe and Markstrom (2007). Linear storage is applied in order to account for the recession of runoff typically related to groundwater processes.

The spatial resolution of the Inn headwaters setup in the AWARE model is 1000 m. Besides a grid-based model domain, AWARE assumes a baseline (reference) meteorological data set for calibration, which is shown in Fig. 2 using the HISTALP data from 1996 to 2009 as the reference period (this run is herein referred to as HISTALP-AWARE). The Nash–Sutcliffe model efficiency (NSE) amounts to $E = 0.92$ which could be considered very good model performance. As suggested by Schaefli and Gupta (2007), the benchmark Nash–Sutcliffe model efficiency is computed as well ($E_b = 0.45$). This benchmark NSE value accounts for strong effects of seasonality (Eq. A1 in Sect. A in the Appendix). While the standard NSE indicates if a model is better than the average of observed values, the benchmark NSE proves if the model performance exceeds the corresponding value of a simple model that simply predicts long-term averages for each month. Since the benchmark NSE is also greater than 0, the model is more skilful than applying long-term averages. According to Klemeš (1986) a split sample test is applied including an independent validation period ranging from 1984 to 1995. The corresponding NSE and benchmark NSE are $E = 0.91$ and $E_b = 0.25$ respectively. A possible reason for the lower E_b value might be the fact that the validation period has seen an advancing of glaciers due to positive glacier mass balances. In contrast, the calibration period is characterized by a shrinkage of glacier volumes. Both processes are not incorporated in the model so far. However, as the model performance of the validation period is still comparable to the calibration period, the model is found to be suitable for prediction. The mismatch of runoff simulations in winter, especially in March, can be attributed to the effects of reservoirs on river flow in the catchment area which are not represented in the model so far. In this period water is released from seasonal storage filled in summer.

Another advantage of the 1-month time step is the lower complexity with respect to downscaling of climate model data. Current approaches focus on statistical (e.g. Crochemore et al., 2016) or dynamical downscaling (e.g. Förster et al., 2014) of coarse atmospheric data fields (e.g. derived by climate models). AWARE builds upon a simple and robust approach which is based on anomalies. For instance, Marzeion et al. (2012) successfully add anomalies from other data sets to a reference climatology to compute glacier mass balances at the global scale. In order to account for different spreads of distributions, standardized anomalies are considered in our study. According to Wilks (2006) this approach is feasible when "working simultaneously with batches of data that are related, but not strictly comparable". This is a typical situation for observational data and re-forecasts. Standardized anomalies z_x are simply computed for a variable x, taking into consideration its long-term mean $\overline{x}$ for a given month and the corresponding empirical standard deviation $\widetilde{s}_x$ (Wilks, 2006):

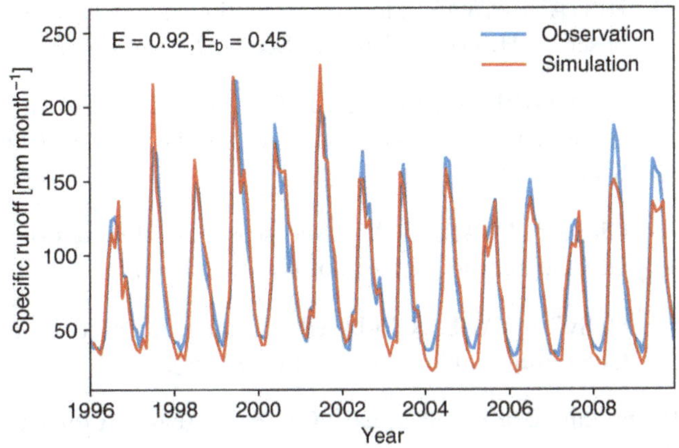

Figure 2. Specific runoff hydrographs for observed and modelled runoff at Kirchbichl gauging station and the corresponding performance measures E (Nash–Sutcliffe model efficiency, NSE) and E_b (benchmark NSE).

$$z_x = \frac{x - \overline{x}}{\widetilde{s}_x}. \quad (1)$$

Given that two data sets x and y are comparable (e.g. reference climatology and the climatology of re-forecasts), their standardized anomalies z_x and z_y could be comparable as well. Based on the assumption that $z_x = z_y$, Eq. (1) can be rearranged to

$$x = \frac{y - \overline{y}}{\widetilde{s}_y} \cdot \widetilde{s}_x + \overline{x}. \quad (2)$$

Anomalies of the climate model (i.e. $y - \overline{y}$) can easily be transformed to the climatology of the reference data set (i.e. x). Mean values and standard deviations are computed separately for each month and climate data set including HISTALP, CFSv2, and GloSea5. In this way, anomalies predicted by the climate models can be reliably transformed to typical anomalies of the observational data.

2.4 Model experiment for analysing the predictability of snow accumulation

The long-term simulations of the water balance provide monthly snapshots of valid system states for each state variable at any point in time. For each CM-based seasonal prediction run starting in November, system states for SWE, soil moisture, and groundwater storage computed for October are defined as initial states. In total four AWARE runs driven with different forcing data sets are available for each winter season between 1997 and 2009 (November to February, NDJF):

1. HISTALP-AWARE: long-term continuous run based on observed HISTALP data (see Sect. 2.2.1 and Fig. 2).
2. CF-AWARE: climatological forecasts (CF) with average conditions computed using HISTALP (see Sect. 2.2.2).
3. GloSea5-AWARE: CM-based seasonal forecast using GloSea5 (ensemble mean of nine members).
4. CFSv2-AWARE: CM-based seasonal forecast using CFSv2 (ensemble mean of 24 members).

The ensemble provided by each CM-based seasonal forecast of meteorological quantities is averaged prior to the water balance simulations. In general, ensemble seasonal predictions are subject to low signal-to-noise ratios. The signal in the ensemble mean is small in most cases and using members individually will mask the signal (Scaife et al., 2014; Eade et al., 2014). In general, each ensemble member of input data is individually processed in hydrological forecasting, which is why the averaging is typically implemented afterwards. However, a skill improvement has been reported in recent seasonal prediction studies (e.g. Bell et al., 2017), in which the concept of averaging is applied prior to hydrological simulations. This approach seems feasible given that the time step of hydrological simulations is 1 month. Although the hydrological model is a conceptual model that mimics the basic physical principles, the temporal scale does not allow for capturing the full dynamics of hydrological processes that are typically studied on smaller scales. Thus, the coarse temporal resolution of the modelling approach is to a certain degree "statistical" in nature which justifies the application of mean ensemble inputs. Moreover, the utilization of standardized anomalies applied to CM-based seasonal forecasts in the AWARE model accounts for variance corrections to the ensemble mean values as suggested by Eade et al. (2014). Appropriate uncertainty can also be added to the predictions to ensure reliable probabilistic forecasts.

The basin-average time series of these water balance simulations are directly comparable. While the continuous long-term simulation represents a reference run (#1) serving as benchmark for seasonal predictions, the climatological forecasts (#2) help to judge whether anomalies will be above or below average. Correlations between the reference run and the water balance simulations forced by CM-based forecasts (#3 and #4) are computed to assess the predictive skill. Moreover, the tendency or sign of anomalies is compared through counting the coincidence of above- or below-average anomalies in the reference run and the seasonal predictions.

A set of skill measures is used throughout the study in order to quantify the model skill of the different forecasts (CF-AWARE, GloSea5-AWARE, CFSv2-AWARE). Besides correlation and hit rate (i.e. the number of correctly predicted states divided by the total number of winters), other measures to assess the skill of the models are considered. For instance, the standard deviation of a single time series is a measure used to compare the variability of forecasts. In contrast, the root mean square error (RMSE) also involves observed time series and provides insight into the absolute difference between time series. Since quadratic differences are summarized, a greater weight is assigned to larger differences, thus making RMSE sensitive to greater mismatches. In order to show the accuracy of the models in predicting the tendencies of anomalies (hit rate), the Brier skill score (BSS) is also computed (see Eqs. A2 and A4 in Sect. A along with a brief description in the Appendix). In general, a skill score judges the improvement of a forecast system relative to a reference (climatology). A value of 0 indicates that the forecast system is not better than the reference. In contrast, a value of 1 indicates a perfect match of forecasts and observations. The BSS is related to the hit rate which has already been defined (higher hit rates go in hand with higher BSS). Finally, the mean absolute error (MAE, Eq. A3) skill is comparable to RMSE but does not account for quadratic weighting of differences. Like BSS, MAE can be computed as a skill score (MAESS; Eq. A4), which is a measure for the differences in absolute terms. In this way, it is less sensitive than RMSE to large differences but rather includes a reference run.

3 Results and discussion

3.1 Long-term simulations and climatological forecast of SWE

While the applicability of AWARE to reconstruct the water balance in terms of observed runoff time series was demonstrated in Sect. 2.3, it is necessary to evaluate the model experiments HISTALP-AWARE and CF-AWARE with respect to SWE prior to the analyses of CM-based SWE forecasts. Figure 3a demonstrates the annual cycle of modelled SWE. The black dashed line is the mean value of all years computed using the reference run (HISTALP-AWARE). It compares well with the black bold line which represents the climatological simulations based on AWARE using average air temperature and precipitation depth for each month (CF-AWARE). Thus, a climatological forecast is suitable to compute average snow conditions. Figure 3c shows the spatial distribution of average SWE in February. The averages of SWE on the model highlights the typical snow distribution with highest values on the mountains and lower values in the valleys. Full time series are shown in Fig. 3b. The boundary conditions of the climatological forecast are equal in each year. However, the initial conditions differ according to the initialization each year in October which is obtained from the long-term run. Figure 3d depicts a subset of SWE observations compiled by Schöber et al. (2016). In contrast to the cited study, which explains the methodology of SWE sampling in detail, here only stations above 1400 m a.s.l. have been selected in order to better match the average catch-

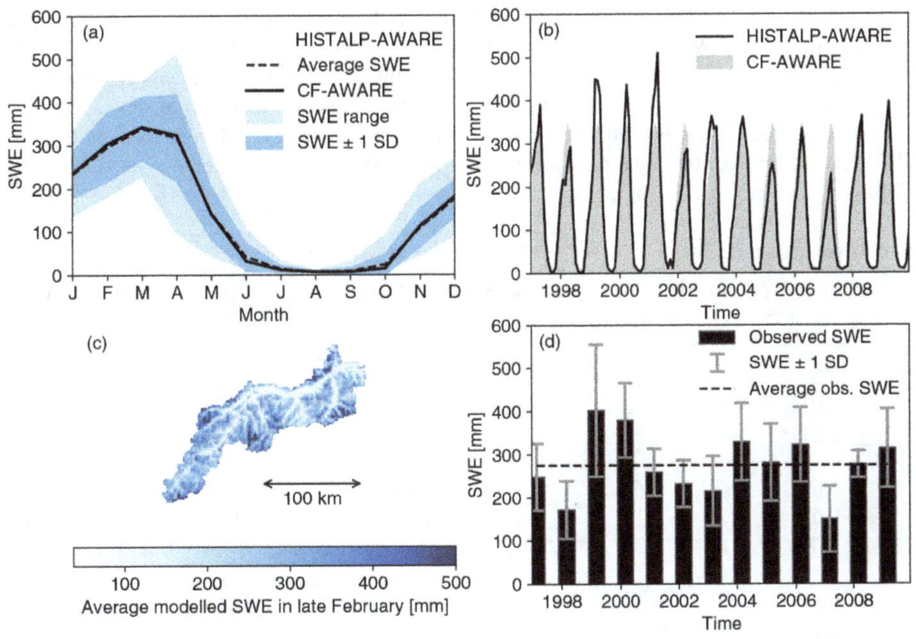

Figure 3. Observed and simulated snow states. **(a)** Range and standard deviation of SWE time series computed for single years and the corresponding mean monthly averages derived from AWARE using HISTALP data (HISTALP-AWARE). Average SWE conditions computed using climatological forecasts (CF-AWARE) are displayed as well. **(b)** This plot shows the same data but as time series between 1997 and 2009. **(c)** Map showing the average spatial SWE distribution computed for February. **(d)** SWE observations in February collected from stations in the study area above 1400 m a.s.l. (Schöber et al., 2016). The error bars show the ±1 SD (standard deviation) of observations for each year (see Schöber et al., 2016, for further details on SWE sampling).

ment elevation (Sect. 2.1). The correlation between computed SWE in February and the SWE observations in February is $r = 0.65$ (Fig. 3b vs. Fig. 3d). This comparison should be interpreted with caution. First, despite the fact that a subselection of stations that better match the mean elevation of the catchment has been chosen for this analysis, the full range of elevation bands in the basin is not fully covered by the observational data set. Moreover, scaling issues limit spatial and temporal representativeness, since averaged point-scale measurements recorded on a weekly scale are compared to basin-scale water balance simulation with 1-month time step. However, observed and computed SWE compare reasonably well. This underlines the applicability of AWARE to predict SWE.

3.2 CM-based seasonal predictions using AWARE

In the next step, anomalies computed using AWARE forced by CM-based seasonal forecasts are compared to the corresponding values of the reference run (HISTALP-AWARE, #1). This evaluation is demonstrated in Fig. 4 for temperature, precipitation depth, and SWE in February. Anomalies in temperature and precipitation depth refer to the period November to February (NDJF) in each winter and represent average values at the basin scale (i.e. the mean of all grid points of the meteorological fields in AWARE). In this way, the values are subject to the statical transformations and elevation-dependent redistributions as outlined in Sect. 2.3. The anomalies of the reference AWARE run driven by HISTALP are shown in the top panels of Fig. 4 (HISTALP-AWARE). Their correlation is set to 1 by definition since this run is viewed as a reference run. The seasonal forecasts computed using AWARE driven by GloSea5-AWARE (centre) and CFSv2-AWARE (bottom) are also displayed. In addition, Table 1 (first model experiment column) provides a summary of skill measures for temperature, precipitation, and SWE.

Correlation coefficients computed for NDJF temperature anomalies range from $r = 0.17$ (CFSv2-AWARE) to $r = 0.32$ (GloSea5-AWARE). Tendencies in anomalies (i.e. the prediction of correct signs of anomalies) also vary between the models. This becomes obvious when counting the shaded areas indicating a mismatch between the seasonal forecast and the reference run. While GloSea5-AWARE correctly predicted the sign of temperature anomalies in 9 of 13 winters, the hit rate achieved for CFSv2-AWARE only amounts to 8 of 13 (see Table 1). The differences between GloSea5-AWARE and CFSv2-AWARE in terms of standard deviation are small. Hence, both model settings show a similar variability of forecasts which can be attributed to the standardized anomaly approach. GloSea5-AWARE shows a smaller RMSE than CFSv2-AWARE does. A similar ranking of skill is obvious when considering BSS and

Table 1. Skill measures for basin-scale averages of NDJF temperature, NDJF precipitation, and SWE in February. For each experiment, results for GloSea5-AWARE and CFSv2-AWARE are summarized. The first experiment (full dynamical model runs) refers to the standard setting of CM-based seasonal forecasts using AWARE (Sect. 3.2), while the other two experiments replace CM-based time series with climatology (Sect. 3.3). Skill measures: *r*, Pearson correlation coefficient; SD, standard deviation of time series; RMSE, root mean square error; hit rate, number of correctly predicted states (out of 13); BSS, Brier skill score; MAESS, mean absolute error skill score. Please note that the units of SD and RMSE are those of the input time series. HISTALP-AWARE is the reference used in the calculations of *r*, RMSE, hit rate, BSS, and MAESS.

	Full dynamical model runs		Temperature from climatology		Precipitation from climatology	
	GloSea5-AWARE	CFSv2-AWARE	GloSea5-AWARE	CFSv2-AWARE	GloSea5-AWARE	CFSv2-AWARE
Temperature						
r [–]	0.32	0.17	–	–	0.32	0.17
SD [K]	1.03	0.86	0.00	0.00	1.03	0.86
RMSE [K]	1.29	1.32	1.16	1.16	1.29	1.32
Hit rate [–]	9	8	–	–	9	8
BSS [–]	0.69	0.62	0.00	0.00	0.69	0.62
MAESS [–]	−0.02	−0.12	0.00	0.00	−0.02	−0.12
Precipitation						
r [–]	0.61	0.31	0.61	0.31	–	–
SD [mm]	64.77	70.95	64.77	70.95	0.00	0.00
RMSE [mm]	55.97	78.66	55.97	78.66	62.20	62.20
Hit rate [–]	9	7	9	7	–	–
BSS [–]	0.69	0.54	0.69	0.54	0.00	0.00
MAESS [–]	0.19	−0.29	0.19	−0.29	0.00	0.00
SWE						
r [–]	0.57	0.28	0.44	0.25	0.51	0.16
SD [mm]	72.57	71.91	62.69	68.03	23.01	20.32
RMSE [mm]	65.27	88.02	68.88	90.03	59.32	66.99
Hit rate [–]	11	7	10	7	8	6
BSS [–]	0.85	0.54	0.77	0.54	0.62	0.46
MAESS [–]	0.14	−0.23	0.06	−0.19	0.13	−0.05

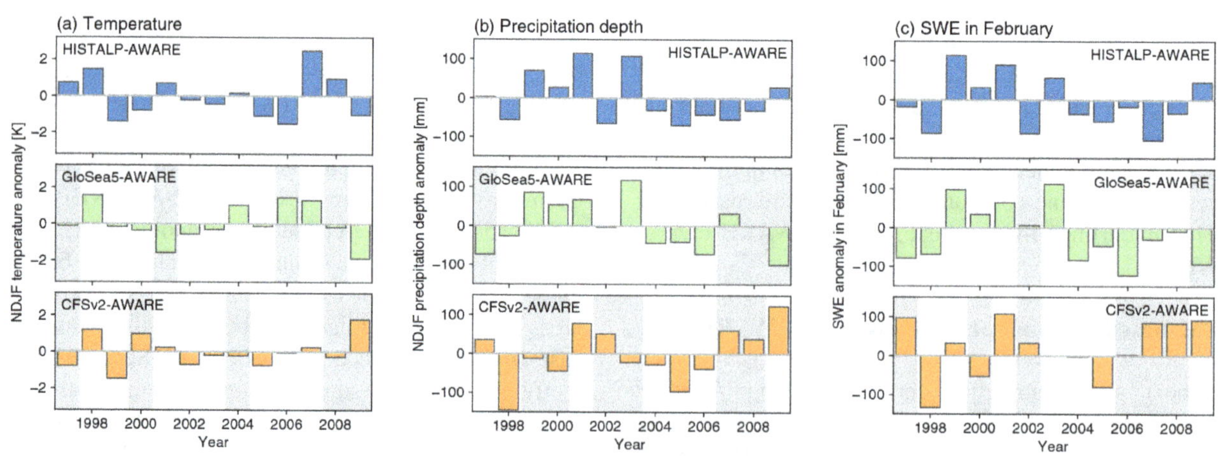

Figure 4. Anomalies of **(a)** basin-scale NDJF temperature, **(b)** NDJF areal precipitation depth, and **(c)** snow accumulation in February. Water balance simulations driven by CM-based seasonal forecasts are compared to water balance simulation driven by HISTALP. Shaded areas indicate years in which the sign of anomalies does not match the reference run.

MAESS. The latter suggests that both model runs (GloSea5-AWARE and CFSv2-AWARE) are less skilful than climatology (MAESS < 0). However, the positive BSS values highlight the capability of predicting the tendency of temperature anomalies.

In the case of GloSea5-AWARE, the hit rate of correctly predicted anomalies regarding precipitation is 9 of 13 ($r = 0.61$). As for temperature, the model skill of precipitation predictions computed by CFSv2-AWARE is also lower (hit rate 7 of 13, $r = 0.31$). This finding holds also true for the other skill measures, namely RMSE, BSS, and MAESS. However, the number of correctly predicted tendencies achieved using GloSea5-AWARE could be considered to be a good result since the seasonal forecasts include a lead time of 4 months. Single months show lower scores, suggesting that temporal integration improves the robustness of results consistent with our approach using hydrological storage rather than fluxes. In our study, we found monthly correlations computed for precipitation forecasts ranging from -0.29 to 0.30 (GloSea5-AWARE) and -0.11 to 0.15 (CFSv2-AWARE). These are generally lower than the corresponding values achieved for the averaged NDJF forecasts (GloSea5-AWARE: 0.61; CFSv2-AWARE: 0.31). Similar values of the same order have been observed for SWE forecasts (GloSea5-AWARE: 0.57; CFSv2-AWARE: 0.28).

Given the skill measures from Table 1 (first column) and the coincidence of anomalies highlighted in Fig. 4c the predictive skill achieved for precipitation depth also prevails for SWE in February. Even though correlation coefficients are slightly lower compared to precipitation depth (GloSea5-AWARE: $r = 0.57$; CFSv2-AWARE: $r = 0.28$), SWE values in February computed by AWARE driven by CM-based forecasts compare well to those of the reference run (HISTALP-AWARE). The hit rate achieved using GloSea-AWARE even reaches 11 of 13 while the hit rate of CFSv2-AWARE remains at the level of 7 of 13. An increase in skill in terms of RMSE, BSS, and MAESS is also at least partially obvious for both models, indicating that some skill measures suggest that SWE predictions are more robust than precipitation predictions.

A Bernoulli experiment helps to judge whether these hit rates differ from the performance of a "fair coin" for predicting above- and below-average conditions. The null hypothesis states that the hit rate of the seasonal forecasts does not differ from a random 50 : 50 probability (binomial test). Given the total number of winters $n = 13$ and a level of significance of $\alpha = 0.05$, the null hypothesis is rejected for hit rates above 9 of 13. This means that according to the results shown in Fig. 4 and Table 1, for seasonal predictions of SWE using GloSea5 this test rejects the null hypothesis, indicating significant skill. In contrast, the scores for CFSv2 are not significant.

Regardless the limitations discussed with respect to observed SWE, the correlations are much lower if the observations from Fig. 3d are involved in skill computations. The correlation between observed anomalies and GloSea5 is $r = 0.21$, while the corresponding value achieved using CFSv2 is only $r = 0.11$. These values are much lower than the correlations achieved using the reference run (HISTALP-AWARE). This finding might also be related to possible mismatches in representativeness between observations and simulations. However, the comparison between HISTALP-AWARE and the CM-based seasonal forecasts highlights GCM forecast skill and acknowledges the fact that the water balance model is never perfect since it introduces uncertainties into hydrological forecasts, too. Due to the reasonably good agreement between seasonal forecasts and the reference run, the skill of CM-based forecasts is considered promising.

Figure 5 depicts time series of the water balance of the snow storage for each year and each AWARE model run. Monthly precipitation (divided into rainfall and snowfall), cumulative snowmelt, and SWE are plotted. Moreover, the snow accumulation of the reference run (HISTALP-AWARE, #1) and the climatological forecast (CF-AWARE, #2) are displayed. The latter is subject to the same forcing in each year but is initialized according to the system states of AWARE in late autumn. If the SWE computed by HISTALP-AWARE exceeds the corresponding value of CF-AWARE, above-average snow accumulation prevails. Accordingly, the opposite is true for below-average conditions. A similar comparison is possible for the predictions of GloSea5-AWARE and CFSv2-AWARE. If the CM-based forecast and HISTALP-AWARE simultaneously indicate either above- or below-average conditions, the label "HIT" is added to the corresponding seasonal forecast. The overall hit rate can be seen in Table 1. Even though monthly precipitation depth differs between HISTALP-AWARE and the CM-based forecasts, the NDJF precipitation totals might compensate for this monthly scale differences resulting in a good match of SWE values in February. This is obvious for many of the winter seasons shown in Fig. 5 (e.g. 1998/1999 and 2000/2001) and confirms the previous finding that improved model skill is possible when storage instead of instantaneous fluxes is considered.

3.3 The role of temperature and precipitation for SWE forecasts

In order to show the importance of both temperature and precipitation in SWE forecasting, Table 1 summarizes the skill measure previously introduced for two other model experiments in which either temperature or precipitation is replaced by climatological forecasts: (i) temperature from climatology is combined with precipitation forecasts from the climate models (second column of Table 1) or (ii) precipitation from climatology is combined with temperature forecasts from the climate models (third column of Table 1). If one variable is

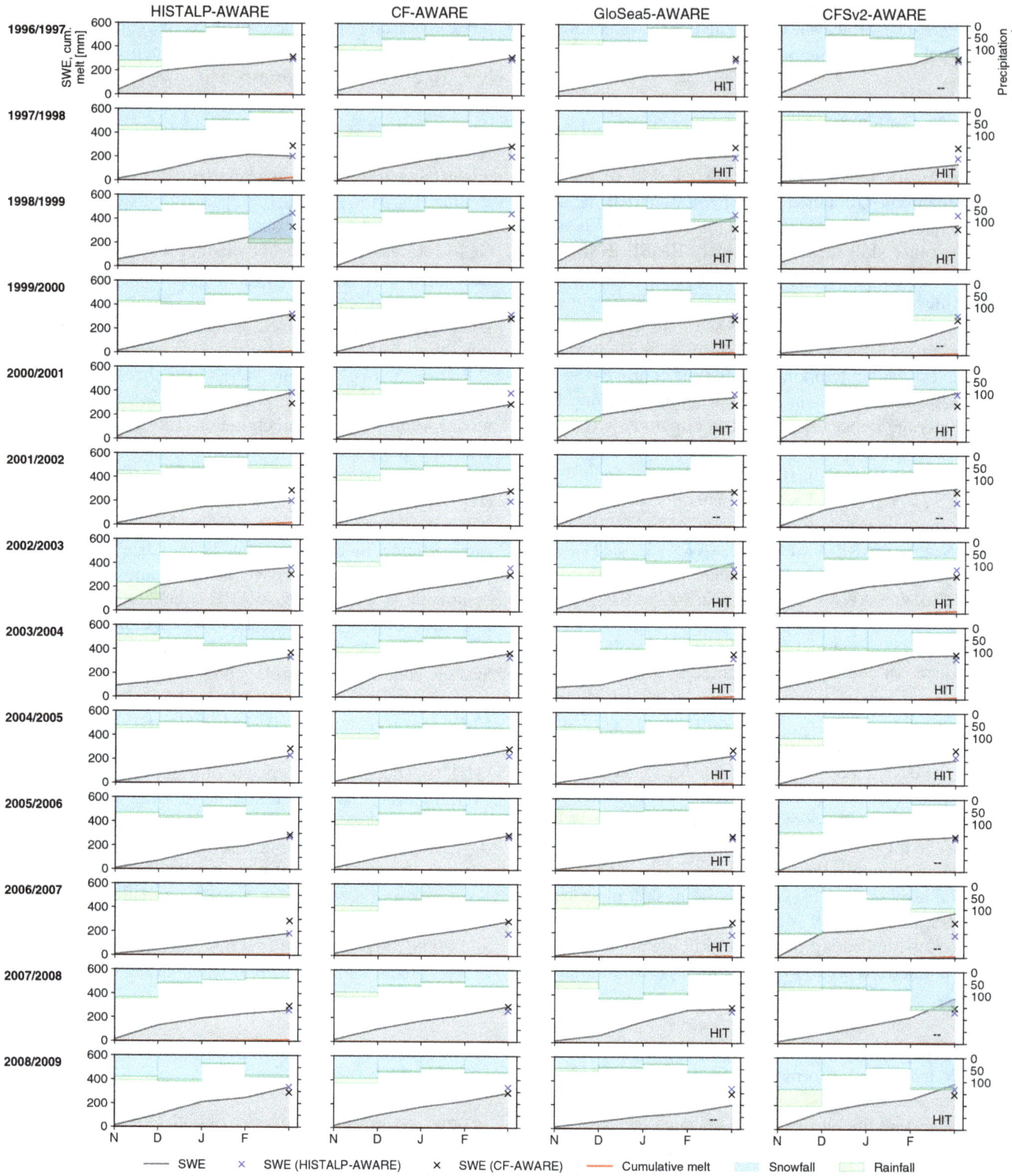

Figure 5. Water balance of the basin-scale snow storage for each year and each forcing data set used for AWARE simulations. CF is the climatological forecast (long-term averages of HISTALP) which can be viewed as forecast yielding average conditions. The evolution of snow accumulation is categorized either "HIT" or "–" if the sign of anomalies obtained from HISTALP-AWARE and CM-based AWARE runs matches or mismatches respectively.

replaced by climatology the standard deviation of anomalies is 0 since the climatological forecasts have no deviations from climatology. This is in line with zero skill in terms of BSS and MAESS (see temperature skills in the second column and precipitation skills in the third column). The skill measures of the respective variable that has not changed in this way are subject to the same characteristics as in the full dynamical run (first column). For instance, if temperature is replaced by climatology, precipitation skills are equal to those in the full dynamic run (e.g. compare the first and second columns for precipitation).

In the case of SWE, the effects of replacing either temperature or precipitation differ in terms of model skill. First, a drop in correlation is obvious in both cases. If temperature is replaced by climatology the hit rate of GloSea5-AWARE decreases only slightly to 10 but remains at 7 for CFSv2-AWARE. If precipitation is replaced by climatology hit rates decrease in both cases and the standard deviation is much lower than in the full dynamic run. This indicates that the variability in SWE forecasts is mainly prescribed by precipitation in the current study setup. However, the influence of temperature would likely increase for predictions of SWE in the ablation season.

Surprisingly, the RMSE in terms of SWE re-forecasts is lowest in the model run in which precipitation is replaced by climatology. Since this finding is confirmed neither by comparing MAESS (which computes similar error statistics but with linear instead of quadratic weighting of errors) values nor by considering any of the other skill measures, it is likely that this effect is explained by the low variability of SWE in this experiment combined with the quadratic weighting of errors in RMSE computations. This comparison underlines the need for different skill measures in the process of evaluating forecasts.

3.4 Model skill and its relation to other studies

Compared to findings reported in the literature, the results achieved in this study are promising given that the skill for Europe is generally found to be low. For instance, according to Weisheimer and Palmer (2014) the skill of DJF temperature is "marginally useful" using ECMWF's System4. Even the rating for DJF precipitation is found to be "not useful" (see Fig. 5 in Weisheimer and Palmer, 2014). Similarly, Kim et al. (2012) found some skill in terms of correlation for wintertime temperature predictions using System4. However, their study also suggests low absolute correlation coefficients for precipitation forecasts and for both temperature and precipitation forecasts achieved using CFSv2. A direct comparison to the results presented in this study is not possible since GloSea5 was not addressed in these studies. Moreover, given that only one single catchment is considered, a ranking of models is beyond the scope of this article. The predictability for SWE detected in this study could be related both to some amount of skill in precipitation prediction and to previous findings on the persistence of SWE predictions with shorter-term forecast horizons. For instance, in the case of Alpine snow cover, Jörg-Hess et al. (2015) underline the persistence of SWE predictions at least up to a lag of 2 weeks.

4 Conclusions

In this study, a systematic evaluation of CM-based seasonal winter forecasts starting in November has been performed using a water balance model. A new method has been developed focussing on hydrological storage instead of instantaneous hydrological fluxes. SWE was chosen as predictand here, and two independent climate models were used as input data for monthly scale distributed water balance model. A robust approach based on standardized anomalies was applied in order to bridge the gap in scale between GCMs and the water balance model. In this way, basin-scale averages of temperature and precipitation depth are temporally integrated in order to achieve November to February (NDJF) averages and totals respectively. Given a lead time of 4 months, the application of the water balance model then allows predicting SWE in February, which is relevant for many sectors like water management or hydropower generation. Based on year-by-year evaluation of re-forecasts using different skill measures and a binomial test, the results achieved using GloSea5-AWARE and CFSv2-AWARE indicate that dynamical (CM-based) seasonal forecasts can provide skill. A sensitivity analysis using different configurations of input data sets showed that SWE forecasts benefit from the skill in precipitation forecasts, especially in terms of variability and hit rate/Brier score. These findings might be related to the hydro-climatological characteristics of the study area where snow accumulation is the major process during winter, while snowmelt as a strong temperature-dependent process is less important in this time (Fig. 5). In other environments the relative role of temperature and precipitation might look different.

Regarding predictability, the location of the study area is also of particular interest in the process of interpreting the results. Due to the fact that the Alps are situated in a transition zone between northern and southern Europe, the influence of large-scale climate patterns, such as the NAO, should be analysed in more detail in the future. It is also known that ENSO impacts the climate in Europe in late winter and sudden stratospheric warming is also important (Ineson and Scaife, 2008; Scaife et al., 2016). The first assessment of possible connections between the NAO and snow- and glacier-related states only resulted in low correlations (cf. Beniston and Jungo, 2002; Scherrer et al., 2004; Bartolini et al., 2009; Marzeion and Nesje, 2012). However, in the southern and western parts of the Alps this relationship between NAO and snow and ice properties might be explained more clearly. Recent improvements regarding CM-based seasonal predictions might explain our detectable skill (Scaife et al., 2014). Future work should address climatological processes that are related to model skill and involve other basins in different parts of the Alps.

Besides studying the climatological perspective of predictability, the results also revealed uncertainties involved in hydrological modelling using the water balance model and scaling issues regarding the representativeness of point-scale SWE observations. These findings also suggest improvements regarding both the provision of basin-scale SWE observations and the water balance model to be considered in future work. Low-flow conditions in March might be better predicted if the model accounted for artificial reservoirs in the study area. Moreover, better representation of changes in glaciated area is currently being investigated through coupling AWARE with a glacier evolution model developed by Marzeion et al. (2012). These features will be added to the model in the future.

However, the results of this study show that it is possible to detect skilful signals from dynamical (CM-based) seasonal predictions of hydrological storage in Europe, where seasonal prediction is still challenging. The results suggest that seasonal prediction of hydrological model storage tendencies is possible, although the skill of such predictions is in many cases low in Europe. Overall this study suggests that focussing on hydrological storage rather than hydrological fluxes might help in exploiting seasonal predictions. The first results of the methodology are promising for practical application, with hit rates above 70 % seen as a reasonable target accuracy. Since snowmelt predictions are of particular interest in the study area, a similar approach could be applied to CM-based seasonal forecasts initialized in May. Future research should also address predictability studies in other regions. Moreover, it would be interesting to study the predictability of other types of hydrological storage such as glaciers, lakes, or groundwater, as well as exploring probabilistic forecasting.

Appendix A: Model performance and skill measures

The definition of the Nash–Sutcliffe model efficiency E and the benchmark Nash–Sutcliffe model efficiency E_b (Schaefli and Gupta, 2007) reads as follows:

$$E_\mathrm{b} = 1 - \frac{\sum_{t=1}^{N}\left[q_\mathrm{obs}(t) - q_\mathrm{sim}(t)\right]^2}{\sum_{t=1}^{N}\left[q_\mathrm{obs}(t) - q_\mathrm{bench}(t)\right]^2}. \tag{A1}$$

In this equation, time series of observed q_obs and modelled q_sim quantities are considered for all time steps t. $q_\mathrm{bench}(t)$ is the time-dependent benchmark value at timestep t. $q_\mathrm{bench}(t)$ is a long-term average computed for the month of time step t. The original definition of Schaefli and Gupta (2007) refers to daily series for which the long-term average for a specific calendar day is applied. According to Schaefli and Gupta (2007) E_b indicates if the model "has greater explanatory power than already contained in the seasonality of the driving forces (the climate)". If $q_\mathrm{bench}(t) = \overline{q}_\mathrm{obs}$ is assumed, E_b is equal to the Nash–Sutcliffe model efficiency E. In contrast to E, E_b presumes the climatological mean of each time step as a benchmark against which all elements of the time series are compared. Since seasonality is inherent in many time series, generally $E_\mathrm{b} \leq E$ holds.

A widely used measure to evaluate the accuracy of forecasts is the Brier score B (Wilks, 2006):

$$B = \frac{1}{N}\sum_{i=1}^{N}(f_i - o_i)^2. \tag{A2}$$

This is a special case of the ranked probability score (RPS; see, e.g., Hersbach, 2000) that restricts the evaluation of forecasts to two categories (e.g. above or below average). The forecast f_i computed for each year i is compared to the corresponding state observed in that year o_i, whereby f_i and o_i are dichotomous states (0 or 1). The Brier score B is the average of the squared differences between f_i and o_i. The average refers to a range of N years. The best value that can be achieved in this way is 0, indicating a perfect forecast skill. In contrast, 1 indicates that all forecasts are wrong.

Another skill measure for forecasts is the MAE which characterizes, similar to RMSE, differences between the forecasted value q_f and the observed value q_o (in units of the underlying time series):

$$M = \frac{1}{N}\sum_{i=1}^{N}\left|q_{\mathrm{f},i}(t) - q_{\mathrm{o},i}(t)\right|. \tag{A3}$$

If $|q_{\mathrm{f},i}(t) - q_{\mathrm{o},i}(t)|$ is replaced by $(q_{\mathrm{f},i}(t) - q_{\mathrm{o},i}(t))^2$ and the square root is calculated from Eq. (A3), this equation yields the RMSE. In contrast to RMSE, MAE is less sensitive to larger differences between $q_{\mathrm{f},i}$ and $q_{\mathrm{o},i}$. Moreover, the MAE is comparable to the continuous ranked probability score (CRPS) used for probabilistic forecasts (Hersbach, 2000; Trinh et al., 2013) and can be used for single-value (deterministic) forecasts.

In order to compare these skill measures computed for different forecasts to a reference forecast (i.e. climatology), a skill score S measure is typically calculated. For instance, the MAESS (S_MAESS) can be derived using

$$S_\mathrm{MAESS} = 1 - \frac{M_\mathrm{forecast}}{M_\mathrm{reference}}, \tag{A4}$$

where M_forecast is the MAE of the forecast system and $M_\mathrm{reference}$ is the MAE of the climatological forecast. Similarly, Eq. (A4) can be applied to derive a BSS S_BSS through replacing M by B.

Author contributions. KF prepared the manuscript with contributions from all co-authors, designed the study, and performed the water balance simulations and predictability analyses. KF and FH developed the AWARE model, which was designed for this kind of study. ES contributed to downscaling of climate model output and reviewed the literature with respect to connections between snow and glacier observations and the NAO index. AAS and CM computed and provided the GloSea5 re-forecasts and helped with data usage, interpretation of the results, and improving the methodology. Snow observations in the study area were evaluated by JS, who also contributed to interpreting anomalies in SWE. MH coordinated the project. SA and US were the key researchers of the project. They supervised the scientific work and helped discuss the results and improve the methodology.

Competing interests. The authors declare that they have no conflict of interest.

Special issue statement. This article is part of the special issue "Sub-seasonal to seasonal hydrological forecasting". It is not associated with a conference.

Acknowledgements. This work was carried out as part of the W01 MUSICALS II – Multiscale Snow/Ice Melt Discharge Simulation for Alpine Reservoirs project at alpS – Centre for Climate Change Adaptation in Innsbruck, Austria. The K1-Centre alpS is funded through the Federal Ministry of Transport, Innovation and Technology (BMVIT), the Federal Ministry of Science, Research and Economy (BMWFW), and the Austrian federal states of Tyrol and Vorarlberg within the scope of COMET – Competence Centers for Excellent Technologies. The COMET programme is managed by the Austrian Research Promotion Agency (FFG). We want to thank Tiroler Wasserkraft AG (TIWAG) for the collaboration and for co-funding the project. Additional thanks go to the NOAA (National Oceanic and Atmospheric Administration) National Centers for Environmental Prediction (NCEP) for the provision the CFSv2 data. The retrospective forecasts of the GloSea5 model were kindly provided by the SPECS project (Seasonal-to-decadal climate Prediction for the improvement of European Climate Services; http://www.specs-fp7.eu/). We would like to thank Felix Oesterle, who wrote the script to automatically retrieve and convert CFSv2 data. Assistance with HISTALP data provided by Anna-Maria Tilg and Barbara Chimani is greatly appreciated. Adam A. Scaife and Craig MacLachlan were supported by the joint DECC/Defra MetOffice Hadley Centre Programme (GA01101). The publication of this article was funded by the Open Access fund of Leibniz Universität Hannover. We wish to thank two anonymous reviewers for their helpful comments that helped to improve the manuscript.

Edited by: Quan J. Wang

References

Abegg, B., Steiger, R., and Walser, R.: Herausforderung Klimawandel: Chancen und Risiken für den Tourismus in Graubünden, Amt für Wirtschaft und Tourismus, Chur, Innsbruck, 2013.

Achleitner, S., Schöber, J., Rinderer, M., Leonhardt, G., Schöberl, F., Kirnbauer, R., and Schönlaub, H.: Analyzing the operational performance of the hydrological models in an alpine flood forecasting system, J. Hydrol., 412–413, 90–100, https://doi.org/10.1016/j.jhydrol.2011.07.047, 2012.

Anghileri, D., Voisin, N., Castelletti, A., Pianosi, F., Nijssen, B., and Lettenmaier, D. P.: Value of long-term streamflow forecasts to reservoir operations for water supply in snow-dominated river catchments, Water Resour. Res., 52, 4209–4225, https://doi.org/10.1002/2015WR017864, 2016.

Auer, I., Böhm, R., Jurkovic, A., Lipa, W., Orlik, A., Potzmann, R., Schöner, W., Ungersböck, M., Matulla, C., Briffa, K., Jones, P., Efthymiadis, D., Brunetti, M., Nanni, T., Maugeri, M., Mercalli, L., Mestre, O., Moisselin, J.-M., Begert, M., Müller-Westermeier, G., Kveton, V., Bochnicek, O., Stastny, P., Lapin, M., Szalai, S., Szentimrey, T., Cegnar, T., Dolinar, M., Gajic-Capka, M., Zaninovic, K., Majstorovic, Z., and Nieplova, E.: HISTALP – historical instrumental climatological surface time series of the Greater Alpine Region, Int. J. Climatol., 27, 17–46, https://doi.org/10.1002/joc.1377, 2007.

Barnett, T. P., Adam, J. C., and Lettenmaier, D. P.: Potential impacts of a warming climate on water availability in snow-dominated regions, Nature, 438, 303–309, https://doi.org/10.1038/nature04141, 2005.

Bartolini, E., Claps, P., and D'Odorico, P.: Interannual variability of winter precipitation in the European Alps: relations with the North Atlantic Oscillation, Hydrol. Earth Syst. Sci., 13, 17–25, https://doi.org/10.5194/hess-13-17-2009, 2009.

Bell, V. A., Davies, H. N., Kay, A. L., Brookshaw, A., and Scaife, A. A.: A national-scale seasonal hydrological forecast system: development and evaluation over Britain, Hydrol. Earth Syst. Sci., 21, 4681–4691, https://doi.org/10.5194/hess-21-4681-2017, 2017.

Beniston, M. and Jungo, P.: Shifts in the distributions of pressure, temperature and moisture and changes in the typical weather patterns in the Alpine region in response to the behavior of the North Atlantic Oscillation, Theor. Appl. Climatol., 71, 29–42, https://doi.org/10.1007/s704-002-8206-7, 2002.

Bock, A. R., Hay, L. E., McCabe, G. J., Markstrom, S. L., and Atkinson, R. D.: Parameter regionalization of a monthly water balance model for the conterminous United States, Hydrol. Earth Syst. Sci., 20, 2861–2876, https://doi.org/10.5194/hess-20-2861-2016, 2016.

Bruno Soares, M. and Dessai, S.: Exploring the use of seasonal climate forecasts in Europe through expert elicitation, Clim. Risk Manage., 10, 8–16, https://doi.org/10.1016/j.crm.2015.07.001, 2015.

Butler, A. H., Arribas, A., Athanassiadou, M., Baehr, J., Calvo, N., Charlton-Perez, A., Déqué, M., Domeisen, D. I. V., Fröhlich, K., Hendon, H., Imada, Y., Ishii, M., Iza, M., Karpechko, A. Y., Kumar, A., MacLachlan, C., Merryfield, W. J., Müller, W. A., O'Neill, A., Scaife, A. A., Scinocca, J., Sigmond, M., Stockdale, T. N., and Yasuda, T.: The Climate-system Historical Forecast Project: do stratosphere-resolving models make better seasonal climate predictions in boreal winter?, Q. J. Roy. Meteorol. Soc., 142, 1413–1427, https://doi.org/10.1002/qj.2743, 2016.

Chahine, M. T.: GEWEX: The global energy and water cycle experiment, Eos Trans. Am. Geophys. Un., 73, 9–14, 1992.

Chimani, B., Matulla, C., Böhm, R., and Hofstätter, M.: A new high resolution absolute temperature grid for the Greater Alpine Region back to 1780, Int. J. Climatol., 33, 2129–2141, https://doi.org/10.1002/joc.3574, 2013.

Crochemore, L., Ramos, M.-H., and Pappenberger, F.: Bias correcting precipitation forecasts to improve the skill of seasonal streamflow forecasts, Hydrol. Earth Syst. Sci., 20, 3601–3618, https://doi.org/10.5194/hess-20-3601-2016, 2016.

Day, G. N.: Extended Streamflow Forecasting Using NWS-RFS, J. Water Resour. Pl. Manage., 111, 157–170, https://doi.org/10.1061/(ASCE)0733-9496(1985)111:2(157), 1985.

Doblas-Reyes, F. J., García-Serrano, J., Lienert, F., Biescas, A. P., and Rodrigues, L. R. L.: Seasonal climate predictability and forecasting: Status and prospects, WIREs Clim. Change, 4, 245–268, https://doi.org/10.1002/wcc.217, 2013.

Domeisen, D. I. V., Butler, A. H., Fröhlich, K., Bittner, M., Müller, W. A., and Baehr, J.: Seasonal Predictability over Europe Arising from El Niño and Stratospheric Variability in the MPI-ESM Seasonal Prediction System, J. Climate, 28, 256–271, https://doi.org/10.1175/JCLI-D-14-00207.1, 2015.

Eade, R., Smith, D., Scaife, A., Wallace, E., Dunstone, N., Hermanson, L., and Robinson, N.: Do seasonal-to-decadal climate predictions underestimate the predictability of the real world?, Geophys. Res. Lett., 41, 5620–5628, https://doi.org/10.1002/2014gl061146, 2014.

Finger, D., Heinrich, G., Gobiet, A., and Bauder, A.: Projections of future water resources and their uncertainty in a glacierized catchment in the Swiss Alps and the subsequent effects on hydropower production during the 21st century, Water Resour. Res., 48, W02521, https://doi.org/10.1029/2011WR010733, 2012.

Förster, K., Meon, G., Marke, T., and Strasser, U.: Effect of meteorological forcing and snow model complexity on hydrological simulations in the Sieber catchment (Harz Mountains, Germany), Hydrol. Earth Syst. Sci., 18, 4703–4720, https://doi.org/10.5194/hess-18-4703-2014, 2014.

Förster, K., Oesterle, F., Hanzer, F., Schöber, J., Huttenlau, M., and Strasser, U.: A snow and ice melt seasonal prediction modelling system for Alpine reservoirs, Proc. Int. Assoc. Hydrol. Sci., 374, 143–150, https://doi.org/10.5194/piahs-374-143-2016, 2016.

Fundel, F., Jörg-Hess, S., and Zappa, M.: Monthly hydrometeorological ensemble prediction of streamflow droughts and corresponding drought indices, Hydrol. Earth Syst. Sci., 17, 395–407, https://doi.org/10.5194/hess-17-395-2013, 2013.

Greatbatch, R. J., Gollan, G., Jung, T., and Kunz, T.: Factors influencing Northern Hemisphere winter mean atmospheric circulation anomalies during the period 1960/61 to 2001/02, Q. J. Roy. Meteorol. Soc., 138, 1970–1982, https://doi.org/10.1002/qj.1947, 2012.

Greenslade, D. and Berkhout, F.: Future Earth-Research for Global Sustainability, in: EGU General Assembly Conference Abstracts, vol. 16, 27 April–2 May 2014, Vienna, Austria, 2014.

Hanzer, F., Marke, T., and Strasser, U.: Distributed, explicit modeling of technical snow production for a ski area in the Schladming region (Austrian Alps), Cold Reg. Sci. Technol., 108, 113–124, https://doi.org/10.1016/j.coldregions.2014.08.003, 2014.

Hanzer, F., Helfricht, K., Marke, T., and Strasser, U.: Multilevel spatiotemporal validation of snow/ice mass balance and runoff modeling in glacierized catchments, The Cryosphere, 10, 1859–1881, https://doi.org/10.5194/tc-10-1859-2016, 2016.

Hanzer, F., Förster, K., Nemec, J., and Strasser, U.: Projected cryospheric and hydrological impacts of 21st century climate change in the Ötztal Alps (Austria) simulated using a physically based approach, Hydrol. Earth Syst. Sci. Discuss., https://doi.org/10.5194/hess-2017-309, in review, 2017.

Hersbach, H.: Decomposition of the Continuous Ranked Probability Score for Ensemble Prediction Systems, Weather Forecast., 15, 559–570, https://doi.org/10.1175/1520-0434(2000)015<0559:DOTCRP>2.0.CO;2, 2000.

Ineson, S. and Scaife, A. A.: The role of the stratosphere in the European climate response to El Niño, Nat. Geosci., 2, 32–36, https://doi.org/10.1038/ngeo381, 2008.

Jörg-Hess, S., Griessinger, N., and Zappa, M.: Probabilistic Forecasts of Snow Water Equivalent and Runoff in Mountainous Areas, J. Hydrometeorol., 16, 2169–2186, https://doi.org/10.1175/JHM-D-14-0193.1, 2015.

Kang, D., Lee, M.-I., Im, J., Kim, D., Kim, H.-M., Kang, H.-S., Schubert, S. D., Arribas, A., and MacLachlan, C.: Prediction of the Arctic Oscillation in boreal winter by dynamical seasonal forecasting systems, Geophys. Res. Lett., 41, 3577–3585, https://doi.org/10.1002/2014GL060011, 2014.

Kaser, G., Großhauser, M., and Marzeion, B.: Contribution potential of glaciers to water availability in different climate regimes, P. Natl. Acad. Sci. USA, 107, 20223–20227, https://doi.org/10.1073/pnas.1008162107, 2010.

Kim, H.-M., Webster, P. J., and Curry, J. A.: Seasonal prediction skill of ECMWF System 4 and NCEP CFSv2 retrospective forecast for the Northern Hemisphere Winter, Clim. Dynam., 39, 2957–2973, https://doi.org/10.1007/s00382-012-1364-6, 2012.

Kirtman, B. and Pirani, A.: The State of the Art of Seasonal Prediction: Outcomes and Recommendations from the First World Climate Research Program Workshop on Seasonal Prediction, B. Am. Meteorol. Soc., 90, 455–458, https://doi.org/10.1175/2008BAMS2707.1, 2009.

Kirtman, B., Power, S. B., Adedoyin, J. A., Boer, G. J., Camilloni, I., Doblas-Reyes, F. J., Fiore, A. M., Kimoto, M., Meehl, G. A., Prather, M., Sarr, A., Schar, C., Sutton, R., van Oldenborgh, G. J., Vecchi, G., and Wang, H. J.: Near-term climate change: projections and predictability, in: Climate change 2013: the physical science basis: contribution to the Fifth Assessment Report of the Intergovernmental Panel on Climate Change, edited by: Stocker, T. F., Qin, D., Plattner, G.-K., Tignor, M. M., Allen, S. K., Boschung, J., Nauels, A., Xia, Y., Bex, V., and Midgley, P. M., Cambridge University Press, Cambridge, 2014.

Klemeš, V.: Operational testing of hydrological simulation models, Hydrolog. Sci. J., 31, 13–24, https://doi.org/10.1080/02626668609491024, 1986.

Kling, H., Fuchs, M., and Paulin, M.: Runoff conditions in the upper Danube basin under an ensemble of climate change scenarios, J. Hydrol., 424-425, 264–277, https://doi.org/10.1016/j.jhydrol.2012.01.011, 2012.

Krajči, P., Kirnbauer, R., Parajka, J., Schöber, J., and Blöschl, G.: The Kühtai data set: 25 years of lysimetric, snow pillow, and meteorological measurements, Water Resour. Res., 53, 5158–5165, https://doi.org/10.1002/2017WR020445, 2017.

Kumar, A., Chen, M., and Wang, W.: Understanding Prediction Skill of Seasonal Mean Precipitation over the Tropics, J. Climate, 26, 5674–5681, https://doi.org/10.1175/JCLI-D-12-00731.1, 2013.

Mackay, J., Jackson, C., Brookshaw, A., Scaife, A., Cook, J., and Ward, R.: Seasonal forecasting of groundwater levels in principal aquifers of the United Kingdom, J. Hydrol., 530, 815–828, https://doi.org/10.1016/j.jhydrol.2015.10.018, 2015.

MacLachlan, C., Arribas, A., Peterson, K. A., Maidens, A., Fereday, D., Scaife, A. A., Gordon, M., Vellinga, M., Williams, A., Comer, R. E., Camp, J., Xavier, P., and Madec, G.: Global Seasonal forecast system version 5 (GloSea5): A High-Resolution Seasonal Forecast System, Q. J. Roy. Meteorol. Soc., 141, 1072–1084, https://doi.org/10.1002/qj.2396, 2015.

Marke, T., Strasser, U., Hanzer, F., Stötter, J., Wilcke, R. A. I., and Gobiet, A.: Scenarios of Future Snow Conditions in Styria (Austrian Alps), J. Hydrometeorol., 16, 261–277, https://doi.org/10.1175/JHM-D-14-0035.1, 2015.

Marzeion, B. and Nesje, A.: Spatial patterns of North Atlantic Oscillation influence on mass balance variability of European glaciers, The Cryosphere, 6, 661–673, https://doi.org/10.5194/tc-6-661-2012, 2012.

Marzeion, B., Jarosch, A. H., and Hofer, M.: Past and future sea-level change from the surface mass balance of glaciers, The Cryosphere, 6, 1295–1322, https://doi.org/10.5194/tc-6-1295-2012, 2012.

McCabe, G. J. and Markstrom, S. L.: A Monthly Water-Balance Model Driven By a Graphical User Interface, US Geological Survey Open-File report, https://pubs.usgs.gov/of/2007/1088/ (last access: 22 February 2017), 2007.

Molteni, F., Stockdale, T. N., and Vitart, F.: Understanding and modelling extra-tropical teleconnections with the Indo-Pacific region during the northern winter, Clim. Dynam., 45, 3119–3140, https://doi.org/10.1007/s00382-015-2528-y, 2015.

Montanari, A., Young, G., Savenije, H., Hughes, D., Wagener, T., Ren, L. L., Koutsoyiannis, D., Cudennec, C., Toth, E., Grimaldi, S., Blöschl, G., Sivapalan, M., Beven, K., Gupta, H., Hipsey, M., Schaefli, B., Arheimer, B., Boegh, E., Schymanski, S. J., Di Baldassarre, G., Yu, B., Hubert, P., Huang, Y., Schumann, A., Post, D. A., Srinivasan, V., Harman, C., Thompson, S., Rogger, M., Viglione, A., McMillan, H., Characklis, G., Pang, Z., and Belyaev, V.: Panta Rhei – Everything Flows: Change in hydrology and society – The IAHS Scientific Decade 2013–2022, Hydrolog. Sci. J., 58, 1256–1275, https://doi.org/10.1080/02626667.2013.809088, 2013.

Olefs, M., Fischer, A., and Lang, J.: Boundary Conditions for Artificial Snow Production in the Austrian Alps, J. Appl. Meteorol. Clim., 49, 1096–1113, https://doi.org/10.1175/2010JAMC2251.1, 2010.

Pagano, T., Garen, D., and Sorooshian, S.: Evaluation of official western US seasonal water supply outlooks, 1922–2002, J. Hydrometeorol., 5, 896–909, https://doi.org/10.1175/1525-7541(2004)005<0896:EOOWUS>2.0.CO;2, 2004.

Painter, T. H., Berisford, D. F., Boardman, J. W., Bormann, K. J., Deems, J. S., Gehrke, F., Hedrick, A., Joyce, M., Laidlaw, R., Marks, D., Mattmann, C., McGurk, B., Ramirez, P., Richardson, M., Skiles, S. M., Seidel, F. C., and Winstral, A.: The Airborne Snow Observatory: Fusion of scanning lidar, imaging spectrometer, and physically-based modeling for mapping snow water equivalent and snow albedo, Remote Sens. Environ., 184, 139–152, https://doi.org/10.1016/j.rse.2016.06.018, 2016.

Pomeroy, J., Bernhardt, M., and Marks, D.: Water resources: Research network to track alpine water, Nature, 521, 32–32, https://doi.org/10.1038/521032c, 2015.

Reid, W. V., Chen, D., Goldfarb, L., Hackmann, H., Lee, Y. T., Mokhele, K., Ostrom, E., Raivio, K., Rockstrom, J., Schellnhuber, H. J., and Whyte, A.: Earth System Science for Global Sustainability: Grand Challenges, Science, 330, 916–917, https://doi.org/10.1126/science.1196263, 2010.

Riddle, E. E., Butler, A. H., Furtado, J. C., Cohen, J. L., and Kumar, A.: CFSv2 ensemble prediction of the wintertime Arctic Oscillation, Clim. Dynam., 41, 1099–1116, https://doi.org/10.1007/s00382-013-1850-5, 2013.

Robertson, D. E. and Wang, Q. J.: A Bayesian Approach to Predictor Selection for Seasonal Streamflow Forecasting, J. Hydrometeorol., 13, 155–171, https://doi.org/10.1175/JHM-D-10-05009.1, 2012.

Robertson, D. E., Pokhrel, P., and Wang, Q. J.: Improving statistical forecasts of seasonal streamflows using hydrological model output, Hydrol. Earth Syst. Sci., 17, 579–593, https://doi.org/10.5194/hess-17-579-2013, 2013.

Saha, S., Moorthi, S., Wu, X., Wang, J., Nadiga, S., Tripp, P., Behringer, D., Hou, Y.-T., Chuang, H.-Y., Iredell, M., Ek, M., Meng, J., Yang, R., Mendez, M. P., van den Dool, H., Zhang, Q., Wang, W., Chen, M., and Becker, E.: The NCEP Climate Forecast System Version 2, J. Climate, 27, 2185–2208, https://doi.org/10.1175/JCLI-D-12-00823.1, 2014.

Scaife, A. A., Arribas, A., Blockley, E., Brookshaw, A., Clark, R. T., Dunstone, N., Eade, R., Fereday, D., Folland, C. K., Gordon, M., Hermanson, L., Knight, J. R., Lea, D. J., MacLachlan, C., Maidens, A., Martin, M., Peterson, A. K., Smith, D., Vellinga, M., Wallace, E., Waters, J., and Williams, A.: Skillful long-range prediction of European and North American winters, Geophys. Res. Lett., 41, 2514–2519, https://doi.org/10.1002/2014GL059637, 2014.

Scaife, A. A., Karpechko, A. Y., Baldwin, M. P., Brookshaw, A., Butler, A. H., Eade, R., Gordon, M., MacLachlan, C., Martin, N., Dunstone, N., and Smith, D.: Seasonal winter forecasts and the stratosphere, Atmos. Sci. Lett., 17, 51–56, https://doi.org/10.1002/asl.598, 2016.

Scaife, A. A., Comer, R. E., Dunstone, N. J., Knight, J. R., Smith, D. M., MacLachlan, C., Martin, N., Peterson, K. A., Rowlands, D., Carroll, E. B., Belcher, S., and Slingo, J.: Tropical rainfall, Rossby waves and regional winter climate predictions, Q. J. Roy. Meteorol. Soc., 143, 1–11, https://doi.org/10.1002/qj.2910, 2017.

Schaefli, B. and Gupta, H. V.: Do Nash values have value?, Hydrol. Process., 21, 2075–2080, https://doi.org/10.1002/hyp.6825, 2007.

Schaefli, B., Hingray, B., and Musy, A.: Climate change and hydropower production in the Swiss Alps: quantification of potential impacts and related modelling uncertainties, Hydrol. Earth Syst. Sci., 11, 1191–1205, https://doi.org/10.5194/hess-11-1191-2007, 2007.

Schattan, P., Baroni, G., Oswald, S. E., Schöber, J., Fey, C., Kormann, C., Huttenlau, M., and Achleitner, S.: Continuous monitoring of snowpack dynamics in alpine terrain by aboveground neutron sensing, Water Resour. Res., 53, 3615–3634, https://doi.org/10.1002/2016WR020234, 2017.

Scherrer, S. C., Appenzeller, C., and Laternser, M.: Trends in Swiss Alpine snow days: The role of local- and large-scale climate variability, Geophys. Res. Lett., 31, L13215, https://doi.org/10.1029/2004GL020255, 2004.

Schick, S., Rössler, O., and Weingartner, R.: Saisonale Abflussprognosen für mittelgroße Einzugsgebiete in der Schweiz: Möglichkeiten und Grenzen hydrologischer Persistenz (Seasonal runoff predictions for mesoscale catchments in Switzerland – Potentials and limitations of hydrologic persistence), Hydrol. Wasserbewirtsch., 59, 59–67, https://doi.org/10.5675/HyWa_2015,2_2, 2015.

Schöber, J., Achleitner, S., Kirnbauer, R., Schöberl, F., and Schönlaub, H.: Impact of snow state variation for design flood simulations in glacierized catchments, Adv. Geosci., 31, 39–48, https://doi.org/10.5194/adgeo-31-39-2012, 2012.

Schöber, J., Schneider, K., Helfricht, K., Schattan, P., Achleitner, S., Schöberl, F., and Kirnbauer, R.: Snow cover characteristics in a glacierized catchment in the Tyrolean Alps – Improved spatially distributed modelling by usage of Lidar data, J. Hydrol., 519, 3492–3510, https://doi.org/10.1016/j.jhydrol.2013.12.054, 2014.

Schöber, J., Achleitner, S., Bellinger, J., Kirnbauer, R., and Schöberl, F.: Analysis and modelling of snow bulk density in the Tyrolean Alps, Hydrol. Res., 47, 419–441, https://doi.org/10.2166/nh.2015.132, 2016.

Sivapalan, M., Savenije, H. H. G., and Blöschl, G.: Socio-hydrology: A new science of people and water, Hydrol. Process., 26, 1270–1276, https://doi.org/10.1002/hyp.8426, 2012.

Smith, D. M., Scaife, A. A., and Kirtman, B. P.: What is the current state of scientific knowledge with regard to seasonal and decadal forecasting?, Environ. Res. Lett., 7, 015602, https://doi.org/10.1088/1748-9326/7/1/015602, 2012.

Svensson, C., Brookshaw, A., Scaife, A. A., Bell, V. A., Mackay, J. D., Jackson, C. R., Hannaford, J., Davies, H. N., Arribas, A., and Stanley, S.: Long-range forecasts of UK winter hydrology, Environ. Res. Lett., 10, 064006, https://doi.org/10.1088/1748-9326/10/6/064006, 2015.

Thornthwaite, C. W.: An Approach toward a Rational Classification of Climate, Geogr. Rev., 38, 55–94, https://doi.org/10.2307/210739, 1948.

Trinh, B. N., Thielen-del Pozo, J., and Thirel, G.: The reduction continuous rank probability score for evaluating discharge forecasts from hydrological ensemble prediction systems, Atmos. Sci. Lett., 14, 61–65, https://doi.org/10.1002/asl2.417, 2013.

Twedt, T. M., Schaake Jr., J. C., and Peck, E. L.: National Weather Service extended streamflow prediction, in: Proceedings Western Snow Conference, 18–21 April 1977, Albuquerque, New Mexico, 52–57, 1977.

Vaughan, C. and Dessai, S.: Climate services for society: origins, institutional arrangements, and design elements for an evaluation framework, Wiley Interdiscip. Rev. Clim. Change, 5, 587–603, https://doi.org/10.1002/wcc.290, 2014.

Viviroli, D., Weingartner, R., and Messerli, B.: Assessing the hydrological significance of the world's mountains, Mt. Res. Dev., 23, 32–40, 2003.

Warner, T. T.: Numerical weather and climate prediction, Cambridge University Press, Cambridge, New York, 2011.

Weisheimer, A. and Palmer, T. N.: On the reliability of seasonal climate forecasts, J. Roy. Soc. Interf., 11, 20131162, https://doi.org/10.1098/rsif.2013.1162, 2014.

Wilks, D. S.: Statistical methods in the atmospheric sciences, in: no. v. 91 in International geophysics series, 2nd Edn., Academic Press, Amsterdam, Boston, 2006.

Wood, A. W. and Lettenmaier, D. P.: An ensemble approach for attribution of hydrologic prediction uncertainty, Geophys. Res. Lett., 35, L14401, https://doi.org/10.1029/2008GL034648, 2008.

Wood, A. W., Hopson, T., Newman, A., Brekke, L., Arnold, J., and Clark, M.: Quantifying Streamflow Forecast Skill Elasticity to Initial Condition and Climate Prediction Skill, J. Hydrometeorol., 17, 651–668, https://doi.org/10.1175/JHM-D-14-0213.1, 2016.

Yuan, X., Wood, E. F., Roundy, J. K., and Pan, M.: CFSv2-Based Seasonal Hydroclimatic Forecasts over the Conterminous United States, J. Climate, 26, 4828–4847, https://doi.org/10.1175/JCLI-D-12-00683.1, 2013.

Yuan, X., Wood, E. F., and Ma, Z.: A review on climate-model-based seasonal hydrologic forecasting: Physical understanding and system development, WIREs Water, 2, 523–536, https://doi.org/10.1002/wat2.1088, 2015.

5

The development and evaluation of a hydrological seasonal forecast system prototype for predicting spring flood volumes in Swedish rivers

Kean Foster[1,2], **Cintia Bertacchi Uvo**[2], **and Jonas Olsson**[1]

[1]Research and Development (hydrology), Swedish Meteorological and Hydrological Institute, 601 76 Norrköping, Sweden
[2]Department of Water Resources Engineering, Lund University, Box 118, 221 00 Lund, Sweden

Correspondence: Kean Foster (kean.foster@smhi.se)

Abstract. Hydropower makes up nearly half of Sweden's electrical energy production. However, the distribution of the water resources is not aligned with demand, as most of the inflows to the reservoirs occur during the spring flood period. This means that carefully planned reservoir management is required to help redistribute water resources to ensure optimal production and accurate forecasts of the spring flood volume (SFV) is essential for this. The current operational SFV forecasts use a historical ensemble approach where the HBV model is forced with historical observations of precipitation and temperature. In this work we develop and test a multi-model prototype, building on previous work, and evaluate its ability to forecast the SFV in 84 sub-basins in northern Sweden. The hypothesis explored in this work is that a multi-model seasonal forecast system incorporating different modelling approaches is generally more skilful at forecasting the SFV in snow dominated regions than a forecast system that utilises only one approach. The testing is done using cross-validated hindcasts for the period 1981–2015 and the results are evaluated against both climatology and the current system to determine skill. Both the multi-model methods considered showed skill over the reference forecasts. The version that combined the historical modelling chain, dynamical modelling chain, and statistical modelling chain performed better than the other and was chosen for the prototype. The prototype was able to outperform the current operational system 57 % of the time on average and reduce the error in the SFV by ∼6 % across all sub-basins and forecast dates.

1 Introduction

The spring flood period (sometimes referred to as the spring melt or freshet period in the literature) is of great importance in snow dominated regions like Sweden where hydropower accounts for nearly half of the country's electrical energy production (Statistiska centralbyrån, 2016). Between 55 and 70 % of the annual inflows to reservoirs in the larger hydropower producing rivers occur during this relatively short period, typically from mid-April/early-May to the end of July. This means that the majority of the annual water resources available for hydropower production would only be available to producers during this period if it were not regulated through carefully planned reservoir management. This reservoir management is important as the energy demand is out of phase with the natural availability of the water resources; typically demand is higher during the colder months when the inflows are lower and vice versa. Therefore the goal is to redistribute the availability of these resources from the spring flood period to other times of the year when electricity demand is higher i.e. during the six months of colder winter, while maintaining the balance between a sufficiently large volume of water for optimal production and enough remaining capacity for safe flood risk management (Olsson et al., 2016). The typical strategy for operators in Sweden is to have reservoirs at around 90 % capacity at the end of the spring flood which is then ideally maintained until the beginning of winter. To achieve this, operators require reliable seasonal forecast information to help them in planning the operations both leading up to and during the spring flood period.

The sources of predictability for hydrological seasonal forecasts come from the initial hydrological conditions i.e. information relating to the water stores within the catchment (e.g. Wood and Lettenmaier, 2008; Wood et al., 2015; Yossef et al., 2013), and also from knowledge of the weather during the forecast period i.e. seasonal meteorological forecasts (e.g. Bennet et al., 2016; Doblas-Reyes et al., 2013; Wood et al., 2015; Yossef et al., 2013). Hydrological seasonal forecasts attempt to leverage at least one of these sources of predictability to make skilful predictions of future streamflow.

In practice, there are two predominant approaches to making hydrological forecasts at the seasonal scale: statistical approaches and dynamical approaches (see Sect. 2.1.4 and 2.1.5 for more information regarding how these approaches were implemented in the context of this work). Statistical approaches utilise empirical relationships between predictors and a predictand, typically streamflow or a derivative thereof (e.g. Garen, 1992; Pagano et al., 2009). These predictors can vary greatly in type from local hydrological storage variables like snow and groundwater storage (e.g. Robertson et al., 2013; Rosenberg et al., 2011), to local and regional meteorological variables (e.g. Còrdoba-Machado et al., 2016; Olsson et al., 2016), or even large scale climate data such as ENSO (El Niño Southern Oscillation) indices (e.g. Schepen et al., 2016; Shamir, 2017). All, however, are trying to leverage the predictability in these predictors that originate from one of the two aforementioned sources. Dynamical approaches use a hydrological model, typically initialised with observed data up to the forecast date so that the model state is a reasonable approximation of the initial hydrological conditions, and then force it with either historical observations (called ensemble streamflow prediction or ESP; e.g. Day (1985)) or using data representative of the future meteorological conditions such as general circulation model (GCM) outputs (e.g. Crochemore et al., 2016; Olsson et al., 2016; Yuan et al., 2013, 2015; Yuan, 2016). Attempts to improve these types of approaches have involved bias adjusting the GCM outputs (e.g. Crochemore et al., 2016; Lucatero et al., 2017; Wood et al., 2002; Yuan et al., 2015), bias adjusting the hydrological model outputs (e.g. Lucatero et al., 2017), or a combination of both (e.g. Yuan and Wood, 2012). Another dynamical approach is the well-established ESP method (Day, 1985). This is similar to the previous approach (GCM); however, instead of using GCM outputs to force the hydrological model it uses an ensemble of historical data. This approach is perhaps one of the most widely used methods and is still the subject of new research. Recent work has looked at conditioning the ensembles before using them, which can be done using GCM outputs (e.g. Crochemore et al., 2016), climate indices, and circulation pattern analysis (e.g. Beckers et al., 2016; Olsson et al., 2016; Candogan Yossef et al., 2017).

The current practice at the Swedish Meteorological and Hydrological Institute (SMHI) for seasonal forecasts of reservoir inflows is the ESP approach. It assumes that historical observations of precipitation and temperature are possible representations of future meteorological conditions and are used to force the HBV hydrological model (e.g. Bergström, 1976; Lindström et al., 1997) to give an ensemble forecast that has a climatological evolution from the initial conditions. A number of attempts have been made in the past to improve the performance of these spring flood forecasts with limited success (Arheimer et al., 2011) demonstrating that these seasonal forecasts are already of a high quality. Work by Olsson et al. (2016) on improving these forecasts was able to realise reasonable improvements using a multi-model approach. By combining a statistical approach, dynamical approach and an analogue approach (conditioned ESP) they were able to show a $\sim 4\,\%$ reduction in the forecast error of the spring flood volume (SFV). The purpose of this paper is to continue on and update the work started by Olsson et al. (2016) regarding developing and evaluating a hydrological seasonal forecast system prototype for forecasting the spring flood volumes in Sweden.

This paper is organised as follows. Section 2 outlines the prototype, including the individual model chains, the experimental set-up, the methods and tools implemented, and the study area and data used in this work. Section 3 presents and discusses the cross-validated evaluation scores for the prototype, first with reference to climatology and then with reference to the current operational system that is in use at SMHI. Section 4 concludes with the main findings and a brief outlook for future work.

2 Materials and methods

2.1 The multi-model system and the individual modelling chains

In this section we present the modelling approaches used in this work. These are based on those explored by Olsson et al. (2016) with some modification to facilitate their use in an operational environment. First we briefly present the multi-model prototype (Sect. 2.1.1) followed by a brief overview of the individual modelling chains used in the multi-model and why we chose them (Sect. 2.1.2–2.1.5). For more information regarding the individual modelling chains readers are referred to Olsson et al. (2016) and the accompanying Supplement.

2.2 The multi-model ensemble (ME)

The prototypes developed in this work build on an approach first proposed by Foster et al. (2010), and later improved upon and first tested by Olsson et al. (2016). The aim is to adapt their methodology for use in an operational environment and then evaluate the resulting prototype against the current operational system using cross-validated hindcasts for 84 gauging stations in northern Sweden (see Sect. 2.6). Four different modelling chains were considered when developing the prototype (Sect. 2.1.2–2.1.5). The performances of different combinations of these four were tested. A combination of all four modelling chains was not considered as the analogue model chain is a subset of the historical model chain.

Figure 1 shows the generalised schematic of the two prototypes, ME_{ads} and ME_{hds} (where the subscripts refer to the individual modelling chains making up each multi-model), including where the current methodologies differ significantly from those in previous works. These differences are discussed in the relevant modelling chain sections below. The prototypes are multi-model ensembles of the outputs from the three respective individual modelling chains. These outputs are pooled together rather than using an asymmetric weighting scheme due to the lack of data points, a total of 35 spring flood events (hindcast period was 1981–2015, see Sect. 2.6), from which to derive a robust weighting scheme. The simple weighting scheme used by Olsson et al. (2016) was tested but, other than improving the ensemble sharpness, did not offer an improvement over the pooling approach.

2.2.1 Historical ensemble (HE)

The historical model chain, the dark blue chain third from the left in Fig. 1, is an ensemble forecast made by forcing a rainfall–runoff model with historical observations of precipitation and temperature. This approach is often referred to as ESP (ensemble streamflow prediction) in the literature but we chose not to as we feel our terminology is more descriptive in the context of this work. This is the current operational seasonal forecasting practice at SMHI. The HBV model (Bergström, 1976; Lindström et al., 1997) is initialised by using observed meteorological inputs (P and T) to force the model up to the forecast date so that the model state reflects the current hydro-meteorological conditions. Then, typically all available historical daily P and T series for the period from the forecast issue date to the end of the forecasting period are used as input to HBV, generating an ensemble of forecasts that are climatological in their evolution from the initial conditions. The historical ensemble (HE) is used as the reference ensemble unless otherwise stated.

The HBV is run one river system at a time and the model outputs are later regrouped into three clusters (Sect. 2.6). Typically only historical data prior to the forecast date are used to force the model, however to allow for a more robust cross validation all data including the years after the forecast date were used (excluding the year in question of course). Unfortunately, the scope of this work did not allow for the recalibration of the HBV model before each cross-validated hindcast. This will potentially inflate the performance of the model for the hindcasts of years that were used in the calibration of the model. This will affect the analogue and dynamical model chains too as they also incorporate the HBV model in their set-up.

2.2.2 Analogue ensemble (AE)

The analogue model chain, the light blue chain furthest to the left in Fig. 1, is a subset of the HE. The hypothesis is that it is possible to identify a reduced set of historical years (an analogue ensemble) that describes the weather in the coming forecasting period better than the full historical ensemble used in HE. In this work the circulation pattern approach used by Olsson et al. (2016) was omitted due to data availability issues making it impractical for operational applications. Additionally, their teleconnection approach was revised to take advantage of the findings by Foster et al. (2018) where they identified which teleconnection patterns are related to the SFV and for which period of their persistence prior to the spring flood this connection is strongest. The teleconnection indices they identified are the Arctic Oscillation (AO) and the Scandinavian pattern (SCA) and the periods of persistence for these indices, expressed as the index mean for the identified period, are the seven and eight months leading up to the spring flood respectively.

The persistence for each teleconnection index is calculated from the beginning of the aforementioned period to one month prior to the forecast date (a limitation imposed by data availability), similarly this was done for all years in the climatological ensemble. If the values of these indices are considered to be coordinates in Euclidean space we define analogue years as those years with positions within a distance of 0.2 units in the Euclidean space from the position of the forecast year. The threshold is a compromise between being small enough to ensure that the climate set-up is indeed similar to the year in question and being large enough to actually be able to capture some analogues from the historical ensemble. The selection of the analogues is done at the regional scale, by cluster (Sect. 2.6), and these selections are applied to the associated sub-catchments in turn.

Similar to the HE, the analogue method makes use of years both before and after the hindcast year for the cross-validated hindcasts.

2.2.3 Dynamical modelling ensemble (DE)

The dynamical model chain, the dark red chain furthest to the left in Fig. 1, is similar to the HE; an adequately initialised HBV model is forced by an ensemble of seasonal forecasts of daily P and T from the ECMWF IFS system 4 (Sect. 2.7). A change to previous work has these daily P and T data bias adjusted first before being used to force HBV. The bias adjustment method used is a version of the distribution based scaling approach (DBS; Yang et al., 2010) which has been adapted for use on seasonal forecast data. DBS is a quantile mapping bias adjustment method where meteorological variables are fitted to appropriate parametric distributions (e.g. Berg et al., 2015; Yang et al., 2010). For precipitation, two discrete gamma distributions are used to adjust the daily seasonal forecast values, one for low intensity pre-

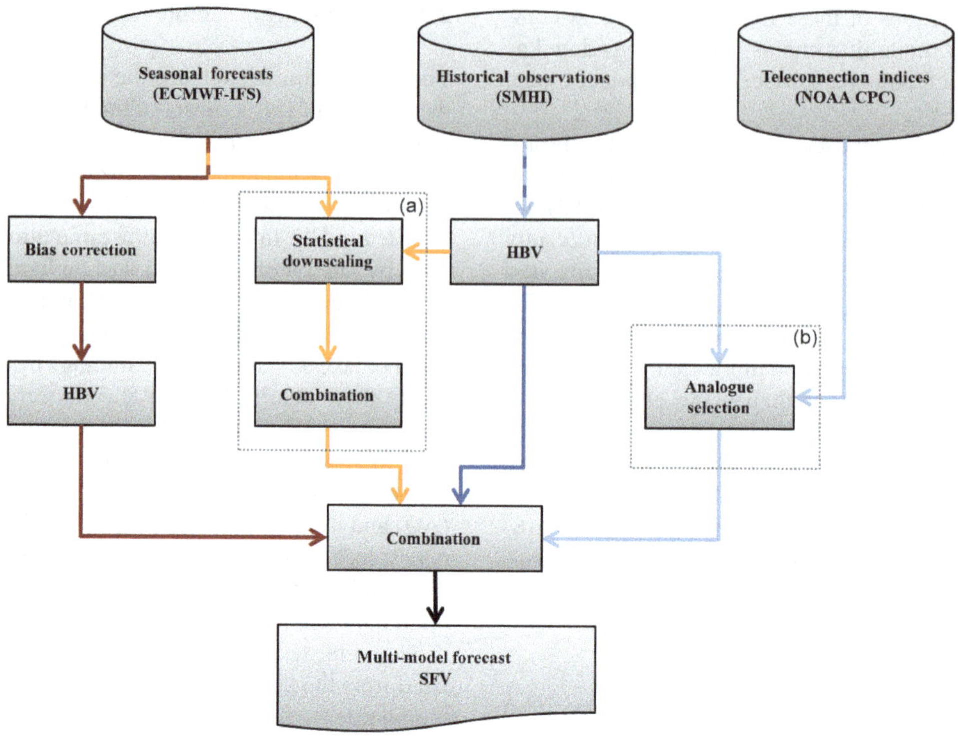

Figure 1. Schematic of the multi-model forecast system. The three individual model chains that are included in the multi-model are (from left to right) the dynamic model chain (red lines), the statistical model chain (orange lines), and the historical (dark blue lines) or analogue (light blue lines) model chain. The dashed boxes labelled **(a)** and **(b)** indicate the parts of the system that have non-trivial changes from the multi-model described in Olsson et al. (2016).

cipitation events ($\leqslant$ 95th percentile) and another for extreme events ($>$ 95th percentile). For temperature, a Gaussian distribution is used to adjust the daily seasonal forecast values.

Observed (Sect. 2.6) and seasonal forecast (Sect. 2.7) time series of P and T spanning the relevant forecast time frame (e.g. January–July for forecasts initialised in January) and for the reference period 1981–2010 are used to derive the adjustment factors to transform the seasonal forecast data to match the observed frequency distributions. First the precipitation data is adjusted and then the temperature data. The latter is done separately for dry and wet days in an attempt to preserve the dependence between P and T (e.g. Olsson et al., 2011; Yang et al., 2010). Adjustment factors are calculated for each calendar month as the distributions can have different shapes depending on the physical characteristics of the precipitation processes that are dominant. It should be emphasised that the adjustment parameters were estimated using much of the same data to which they were applied. Ideally the parameters would be estimated using data that does not overlap the data which is being adjusted. However, this was not possible in the scope of this work.

There have recently been some criticism voiced regarding the applicability of quantile mapping for bias adjusting seasonal data (e.g. Zhao et al., 2017). This criticism points out that although quantile mapping approaches are effective at bias correction they cannot ensure reliability in forecast ensemble spread or guarantee coherence. Unfortunately, the scope of this work did not allow the testing of other bias adjustment methods but the criticism is noted and further work is planned to address these points.

These bias adjusted data are then converted into HBV inputs by mapping them from their native grid onto the HBV sub-catchments. The mapping is done by areal weighting and the resulting sub-catchment average P and T values are then adjusted to represent different altitude fractions within the catchment. These data are consequently used to force the HBV model from the same initial state as that used in the HE procedure.

No changes to this methodology are needed to accommodate the cross-validated hindcasting as done with the other model chains.

2.2.4 Statistical modelling ensemble (SE)

The statistical model chain, the orange chain second from the left in Fig. 1, is an ensemble forecast produced by downscaling forecasted or modelled large-scale variables (predictors) to the SFV for each cluster (predictand). The downscaling is done using an SVD approach (singular variable decomposition). The predictors are three large scale circulation variables (Sect. 2.7) and the modelled snow depths from the

HBV initial conditions. The outputted ensembles of SFV are combined using a simple arithmetic weighting system. The normalised squared covariance between the four predictors and the predictands are ranked for each forecast initialisation date, and weights between 0 and 1 are applied to the different predictors according to their rank. The lowest ranked predictor is assigned a weight of 0.1 ($= 1/10$), the next lowest predictor is assigned a weight of 0.2 ($= 2/10$), and so on until all four have been assigned a weight. The reason that an asymmetric weighting scheme is used here is that there is physical support for it. Early in the season the snowpack, which is the majority contributor to the spring flood volume, is a fraction of what it will be and is still accumulating. Therefore, the coming meteorological conditions, which dictate snowpack evolution, are more important earlier on in the season than they are later giving physical support for asymmetric weighting. Additionally, the relative importance of these meteorological predictors with respect to each other differs with time too.

The relative simplicity of the statistical model chain means that it was possible to retrain the model before each hindcast during the cross-validation calculations allowing for no overlap between the calibration and validation periods.

2.3 Defining the spring flood

In previous works the spring flood period has often been defined in terms of calendar months e.g. May–June–July (Nilsson et al., 2008; Foster and Uvo, 2010; Arheimer et al., 2011; Olsson et al., 2016; Foster et al., 2016). This definition of the spring flood period is not ideal as it does not take the interannual and geographical variations in the timing of the spring flood onset into account. In this work we propose an improvement to this practice where we define the spring flood to be the period from the onset date to the end of July.

We define the onset as the nearest local minima in the hydrograph before the date after which the inflows are above the 90th percentile, with reference to the inflows during the first 80 days of the current year, for a period of at least 30 days (Fig. 2). For forecasts made after January i.e. those made in February, March, April, and May, the missing inflow data between the 1 January and the forecast date are filled with simulated inflow data from the HBV model using observed precipitation and temperatures as input data.

A drawback to this definition is that it is not comprehensive as the end of the spring flood is not defined according to the hydrograph but rather by date. The reason for not defining the end of the spring flood objectively is twofold. Firstly, the forecast horizon for the ECMWF-IFS is seven months which means that forecasts initialised in January may not encompass the entire spring flood period. Secondly, a robust and objective definition of what constitutes the end of the spring flood was difficult to realise within the scope of this work. Further work is needed to resolve this in a more satisfactory manner.

2.4 Experimental set-up

The challenge in this work was to perform a robust evaluation on a limited dataset (35 spring floods, 1981–2015) while minimising the risk of unstable or over fitted statistics. Therefore, a leave-one-out cross-validation (LOOCV) protocol was adopted. Additionally, as it was not practical to recalibrate the HBV model before each step of the LOOCV process; the statistical model uses the same periods for training as those used to calibrate and validate the HBV model. LOOCV is a model evaluation technique that uses $n - 1$ data points to train the models and the data point left out is used for validation. This process is repeated n times to give a validation dataset of length n. This allows for a more robust evaluation with a limited dataset and also gives one the ability to sample more of the variability in the training period than if a traditional validation were performed. The second point is especially advantageous for evaluating the statistical model which is particularly sensitive to situations that were not found within the training period. LOOCV was applied to the individual model chains.

To assess the relative skill for different lead times, we evaluate hindcasts issued on the 1 January (Jan), 1 February (Feb), 1 March (Mar), 1 April (Apr), and 1 May (May) for the spring floods 1981–2015. The evaluation of performance is done in terms of how well the SFV is forecast.

2.5 Evaluation

As it has been mentioned above, we are interested in the ability of a multi-model ensemble's ability to forecast the SFV at differing lead times i.e. forecasts initialised on the first of the month for the months of January through May. It was suggested by Cloke and Pappenberger (2008) that for a rigorous assessment of the quality of a hydrological ensemble prediction system (HEPS) it is not only important to select appropriate verification measures but also to use several different measures so that different properties of the forecast skill can be estimated, resulting in a more comprehensive evaluation.

The evaluations in this paper are designed to answer the following questions:

- Can the forecasts improve on the reference forecast error?
- How often do the forecasts perform better than the reference forecast?
- Are the forecasts better at capturing the interannual variability than the reference forecast?
- Are the forecasts better at discriminating between events and non-events than the reference forecast?
- Are the forecasts sharper than the reference forecast?
- Are the forecasts more sensitive to uncertainty that the reference forecasts?

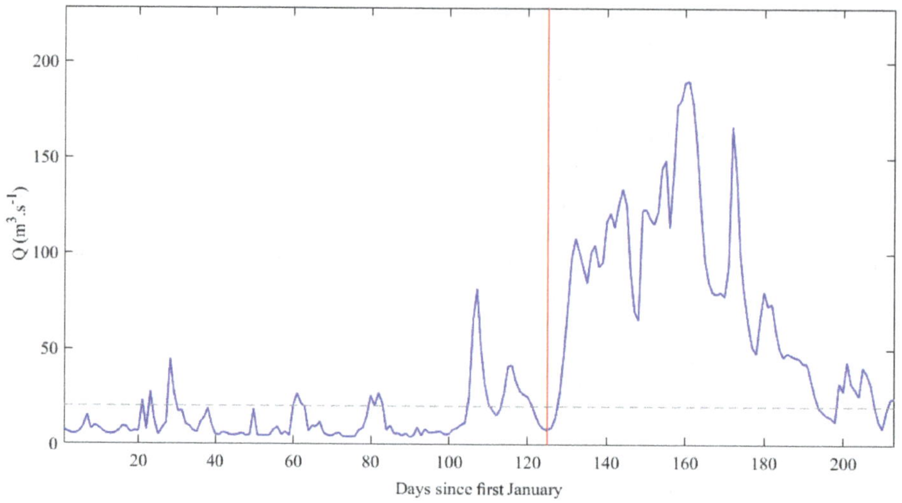

Figure 2. Schematic of how the spring flood is defined. The spring flood is the period between the onset and the last day of July (day 211/212 since the 1 January). The hydrograph from which the spring flood period is to be derived (blue line), the onset date (red line), and the 90th percentile of the inflow for first 80 days (dashed line).

The verification measures used to answer these questions are described below and summarised in Table 1.

2.5.1 Mean absolute error skill score (MAESS)

One of the most commonly published scores, even the recommended method, when evaluating HEPS is the continuous rank probability score (CRPS, Hersbach, 2000). However, since we have a limited number of data points, only 35 cross-validated hindcasts per sub-basin, and the CRPS compares distributions, we deemed its use unsuitable for this work. We instead chose to use the mean absolute error (MAE) to evaluate general forecast performance as the CRPS collapses to the MAE for deterministic forecasts (Hersbach, 2000). Therefore, by assuming the ensemble mean to be the deterministic forecast a MAE skill score (MAESS) can be expressed as

$$\mathrm{MAESS} = 1 - \frac{\mathrm{MAE_f}}{\mathrm{MAE_r}}, \tag{1}$$

where f and r denote forecast and reference, respectively. MAE is furthermore defined as

$$\mathrm{MAE} = \frac{1}{n}\sum_{y=1}^{n}\left|\frac{\mathrm{SFV}_o^y - \mathrm{SFV}_f^y}{\mathrm{SFV}_o^y}\right|, \tag{2}$$

where y denotes year, n denotes the total number of years, and o denotes observations. The MAESS has a range between negative infinity and one with positive values indicating skill over the reference forecast and a value of one a perfect forecast.

2.5.2 Frequency of years (FY^+)

In their work Olsson et al. (2016) proposed FY^+ as a complimentary performance measure to scores such as the MAESS. They are complimentary in that the MAESS is a measure of how much better the forecast is than the reference forecast while FY^+ is the frequency or how often the forecast is better i.e. how often the absolute error is lower. FY^+ scores range from 0 to 100 % where values above 50 % indicate that the multi-model forecast has skill over the reference forecast. By assuming the ensemble mean to be the deterministic forecast FY^+ is expressed as

$$\mathrm{FY}^+ = \frac{100}{n}\sum_{y=1}^{n} H^y, \tag{3}$$

where H is the Heaviside function defined by

$$H^y = \begin{cases} 0, \mathrm{AbsE}_r^y < \mathrm{AbsE}_f^y \\ 1, \mathrm{AbsE}_r^y > \mathrm{AbsE}_f^y \end{cases} \tag{4}$$

and AE is the absolute error.

2.5.3 Nash–Sutcliffe model efficiency (NSE)

The NSE (Nash and Sutcliffe, 1970) is a normalised statistic that determines the relative magnitude of the residual variance compared to the measured data variance. The NSE has a range from negative infinity to one with one being a perfect match and values above zero denoting that the forecast has skill over climatology. For this work it can be interpreted as how well the forecasted SFV matches the observed SFV year on year and as such is complimentary to MAESS and FY^+. By assuming the ensemble mean to be the deterministic fore-

Table 1. The validation metrics used to evaluate the multi-model performance. The threshold for skill is 50 for FY^+ and 0 for all the other metrics.

Name	Equation	Description
Mean absolute error skill score (MAESS)	$\mathrm{MAESS} = 1 - \frac{\mathrm{MAE_f}}{\mathrm{MAE_r}}$	Measure of the model's general performance; it quantifies the relative forecast error against a reference forecast.
Frequency of years (FY^+)	$\mathrm{FY}^+ = \frac{100}{n}\sum_{y=1}^{n} H^y$, where H is the Heaviside function defined by $H^y = \begin{cases} 0, \mathrm{AE}_r^y < \mathrm{AE}_f^y \\ 1, \mathrm{AE}_r^y > \mathrm{AE}_f^y \end{cases}$ AE is the absolute error.	Measure of the model's general performance; it quantifies how often the forecast outperforms a reference forecast.
Nash–Sutcliffe efficiency (NSE)	$\mathrm{NSE} = 1 - \frac{\sum_{y=1}^{n}(\mathrm{SFV}_{\mathrm{obs}}^y - \mathrm{SFV}^y)^2}{\sum_{y=1}^{n}(\mathrm{SFV}_{\mathrm{obs}}^y - \overline{\mathrm{SFV}_{\mathrm{obs}}})^2}$	Measure of the model's general performance; it quantifies the model's residual variance against a reference forecast's variance.
Relative operating characteristic skill score (ROCSS)	$\mathrm{ROCSS} = 2 \cdot \mathrm{AUC} - 1$ where AUC is the area under the curve $\mathrm{AUC} = \sum_{y=1}^{n+1} \frac{(\mathrm{FR}^y - \mathrm{FR}^{y-1})(\mathrm{HR}^y + \mathrm{HR}^{y-1})}{2}$, where FR is the false alarm rate and HR is the hit rate.	Measure of the model's probabilistic performance; it quantifies the model's ability to discriminate between an event and a non-event given a specific threshold.
Interquartile range skill score (IQRSS)	$\mathrm{IQRSS} = 1 - \frac{\mathrm{IQR_f}}{\mathrm{IQR_r}}$, where IQR is the interquartile range.	Measure of the forecast sharpness; it quantifies the relative spread in the forecast against a reference forecast.
Uncertainty sensitivity skill score (USS)	$\mathrm{USS} = \frac{(\rho_r - \rho_f)}{(1-\rho_r)}$, where ρ is the Spearman rank correlation between the IQR and absolute error.	Measure of the model's sensitivity to uncertainty; it quantifies the correlation between forecast sharpness and absolute error

cast the NSE can be expressed as

$$\mathrm{NSE} = 1 - \frac{\sum_{y=1}^{n}\left(\mathrm{SFV}_o^y - \mathrm{SFV}_f^y\right)^2}{\sum_{y=1}^{n}\left(\mathrm{SFV}_o^y - \overline{\mathrm{SFV}_o}\right)^2}. \quad (5)$$

To assess the skill of the multi-model ensemble, with respect to the reference historical ensemble, the difference in their NSE is calculated as

$$\Delta\mathrm{NSE} = \mathrm{NSE_f} - \mathrm{NSE_r}, \quad (6)$$

where $\Delta\mathrm{NSE} > 0$ indicates that the multi-model forecast has skill over the reference forecast.

2.5.4 Relative operating characteristic skill score (ROCSS)

The ROCSS is a skill score based on the area under the curve (AUC) in a relative operating characteristic diagram. ROCSS values below zero indicate the forecast has no skill over climatology while values over zero indicate skill with one being a perfect forecast. The ROC diagram measures the ability of the forecast ensemble to discriminate between an event and a non-event given a specific threshold. For this work the ROCSS were calculated for the upper tercile ($x \geqslant 66.7\,\%$), middle tercile ($66.7\,\% < x \leqslant 33.3\,\%$), and lower tercile ($x < 33.3\,\%$). These scores estimate the skill of ensemble forecasts to distinguish between below normal (BN), near normal (NN) and above normal (AN) anomalies. Hamill and Juras (2006) define the ROC skill score to be

$$\mathrm{ROCSS} = 2 \cdot \mathrm{AUC} - 1, \quad (7)$$

where AUC is the area under the curve when mapping hit rates against false alarm rates

$$\mathrm{AUC} = \sum_{y=1}^{n+1} \frac{\left(\mathrm{FAR}_y - \mathrm{FAR}_{y-1}\right)\left(\mathrm{HR}_y + \mathrm{HR}_{y-1}\right)}{2} \quad (8)$$

and FAR is the false alarm rate and HR is the hit rate. False alarms are defined as both the false positive and false negative forecasts, or type I and type II errors. Hits are defined as correctly forecasted events.

2.5.5 Interquartile range skill score (IQRSS) and uncertainty sensitivity skill score (USS)

Sharpness is an intrinsic attribute to HEPS, giving an indication of how large the ensemble spread is. Forecasts ensembles that are too spread are overly cautious and have limited value for an end user due to the uncertainty of the true magnitude of the SFV; conversely ensembles that are not spread enough are overly confident and may not be a true representation of the uncertainty, thus giving the end user false confidence in the forecast (Gneiting et al., 2007). For this work the sharpness is computed as the difference between the 75th and 25th percentiles of the forecast distribution or the interquartile range (IQR). The IQRSS is skill score based on the IQR and is a measure of how much better, i.e. sharper, the forecast ensemble is over the reference ensemble, values above zero indicate that the forecast ensemble is an improvement over the reference ensemble. The IQRSS is expressed as

$$\mathrm{IQRSS} = 1 - \frac{\mathrm{IQR_f}}{\mathrm{IQR_r}}. \tag{9}$$

As mentioned above, sharpness can be misleading. A well-designed and calibrated ensemble should give the user an idea of the uncertainty of the forecast conveyed through the relative sharpness of the ensemble. Thus it follows that the IQR should be positively correlated to the absolute forecast error; a larger (smaller) IQR would indicate to the user that there is a larger (smaller) uncertainty in the SFV forecast. The uncertainty sensitivity skill score (USS) can be expressed as the skill score of the Spearman rank correlations between the IQR and the absolute deterministic error as follows:

$$\mathrm{USS} = \frac{(\rho_\mathrm{r} - \rho_\mathrm{f})}{(1 - \rho_\mathrm{r})}, \tag{10}$$

where ρ is the Spearman rank correlation.

2.6 Uncertainty estimation

Due to the limited sample size of data available in this work a bootstrap approach is employed to estimate the verification measures and determine whether they are statistically significant. Again due to data limitations a more circumspect significance level is prudent due to the course nature of the resulting statistics; we chose to set the significance level at 0.1 resulting in a 90 % confidence interval between the 5th and 95th percentiles. The cross-validated hindcast ensembles were sampled, allowing for repetition, 10 000 times to calculate the verification measures. We define a result to be statistically significant if the 5th (95th) percentile of the bootstrapped ensemble being evaluated does not overlap the 95th (5th) percentile of the bootstrapped reference ensemble.

2.7 Study area and local data

The sub-basins used in this work are divided into three groups using the clusters defined by Foster et al. (2016), namely clusters, S^1, S^2, and S^3 (Fig. 3). Sweden was divided into five regions of homogeneous streamflow variability: three clusters located in the northern parts of the country, where snow dominates the hydrological processes (northern group); and two located in the southern part, where rain dominates the hydrological processes (southern group). For the purposes of this work we are only interested in the northern group. The numbers of sub-basins in each of these clusters are 25, 19, and 40 respectively. The S in the cluster's designation denotes that the hydrological regimes are dominated by snow processes and the superscripts give the relative strength of the signal from these processes in the hydrological regime. During the winter months most of the precipitation that falls within these basins is stored in the form of a snowpack and does not immediately contribute to streamflow. During the warmer spring months, when the temperatures rise above freezing, these snowpacks begin to melt (typically around mid- to late-April), which results in a period of high streamflow commonly referred to as the spring flood. We focus on forecasts of the accumulated streamflow volume during this period, or SFV.

For this work, 84 sub-basins from seven hydropower producing rivers in northern Sweden (Fig. 3) were used for the development and testing of the multi-model prototype. The sub-basins utilised include those in the current operational seasonal forecast system at SMHI plus the two unregulated sub-basins used by Olsson et al. (2016). Daily reservoir inflows for each sub-basin are available from the SMHI archives starting from 1961 to the end of the last hydrological year; the data used in our work are for the period 1981–2015 due to some of the other datasets used in this study only being available from 1981. These inflows are derived by adding the local streamflow to the change in reservoir storage then subtracting the streamflow from upstream basins as follows:

$$Q_\mathrm{in} = \Delta S + Q_\mathrm{local} - Q_\mathrm{upstream}. \tag{11}$$

Missing inflow data were filled by a multiple linear regression approach using simulated inflows for the sub-basin and observations from the surrounding sub-basins as predictors. Of the 84 sub-basins used in this work 68 had less than 1 % missing data (50 of these had no missing data), four had 1–10 % missing data, five had 20–30 % missing data, four had 30–40 % missing data, and three had 61, 63 and 71 % missing data, respectively. As these sub-basins are a part of the current operational forecast system they were included in the study despite some of them having a significant missing data fraction. The average NSE for the data used for filling was 0.70 (the NSE scores for the intervals above were 0.67, 0.75,

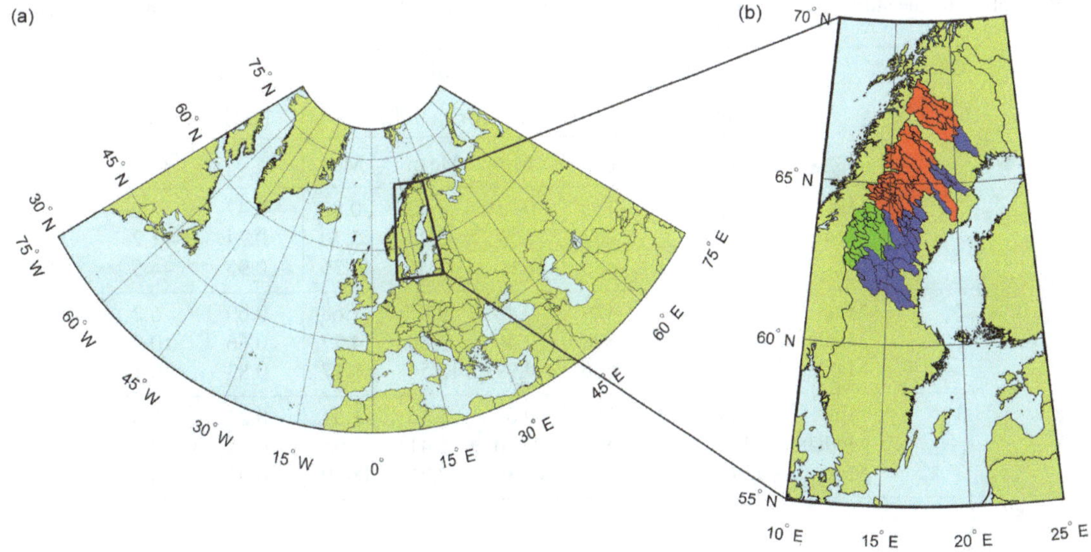

Figure 3. Map showing **(a)** the domain for the predictors used in the SE modelling chain, **(b)** the domain of the seasonal forecast data used in the DE modelling chain, and the location of the forecasts sub-basins used in this work. The sub-basins shown in blue belong to cluster S^1, those shown in green belong to cluster S^2, and those shown in red belong to cluster S^3.

0.77, 0.61, and 0.73, respectively) which suggests that this approach is acceptable.

Daily observations of precipitation and temperature data used in this work were obtained from the PTHBV dataset from SMHI (Johansson, 2002). The PTHBV dataset is a 4×4 km gridded observation dataset of daily precipitation and temperature data that has been created by optimal interpolation with elevation and wind taken into account. These data are available from 1961 to the present.

Table 2 gives a summary of some basic basin characteristics and statistics regarding the SFV as well as selected performance measures for the HBV rainfall–runoff model for the sub-basins in each cluster. Although, the ranges in sub-basin areas in the different clusters are similar, except for the maximum in S^3, the SFV statistics increase with each cluster when looking from cluster S^1 to S^3. This is due to the effects that elevation and latitude have on how much snow processes dominate the hydrological regimes in each cluster. Sub-basins in cluster S^1 are typically at a latitude or elevation lower than those in clusters S^2 and S^3, similarly to the sub-basins in S^2 with respect to those in S^3. The ranges in the NSE and the relative MAE imply that, in general, the HBV model is adequately or well calibrated for most sub-basins, however there are some sub-basins for which the HBV model appears not to be well calibrated and for which there is some scope for improvement. This can be somewhat misleading as these data are a function of three different observations and as such can be subject to noise and uncertainties.

2.8 Driving Data

2.8.1 Teleconnection indices

This work uses monthly indices of the Arctic Oscillation and Scandinavian pattern collected from the Climate Prediction Center (Climate Prediction Center, 2012) for the period October 1960 to May 2015.

2.8.2 Seasonal data

The ECMWF seasonal forecast system model from system 4 (Molteni et al., 2011) is the cycle36r4 version of ECMWF IFS (Integrated Forecast System) coupled with a 1° version of the NEMO ocean model. The seasonal forecasts from the ECMWF IFS were used in the following two different forms: a field of seasonal monthly averages as input to the statistical model, and individual grid points of daily data for input into the HBV model. The choice to use ECMWF data is primarily a practical one. The ECMWF is an established and proven producer of medium range forecasts and SMHI already has operational access to their products.

The seasonal forecast averages are the seasonal means for each ensemble member of the different predictors which had a domain covering 75° W to 75° E and 80 to 30° N (Fig. 2) with a $1° \times 1°$ resolution. For each predictor only the first 15 ensemble members were used in this work. This is because the number of ensemble members available in the ECMWF seasonal forecast is limited to 15 for the hindcast period while the operational seasonal forecast ensemble has 51 members. The predictors considered in this part of the work were the following: 850 hPa geopotential, 850 hPa tem-

Table 2. Basic information on the study area including overall performance of the HBV model for the sub-basins in each cluster.

Cluster		Basin		SFV			HBV	
		Area	Elevation	($m^3 \times 10^8$)			NSE	rMAE
		(km^2)	(m)	Min	Mean	Max		(%)
1	Min	233	135	0.21	0.42	0.82	−0.47	3.9
	Median	1827	282	1.27	3.23	5.30	0.74	11.2
	Max	6258	584	18.95	34.36	44.36	0.95	44.5
2	Min	184	429	0.40	0.81	1.68	−0.69	6.2
	Median	1166	598	2.49	3.80	4.77	0.66	10.2
	Max	4272	666	9.99	13.56	18.17	0.83	70.1
3	Min	270	212	0.67	1.12	1.54	0.22	3.8
	Median	1309	586	2.96	5.41	7.95	0.74	7.3
	Max	13177	776	19.39	37.76	48.50	0.92	20.0

perature, 850 hPa zonal wind component, 850 hPa meridional wind component, 850 hPa specific humidity, surface sensible heat flux, surface latent heat flux, mean sea level pressure, 10 m zonal wind component, 10 m meridional wind component, 2 m temperature, and total precipitation.

The daily time series data are the ECMWF IFS seasonal forecasts of daily values of temperature (2mT) and the accumulated total precipitation (pr). These data have a resolution of 0.5° × 0.5°, span a period from 1981 to 2015, and have a domain covering 11 to 23° E and 55 to 70° N.

3 Results and discussion

The following section outlines and discusses the results from the cross-valuated hindcasts of the different approaches' ability to hindcast the SFV for the period 1981–2015. First this evaluation is done for each system using climatology as a reference to assess their general skill. Following this the more skilful of the two multi-model ensembles is evaluated using the HE as a reference to assess any improved skill, and thus added value, of the multi-model ensemble approach over the current HE approach. The analysis is carried out on the cross-validated hindcasts of the SFV initialised on the 1 January, February, March, April, and May.

3.1 Evaluating the different forecast systems against climatology

The different forecast approaches' general skill to predict the SFV was estimated using MAESS, their skill to reproduce the interannual variability was estimated using NSE, and finally the skill to discriminate between below BN, NN, and AN SFVs was estimated using ROCSS. Table 3 gives an overview of these scores across all sub-basins and clusters for each initialisation month as well as the percentage of sub-basins where the hindcasts outperformed climatology; the values in brackets are the percentage of sub-basins where the hindcasts outperformed climatology and the result is statistically significant.

The performance measures for each of the three approaches are positively related to the relative timing of the hindcasts i.e. hindcasts initialised in any month are generally more (less) skilful than the hindcasts initialised in the preceding (following) months. This can be expected as the further away in time from the spring flood that the hindcast is initialised, increasing the lead time, the less the initial hydrological conditions contribute to predictability and the more uncertain the forcing data become (e.g. Wood et al., 2016; Arnal et al., 2017).

With respect to general skill and the ability to capture the interannual variation shown by the observations, the prototypes tend to perform better than HE with ME_{hds} typically having the best performance. This is especially so when we consider the percentage of the sub-basins where this improved performance is statistically significant. The gap between HE and the two prototypes in MAESS, NSE, and percentage of sub-basins with improved performance over climatology tends to get smaller as the season progresses while the gap in the percentage of sub-basins where improved performance is statistically significant appears to grow, at least early in the season.

However, if we turn our attention to the forecast's ability to discriminate between BN, NN, and AN SFVs then the HE holds an advantage over the two prototypes especially when it comes to identifying NN events from all forecast initialisation dates and, to a lesser extent, BN events for the later forecasts. The proposed prototypes are better at identifying AN events for all forecasts except those initialised in May where the ability of the HE is comparable. The advantage displayed by the HE to identify NN events is to be expected due to its climatological nature while the advantage with respect to BN events can probably be attributed to a cold bias in the historical forcing data caused by climate change. The drop in relative skill by the prototypes in the later forecasts

Table 3. Bootstrapped ($N = 10\,000$) skill scores and the number of sub-basins, as a percentage, where the HEPS performs better than climatology averaged over all 84 sub-basins. The n^+ values in brackets show the percentages of the sub-basins for which these scores are statistically significant at the 0.1 level.

		MAESS		NSE		ROCSS					
						BN		NN		AN	
			n^+ (%)		n^+ (%)		n^+ (%)		n^+ (%)		n^+ (%)
HE	Jan	−0.09	25 (1)	−0.24	17 (0)	0.23	90 (21)	0.07	70 (0)	0.10	68 (11)
	Feb	0.00	51 (6)	−0.07	42 (5)	0.41	99 (52)	0.11	69 (1)	0.26	92 (27)
	Mar	0.09	80 (17)	0.13	77 (23)	0.55	100 (87)	0.10	73 (5)	0.44	99 (56)
	Apr	0.15	85 (35)	0.22	80 (35)	0.62	100 (92)	0.17	85 (7)	0.51	100 (75)
	May	0.21	90 (49)	0.32	90 (49)	0.68	100 (98)	0.23	92 (10)	0.61	100 (92)
ME_{ads}	Jan	0.00	50 (2)	0.00	55 (1)	0.31	99 (31)	−0.01	48 (0)	0.20	80 (18)
	Feb	0.06	73 (23)	0.11	76 (21)	0.39	99 (51)	0.08	74 (0)	0.36	96 (42)
	Mar	0.11	86 (25)	0.20	87 (36)	0.47	100 (76)	0.07	61 (4)	0.47	100 (60)
	Apr	0.20	95 (62)	0.32	94 (64)	0.60	100 (90)	0.16	83 (5)	0.52	100 (79)
	May	0.22	96 (67)	0.36	98 (68)	0.66	100 (94)	0.18	82 (8)	0.57	100 (76)
ME_{hds}	Jan	0.02	60 (6)	0.03	63 (5)	0.32	100 (31)	0.00	51 (0)	0.22	83 (24)
	Feb	0.08	80 (25)	0.14	85 (29)	0.41	99 (57)	0.07	69 (1)	0.38	99 (44)
	Mar	0.14	90 (32)	0.24	92 (45)	0.51	100 (81)	0.07	61 (5)	0.48	100 (64)
	Apr	0.19	94 (56)	0.32	93 (62)	0.60	100 (90)	0.17	88 (5)	0.54	100 (80)
	May	0.24	98 (74)	0.39	96 (76)	0.67	100 (94)	0.18	85 (10)	0.60	100 (88)

is in part due to their sharpness being worse than the HE in the later forecast (Sect. 3.3).

From these results we are now able to make an informed choice as to which prototype to proceed with, ME_{hds} (hereafter referred to as the prototype unless stated otherwise). If we take all the results and rank the performances of the three methods then the prototype would rank first, followed closely by ME_{ads}, and HE would rank third. However, all three forecast methods have been shown to be skilful at forecasting the SFV – albeit a naïve skill.

3.2 Evaluating the prototype against HE

The frequency at which the prototype outperforms HE is estimated using FY^+, its general skill to predict the SFV is estimated using MAESS, its skill to reproduce the interannual variability is estimated using ΔNSE, and finally its skill to discriminate between BN, NN, and AN SFVs is estimated using ΔROCSS. Figure 4 shows the bootstrapped scores for MAESS, FY+, and ΔNSE calculated for each hindcast initialisation month for the sub-basins in cluster S^3. The medians of these bootstrapped scores are presented as a histogram; summary statistics are documented above the histogram. On the left-hand side are the max, mean, and min scores for the cluster i.e. the sub-basins with the highest and lowest scores and the mean for the basin. On the right-hand side are the percentage of sub-basins where the prototype outperformed HE (shows skill over HE (n^+_{abs})), the percentage of sub-basins where the prototype shows statistically significant skill over HE ($n^+_{0.1}$), and the percentage of sub-basins where HE shows statistically significant skill over the prototype ($n^-_{0.1}$).

3.2.1 FY^+

The prototype has a $FY^+ > 50\,\%$ for the majority of the sub-basins in cluster S^3, ranging from 98 % of the sub-basins with a mean FY^+ of 61 % for hindcasts initialised in January down to 73 % of the sub-basins with a mean FY^+ of 56 % for hindcasts initialised in May. These figures are similar, even a little higher, for sub-basins in cluster S^2 while somewhat lower for cluster S^1. The number of sub-basins for which the prototype has a statistically significant $FY^+ > 50\,\%$ ranges between 10 and 28 % in cluster S^3 ($S^2 = 5$–37 %, and $S^1 = 12$–16 %). Whilst the prototype has a statistically significant $FY^+ < 50\,\%$ (performs worse than HE more often than not) for 5 % of sub-basins for hindcasts initialised in April in cluster S^3 and 4 % of sub-basins in hindcasts initialised in May for cluster S^1.

3.2.2 MAESS

The prototype shows skill at improving the volume error hindcast by HE for the majority of the sub-basins, ranging between 65 and 100 % of the sub-basins in cluster S^3 ($S^2 =$ 74–95 %, and $S^1 = 64$–80 %). This improvement tends to be largest for hindcasts initialised in January, mean MAESS of 0.12 ($S^2 = 0.11$, and $S^1 = 0.04$), and lowest for those in May, 0.04 ($S^2 = 0.05$, and $S^1 = 0.02$). The percentage of sub-basins for which MAESS > 0 is statistically significant

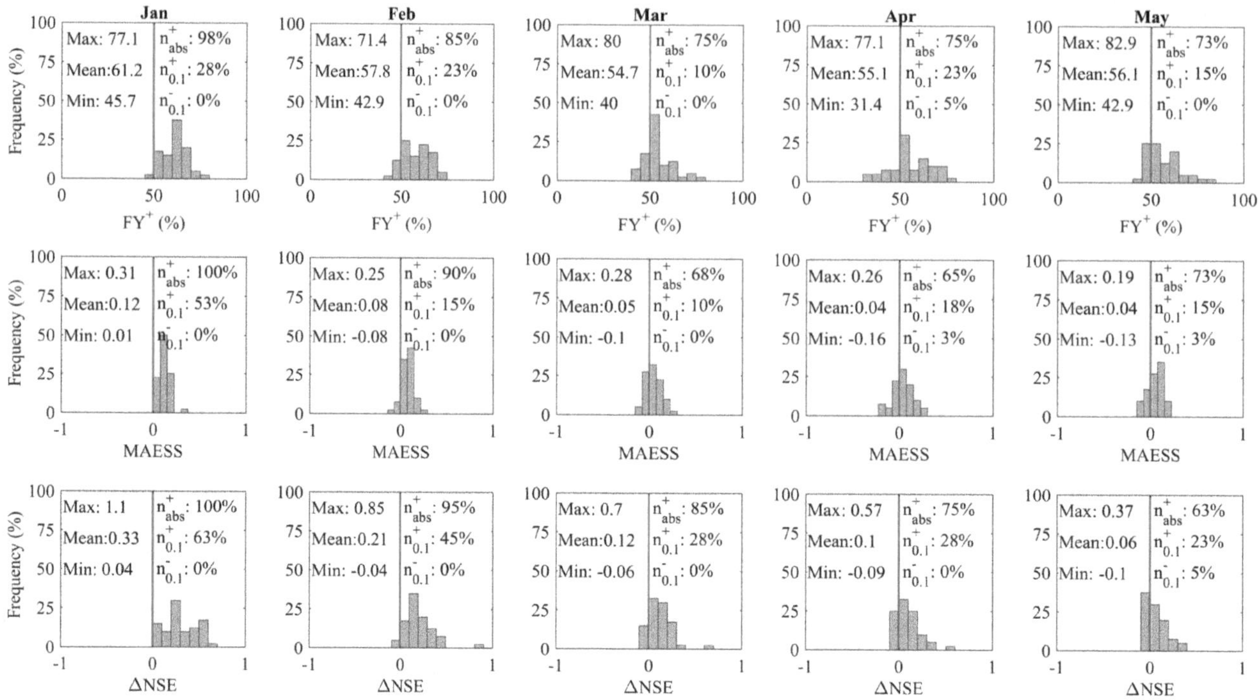

Figure 4. Bootstrapped ($N = 10\,000$) FY+, MAESS, and ΔNSE scores for ME_{hds} with respect to HE for all sub-basins in the cluster S^3. Each subplot is a histogram of the medians of the bootstrapped validation scores for each initialisation month. Above the histograms are six related statistics: (left of the red line) the maximum, mean, and minimum of the validation scores shown in the histograms; (right of the red line) percentages of the sub-basins where ME_{hds} performed better than HE (n^+_{abs}), the percentage of sub-basins where ME_{hds} performed better than HE ($n^+_{0.1}$) at the significance level 0.1, and lastly the percentage of sub-basins where ME_{hds} performed worse than HE ($n^-_{0.1}$) at the 0.1 level.

ranges between 10 and 53 % for all clusters and hindcast initialisations, while the percentage of sub-basins for which MAESS < 0 is statistically significant are 8 and 4 % for hindcasts initialised in March and May in cluster S^1, respectively, and 3 % for hindcasts initialised in both April and May in cluster S^3. These results also show that the prototype generally has a smaller MAE than HE especially for earlier hindcast initialisations and again for clusters S^3 and S^2.

3.2.3 ΔNSE

The prototype shows skill at improving the representation of the interannual variability of the observed SFV again for most of the sub-basins, ranging between 63 and 100 % of sub-basins in cluster S^3 ($S^2 = 74$–100 %, and $S^1 = 76$–92 %), and the mean ΔNSE ranges between 0.06 and 0.33 for sub-basins in cluster S^3 ($S^2 = 0.09$ and 0.32, and $S^1 = 0.06$ and 0.15). The percentage of sub-basins for which ΔNSE > 0 is statistically significant ranges between 16 and 63 % for all clusters and hindcast initialisations, while the percentage of sub-basins for which ΔNSE < 0 is statistically significant are 4 % for hindcasts initialised in January, March, and May in cluster S^1, and 5 % for hindcasts initialised in May in cluster S^3.

3.2.4 ΔROCSS

Figure 5, which has the same information presentation structure as Fig. 4, shows the bootstrapped ΔROCSS for the lower (BN), middle (NN), and upper (AN) terciles calculated for each hindcast initialisation month for the sub-basins in cluster S^3.

The prototype shows skill over HE to discriminate between BN events and non-BN events for the majority of the sub-basins in cluster S^3 for hindcasts initialised in January and February, 95 and 68 % respectively ($S^2 = 63$, 53 % and $S^1 = 68$, 32 %) but this drops to less than half the sub-basins in hindcasts initialised thereafter. The mean ΔROCSS ranges between -0.04 and 0.14 in cluster S^3 ($S^2 = -0.03$ and 0.05, and $S^1 = -0.06$ and 0.02) with only statistically significant results being found in favour of the prototype for 15 and 5 % of sub-basins for hindcasts initialised in January in clusters S^3 and S^2, respectively, and in favour of HE for 4 % of sub-basins for hindcasts initialised in April in cluster S^1.

Out of the three terciles the prototype shows the least skill over HE at discriminating between NN events and non-NN events. The percentage of sub-basins for which the prototype outperforms HE ranges between 23 and 57 % ($S^2 = 37$–63 %, and $S^1 = 24$–60 %) with mean ΔROCSS ranges between -0.07 and 0.01 ($S^2 = -0.06$ and 0, and $S^1 = -0.05$

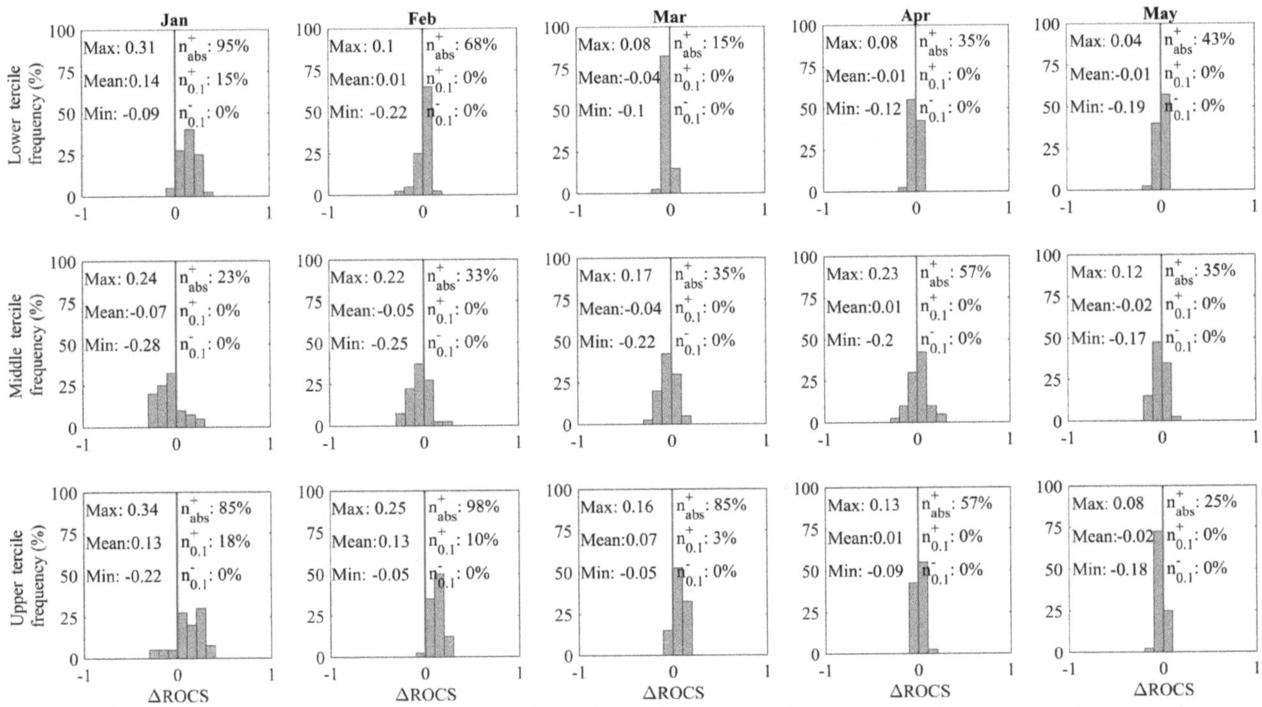

Figure 5. Bootstrapped ($N = 10\,000$) ΔROCSS for the lower, middle, and upper terciles between the ME_{hds} and HE for sub-basins in the cluster S^3. Each subplot is a histogram of the medians of the bootstrapped validation score's ensembles for each initialisation month. Above the histograms are six related statistics: (left of the red line) the maximum, mean, and minimum of the validation scores shown in the histograms; (right of the red line) percentages of the sub-basins where ME_{hds} performed better than HE (n^+_{abs}), the percentage of sub-basins where ME_{hds} performed better than HE ($n^+_{0.1}$) at the significance level 0.1, and lastly the percentage of sub-basins where ME_{hds} performed worse than HE ($n^-_{0.1}$) at the 0.1 level.

and 0.02) and no statistically significant results for any sub-basins, both in favour of or against the prototype.

The prototype shows the best performance when discriminating between AN events and non-AN events. The percentage of sub-basins for which the prototype shows skill over HE in the upper tercile ranges between 85 and 98 % for hindcasts initialised in the first three months ($S^2 = 47$–$89\,\%$, and $S^1 = 48$–$84\,\%$) then 57 and 25 % for the last two months, respectively ($S^2 = 63$ and 53 %, and $S^1 = 80$ and 32 %). The mean ΔROCSS ranges between −0.02 and 0.13 ($S^2 = -0.01$ and 0.14, and $S^1 = -0.01$ and 0.07). The percentage of sub-basins in cluster S^3 for which ΔROCSS > 0 is statistically significant are 18, 10, and 3 % for hindcasts initialised in January, February, and March respectively, and 16 % for forecasts initialised in January and February in cluster S^2. There are no statistically significant results in favour of HE.

3.3 Analysis of the forecast ensemble sharpness

Figure 6 shows the cross-validated hindcasts by the prototype initialised in January (top panel) and May (bottom panel) for Göuta-Ajaure, a cluster S^3 sub-basin in upper reaches of the Ume River system. This basin was chosen as an example of a where the prototype showed typical performance results i.e. neither the best nor the worst. The total ensemble spread (the whiskers) of the forecasts initialised in January remains somewhat consistent from year to year while the IQR (the blue boxes) displays a more pronounced variation. The lack of variation in total spread is primarily the result of the climatological nature of the HE component which tends to have a larger and more consistent spread than that for DE and SE at longer lead times. The greater variation exhibited by the IQR is mostly due to the "true" forecast nature of the DE and SE components in the multi-model ensemble. If we turn our attention to the forecasts initialised in May we see a more pronounced variation in both the total spread and the IQR. This is because the spread in the DE and SE components is now comparable to and often larger than the spread in the HE component. Table 4 shows how the IQRSS drops as the spring flood season approaches. It can also be seen in Fig. 6 that the ensemble median (red line) is more responsive to the year on year variation in SFV in the May forecasts than in the January forecasts. This is because the relative contribution to predictability by the initial conditions is greater than the contribution from the meteorological drivers closer to the spring flood period. These patterns are generally true for both the

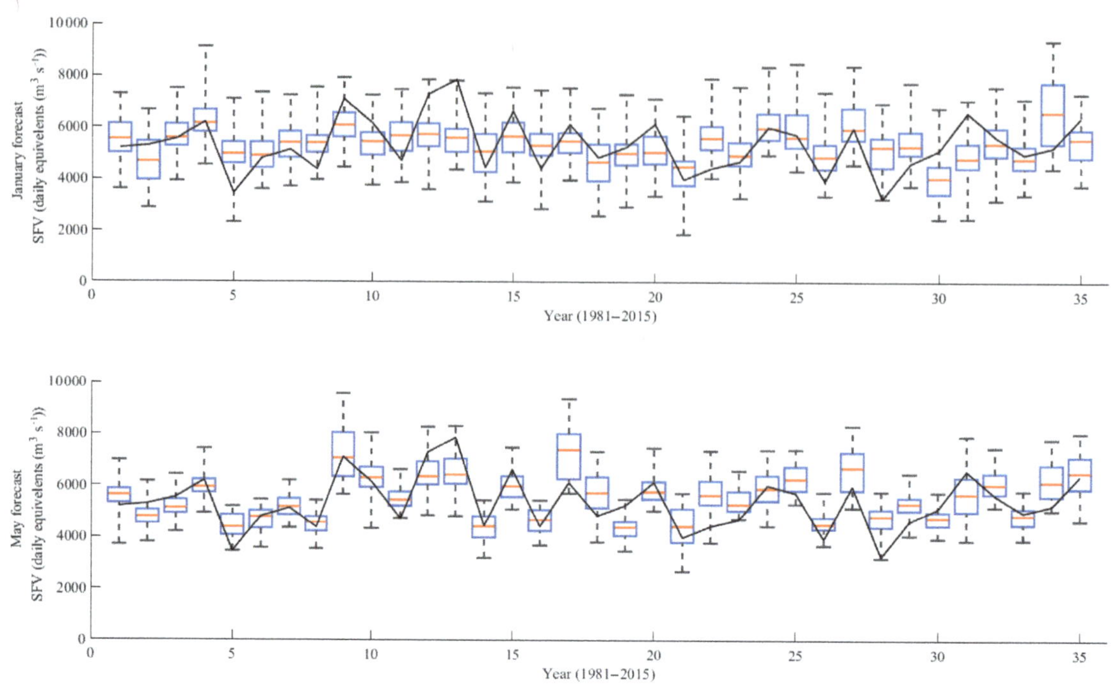

Figure 6. The cross-validated hindcasts of the SFV for a sub-basin in cluster S^3 made by ME_{hds} (boxplots) together with the observed SFV (black line). The box plots represent the entire forecast ensemble, the red lines represent the ensemble medians, the blue boxes the 25th and 75th percentiles (IQR), and the feelers represent the 0th and 100th percentiles.

forecasts initialised in the intermediate months and for the other sub-basins.

If we assume that the more sensitive an ensemble is to uncertainty the more the forecast sharpness will vary. We would therefore expect the USS values to generally be positive i.e. that the forecast sharpness of prototype is better correlated with the forecast error than for HE. This is largely supported by the USS values in Table 4 where only three values are negative, the January forecast in cluster S^2 and the April forecasts in both clusters S^1 and S^2, and even then not by very much. This suggests that at least one but probably both of the DE and SE ensembles are responsible for this improvement due to their variability. There is a general decreasing trend with initialisation date in the USS values in clusters S^1 and S^2 (if we ignore the value for January in S^2) while the values are more consistent in cluster S^3. All the uncertainty correlation values for both the HE and the prototype are significant at the 0.1 level (not shown for brevity) suggesting that both exhibit sensitivity to uncertainty to some extent, however the prototype is more sensitive to uncertainty which should instil more confidence in the forecast for the users.

The IQRSS values show that the prototype tends to produce sharper forecasts than HE early in the season i.e. for forecasts initialised in January and February in cluster S^1 and forecasts initialised in January, February, and March in clusters S^2 and S^3. This is reversed for the remaining initialisation dates where HE tends to produce sharper forecasts than the prototype. This is probably due to the climatological nature of the HE having less of an impact on forecast sharpness as the initialisation date approaches the spring flood period in combination with the uncertainties and biases in the other individual ensembles, which further exacerbate the situation.

3.4 Spatial and temporal variations and transferability of the prototype

Both multi-model ensembles show skill at forecasting SFV with respect to forecast error, ability to reproduce the interannual variability in SFV, and the ability to discriminate between BN, NN, and AN events. The prototype, in particular, is at worst comparable to the HE and at best clearly more skilful. This relative performance of the prototype varies both in space and time. Figure 7 shows maps of the median bootstrapped FY^+ values. For hindcasts initialised in January the spatial pattern in the FY+ scores show that the prototype tends to outperform HE more in sub-basins that have a higher latitude or elevation. However, as the initialisation date approaches the spring flood period this pattern becomes less and less coherent. This general pattern is also true for MAESS scores. This suggests that the change in the performances of the prototype and HE, as a function of initialisation date, are not always similar for sub-basins that are near one another. Further work would be needed to find out what the underlying reason for this is.

Data availability is the biggest limiting factor to the transferability of this approach to other areas. The HE, AE, and SE approaches are all dependant on good quality observation

Table 4. Bootstrapped ($N = 10\,000$) USS and IQRSS for ME_{hds} using HE as a reference, and the percentage of sub-basins where ME_{hds} performed better than HE. The n^+ values in brackets show the percentages of the sub-basins for which these scores are statistically significant at the 0.1 level.

		USS					IQRSS				
		Jan	Feb	Mar	Apr	May	Jan	Feb	Mar	Apr	May
s^1	SS	0.21	0.04	0.02	−0.02	0.00	0.01	0.05	−0.02	−0.08	−0.18
	n^+ (%)	80 (16)	64 (0)	56 (8)	48 (4)	52 (8)	56 (24)	72 (20)	40 (8)	16 (4)	16 (4)
s^2	SS	−0.05	0.17	0.07	−0.10	0.17	0.01	0.06	0.05	−0.13	−0.15
	n^+ (%)	48 (4)	80 (4)	60 (4)	40 (0)	68 (4)	52 (36)	68 (40)	64 (16)	16 (4)	12 (0)
s^3	SS	0.05	0.06	0.11	0.07	0.06	0.09	0.08	0.02	−0.05	−0.03
	n^+ (%)	60 (12)	65 (10)	65 (15)	58 (10)	60 (12)	70 (52)	80 (48)	60 (25)	32 (8)	48 (18)

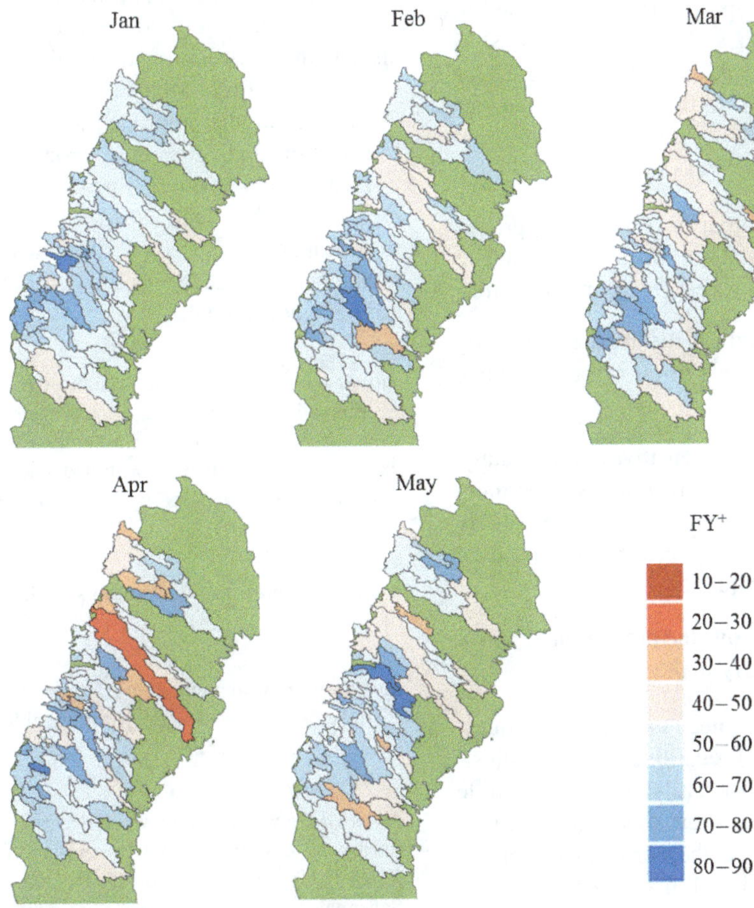

Figure 7. Maps of the median bootstrapped FY^+ values for each of the initialisation dates.

time series. Additionally, the skill of all three of these approaches would be expected to be affected by the length of these time series. They length of the time series should be long enough to be a good representative sample of the climatology otherwise the forecasts would be biased in favour of the climate represented in the data and not the true climatology.

The SE and AE approaches require an understanding of how the variability in the local hydrology is affected by large scale circulation phenomena such as teleconnection patterns to help select predictors and teleconnection indices for inputs to each approach, respectively. The hydrological rainfall–runoff model used in the prototype should not pose a problem, although HBV has been successfully set up for snow dominated catchments outside of Sweden (e.g. Seibert et al.,

2010; Okkonen and Kløve, 2011), any sufficiently well calibrated rainfall–runoff model would suffice.

We believe that, if the above requirements are met, a seasonal hydrological forecast system similar to the prototype can be set up in other snow dominated regions around the world.

4 Conclusions

In this paper we present the development and evaluation of a hydrological seasonal forecast system prototype for predicting the SFV in Swedish rivers. Initially, two versions of the prototype, ME_{ads} and ME_{hds}, were evaluated together using the HE using climatology as a reference for both, to help select which version of the prototype to proceed with and to get a general impression of their skill to forecast the SFV. Thereafter the chosen prototype was evaluated using HE as a reference and finally the sharpness of the hindcast ensembles were analysed.

The main findings are summarised as follows:

- The prototype is able to outperform the HE approach 57 % of the time on average. It is at worst comparable to the HE in forecast skill and at best clearly more skilful.
- The prototype is able to reduce the forecast error by 6 % on average. This translates to an average volume of $9.5 \times 10^6\,m^3$.
- The prototype is generally more sensitive to uncertainty, that is to say that the ensemble spread tends to be more correlated with the forecast error. This is potentially useful to users as the ensemble spread could be used as a measure of the forecast quality.
- The prototype is able to improve the prediction of above and below normal events early in the season.

Looking forward, future studies need to address the questions raised by Zhao et al. (2017) regarding the bias adjustment of meteorological seasonal forecast data using quantile mapping. Results from this study show that while the seasonal forecasts were bias adjusted the performance of the DE was disappointing; however it still had value within the multi-model setting suggesting that it has more of a modulating roll on the other modelling chains as opposed to contributing directly to predictability.

How the individual model ensembles are combined to give the multi-model output needs to be revisited. When we applied the asymmetric weighting scheme proposed by Olsson et al. (2016) we did not find that it improved the multi-model performance in general across all stations and forecasts and so we did not use it. However, we do believe that more work should be done to find a more appropriate weighting scheme than simple pooling. Perhaps better understanding how the performance of the different modelling chains are affected by the initial conditions and lead time will shed more light on how to best approach this issue. Further development and testing along these lines are planned for the future.

The AE approach did not exhibit the promising performance found by Olsson et al. (2016) using circulation pattern analysis to select the analogues. A part of the explanation for this poor performance is that the teleconnection information used to select the analogues only partially span the full periods identified by Foster et al. (2016), from October/November to the beginning of the spring flood. The missing data could be filled by making forecasts of the indices. Another approach would be to revisit the circulation pattern analysis based approach now that data inhomogeneity issues are largely addressed by the new ERA5 reanalysis data that is becoming available (http://climate.copernicus.eu/products/climate-reanalysis, last access: 15 September 2017). Yet another approach would be to use GCM forecasts to select the analogues (e.g. Crochemore et al., 2016).

Lastly, the post processing of model outputs (e.g. Lucatero et al., 2017) has been shown to be beneficial, the incorporation of a simple approach like linear scaling is possibly the most appealing due to its ease of implementation in an operational environment.

Author contributions. The AE and SE approaches were designed by KF and CBU and were implemented by KF. The DE approach was designed by JO and implemented by KF. The multi-model experimental set-up was designed and implemented by KF. The manuscript was prepared by KF with contributions from CBU and JO.

Competing interests. The authors declare that they have no conflict of interest.

Special issue statement. This article is part of the special issue "Sub-seasonal to seasonal hydrological forecasting". It does not belong to a conference.

Acknowledgements. This work was supported by research projects funded by Energiforsk AB (formerly Elforsk AB) and the EUPORIAS (EUropean Provision Of Regional Impacts Assessments on Seasonal and decadal timescales) project funded by the Seventh Framework Programme for Research and Technological Development (FP7) of the European Union (grant agreement no. 308291). Many thanks go to Johan Södling, Jonas German, and Barbro Johansson for technical assistance as well as fruitful discussions. Constructive and detailed reviews of the original manuscript by Joost Beckers, Fernando Mainardi Fan, Hannes Tobler, and Sebastian Röthlin are gratefully acknowledged.

Edited by: Albrecht Weerts

References

Arheimer, B., Lindström, G., and Olsson, J.: A systematic review of sensitivities in the Swedish flood-forecasting system, Atmos. Res., 100, 275–284, 2011.

Arnal, L., Wood, A. W., Stephens, E., Cloke, H. L., and Pappenberger, F.: An efficient approach for estimating streamflow forecast skill elasticity, J. Hydrometeorol., 18, 1715–1729, 2017.

Beckers, J. V. L., Weerts, A. H., Tijdeman, E., and Welles, E.: ENSO-conditioned weather resampling method for seasonal ensemble streamflow prediction, Hydrol. Earth Syst. Sci., 20, 3277–3287, https://doi.org/10.5194/hess-20-3277-2016, 2016.

Bennett, J. C., Wang, Q. J., Li, M., Robertson, D. E., and Schepen, A.: Reliable long-range ensemble streamflow forecasts: Combining calibrated climate forecasts with a conceptual runoff model and a staged error model, Water Resour. Res., 52, 8238–8259, 2016.

Berg, P., Bosshard, T., and Yang, W.: Model consistent pseudo-observations of precipitation and their use for bias correcting regional climate models, Climate, 3, 118–132, 2015.

Bergström, S.: Development and application of a conceptual runoff model for Scandinavian catchments, SMHI Reports RHO No. 7, SMHI, Norrköping, Sweden, 1976.

Candogan Yossef, N., van Beek, R., Weerts, A., Winsemius, H., and Bierkens, M. F. P.: Skill of a global forecasting system in seasonal ensemble streamflow prediction, Hydrol. Earth Syst. Sci., 21, 4103–4114, https://doi.org/10.5194/hess-21-4103-2017, 2017.

Climate Prediction Center: Teleconnections, Scandinavia (SCAND), available at: http://www.cpc.ncep.noaa.gov/data/teledoc/scand.shtml (last access: 12 November 2016), 2012.

Climate Prediction Center: AO index, available at: http://www.cpc.ncep.noaa.gov/products/precip/CWlink/daily_ao_index/monthly.ao.index.b50.current.ascii, last access: 16 September 2017a.

Climate Prediction Center: Scand index, available at: ftp://ftp.cpc.ncep.noaa.gov/wd52dg/data/indices/scand_index.tim, last access: 16 September 2017b.

Cloke, H. L. and Pappenberger, F.: Evaluating forecasts for extreme events for hydrological applications: an approach for screening unfamiliar performance measures, Meteorol. Appl., 15, 181–197, 2008.

Córdoba-Machado, S., Palomino-Lemus, R., Gámiz-Fortis, S. R., Castro-Díez, Y., and Esteban-Parra, M. J.: Seasonal streamflow prediction in Colombia using atmospheric and oceanic patterns, J. Hydrol., 538, 1–12, https://doi.org/10.1016/j.jhydrol.2016.04.003, 2016.

Crochemore, L., Ramos, M.-H., and Pappenberger, F.: Bias correcting precipitation forecasts to improve the skill of seasonal streamflow forecasts, Hydrol. Earth Syst. Sci., 20, 3601–3618, https://doi.org/10.5194/hess-20-3601-2016, 2016.

Day, G. N.: Extended streamflow forecasting using NWSRFS, J. Water Res. Plan. Man., 111, 157–170, https://doi.org/10.1061/(ASCE)0733-9496(1985)111:2(157), 1985.

Doblas-Reyes, F. J., García-Serrano, J., Lienert, F., Biescas, A. P., and Rodrigues, L. R.: Seasonal climate predictability and forecasting: status and prospects, Wires Clim. Change, 4, 245–268, 2013.

Foster, K., Olsson, J., and Uvo, C. B.: New approaches to spring flood forecasting in Sweden, Vatten, 66, 193–198, 2010.

Foster, K., Uvo, C. B., and Olsson, J.: The spatial and temporal effect of selected teleconnection phenomena on Swedish hydrology, Clim. Dynam., submitted, 2018.

Foster, K. L. and Uvo, C. B.: Seasonal streamflow forecast: a GCM multimodel downscaling approach, Hydrol. Res., 41, 503–507, 2010.

Garen, D.: Improved techniques in regression-based streamflow volume forecasting, J. Water Res. Plan. Man., 118, 654–670, 1992.

Gneiting, T., Balabdaoui, F., and Raftery, A. E.: Probabilistic forecasts, calibration and sharpness, J. Roy. Stat. Soc. B, 69, 243–268, https://doi.org/10.1111/j.1467-9868.2007.00587.x, 2007.

Hamill, T. M. and Juras, J.: Measuring forecast skill: is it real skill or is it the varying climatology?, Q. J. Roy. Meteor. Soc., 132, 2905–2923, 2006.

Hersbach, H.: Decomposition of the continuous ranked probability score for ensemble prediction systems, Weather Forecast., 15, 559–570, 2000.

Johansson, B.: Estimation of areal precipitation for hydrological modelling, PhD thesis, Earth Sciences Centre, Report no. A76, Göteborg University, Göteborg, 2002.

Lindström, G., Johansson, B., Persson, M., Gardelin, M., and Bergström, S.: Development and test of the distributed HBV-96 hydrological model, J. Hydrol., 201, 272–288, 1997.

Lucatero, D., Madsen, H., Refsgaard, J. C., Kidmose, J., and Jensen, K. H.: Seasonal streamflow forecasts in the Ahlergaarde catchment, Denmark: effect of preprocessing and postprocessing on skill and statistical consistency, Hydrol. Earth Syst. Sci. Discuss., https://doi.org/10.5194/hess-2017-379, in review, 2017.

Molteni, F., Stockdale, T., Balmaseda, M., Balsamo, G., Buizza, R., Ferranti, L., Magnusson, L., Mogensen, K., Palmer, T., and Vitart, F.: The new ECMWF seasonal forecast system (System 4), ECMWF Tech. Memo., 656, 49 pp., available at: http://www.ecmwf.int/sites/default/files/elibrary/2011/11209-new-ecmwf-seasonal-forecast-system-system-4.pdf (last access: 29 August 2016), 2011.

Nash, J. E. and Sutcliffe, J. V.: River flow forecasting through conceptual models part I — A discussion of principles, J. Hydrol., 10, 282–290, https://doi.org/10.1016/0022-1694(70)90255-6, 1970.

Nilsson, P., Uvo, C. B., Landman, W. A., and Nguyen, T. D.: Downscaling of GCM forecasts to streamflow over Scandinavia, Hydrol. Res., 39, 17–26, https://doi.org/10.1016/j.jhydrol.2005.08.007, 2008.

Okkonen, J. and Kløve, B.: A sequential modelling approach to assess groundwater–surface water resources in a snow dominated region of Finland, J. Hydrol., 411, 91–107, 2011.

Olsson, J., Yang, W., Graham, L. P., Rosberg, J., and Andréasson J.: Using an ensemble of climate projections for simulating recent and near-future hydrological change to Lake Vänern in Sweden, Tellus, 63A, 126–137, https://doi.org/10.1111/j.1600-0870.2010.00476.x, 2011.

Olsson, J., Uvo, C. B., Foster, K., and Yang, W.: Technical Note: Initial assessment of a multi-method approach to spring-flood forecasting in Sweden, Hydrol. Earth Syst. Sci., 20, 659–667, https://doi.org/10.5194/hess-20-659-2016, 2016.

Pagano, T. C., Garen, D. C., Perkins, T. R., and Pasteris, P. A.: Daily updating of operational statistical seasonal water supply forecasts for the western U.S., J. Am. Water Resour. As., 45, 767–778, 2009.

Robertson, D. E., Pokhrel, P., and Wang, Q. J.: Improving statistical forecasts of seasonal streamflows using hydrological model output, Hydrol. Earth Syst. Sci., 17, 579–593, https://doi.org/10.5194/hess-17-579-2013, 2013.

Rosenberg, E. A., Wood, A. W., and Steinemann, A. C.: Statistical applications of physically based hydrologic models to seasonal streamflow forecasts, Water Resour. Res., 47, https://doi.org/10.1029/2010WR010101, 2011.

Schepen, A., Zhao, T., Wang, Q. J., Zhou, S., and Feikema, P.: Optimising seasonal streamflow forecast lead time for operational decision making in Australia, Hydrol. Earth Syst. Sci., 20, 4117–4128, https://doi.org/10.5194/hess-20-4117-2016, 2016.

Seibert, J., McDonnell, J. J., and Woodsmith, R. D.: Effects of wildfire on catchment runoff response: a modelling approach to detect changes in snow-dominated forested catchments, Hydrol. Res., 41, 378–390, 2010.

Shamir, E.: The value and skill of seasonal forecasts for water resources management in the Upper Santa Cruz River basin, southern Arizona, J. Arid Environ., 137, 35–45, https://doi.org/10.1016/j.jaridenv.2016.10.011, 2017.

Statistiska centralbyrån: Electricity supply, district heating and supply of natural gas 2015, Report EN11SM1601, Statistiska centralbyrån, Stockholm, Sweden, 2016.

Wood, A. W. and Lettenmaier, D. P.: An ensemble approach for attribution of hydrologic prediction uncertainty, Geophys. Res. Lett., 35, L1401, https://doi.org/10.1029/2008GL034648, 2008.

Wood, A. W., Maurer, E. P., Kumar, A., and Lettenmaier, D.: Longrange experimental hydrologic forecasting for the eastern United States, J. Geophys. Res., 107, 4429, https://doi.org/10.1029/2001JD000659, 2002.

Wood, A. W., Hopson, T., Newman, A., Brekke, L., Arnold, J., and Clark, M.: Quantifying Streamflow Forecast Skill Elasticity to Initial Condition and Climate Prediction Skill, J. Hydrometeorol., 17, 651–668, 2015.

Wood, A. W., Hopson, T., Newman, A., Brekke, L., Arnold, J., and Clark, M.: Quantifying streamflow forecast skill elasticity to initial condition and climate prediction skill, J. Hydrometeorol., 17, 651–668, https://doi.org/10.1175/JHM-D-14-0213.1, 2016.

Yang, W., Andréasson, J., Graham, L. P., Olsson, J., Rosberg, J., and Wetterhall, F.: Distribution-based scaling to improve usability of regional climate model projections for hydrological climate change impacts studies, Hydrol. Res., 41, 211–229, 2010.

Yossef, N. C., Winsemius, H., Weerts, A., van Beek, R., and Bierkens, M. F. P.: Skill of a global seasonal streamflow forecasting system, relative roles of initial conditions and meteorological forcing, Water Resour. Res., 49, 4687–4699, https://doi.org/10.1002/wrcr.20350, 2013.

Yuan, X.: An experimental seasonal hydrological forecasting system over the Yellow River basin – Part 2: The added value from climate forecast models, Hydrol. Earth Syst. Sci., 20, 2453–2466, https://doi.org/10.5194/hess-20-2453-2016, 2016.

Yuan, X. and Wood, E. F.: Downscaling precipitation or bias-correcting streamflow? Some implications for coupled general circulation model (CGCM)-based ensemble seasonal hydrologic forecast, Water Resour. Res., 48, 1–7, https://doi.org/10.1029/2012WR012256, 2012.

Yuan, X., Wood, E. F., Roundy, J. K., and Pan, M.: CFSv2-Based seasonal hydroclimatic forecasts over the conterminous United States, J. Climate, 26, 4828–4847, https://doi.org/10.1175/JCLI-D-12-00683.1, 2013.

Yuan, X., Roundy, J. K., Wood, E. F., and Sheffield, J.: Seasonal forecasting of global hydrologic extremes?: System development and evaluation over GEWEX basins, B. Am. Meteorol. Soc., 96, 1895–1912, https://doi.org/10.1175/BAMS-D-14-00003.1, 2015.

Zhao, T., Bennett, J. C., Wang, Q. J., Schepen, A., Wood, A. W., Robertson, D. E., and Ramos, M.-H.: How suitable is quantile mapping for postprocessing GCM precipitation forecasts?, J. Climate, 30, 3185–3196, https://doi.org/10.1175/JCLI-D-16-0652.1, 2017.

6

The WACMOS-ET project – Tower-scale evaluation of four remote-sensing-based evapotranspiration algorithms

D. Michel[1], C. Jiménez[2,3], D. G. Miralles[4,5], M. Jung[6], M. Hirschi[1], A. Ershadi[7], B. Martens[5], M. F. McCabe[7], J. B. Fisher[8], Q. Mu[9], S. I. Seneviratne[1], E. F. Wood[10], and D. Fernández-Prieto[11]

[1]Institute for Atmospheric and Climate Science, ETH Zürich, Zürich, Switzerland
[2]Estellus, Paris, France
[3]LERMA, Paris Observatory, Paris, France
[4]Department of Earth Sciences, VU University Amsterdam, Amsterdam, the Netherlands
[5]Laboratory of Hydrology and Water Management, Ghent University, Ghent, Belgium
[6]Max Planck Institute for Biogeochemistry, Jena, Germany
[7]Division of Biological and Environmental Sciences and Engineering, King Abdullah University of Science and Technology, Thuwal, Saudi Arabia
[8]Jet Propulsion Laboratory, California Institute of Technology, Pasadena, California, USA
[9]Department of Ecosystem and Conservation Sciences, University of Montana, Missoula, Montana, USA
[10]Department of Civil and Environmental Engineering, Princeton University, Princeton, New Jersey, USA
[11]ESRIN, European Space Agency, Frascati, Italy

Correspondence to: D. Michel (dominik.michel@env.ethz.ch)

Abstract. The WAter Cycle Multi-mission Observation Strategy – EvapoTranspiration (WACMOS-ET) project has compiled a forcing data set covering the period 2005–2007 that aims to maximize the exploitation of European Earth Observations data sets for evapotranspiration (ET) estimation. The data set was used to run four established ET algorithms: the Priestley–Taylor Jet Propulsion Laboratory model (PT-JPL), the Penman–Monteith algorithm from the MODerate resolution Imaging Spectroradiometer (MODIS) evaporation product (PM-MOD), the Surface Energy Balance System (SEBS) and the Global Land Evaporation Amsterdam Model (GLEAM). In addition, in situ meteorological data from 24 FLUXNET towers were used to force the models, with results from both forcing sets compared to tower-based flux observations. Model performance was assessed on several timescales using both sub-daily and daily forcings. The PT-JPL model and GLEAM provide the best performance for both satellite- and tower-based forcing as well as for the considered temporal resolutions. Simulations using the PM-MOD were mostly underestimated, while the SEBS performance was characterized by a systematic overestimation. In general, all four algorithms produce the best results in wet and moderately wet climate regimes. In dry regimes, the correlation and the absolute agreement with the reference tower ET observations were consistently lower. While ET derived with in situ forcing data agrees best with the tower measurements ($R^2 = 0.67$), the agreement of the satellite-based ET estimates is only marginally lower ($R^2 = 0.58$). Results also show similar model performance at daily and sub-daily (3-hourly) resolutions. Overall, our validation experiments against in situ measurements indicate that there is no single best-performing algorithm across all biome and forcing types. An extension of the evaluation to a larger selection of 85 towers (model inputs resampled to a common grid to facilitate global estimates) confirmed the original findings.

1 Introduction

Research on climate variability and the development of predictive capabilities relies largely on globally available reference data time series of the various components of the energy and water cycles. Turbulent fluxes of sensible and latent heat determine the development of the planetary boundary layer and thus govern the interactions between the Earth surface and the atmosphere. Evapotranspiration (ET) represents the time-integrated flux of latent heat and is an essential component of the energy and water cycle, playing a key role in meteorology and climate as well as agriculture (see, e.g., Ershadi et al., 2014).

Historically, there has been a lack of reliable estimates of turbulent fluxes, since the partitioning of the available energy at the Earth's surface into these fluxes is complex and characterized by large spatial and temporal variability. Also, these components of the energy balance cannot be monitored directly on a global scale by remote sensing techniques. Thus, efforts to produce satellite-based estimates typically involve combining multi-sensor data sets with predictive formulations of varying complexity, ranging from relatively simple empirical formulations to more complex modeling approaches (see, e.g., Courault et al. (2005), Kalma et al. (2008) and Wang and Dickinson (2012) for comprehensive reviews). In recent years, such efforts have generated global ET products (Mu et al., 2007; Fisher et al., 2008; Jung et al., 2010; Zhang et al., 2010; Vinukollu et al., 2011; Miralles et al., 2011a) that have typically been evaluated by comparing them individually to in situ data or by inter-comparing them with other existing global heat flux estimates. For example, within the Evaluation and inter-comparison of existing land evapotranspiration products (Landflux-Eval initiative) initiative of the Global Energy and Water cycle Exchanges (GEWEX) Data and Assessments Panels (GDAP), Jiménez et al. (2011) investigated 3 years (1993–1995) of global sensible and latent fluxes from a selection of 12 products, including satellite-based estimates, atmospheric reanalyses and offline land surface model simulations, while Mueller et al. (2011) examined a total of 30 observation-based ET estimates from similar sources over the longer period of 1989–1995, while also providing a comparison with global climate model simulations contributing to the Intergovernmental Panel on Climate Change (IPCC) fourth assessment report. More recently, Mueller et al. (2013) extended the previous LandFlux-EVAL evaluations and presented two monthly global ET synthesis products, merged from individual data sets spanning the periods 1989–1995 and 1989–2005. Based on the Mueller et al. (2013) methodology, Mao et al. (2015) synthesized a global ET time series for the period 1982–2010 based on a set of diagnostic ET products (including data sets produced with PT-JPL (Priestley–Taylor Jet Propulsion Laboratory model) and GLEAM (Global Land Evaporation Amsterdam Model)) to investigate the role of anthropogenic and climatic controls on ET trends. For the period 1982–2013 Zhang et al. (2015) produced a satellite-based global ET data set using remote sensing data and daily surface meteorology records for the investigation of multidecadal changes in ET, which was validated against precipitation and discharge records.

The GEWEX-LandFlux initiative is currently working towards producing an observation-based data set of heat fluxes that can be used together with related GDAP products to enable a joint analysis of the water and energy cycles (Jimenez et al., 2012). To contribute towards that goal, the European Space Agency (ESA) has conducted the Water Cycle Multi-mission Observation Strategy EvapoTranspiration project (WACMOS-ET), aimed at the identification of appropriate algorithms to develop regional and global ET products. WACMOS-ET efforts, which aimed at maximizing the use of European Earth Observation assets, have also included the compilation of a multi-sensor data set to run the ET methodologies for a 3-year period (2005-2007).

In WACMOS-ET, the methodologies by Su (2002) (Surface Energy Balance System, hereafter referred to as SEBS), Mu et al. (2011) (Penman–Monteith algorithm from the MODerate resolution Imaging Spectroradiometer (MODIS) evaporation product, PM-MOD), Fisher et al. (2008) (Priestley–Taylor Jet Propulsion Laboratory model, PT-JPL) and Miralles et al. (2011a, b) (Global Land Evaporation Amsterdam Model, GLEAM) were selected to produce ET estimates on different temporal and spatial scales. The same algorithms have also been examined at a selection of different tower sites in a recent paper by McCabe et al. (2015) in preparation for the GEWEX-LandFlux product. In McCabe et al. (2015) the algorithms are run at 3-hourly time steps with both point-scale inputs (from tower meteorological observations) and gridded inputs (from the GEWEX-LandFlux global forcing data set) over a longer time period. Here, the ET algorithms are run with the WACMOS-ET forcings (see Sect. 2.2) and the analyses of model performance are extended to evaluate different timescales (3-hourly and daily) and time integrations (nighttime, daytime and full day). In a companion paper, Miralles et al. (2016) present the second part of the WACMOS-ET study, in which PT-JPL, GLEAM and PM-MOD are evaluated on the global scale.

The paper is structured as follows. First, the WACMOS-ET input data set is described in detail, together with the tower flux data used for driving and evaluating the ET models. This is followed by an evaluation of ET model performance on the tower scale using the tower eddy-covariance (EC) fluxes as the reference data set. The model evaluation is first performed over a selection of 24 stations covering nine biomes in three continents (Europe, North America and Australia), in which models are run based on in situ and remote sensing forcing. Then the validation is extended to embrace a larger sample of 85 towers, in which models are driven only by satellite data resampled to a common grid. Finally, the main findings of our model evaluation on the pixel scale are summarized.

2 Methods and data

Here the ET methodologies comprising the WACMOS-ET effort together with the input data sets that have been compiled to run the models and evaluate the ET estimates are presented. A summary of the data sets and the model-specific forcing requirements is provided in Table 1.

2.1 ET models

The four algorithms selected here estimate the fraction of the available energy at the surface used by the soil and canopy evaporation processes. Therefore, the available energy (i.e., the difference between the surface net radiation and the ground heat flux) is a key input for all algorithms. However, this evaporative fraction is parameterized differently by each model. SEBS is an energy balance model (Su, 2002) based on a detailed parameterization of the sensible heat flux at the surface, where ET is estimated as the residual of the surface energy balance once the sensible heat flux is calculated. Therefore, key inputs are the surface temperature and the temperature gradient between the surface and the air, and a key component of the model is the aerodynamic resistance to sensible heat transfer. PM-MOD (Mu et al., 2011) derives ET directly based on the Penman–Monteith equation (Monteith, 1965), which relates the latent heat flux to the vapor pressure deficit between the surface and the overlying air and requires a resistance parameter to characterize the canopy transpiration. PT-JPL (Fisher et al., 2008) and GLEAM (Miralles et al., 2011a) are based on first determining the potential evaporation by applying the Priestley and Taylor equation (Priestley and Taylor, 1972), followed by reducing the potential evaporation to actual evaporation with a number of evaporative stress factors. The derivation of these stress factors is different between both models. PT-JPL requires the vapor pressure deficit, relative humidity and visible-infrared-related vegetation indexes, while GLEAM combines microwave data of vegetation optical depth and soil moisture. A more detailed description of the input requirements for each model can be found in Table 1, while a more comprehensive description of each individual model is given in the following sections.

2.1.1 SEBS

SEBS (Su, 2002) is a one-source energy balance algorithm that is arguably one of the most widely used energy balance approaches to derive turbulent fluxes. The SEBS model calculates the sensible heat flux (H) based on the Monin and Obukhov theory (Monin and Obukhov, 1954):

$$\theta_0 - \theta_a = \frac{H}{k u_* \rho c_p}\left[\ln\left(\frac{z-d_0}{z_{0h}}\right) - \Gamma_h\left(\frac{z-d_0}{L}\right) + \Gamma_h\left(\frac{z_{0h}}{L}\right)\right], \quad (1)$$

$$u = \frac{u_*}{k}\left[\ln\left(\frac{z-d_0}{z_{0m}}\right) - \Gamma_m\left(\frac{z-d_0}{L}\right) + \Gamma_m\left(\frac{z_{0h}}{L}\right)\right], \quad (2)$$

$$L = -\frac{\rho c_p u_*^3 \theta_v}{k g H}, \quad (3)$$

where u is the wind speed, u_* is the friction velocity, k is the von Kármán constant, z is the height above the surface, d_0 is the zero plane displacement height, z_{0m} and z_{0h} are the roughness heights for momentum and heat transfer, and Γ_m and Γ_h are the stability correction functions for momentum and sensible heat transfer, respectively. L refers to the Obukhov length, ρ is the air density, θ_0 is potential land surface temperature, θ_a is the potential air temperature at height z, g is the gravity acceleration and θ_v is the potential virtual air temperature at level z. When the suitable data are available, the only unknowns are H, u_* and L. This allows the calculation of H and the further estimation of ET based on closing the energy balance at the surface, i.e., ET is estimated as the difference between net radiation (R_n) and the sum of the calculated H and ground flux (G).

Additionally, in order to constrain H estimates, two limiting cases are considered that set upper and lower bounds for the evaporative fraction. Under very dry conditions, ET becomes 0 and H is at its maximum, set by $R_n - G$. Under wet conditions, ET occurs at potential rates and therefore H is at its minimum. In this wet case, H is calculated via reverse application of the Penman–Monteith equation (see Sect. 2.1.2) assuming that the surface resistance is zero.

SEBS has been extensively validated against tower measurements and has proved to estimate realistic evaporation rates on a variety of scales, ranging from local to regional (Jia et al., 2003; Su et al., 2005; McCabe and Wood, 2006). As an example, Chen et al. (2015) recently reported an average correlation of ~ 0.8 and root mean square difference (RMSD) of 0.7 mm day^{-1} against eddy-covariance measurements in a validation of monthly SEBS ET aggregates over China.

As a one-source energy model, SEBS does not separate the contributing components of ET (i.e., transpiration, interception loss, bare-soil evaporation), unlike the other models studied in WACMOS-ET, which provide independent estimates of these vapor sources. Although not examined here, we note that two-source energy balance models can also treat the soil and vegetation components separately (e.g., Kustas and Norman, 2000; Anderson et al., 2007) but have had limited application on the global scale.

2.1.2 PM-MOD

PM-MOD (Mu et al., 2011) is based on the Penman–Monteith equation. It estimates ET as the sum of interception loss (I), transpiration (ET_t) and evaporation from the soil (ET_s). The interception loss is modeled as

Table 1. Table summarizing the model inputs. Listed are the main inputs, product selected, original temporal and spatial resolutions, and the satellite sensors used to derive the product.

Variable	Models	Product	Resolution	Sensors
Surface radiation	All	SRB	3-hourly, 100 km	Several VIS-IR sensors
Surface temperature	SEBS	IPMA		
polar			twice daily, 1 km	AATSR
geostationarity			hourly, 5 km	MSG-2, MT-SAT, GOES-12
Surface meteorology		ERA-Interim	3-hourly, 75 km	Assimilation of satellite and other meteo observations
temperature	All			
humidity	SEBS, PM-MOD, PT-JPL			
wind	SEBS			
*f*APAR, LAI	SEBS, PM-MOD, PT-JPL	From ESA GlobAlbedo	8 days, 1 km	VEGETATION, MERIS, MODIS
Soil moisture	GLEAM	ESA-CCI	daily, 25 km	SSM/I, TMI, AMSR-E, ASCAT
Precipitation	GLEAM	CMORPH	30 min, 15 km	AMSU-B, AMSR-E, TMI
Snow water	GLEAM	ESA	daily, 1 km	AMSR-E

*f*APAR: fraction of absorbed photosynthetically active radiation; VIS-IR:visible and infrared; IPMA: Portuguese Sea and Atmosphere Institute; AATSR: Advanced Along-Track Scanning Radiometer; MSG-2: Meteosat Second Generation 2; MT-SAT: Multi-functional Transport Satellites; GOES-12: Geostationary Operational Environmental Satellite 12; VEGETATION: SPOT (Satellite Pour l'Observation de la Terre) vegetation; MERIS: Medium Resolution Imaging Spectrometer; SSM/I: Special Sensor Microwave Imager; TMI: TRMM (Tropical Rainfall Measuring Mission) Microwave Imager; AMSR-E: Advanced Microwave Scanning Radiometer – Earth Observing System; ASCAT: Advanced Scatterometer; AMSU-B: Advanced Microwave Sounding Unit-B.

$$I = f_{\mathrm{wet}} f_{\mathrm{c}} \frac{\Delta\left(R_{\mathrm{n}}-G\right)+\rho c_p \mathrm{VPD}/r_{\mathrm{a}}^{\mathrm{wc}}}{\lambda\left(\Delta+\gamma \frac{r_{\mathrm{s}}^{\mathrm{wc}}}{r_{\mathrm{a}}^{\mathrm{wc}}}\right)}, \tag{4}$$

where Δ is the slope of the curve relating saturated water vapor pressure to temperature, VPD is the vapor pressure deficit, γ is the psychrometric constant, f_{c} is the canopy fraction, f_{wet} is the wet cover fraction based on the derivation by Fisher et al. (2008) (see Eq. (9) in Sect. 2.1.3), and $r_{\mathrm{a}}^{\mathrm{wc}}$ and $r_{\mathrm{s}}^{\mathrm{wc}}$ are aerodynamic and surface resistances against evaporation of intercepted water (calculated as a function of air temperature and leaf area index, LAI).

Canopy transpiration is estimated as

$$\mathrm{ET}_{\mathrm{t}} = \left(1-f_{\mathrm{wet}}\right) f_{\mathrm{c}} \frac{\Delta\left(R_{\mathrm{n}}-G\right)+\rho c_p \mathrm{VPD}/r_{\mathrm{a}}^{\mathrm{t}}}{\lambda\left(\Delta+\gamma \frac{r_{\mathrm{s}}^{\mathrm{t}}}{r_{\mathrm{a}}^{\mathrm{t}}}\right)}, \tag{5}$$

where $r_{\mathrm{a}}^{\mathrm{t}}$ and $r_{\mathrm{s}}^{\mathrm{t}}$ are the aerodynamic and surface resistances against transpiration. $r_{\mathrm{a}}^{\mathrm{t}}$ is determined in a similar way to $r_{\mathrm{a}}^{\mathrm{wc}}$, and $r_{\mathrm{s}}^{\mathrm{t}}$ is a function of stomatal conductance, biome-constant values of cuticular conductance and canopy boundary-layer conductance. The values of stomatal conductance are a function of air temperature, VPD and LAI.

Evaporation from the soil surface is the sum of evaporation from wet soil and evaporation from saturated soil, which are both calculated separately based on Eq. (7) with specific values of aerodynamic and surface resistances for bare soils and a soil moisture constraint (f_{sm}) depending on relative humidity (taken from Fisher et al. (2008), see Sect. 2.1.3).

The Mu et al. (2011) daily ET estimates have been previously validated against EC measurements from 46 FLUXNET towers in North America, reporting for the daily estimates an average RMSD of $\sim 0.9\,\mathrm{mm\,day}^{-1}$ and a ~ 0.6 average correlation coefficient.

2.1.3 PT-JPL

PT-JPL (Fisher et al., 2008) is based upon the Priestley and Taylor equation. As in PM-MOD, ET is estimated as the sum of interception loss I, transpiration ET_{t} and evaporation from the soil ET_{s}. The driving equations in the model are

$$\mathrm{ET}_{\mathrm{t}} = \left(1-f_{\mathrm{wet}}\right) f_{\mathrm{g}} f_T f_{\mathrm{M}} \alpha \frac{\Delta}{\Delta+\gamma} R_{\mathrm{n}}^{\mathrm{c}}, \tag{6}$$

$$\mathrm{ET}_{\mathrm{s}} = f_{\mathrm{wet}} + f_{\mathrm{sm}}\left(1-f_{\mathrm{wet}}\right) \alpha \frac{\Delta}{\Delta+\gamma}\left(R_{\mathrm{n}}^{\mathrm{s}}-G\right), \tag{7}$$

$$I = f_{\mathrm{wet}} \alpha \frac{\Delta}{\Delta+\gamma} R_{\mathrm{n}}^{\mathrm{c}}, \tag{8}$$

where α is known as the Priestley and Taylor (PT) coefficient and is considered here as a constant value (1.26) (Priestley and Taylor, 1972) that aims to summarize the atmospheric term in the Penman–Monteith equation (Eq. 5), λ is the latent heat of vaporization and f_{wet}, f_{g}, f_{M}, f_{sm} and f_T are ecophysiological constraint functions with values between 0 and 1 referred to as f functions. The f functions are given by

$$f_{\mathrm{wet}} = \mathrm{RH}^4, \tag{9}$$

$$f_{\mathrm{g}} = f\mathrm{APAR}/f\mathrm{IPAR}, \tag{10}$$

$$f_{\mathrm{M}} = f\mathrm{APAR}/f\mathrm{APAR}_{\max}, \tag{11}$$

$$f_{\mathrm{sm}} = \mathrm{RH}^{\mathrm{VPD}}, \tag{12}$$

$$f_T = e^{-\left(\frac{T_{\mathrm{s}}-T_{\mathrm{opt}}}{T_{\mathrm{opt}}}\right)^2}, \tag{13}$$

where f_{wet} is the relative surface wetness, f_{g} is green canopy fraction, $f\mathrm{APAR}$ ($f\mathrm{IPAR}$) is the fraction of absorbed (intercepted) photosynthetically active radiation, f_{M} is a plant moisture constraint, $f\mathrm{APAR}_{\max}$ is the maximum of $f\mathrm{APAR}$, f_{sm} is a soil moisture constraint, f_T is a plant temperature constraint and T_{opt} is the optimum plant growth temperature, estimated as the air temperature at the time of peak canopy activity when the highest $f\mathrm{APAR}$ and minimum VPD occur. Note that as the input data set does not include $f\mathrm{IPAR}$; $f\mathrm{IPAR}$ is derived from the rescaled project LAI by inverting the model original relationships between LAI and $f\mathrm{IPAR}$.

Using this methodology, monthly estimates of ET were tested against EC measurements from 16 FLUXNET towers worldwide (Baldocchi et al., 2001) with a reported average RMSD of $\sim 0.4\,\mathrm{mm\,day^{-1}}$ and a ~ 0.9 average correlation coefficient (Fisher et al., 2008). Note that unlike the above statistics reported for SEBS and PM-MOD, these numbers come from the model run with the tower meteorology, instead of global forcings.

2.1.4 GLEAM

GLEAM (Miralles et al., 2011a, b) calculates ET via the PT equation, a soil moisture stress computation and a Gash analytical model of rainfall interception loss (Gash, 1979). In the absence of snow, evaporation from land is calculated as

$$\mathrm{ET} = \mathrm{ET}_{\mathrm{tc}} + \mathrm{ET}_{\mathrm{sc}} + \mathrm{ET}_{\mathrm{s}} + \beta I, \tag{14}$$

where $\mathrm{ET}_{\mathrm{tc}}$ is transpiration from tall canopy, $\mathrm{ET}_{\mathrm{sc}}$ is transpiration from short vegetation, ET_{s} is soil evaporation and I is tall canopy interception loss. β is a constant used to account for the times at which vegetation is wet; thus, transpiration occurs at lower rates ($\beta = 0.93$) (Gash and Stewart, 1977).

The first three terms in Eq. (14) are derived using the Priestley and Taylor equation, so ET becomes

$$\mathrm{ET} = \frac{\Delta\left[f_{\mathrm{tc}} S_{\mathrm{tc}} \alpha_{\mathrm{tc}} \left(R_{\mathrm{n}}^{\mathrm{tc}} - G_{\mathrm{tc}}\right) + f_{\mathrm{sc}} S_{\mathrm{sc}} \alpha_{\mathrm{sc}} \left(R_{\mathrm{n}}^{\mathrm{sc}} - G_{\mathrm{sc}}\right) + f_{\mathrm{s}} S_{\mathrm{s}} \alpha_{\mathrm{s}} \left(R_{\mathrm{n}}^{\mathrm{s}} - G_{\mathrm{s}}\right)\right]}{\lambda(\Delta + \gamma)} + \beta I,, \tag{15}$$

where the subscripts "tc", "sc" and "s" correspond to tall vegetation, short vegetation and bare soil (respectively) and the fraction of each of these three cover types per pixel is represented by f. Different cover types have different values of α and parameterizations of G; additionally, R_{n} is distributed within the cover fractions using average albedo ratios from the literature. S represents the evaporative stress due to soil moisture deficit and vegetation phenology. Soil moisture deficit is estimated using a multilayer running water balance to describe the infiltration of observed precipitation through the vertical soil profile. Microwave soil moisture observations are assimilated into the soil profile (Martens et al., 2016). In vegetated land covers, phenology effects on ET are based on microwave observations of vegetation optical depth, used as a proxy of vegetation water content.

I is independently derived using a Gash analytical model (Gash, 1979), in which a running water balance for canopies and trunks is driven by observations of precipitation. The derivation of the parameters and the global implementation and validation of this I model are described in Miralles et al. (2010). For regions covered by ice and snow, sublimation is calculated based on a PT equation parameterized for ice and supercooled waters (Murphy and Koop, 2005).

The ET estimates from GLEAM have been validated against eddy covariance towers worldwide; reported average correlations are 0.83 and 0.90 for daily and monthly estimates, respectively, with an average RMSD of $\sim 0.3\,\mathrm{mm\,day^{-1}}$, based on a sample of 43 towers (Miralles et al., 2011a), and correlations of 0.71–0.75 and 0.81–0.86 for daily and monthly estimates, respectively, based on a sample of 163 towers and different satellite products as forcing (Miralles et al., 2016).

2.2 Model inputs

2.2.1 Surface meteorology

The European Centre for Medium-range Weather Forecasts (ECMWF) Re-Analysis-Interim (ERA-Interim) (Dee et al., 2011) was selected to provide the near-surface meteorology every 3 h at a spatial resolution of ~ 75 km. The use of reanalysis data is necessary as satellite observations are generally unable to retrieve the surface variables needed, such as temperature, humidity and wind speed, with sufficient accuracy or at a suitable sub-daily temporal resolution. Although products of near-surface air temperature and humidity derived from satellite sounders exist (Ferguson and Wood, 2010), atmospheric reanalyses have the advantage of providing regular sub-daily estimates for all weather conditions. ERA-Interim is also used in the derivation of the land surface temperature product (see Sect. 2.2.2), to ensure inter-product consistency between air and surface temperatures. In terms of accuracy, ERA-Interim data have been evaluated through comparison with other reanalyses and weather stations over specific areas, showing a good general performance (e.g., Mooney et al., 2011; Szczypta et al., 2011).

2.2.2 Land surface temperature

Land surface temperature (LST) estimates have been internally generated by the project from level 1 radiances from the Advanced Along-Track Scanning Radiometer (AATSR) onboard ESA's Environmental Satellite (EnviSat) polar-orbiting satellite, from Multi-functional Transport Satellites (MT-SAT) 2 (over Australia), Meteosat Second Generation (MSG) 2 and Geostationary Operational Environmental Satellite (GOES) 12. The data sets are provided over a sinusoidal grid with 1 km resolution for AATSR at the two satellite overpasses per day (~ 10:00 LT descending node) and 5 km for the remaining sensors (1-hourly estimates for MSG and MT-SAT, 3-hourly for GOES). Ancillary atmospheric information for the inversion of the L1 radiances comes from ERA-Interim. Estimates of surface emissivity are taken from the Global Infrared Land Surface Emissivity UW-Madison Baseline Fit Emissivity Database developed by Seemann et al. (2008).

2.2.3 Surface radiation

The National Aeronautics and Space Administration (NASA)–GEWEX Surface Radiation Budget (SRB) satellite product version 3.1 (Stackhouse et al., 2004) is used to provide the surface net radiation input to the ET models. The SRB product is used by a large number of global ET algorithms to characterize the radiation at the surface, given its relatively long data record and sub-daily temporal resolution. SRB data sets include global 3-hourly averages of surface and top-of-atmosphere longwave and shortwave radiative parameters on a ~ 100 km grid.

2.2.4 LAI and *f*APAR

To characterize the vegetation state using visible and near-infrared wavebands, estimates of LAI and *f*APAR have been derived by applying the Joint Research Centre (JRC) two-stream inversion package (hereafter TIP) (Pinty et al., 2007, 2011a, b) to the ESA GlobAlbedo bihemispherical reflectances. Here, LAI is defined as the one-sided leaf area per unit ground area and *f*APAR as the fraction of absorbed photosynthetically active radiation in the 400–700 nm region.

The application of the TIP LAI and *f*APAR with our ET models required some LAI and *f*APAR calibration. The TIP LAI is a one-dimensional (1-D) equivalent LAI for solving the radiative transfer in a three-dimensional (3-D) medium, and it is consistent with the fluxes from which it is derived. It is not consistent with LAI derived using a 3-D radiative scheme that allows some form of horizontal canopy clumping (e.g., the MODIS MOD15A2 LAI product). In practical terms, this means that if an ET model was constructed to use a MODIS-like LAI and *f*APAR, a straight use of the project LAI and *f*APAR will result in the ET model producing lower values than expected for those biomes where horizontal clumping is significant (e.g., for forests). While the ET dynamics may not be highly affected (there is a high degree of correlation between different LAI and *f*APAR estimates), the absolute values would be. As the SEBS, PM-MOD and PT-JPL models have typically been used with the MODIS vegetation product, a rescaling of our TIP-derived LAI and *f*APAR products against the MODIS product has been undertaken. For running the models on the tower scale, a local rescaling is conducted by a linear regression between the MOD15A2 and the TIP values co-registered at each tower. For global model simulations, individual rescaling per biome or climate classification is conducted. For PT-JPL, given the model internal relationships between these variables and the vegetation indexes used as model inputs (see Table 1 in Fisher et al., 2008), it can be discussed whether the original TIP LAI and *f*APAR or the rescaled LAI and *f*APAR are the most appropriate to be used as model inputs. For simplicity we will apply also the rescaled LAI and *f*APAR, but this choice will be further evaluated in future applications of the model with the TIP LAI and *f*APAR inputs.

Figure 1 illustrates an example of both products at two towers. The station Quebec – Eastern Boreal, Mature Black Spruce (CA-Qfo) is located in an evergreen needleleaf forest and shows that the MOD15A2 and WACMOS-ET LAI and *f*APAR absolute values differ considerably. This is expected, as allowing some form of horizontal clumping (MODIS 3-D radiative transfer scheme) or not (TIP 1-D) can result in large differences in the estimated LAI and *f*APAR in forests. It can be seen that the local calibration of the MODIS-like product retains the dynamics of the WACMOS-ET product, while adding absolute values close to the MODIS product. The station Brookings (US-Bkg) is situated in a cropland area, where the effects of clumping are much less severe, and the different LAI and *f*APAR values are much closer.

2.2.5 Vegetation height

Vegetation height on the global scale is required by SEBS. For shrubland and forest biomes the product developed by Simard et al. (2011) was used as static canopy height cover. For grassland and cropland biomes, where the temporal dynamics of canopy height can be more considerable, we approximated canopy height with the method by Chen et al. (2015), with the minimum and maximum canopy height obtained from the static vegetation table of the North American Land Data Assimilation System (NLDAS).

2.2.6 Soil moisture and vegetation optical depth

A soil moisture product combining observations from active and passive microwave sensors has been developed as part of the ESA Climate Change Initiative (CCI) and is adopted here to provide the surface soil moisture data that are assimilated into GLEAM. Details on the product algorithm and evaluation can be found in Liu et al. (2011b). The data are provided on a regular grid with a resolution of $0.25° \times 0.25°$. The product performs well in moderately vegetated regions but shows higher uncertainties in densely vegetated regions (as vegetation attenuates the microwave signal from the ground)

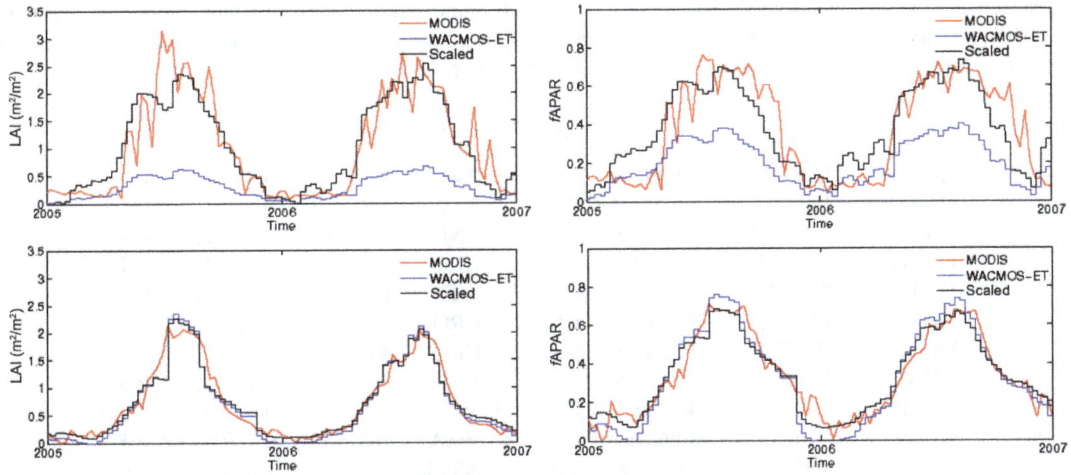

Figure 1. Time series for 2005–2006 of MODIS MOD15A2 LAI and fAPAR, WACMOS-ET LAI and fAPAR, and the MODIS-like LAI and fAPAR (referred to as "scaled" in the figures) at the tower stations CA-Qfo (top panels) and US-Bkg (bottom panels).

and mountainous areas (due to the high surface roughness) (Liu et al., 2011b).

The retrieval of soil moisture from passive sensors discussed in Sect. 2.2.6 can be accompanied by an estimation of the vegetation optical depth (VOD). VOD can be used to account for the development of vegetation over the year as it is a good proxy of vegetation water content (Liu et al., 2015). Although most ET models traditionally use parameters derived from visible and near-infrared wavelengths, microwave VOD is used by GLEAM. Here the long-term record by Liu et al. (2011a) based on the application of the Land Parameter Retrieval Model by Owe et al. (2001) is used by GLEAM.

2.2.7 Precipitation and snow

Observations of precipitation and snow water equivalent are also required by GLEAM only. Precipitation is used both to estimate the effects of soil water limitations on ET and to calculate interception loss. To run the model on the tower scale we use the Climate Prediction Center (CPC) Morphing Technique (CMORPH) (Joyce et al., 2004). CMORPH transports the features of precipitation estimates derived from low orbiter satellite microwave observations using information from geostationary satellite infrared (IR) data. Precipitation estimates are available every 30 min on a grid with a spacing of 8 km at the Equator, although the resolution of the individual satellite-derived estimates is coarser at $\sim 12\,\text{km} \times 15\,\text{km}$. The spatial coverage ranges from 60° N–60° S. To run the model globally, we use the National Centers for Environmental Prediction (NCEP) Climate Forecast System Reanalysis (CFSR) land precipitation estimates (Coccia and Wood, 2015). These precipitation estimates come from the hourly CFSR output (Saha et al., 2010) but are corrected using the observation-based data sets of the CPC (Xie and Arkin, 1997) and the Global Precipitation Climatology Project (GPCP) (Adler et al., 2003). Finally, snow water equivalent estimates come from ESA GlobSnow. Since GlobSnow covers the Northern Hemisphere only, data from the National Snow and Ice Data Center (NSIDC) are used in snow-covered regions of the Southern Hemisphere (Kelly et al., 2003). The product combines satellite passive microwave measurements with ground-based weather station data in a data assimilation scheme (Luojus and Pulliainen, 2010). The products exist at a daily resolution and a spatial resolution of 25 km.

2.3 Tower data

2.3.1 Tower selection

Model simulations are evaluated by comparison with the turbulent latent fluxes measured by the eddy-covariance technique at a selection of tower sites from FLUXNET (Baldocchi et al., 2001). A first sample of towers was compiled by selecting those stations from the FLUXNET La Thuile synthesis data set which contain latent flux measurements in the 2005–2007 period, as well as the meteorological and radiation inputs required to run the ET models at the towers' locations. The 24 selected stations are described in Table 2 and their geographical location is displayed in Fig. 2. While some meteorological variables such as near-surface air temperature or humidity are measured at nearly all towers, other inputs such as the surface net radiation or the ground heat flux are measured at only a few towers. Some stations that were very close to the shore or in places with regular flooding were discarded. The final selection of 24 towers represents a significant number of biomes and a reasonable sample of dry and wet climate regimes.

In a later step, by removing the constraint of requiring local measurements of all the model inputs, the first selection of 24 towers is extended to a total of 85 stations. This second selection is used to evaluate model performance when

Table 2. Stations selected to run the models with tower inputs. From left to right: station name; longitude; latitude; Köppen–Geiger Climate Classification (KGCC); International Geosphere–Biosphere International Programme (IGBP) land cover; total number of days with data, no precipitation number of days with data; evaporative fraction (EF) for the DJF, MAM, JJA, SON 3-month periods.

Name	Long	Lat	KGCC	IGBP	Days	EF
AU-How	131.15° E	12.49° S	Aw	SV	114, 100	0.7, 0.7, 0.5, 0.3
CA-Ojp	104.69° W	53.92° N	Dfc	ENF	126, 101	0.2, 0.1, 0.3, 0.5
CA-Qfo	74.34° W	49.69° N	Dfc	ENF	253, 166	0.1, 0.1, 0.4, 0.4
DE-Geb	10.91° E	51.1° N	Cfb	CRO	188, 113	0.0, 0.4, 0.5, 0.7
DE-Har	7.60° E	47.93° N	Cfb	MF	105, 88	1.0, 0.5, 0.5, 0.7
DE-Kli	13.52° E	50.89° N	Cfb	CRO	275, 98	0.0, 0.5, 0.5, 0.0
DE-Meh	10.66° E	51.28° N	Cfb	CRO	444, 269	0.0, 0.3, 0.5, 0.5
DE-Wet	11.46° E	50.45° N	Cfb	ENF	384, 182	0.1, 0.3, 0.4, 0.6
IT-MBo	11.08° E	46.03° N	Dfb	MF	149, 126	0.0, 0.6, 0.7, 0.9
IT-Noe	8.15° E	40.6° N	Csa	WSA	182, 182	0.7, 0.3, 0.2, 0.3
NL-Ca1	4.93° E	51.97° N	Cfb	NVM	38, 22	1.0, 0.6, 0.6, 0.9
PT-Mi2	8.02° W	38.48° N	Csa	SV	275, 221	0.5, 0.4, 0.3, 0.4
RU-Fyo	32.92° E	56.46° N	Dfb	MF	374, 216	0.0, 0.4, 0.5, 0.4
US-ARM	97.49° W	36.61° N	Cfa	CRO	159, 131	0.4, 0.5, 0.3, 0.3
US-Aud	110.51° W	31.59° N	BSk	OSH	219, 219	0.5, 0.2, 0.3, 0.5
US-Bkg	96.84° W	44.35° N	Dfa	CRO	174, 172	0.6, 0.7, 0.9, 1.0
US-Bo2	88.29° W	40.01° N	Dfa	CRO	192, 192	0.3, 0.3, 0.6, 0.3
US-FPe	105.1° W	48.31° N	BSk	GRA	184, 184	1.0, 0.4, 0.6, 0.4
US-Goo	89.87° W	34.25° N	Cfa	NVM	183, 179	0.7, 0.6, 0.5, 0.6
US-MOz	92.2° W	38.74° N	Cfa	DBF	252, 252	0.3, 0.4, 0.5, 0.5
US-SRM	110.87° W	31.82° N	BSk	OSH	139, 137	0.2, 0.1, 0.3, 0.4
US-WCr	90.08° W	45.81° N	Dfb	DBF	338, 239	0.1, 0.2, 0.6, 0.4
US-Wkg	109.94° W	31.74° N	BSk	GRA	137, 137	0.2, 0.1, 0.2, 0.3
US-Wrc	121.95° W	45.82° N	Csb	ENF	146, 107	0.4, 0.3, 0.3, 0.5

KGCC abbreviations: Aw – tropical wet and dry; BSk – semiarid midlatitudes; Cfa – humid subtropical; Cfb – marine, mild winter; Csa – interior Mediterranean; Csb – coastal Mediterranean; Dfa – humid continental hot summer, wet all year; Dfb: humid continental mild summer, wet all year; Dfc – subarctic with cool summer, wet all year. IGBP abbreviations: SV – savannas; ENF – evergreen needleleaf forests; CRO – croplands; MF – mixed forests; WSA – woody savannas; OSH – open shrublands; GRA – grasslands; DBF – deciduous broadleaf forests; NVM – natural vegetation mosaic.

the models are run with the satellite data used for the global runs.

2.3.2 In situ surface energy balance

While, in principle, the surface energy balance should close at the tower, this is rarely the case: a lack of closure in the surface energy balance of about 10–30 % is commonly found when comparing the EC measurements against the energy balance residual (ER) term, i.e., the difference between net radiation and the sum of the sensible, latent and ground fluxes (e.g., Foken et al., 2006). Consequently, throughout the paper the model evaluation is discussed by comparing it with both the EC measurements and the in situ ER estimates.

2.3.3 In situ LST

To run SEBS, the broadband longwave radiometer measurements need to be converted into LST estimates. This is done by inverting the equation relating the upwelling spectral radiance measured by the radiometer and the LST. Broadband emissivity is required, and it is estimated from the MODIS-based Global Infrared Land Surface Emissivity Database (Seemann et al., 2008) operated by the Cooperative Institute for Meteorological Satellite Studies (CIMSS). The estimates are calculated by following the approach suggested by Wang et al. (2005) using a linear combination of narrowband emissivities at 8.5, 11 and 12 μm.

2.3.4 In situ vegetation height

SEBS also requires vegetation height to derive the surface roughness values. In most cases a mean annual value can be obtained from the tower metadata, and this value is adopted here as vegetation height at the tower. However, a clear limitation in this assumption is that it does not include dynamic changes in vegetation height over time. As discussed in Sect. 2.2, the importance of neglecting the temporal variability in height is biome-dependent; for instance, in forests the mean vegetation height is typically more constant than in, e.g., croplands, where the changes derived from agricultural practices can be large.

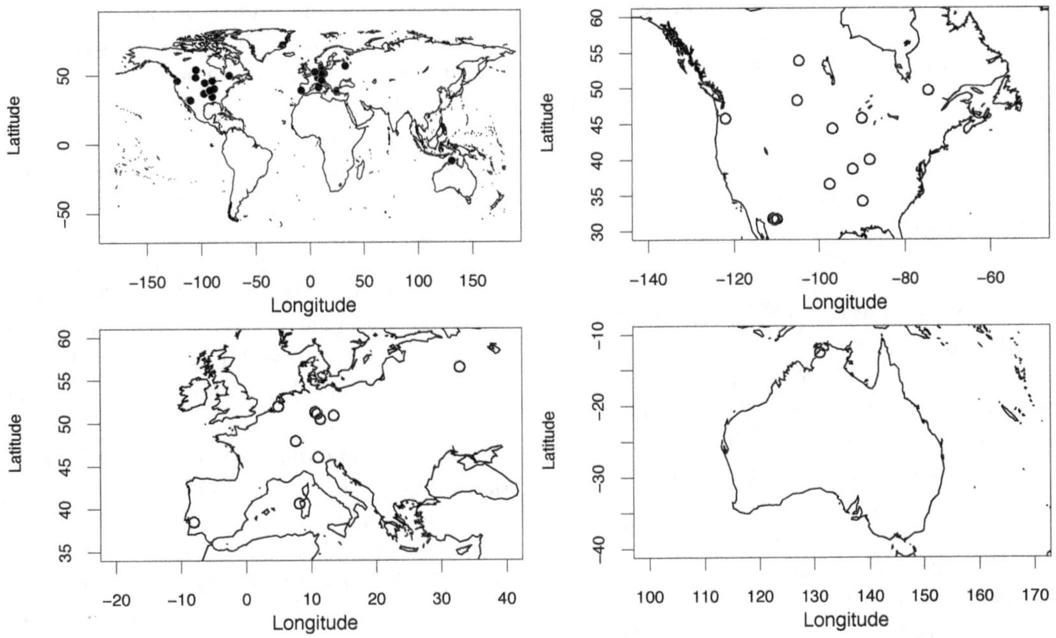

Figure 2. Location of the 24 FLUXNET stations used for the main analysis of the study. They are located on three different continents, encompassing nine different biomes, i.e., vegetation mosaic, croplands, mixed forests, deciduous forest, savanna, evergreen needleleaf forests, grasslands, woody savanna and shrublands.

2.4 ET experiments

2.4.1 Evaluation times

The model performance is investigated on sub-daily and daily timescales. The tower data are available at 0.5 h intervals and have been time-integrated to 3 hours in order to run the ET models at that sub-daily resolution. The satellite data have been time-matched to the 3-hourly or daily resolutions from their native resolution in different ways (see below), depending on the type of data and original resolution. The 3-hourly inputs were then aggregated to daily values in order to run the models with tower-based daily inputs. The tower data record is not always time-continuous, as in some instances there are gaps in the record. This is not a problem for the PM-MOD, PT-JPL and SEBS models because the ET estimates depends only on the instantaneous atmospheric or surface state. When inputs to the models and/or ET for the evaluation are missing, those three models are not run. Conversely, GLEAM requires continuous data records to update the soil moisture state variable. To facilitate running GLEAM with tower inputs, the tower measurements are gap-filled with the corresponding pixel data (see Sect. 2.2). ET estimates from these periods are removed after the runs, so, as before, only the time steps where tower forcing data are available are used for model evaluation.

The models are validated against the tower ET only under dry (non-raining) conditions, as EC gas analyzers are not reliable during rain events due to disturbance of the infrared signal by droplets on the sensor (Burba et al., 2010; Hirschi et al., 2016). Therefore, any days with precipitation as indicated by the tower or satellite precipitation are removed from the validation, and the interception component from PM-MOD, PT-JPL and GLEAM is not considered in the validations.

2.4.2 Nighttime ET

Only PM-MOD and GLEAM specifically deal with nighttime evaporation. Nevertheless, nighttime values are required from all models to integrate the 3-hourly ET estimates to daily values. For SEBS and PT-JPL negative nighttime estimates are set to 0 to allow the daily integration for those models. To separate day and night, daylight times are identified by calculating the solar zenith angle. Time intervals, where the cosine of the zenith angle is larger than 0.2, are kept as day values. This day and night separation may be less accurate than using a solar downward radiation threshold, but it allows a day–night flag for those stations without solar radiation measurements. The impact of setting ET from SEBS and PT-JPL to 0, as these models cannot specifically simulate nighttime conditions, is addressed in Sects. 3.1 and 3.2, where sub-daily periods, including daytime, are investigated.

2.4.3 ET production

The following ET estimates are generated to evaluate model performance.

- Tower-based ET: ET generated by the four models using the 3-hourly or daily in situ data (surface radiation, LST, air temperature, air humidity, and wind speed and precipitation), the scaled WACMOS-ET LAI and fAPAR and gridded soil moisture and VOD data. Note that the 3-hourly tower-based ET estimates are also time-integrated to daily values, so daily ET estimates exist both from the runs with daily inputs and from the integration of the 3-hourly ET estimates.
- Original-resolution satellite-based ET: ET generated by the four models using the 3-hourly or daily satellite data (SRB surface radiation, ERA-Interim air temperature, humidity, wind speed, CMORPH precipitation, ESA-CCI soil moisture, scaled WACMOS-ET LAI and fAPAR) at their original resolutions. In situ LST is still used here in order to have the same number of SEBS estimates as in the tower-based ET (cloudiness, satellite overpass time and revisiting times would have notably reduced the number of SEBS estimates if the satellite LST had been used). As for the tower-based ET, 3-hourly tower-based ET estimates are also time-integrated to daily values.
- Common-grid satellite-based ET: ET generated by the four models using the 3-hourly satellite data resampled to a common grid. In contrast to the previous runs, the satellite data are not applied at their original resolutions but after resampling them to the sinusoidal grid at ~ 25 km, adopted to produce the global model runs. Note that the CMORPH precipitation is replaced by the CFSR-Land product in order to have global coverage and that the LST is based on the AATSR observations.

3 Results and discussion

Here we look at the model performance against the in situ measurements, when the models are run with tower-based and satellite inputs. This section is divided into the three subsections, each of them dealing with one of the three experiments introduced in the previous Sect. 2.4.3. First, the 3-hourly and daily runs based on in situ forcing at 24 FLUXNET stations (see Table 2) are investigated. In the second part we look at the model performance at the same stations using 3-hourly and daily resolution satellite forcing. Finally, the ET estimates from the run using 3-hourly common-grid satellite forcing are compared to the in situ measurements at 85 FLUXNET stations.

3.1 Three-hourly and daily tower-based ET

The agreement of modeled evaporative fraction (EF) – defined here as $\lambda E/R_n$, using modeled λE and the net radiation from the respective forcing – with the measured EF (i.e., based on tower measurements of λE and R_n) gives an indication of the algorithm skill to model evaporative stress. Figure 3 (top panel) illustrates the agreement of modeled evaporative fraction with in situ measurements (derived using both EC and ER measurements of evapotranspiration; see Sect. 2.3.2), when models are run with tower inputs. GLEAM generally ranges between the EC and ER measurements, even at dry stations in open shrubland (OSH), woody savannas (WSA) and grassland (GRA) biomes, e.g., Sardinia/Arca di Noe (IT-Noe), Audubon Research Ranch (US-Aud), Santa Rita Mesquite (US-SRM) and Walnut Gulch Kendall Grasslands (US-Wkg). Only in the evergreen needleleaf forests (ENF) does GLEAM exceed the range of in situ measurements. PT-JPL mostly agrees with the reference as well, although it presents positive biases at some dry sites, like Wind River Crane (US-Wrc) and IT-Noe. PM-MOD underestimates EF for most stations (but it is very close to the EC measurements at six stations), while SEBS is characterized by an overall overestimation (for six stations SEBS EF is within the tower EC–ER range). In terms of the model performance per biome type, it can be stated that models generally perform the best in croplands (CRO) and deciduous broadleaf forest (DBF); at least this is the case for PT-JPL, PM-MOD and GLEAM. SEBS seems to perform better in grassland and savanna biomes (SAV). It is, however, difficult to derive robust conclusions on the model performance as a function of biome due to the low number of stations per biome type.

As the surface meteorology plays an important role in the ET production, we also compare the point-scale model performance with the gridded ERA-Interim ET data set (ERA) in Fig. 3 (top panel). ERA-Interim estimates are mostly within the range of EF measurements. The good agreement between ERA EF and the in situ measurements indicates that the ERA-Interim meteorology reliably captures the station conditions. It can also be stated that the point-scale tower-forced EF derived with PT-JPL and GLEAM match the ERA-Interim product based on a ~ 75 km resolution.

A statistical assessment of the model performance is given in Fig. 4, which shows the correlation (R^2), the RMSD and the average of the bias normalized by the reference (MBD) between modeled ET and tower measurement of ET (i.e., using the EC approach). In the left column of Fig. 4 the station averages of the statistical inferences are shown according to measured EF, i.e., from wet to dry. In general, the correlation to in situ data is high in wet and in moderately wet biomes for most sites and for all models. This is also true for SEBS, despite its substantial overestimation of EF (see Fig. 3). However, there seems to be a distinct decrease in R^2 from wet to dry biomes for all models; this decrease in performance is lower for GLEAM and higher for PM-MOD, which presents correlations ($R^2 < 0.4$) at dry sites. PT-JPL stands out amongst the ensembles with the highest correlation at most sites and especially in dry conditions. In comparison to the mostly underestimated evaporative fraction derived with PM-MOD (see Fig. 3), the RMSD of PM-MOD ET corre-

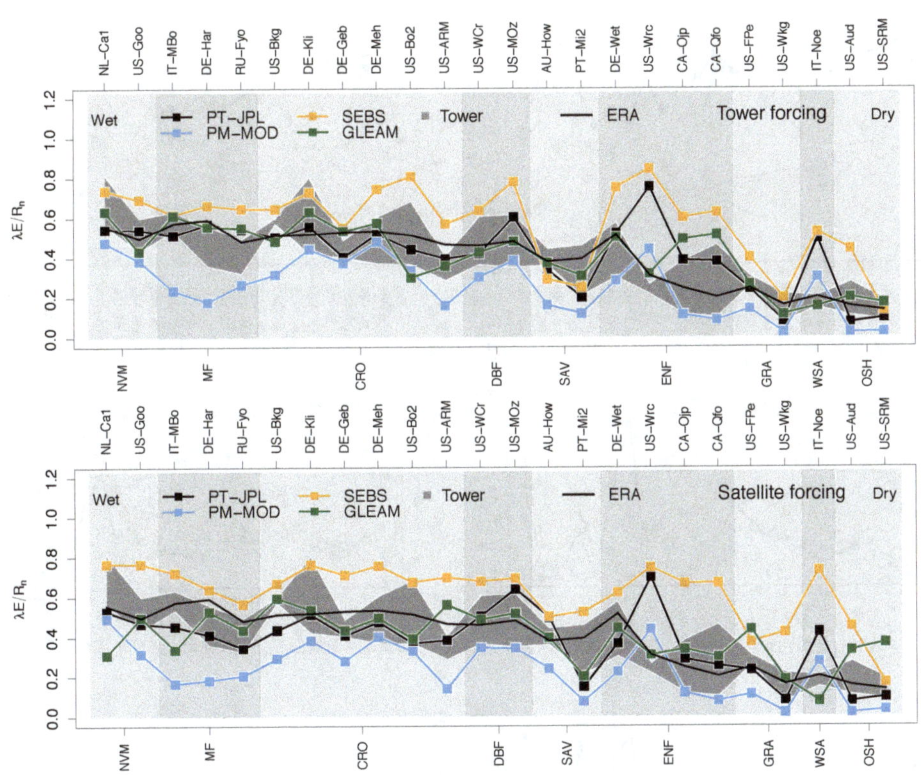

Figure 3. Station means of 3-hourly EC-observed and tower-forced (top panel) and satellite-forced (bottom panel) evaporative fraction against tower reference, as function of biomes, sorted from wet to dry (based on the biome average). The grey area denotes the range of evaporative fraction between EC and ER measurements. The black line denotes EF derived from ERA-Interim ET (ERA) and R_n.

sponds to PT-JPL and GLEAM and even produces the lowest maximum value ($0.13\,\mathrm{mm\,h^{-1}}$), followed by GLEAM ($0.17\,\mathrm{mm\,h^{-1}}$). Note that the large positive MBD values of PT-JPL and SEBS (> 200 %) may partly result from forcing ET to 0 during nighttime (see Sect. 2.4.2), when tower ET is negative, and thus leading to large relative errors, even for small negative reference ET values.

In order to evaluate the impact of using EC measurements as reference (in contrast to the ER method in Fig. 4), Table 3 shows the overall average 3-hourly model performance (i.e., the average of all station statistics) using both EC and ER data as reference. Overall, the average statistics of PT-JPL and GLEAM appear more favorable than those of SEBS and PM-Mu, although the RMSD and MBD of PM-MOD and the R^2 of SEBS are in general comparable to those of GLEAM and PT-JPL. This is again to a large extent affected by the overall overestimation and underestimation by SEBS and PM-MOD, respectively. The RMSD of SEBS is significantly smaller when using the ER method as reference ($0.10\,\mathrm{mm\,h^{-1}}$) as opposed to using EC ($0.13\,\mathrm{mm\,h^{-1}}$); on the other hand, the RMSD of PM-MOD is larger compared to ER ($0.12\,\mathrm{mm\,h^{-1}}$) than compared to EC (from $0.06\,\mathrm{mm\,h^{-1}}$). Note that the transpiration resistances in PM-MOD are calibrated based on a biome-dependent annual ET derived from EC observations, which may explain the smaller RMSD and MBD when using EC as a reference. Finally, the RMSD station averages are similar to both in situ references for by PT-JPL ($0.08, 0.09\,\mathrm{mm\,h^{-1}}$) and GLEAM ($0.08, 0.08\,\mathrm{mm\,h^{-1}}$).

Here the skill of models at representing ET at specific times of the day is examined. Note that small nighttime ET values from models and measurements may produce small absolute errors and thus can improve the overall full-day model performance in comparison to daytime periods, even if the relative bias is large.

The Taylor diagrams in Fig. 5 show that correlations to in situ observations (using the EC method) considering the entire daily cycle (left panel) are very similar compared to those considering daytime values only (left panel, top row) or nighttime values only (right panel, top row). The overall R^2 with tower forcing including all models is 0.67 for full-day as well as daytime evaluation and 0.68 for nighttime evaluation; this indicates that the results are independent of the timescale. Note that nighttime is identified as cases when the cosine of the zenith angle is < 0.2.

We can see (right panel, top row) that forcing negative nighttime ET values of PT-JPL and SEBS to 0 (in contrast to specific negative ET produced by PM-MOD and GLEAM; see Sect. 2.4.2) does not have a substantial impact on the overall agreement with tower measurements. However, it should be noted that the uncertainty of nighttime EC mea-

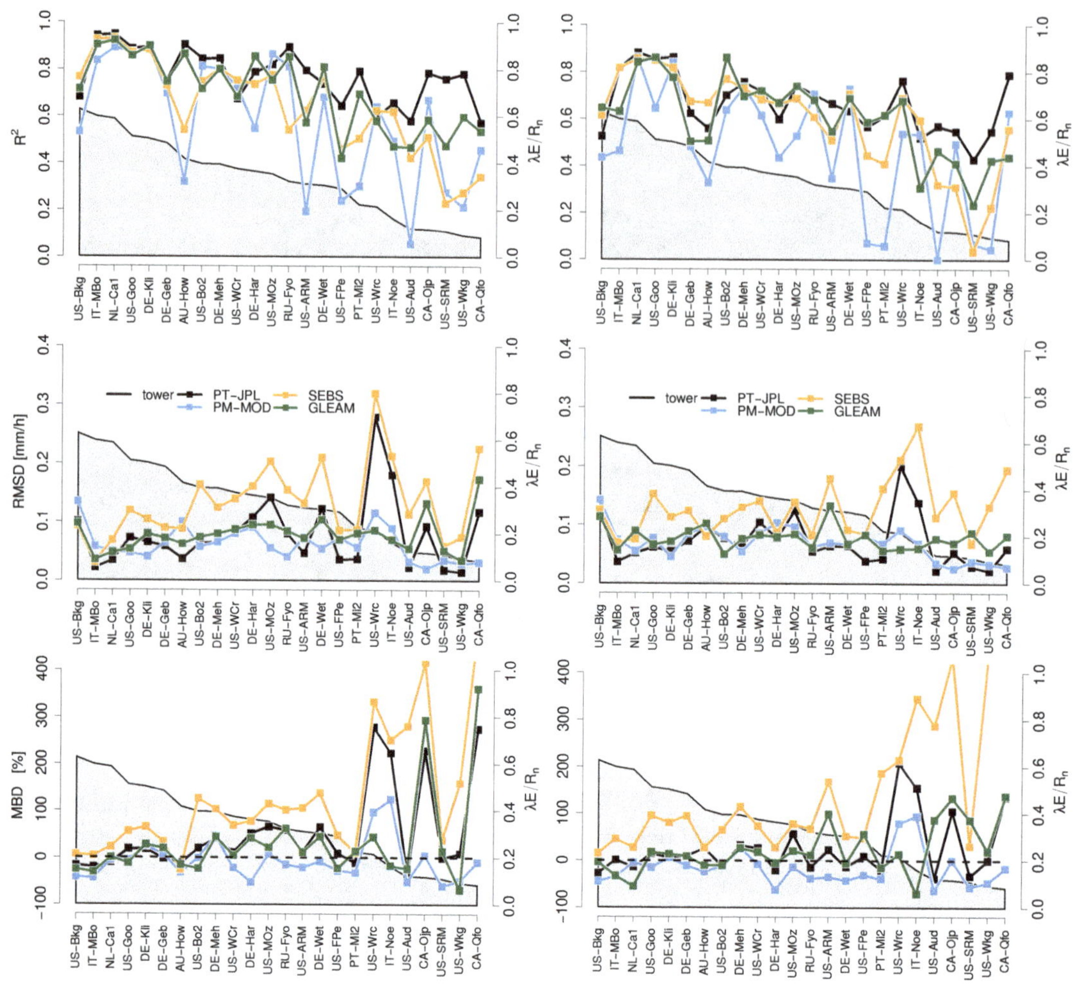

Figure 4. Station mean statistics of 3-hourly model data against EC reference. Left-column panels: tower-forced ET; right-column panels: satellite-forced ET; top-row panels: R^2 correlation coefficient (left y axis); middle-row panels: mean bias deviation (MBD, left y axis); bottom-row panels: root mean square difference (RMSD, left y axis). For all plots the evaporative fraction is given by the grey area (right y axis).

surements is high because of low turbulence. Hence, large nighttime errors can be present not only in the ET simulations but also in the EC data.

Sub-daily resolution is desirable in evaporation modeling, as it allows investigation of the underlying land–atmospheric interactions during the daily cycle of the planetary boundary layer. Given the short timescale of these interactions, one may expect that models that are able to reproduce short-term variability in ET would also be able to provide more reliable aggregates on daily timescales. Therefore, we investigate whether the model performance would benefit from solving evaporation at a 3-hourly resolution and aggregating it to daily values, as opposed to generating the estimates with daily input directly. Figure 5 (bottom row) clearly shows that not much more skill is gained by producing daily ET based on 3-hourly input (i.e., resolved diurnal cycles in the meteorological inputs) as opposed to forcing the models with the original daily input; results are almost identical when using aggregated 3-hourly output (left panel, bottom row) or using daily forcing (right panel, bottom row). In fact, for GLEAM the correlation to the EC reference is slightly higher when daily input is used, even if the standard deviation agrees marginally less well with the reference.

Figure 6 shows the statistics of the models' evaluation after forcing them with daily inputs. As expected, the general correlations become lower when daily (as opposed to 3-hourly) estimates are validated, since the daily cycle no longer plays a role in the enhancement of correlations – this was already highlighted by Table 3. Comparison of Figs. 4 and 6 shows that the decline in average R^2 from wet to dry stations is less evident at a daily resolution. This may be due to the smaller sample size when daily values are analyzed. PM-MOD and SEBS in particular correlate poorly at dry stations (also at other stations, such as the moderately wet AU-How). PT-JPL and GLEAM perform worse (compared with the 3-hourly resolution) at dry stations when they are run at

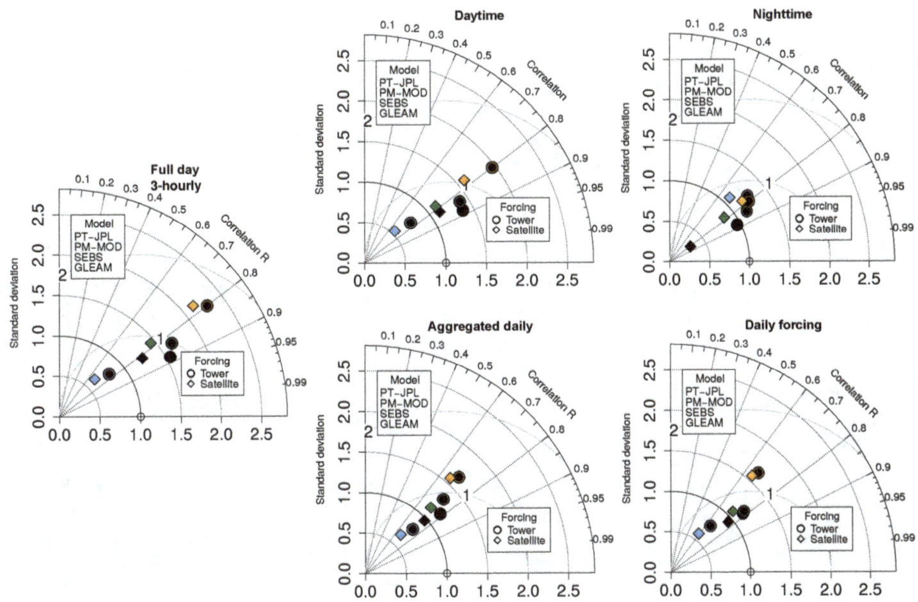

Figure 5. Taylor diagrams of 3-hourly model performance against EC reference in sub-daily periods (top-row panels) and as function of temporal resolution (bottom-row panels). The left panel shows the average model statistics for full-day (compare to top row) and 3-hourly output data (compare to bottom row). Daytime is defined as cases when the cosine of the sun elevation azimuth is > 0.2; nighttime is defined as cases when the cosine of the sun elevation azimuth is < 0.2. Shown are the normalized standard deviation, the normalized RMSD and the correlation coefficient (R).

Table 3. Summary of 24 stations average statistics for 3-hourly and daily tower forcing. EC denotes the model agreement with the evapotranspiration reference from eddy-covariance measurements, and ER is the model agreement with the evapotranspiration reference based on the in situ energy residual. RMSD is given in millimeters per hour for both 3-hourly data (3 h) and daily data (d).

		R^2		RMSD		MBD	(%)
		EC	ER	EC	ER	EC	ER
PT-JPL	3 h	0.77	0.78	0.08	0.09	53.1	37.4
	d	0.61	0.60	0.04	0.05	47.8	21.8
PM-MOD	3 h	0.58	0.55	0.06	0.12	−6.7	−18.2
	d	0.43	0.41	0.04	0.06	3.8	−11.3
SEBS	3 h	0.64	0.78	0.13	0.10	125.9	78.4
	d	0.45	0.50	0.08	0.07	113.5	48.5
GLEAM	3 h	0.70	0.80	0.08	0.08	31.9	−8.7
	d	0.60	0.66	0.03	0.04	15.6	−14.1

daily resolutions. In terms of the RMSD and MBD, the results are quite similar to the 3-hourly findings, but in most cases worst performance at the daily resolution is found at dry stations. An exception is GLEAM, which shows smaller RMSD at the dry stations when using daily rather than 3-hourly resolution.

The change in overall MBD (against the EC reference) from using 3-hourly tower input to using daily tower input is from 53.1 to 47.8 % for PT-JPL, from −6.7 to 3.8 % for PM-MOD, from 125.9 to 113.5 % for SEBS, and from 31.9 to 15.6 % for GLEAM. While the pattern of EF (Fig. 3) and MBD (Fig. 4) indicates a substantial underestimation of 3-hourly ET by PM-MOD, this underestimation is attenuated when daily input is used (−18.2 to −11.3 % compared to the ER reference). Note that even if we employ the term daily input, the PM-MOD model estimates day and night ET separately by using integrated day and night inputs (as opposed to PT-JPL, SEBS and GLEAM, which use daily integrated inputs) and then combines them to provide a daily value. This is how the PM-MOD model was originally used and how it is implemented in this study for daily estimation. The better agreement on a daily scale thus may reflect a more appropriate use of the inputs.

The similarity of the results for different temporal resolutions underlines the robustness of the modeling processes. PT-JPL and GLEAM agree best with the in situ measurements, while SEBS yields a good correlation in comparison to the other models yet produces the largest absolute errors due to its large overestimation. PM-MOD produces the lowest correlation but agrees rather well in terms of absolute deviations.

Table 3 summarizes the main statistics of the model evaluation for the 3-hourly and daily tower inputs.

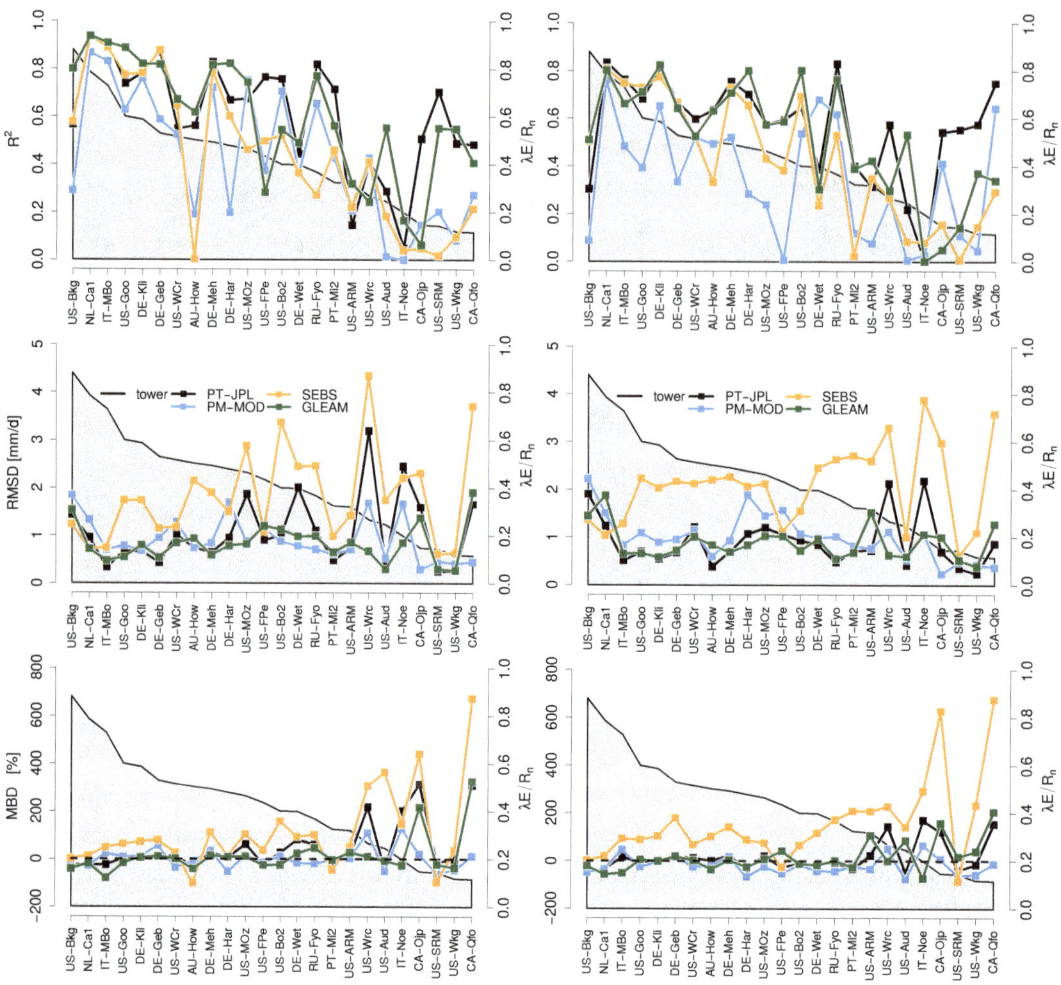

Figure 6. Station mean statistics of daily data from daily input against EC reference. Left-column panels: tower-forced ET; right-column panels: satellite-forced ET; top-row panels: R^2 correlation coefficient (left y axis); middle-row panels: mean bias deviation (MBD, left y axis); bottom-row panels: root mean square difference (RMSD, left y axis). For all plots the evaporative fraction is given by the grey area (right y axis).

3.2 Three-hourly and daily original-resolution satellite ET

In this section we discuss the model performance using 3-hourly and daily satellite forcing with original resolution at the selected 24 FLUXNET stations. The findings are compared to the results of the tower forcing in the previous section in order to allocate model uncertainty to either the algorithms used or the common forcing.

The evaluation of 3-hourly modeled EF using satellite forcing (Fig. 3, bottom panel) shows a very similar picture of agreement with the reference compared to the results of the tower forcing. Note that the satellite EF shown here slightly differs from tower-forced EF, as the data availability of the input time series may be different at some stations. The ET overestimation by SEBS seems to be slightly emphasized when using satellite input in comparison to the tower forcing. Note that the LST used in SEBS is still obtained from the tower measurements, as discussed in Sect. 2.4.1. EF derived with PT-JPL and GLEAM still agrees well with the reference, yet GLEAM overestimates EF in dry biomes when using satellite forcing but is more accurate at needleleaf forest sites. The good model performance of PT-JPL and GLEAM, independent of forcing type, indicates a robust performance of the models on the one hand and a reliable satellite forcing – in the sense of their meteorology comparing well with the in situ tower data – on the other hand.

In Fig. 3 (bottom panel) we also compare the model performance with the gridded ERA-Interim ET data set. Note that while the tower forcing runs (top panel) are independent of ERA-Interim, the satellite runs use ERA-Interim estimates as inputs for the surface meteorology. As shown in Sect. 3.1, the ERA-Interim EF product agrees with the in situ measurements. The correlation of the models to ERA-Interim is not substantially improved with satellite input in comparison to the tower forcing, although the models now use the ERA-Interim meteorology as input.

The station averages of the statistical indices R^2, RMSD and MBD of the models forced with satellite observations (Fig. 4, bottom panel) against the in situ measurements underline the previously reported high similarity of modeled ET based on tower and satellite forcing. Only the RMSD of SEBS is slightly attenuated with remotely sensed forcing. However, the algorithm is still characterized by substantial overestimation.

In the following we compare the model performance with daily satellite forcing to the model performance with daily tower forcing. In accordance with the evaluation of 3-hourly data (see Fig. 4), Fig. 6 indicates that the daily satellite-based ET products also correspond to the tower-based modeled ET. We want to highlight, however, that in contrast to the 3-hourly runs, the RMSD of SEBS substantially increases when satellite input is used. This suggests that the SEBS physical modeling captures the ET processes more accurately with the high temporal resolution inputs (3-hourly vs. daily).

Table 4 provides a summary of the main statistics of the model evaluation for the 3-hourly and daily satellite inputs.

Table 4. Summary of 24 station average statistics for 3-hourly and daily satellite forcing. ERA denotes the agreement of ERA-Interim evapotranspiration with the in situ reference evapotranspiration. For other abbreviations, see Table 3. RMSD is given in millimeters per hour for both 3-hourly data (3 h) daily data (d).

		R^2		RMSD		MBD	(%)
		EC	ER	EC	ER	EC	ER
PT-JPL	3 h	0.67	0.68	0.07	0.11	25.8	14.6
	d	0.57	0.49	0.04	0.05	16.5	−1.7
PM-MOD	3 h	0.47	0.47	0.07	0.14	−16.1	−27.2
	d	0.35	0.29	0.04	0.06	−17.9	−34.2
SEBS	3 h	0.59	0.71	0.13	0.11	145.1	148.6
	d	0.42	0.41	0.09	0.08	160.2	123.9
GLEAM	3 h	0.61	0.72	0.08	0.10	22.7	−4.6
	d	0.52	0.52	0.04	0.05	16.2	−10.6
ERA	3 h	0.51	0.45	0.10	0.14	111.3	87.0
	d	0.62	0.50	0.07	0.07	114.7	74.6

3.3 Three-hourly common-grid satellite ET

Here the ET algorithms are tested against 85 FLUXNET stations using the gridded sinusoidal (~ 25 km) satellite input (as opposed to using their original input resolutions) in order to evaluate the common-gridded global ET estimates on the tower scale. Only the evaluation over the towers is discussed here, with the evaluation on the global scale discussed in the companion paper of Miralles et al. (2016). Note that the spatial mismatch between the tower fetch and the ~ 750 km^2 of the gridded cells is very large, and the agreement between the tower fluxes and the modeled ET certainly depends on the tower conditions being representative of the corresponding gridded pixel. This was also the case for some of the original-resolution satellite inputs used over the 24 stations, such as the SRB radiation or the ERA-Interim meteorology. The results of the satellite runs using common-grid forcing are compared to the results using the tower and satellite inputs on the tower scale presented in Sects. 3.1 and 3.2.

The top panel of Fig. 7 shows the mean 3-hourly EF over 70 stations for PM-MOD, PT-JPL and GLEAM. For 15 of the 85 stations the surface radiation or the ground flux was not available; hence, the ER reference could not be calculated. As the gridded inputs use satellite LST from AATSR, SEBS ET is only estimated at the midmorning AATSR overpass. The bottom panel of Fig. 7 shows the annual midmorning evaporative fraction, this time including SEBS. Due to the 3-day revisiting time of AATSR and the lack of measurements in cloudy conditions, the number of available SEBS ET estimates reduces drastically, compared with the previous simulations using tower LST. The bottom panel of Fig. 7 shows station averages from all models only when SEBS ET is available. Thus, it is based on fewer data and with the number of stations reduced to 67.

The 3-hourly model performances from PM-MOD, PT-JPL and GLEAM correspond closely to the performance in the analysis using the 24 towers and the original-resolution satellite inputs. The EF station averages produced by PT-JPL and GLEAM are very close at all locations and respond well to the hydrological and energetic conditions expected in the respective biome. The overall agreement with the range between EC and ER in situ measurements is comparable to what has previously been found in the smaller sample of stations (see Fig. 3). PM-MOD keeps underestimating ET, except for the cropland biome, where the majority of station averages matches well with the reference. Concerning the midmorning evaporative fractions, the PM-MOD, PT-JPL and GLEAM patterns are all very similar to the case with the full diurnal cycle. SEBS again tends to overestimate over a large number of stations, compared with the in situ measurements. Overall, it can be stated that the model accuracy and inter-model agreement obtained with in situ and satellite forcing on the tower scale could be reproduced with the common-grid satellite forcing.

Figure 8 summarizes the results above by displaying standard deviation, correlation and RMSD of the modeled ET shown in Fig. 7 against the EC reference. The Taylor plots highlight the fact that the variability of PT-JPL, PM-MOD and GLEAM is not substantially influenced by the low sample size for cases when SEBS ET is available. Again, the similarity between Fig. 5 (left panel) for satellite forcing on the tower scale and Fig. 8 for gridded input data confirms the robustness of the analyses independent of tower and time sampling.

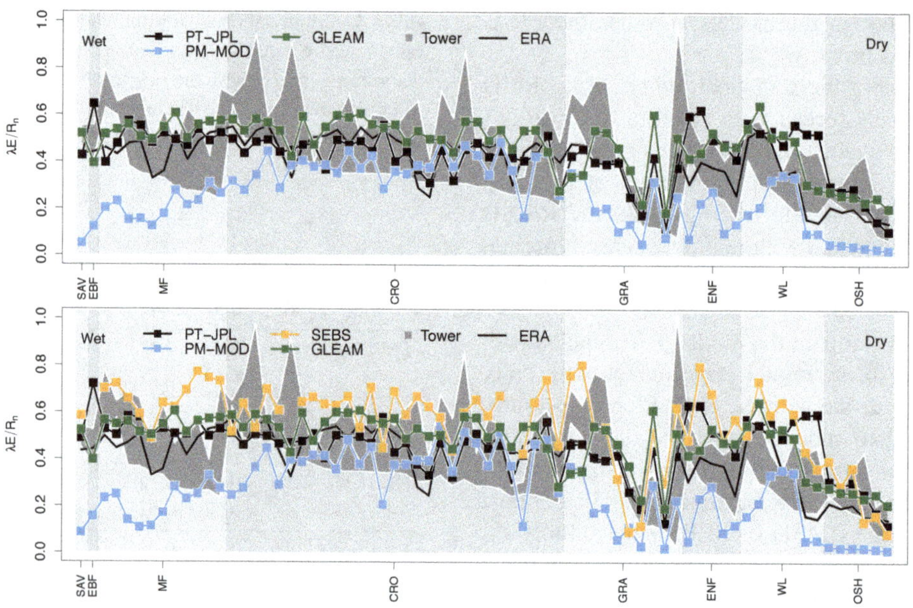

Figure 7. Top panel: station means of 3-hourly sinusoidal gridded satellite-forced evaporative fraction for full days (70 stations) against tower reference, as function of biomes, sorted from wet to dry (based on the biome average). Bottom panel: same as top panel but for midmorning only (from 09:00 to 13:00 LT; 67 stations). The grey area denotes the range of evaporative fraction between EC and ER measurements. The black line denotes EF derived from ERA-Interim ET and R_n.

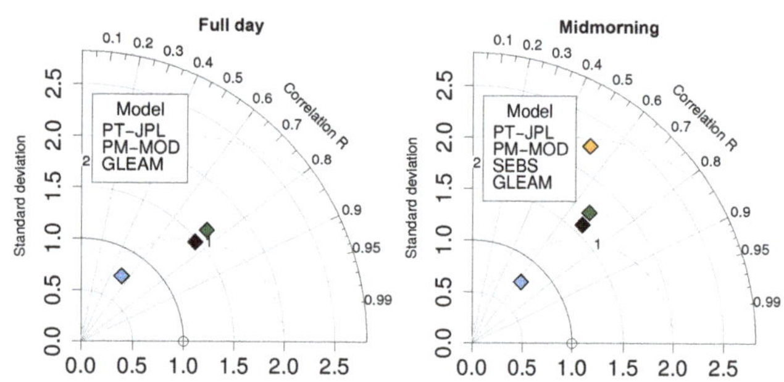

Figure 8. Taylor diagram plots of sinusoidal gridded model data against tower EC reference. Left panel: full-day 3-hourly data compared to 85 stations. Right panel: midmorning data (from 09:00 and 13:00 LT) compared to 82 stations. Shown are the normalized standard deviation, the normalized RMSD and the correlation coefficient (R).

4 Conclusions

In this first part of the WACMOS-ET study, the skill of the PT-JPL, PM-MOD, SEBS and GLEAM ET algorithms has been tested on the tower scale against in situ measurements at 24 FLUXNET sites. The algorithms are forced using in situ meteorological data from these towers, covering the period 2005–2007 on three continents and across nine different biomes, while ensuring spatial consistency between input and reference data. Additionally, the models are run for the same period with reanalysis and satellite forcing of varying spatial resolutions, including ERA-Interim (surface meteorology), SRB (radiation), AATSR (LST), GlobAlbedo (LAI and fAPAR), CMORPH (precipitation) and WACMOS-CCI (soil moisture). The models were run with 3-hourly and daily input to assess the robustness of their performance for sub-daily and daily resolution.

Our analyses have shown that the four models' performance is robust in terms of changes in forcing types and temporal resolutions (i.e., the changes do not alter the model behavior at the selected stations significantly). Against the in situ 3-hourly energy residual estimates at the tower, the tower-based model simulations are ranked (according to station averages) as follows: GLEAM (0.80, 0.08), PT-JPL (0.78, 0.09), SEBS (0.78, 0.10) and PM-MOD (0.55, 0.12). The first value in the brackets denotes R^2 and the second value denotes RMSD in millimeters per hour. Compared to the eddy-covariance measurements, however, the station averages of RMSD do not reflect the same outcome. Due to more substantial overestimation at two stations each, the RMSD of PT-JPL (0.77, 0.08) and GLEAM (0.70, 0.08) are larger than that of PM-MOD (0.58, 0.06). However, correlations are consistently higher for GLEAM and PT-JPL. Thus, over our selection of towers and the reference period (2005–2007), we judge GLEAM and PT-JPL as the algorithms more closely matching the in situ observations. At some stations, PM-MOD and SEBS also agree well with the observations, but in general the PM-MOD and SEBS performance is characterized by under- and overestimation, respectively.

For the satellite forcing, the RMSD between the models and the reference yields very similar numbers as for tower forcing. Correlations are closer but in most situations slightly smaller for the satellite forcings. This can be the result of discrepancies between the spatial resolution of satellite observations and tower measurements, although different inputs errors (in situ vs. satellite) may also play a role. This performance closeness between in situ and satellite-derived values can be an indication of the spatial representativeness of the tower measurements (i.e., reasonable spatial homogeneity around the tower) and the consistency of the input data set across forcing types. This is underlined by a comparison to the ~ 75 km resolved reanalysis ET product of ERA-Interim, which agrees well with the modeled ET across the different biomes.

Regarding the analysis over the 85 stations, a similar overall picture is obtained using the ~ 25 km common-grid ET prepared for the global runs. The evaluations of McCabe et al. (2015) of a different selection of towers (45 stations), over a more extended period (1997–2007) and with different satellite forcings (LandFlux forcings) also results in an overall similar analysis, confirming the robustness of the model performance evaluations.

Using daily input data reduces the RMSD of the models with the tower measurements but results in slightly worse correlations. This is due to the lower variability of daily values in contrast to 3-hourly data (variability accentuated by the diurnal cycle). However, the consistency of the model agreements with the reference with regard to 3-hourly and daily ET estimates highlights the robustness of the integration method applied to the models. This is also underlined by the good agreements of modeled daily ET from aggregated 3-hourly output data with modeled daily ET from daily input.

While GLEAM and PM-MOD can produce negative ET, PT-JPL and SEBS cannot operate under these conditions (mostly at nighttime when the flux of available energy reverses sign) and their negative values are forced to 0. This does not have a large impact on their full-day performance, since these values occur at night, when tower ET is negative and with generally low values. Only for the relative bias is the effect significant, since the two models consequently overestimate ET in these cases.

In terms of high and low temporal input resolution, it was found that using 3-hourly input data does not significantly increase the accuracy of the models for producing daily ET. Hence, it is sufficient to use daily input to achieve a similar result if the intended application of the ET product does not demand a reproduction of the diurnal cycle.

The ET models generally perform best in wet biomes and tend to overestimate values in dry biomes, where ET is constrained by water availability. Focusing on water stress in the model development within the community would thus provide the opportunity to obtain more robust simulations of surface fluxes for global-scale employment.

The conducted analyses based on in situ ET are useful to evaluate model performance, but there are some clear limitations. Our requirements for tower selection resulted in a somewhat limited number of stations, so it would be desirable to extend the evaluations to larger regions in order to better cover different climate and biome conditions. Therefore, in the companion paper of Miralles et al. (2016) our analyses are extended by looking at the global spatiotemporal variability of the modeled ET, the closure of regional water budgets, and the discrete estimation of land evaporation components or sources (i.e., transpiration, interception loss and direct soil evaporation).

Acknowledgements. This study was funded by the European Space Agency (ESA) and conducted as part of the project WACMOS-ET (Contract no. 4000106711/12/I-NB). D. G. Miralles acknowledges the financial support from the Netherlands Organization for Scientific Research through grant 863.14.004 and the Belgian Science Policy Office (BELSPO) in the framework of the STEREO III programme, project SAT-EX (SR/00/306). M. F. McCabe and A. Ershadi acknowledge the support of the King Abdullah University of Science and Technology. The SEBS team is acknowledged for facilitating discussions concerning the implementation of their model. This work used eddy-covariance data acquired by the FLUXNET community and in particular by the following networks: AmeriFlux (US Department of Energy, Biological and Environmental Research, Terrestrial Carbon Program, DE-FG02-04ER63917 and DE-FG02-04ER63911), AfriFlux, AsiaFlux, CarboAfrica, CarboEuropeIP, CarboItaly, CarboMont, ChinaFlux, Fluxnet-Canada (supported by CFCAS, NSERC, BIOCAP, Environment Canada and NRCan), GreenGrass, KoFlux, LBA, NECC, OzFlux, TCOS-Siberia and USCCC. Data and logistical support for the station US-Wrc were provided by the US Forest Service Pacific Northwest Research Station. All WACMOS-ET forcing data and ET estimates are publicly available and can be requested through the project website (http://wacmoset.estellus.eu).

Edited by: P. Gentine

References

Adler, R., Huffman, G., Chang, A., Ferraro, R., Xie, P.-P., Janowiak, J., Rudolf, B., Schneider, U., Curtis, S., Bolvin, D., Gruber, A., Susskind, J., Arkin, P., and Nelkin, E.: The version-2 Global Precipitation Climatology Project (GPCP) monthly precipitation analysis (1970–Present), J. Hydrol., 4, 1147–1167, 2003.

Anderson, M., Norman, J., Mecikalski, J., Otkin, J., and Kustas, W.: A climatological study of evapotranspiration and moisture stress across the continental United States based on thermal remote sensing: 1. Model formulation, J. Geophys. Res., 112, D10117, doi:10.1029/2006JD007506, 2007.

Baldocchi, D., Falge, E., Gu, L., Olson, R., Hollinger, D., Running, S., Anthoni, P., Bernhofer, C., Davis, K., Evans, R., Fuentes, J., Goldstein, A., Katul, G., Law, B., Lee, X., Malhi, Y., Meyers, T., Munger, W., Oechel, W., Paw, K., Pilegaard, K., Schmid, H., Valentini, R., Verma, S., Vesala, T., Wilson, K., and Wofsyn, S.: FLUXNET: a new tool to study the temporal and spatial variability of ecosystem-scale carbon dioxide, water vapor, and energy flux densities, B. Am. Meteorol. Soc., 82, 2415–2434, 2001.

Burba, G., McDermitt, D., Anderson, D., and Furtaw, M. D.: Novel design of an enclosed CO_2/H_2O gas analyser for eddy covariance flux measurements, Tellus, 62, 743–748, 2010.

Chen, X., Su, Z., Ma, Y., Liu, S., Yu, Q., and Xu, Z.: Development of a 10-year (2001–2010) 0.1° data set of land-surface energy balance for mainland China, Atmos. Chem. Phys., 14, 13097–13117, doi:10.5194/acp-14-13097-2014, 2014.

Coccia, G. and Wood, E. F.: CFSR-Land: a new high temporal resolution global land data assimilation product, J. Geophys. Res., in preparation, 2015.

Courault, D., Seguin, B., and Olioso, A.: Review on estimation of evapotranspiration from remote sensing data: from empirical to numerical modeling approaches, Irrig. Drain. Syst., 19, 223–249, 2005.

Dee, D., Uppala, M., S., Simmons, J., A., Berrisford, P., Poli, P., Kobayashi, S., Andrae, U., Balmaseda, A., M., Balsamo, G., Bauer, P., Bechtold, P., Beljaars, M., A. C., van de Berg, L., Bidlot, J., Bormann, N., Delsol, C., Dragani, R., Fuentes, M., Geer, J., A., Haimberger, L., Healy, B., S., Hersbach, H., Hólm, V., E., Isaksen, L., Kallberg, P., Khaler, M., Matricardi, M., McNally, P., A., Monge-Sanz, M., B., Morcrette, J.-J., Park, B.-K., Peubey, C., de Rosnay, P., Tavolato, C., Thapaut, J.-N., and Vitart, F.: The ERA-Interim reanalysis: configuration and performance of the data assimilation system, Q. J. Roy. Meteorol. Soc., 137, 553–597, doi:10.1002/qj.828, 2011.

Ershadi, A., McCabe, M. F., Evans, J. P., Chaney, N. W., and Wood, E. F.: Multi-site evaluation of terrestrial evaporation models using FLUXNET data, Agr. Forest Meteorol., 187, 46–61, 2014.

Ferguson, C. R. and Wood, E. F.: An evaluation of satellite remote-sensing data products for land surface hydrology: Atmospheric Infrared Sounder (AIRS), J. Hydrometeorol., 11, 1234–1262, 2010.

Fisher, J., Tu, K., and Baldocchi, D.: Global estimates of the land-atmosphere water flux based on monthly AVHRR and ISLSCP-II data, validated at 16 FLUXNET sites, Remote Sens. Environ., 112, 901–919, 2008.

Foken, T., Wimmer, F., Mauder, M., Thomas, C., and Liebethal, C.: Some aspects of the energy balance closure problem, Atmos. Chem. Phys., 6, 4395–4402, doi:10.5194/acp-6-4395-2006, 2006.

Gash, J.: An analytical model of rainfall interception by forests, Q. J. Roy. Meteorol. Soc., 105, 43–45, 1979.

Gash, J. and Stewart, J.: The evaporation from Thetford Forest during 1975, J. Hydrol., 35, 385–396, 1977.

Hirschi, M., Michel, D., Lehner, I., and Seneviratne, S. I.: A site-level comparison of lysimeter and eddy-covariance flux measurements of evapotranspiration, Hydrol. Earth Syst. Sci., in preparation, 2016.

Jia, L., Su, Z., van den Hurk, B., Menenti, M., Moene, A., De Bruin, H. A. R., and Yrisarry, J. J. B.: Estimation of sensible heat flux using the Surface Energy Balance System (SEBS) and ATSR measurements, Phys. Chem. Earth, 28, 75–88, 2003.

Jiménez, C., Prigent, C., Mueller, B., Seneviratne, S. I., McCabe, M. F., Wood, E. F., Rossow, W. B., Balsamo, G., Betts, A. K., Dirmeyer, P. A., Fisher, J. B., Jung, M., Kanamitsu, M., Reichle, R. H., Reichstein, M., Rodell, M., Sheffield, J., Tu, K., and Wang, K.: Global intercomparison of 12 land surface heat flux estimates, J. Geophys. Res., 116, 6809–6866, D02102, doi:10.1029/2010JD014545, 2011.

Jimenez, C., McCabe, M., Seneviratne, S. I., Wood, E., and Rossow, W.: 4th LandFlux Workshop, GEWEX News, 22, 17–18, 2012.

Joyce, R. J., Janowiak, J. E., Arkin, P. A., and Xie, P.: CMORPH: a method that produces global precipitation estimates from passive microwave and infrared data at high spatial and temporal resolution, J. Hydrometeorol., 5, 487–503, 2004.

Jung, M., Reichstein, M., Ciais, P., Seneviratne, S. I., Sheffield, J., Goulden, M. L., Bonan, G., Cescatti, A., Chen, J., and de Jeu, R.: Recent decline in the global land evapotranspiration trend due to limited moisture supply, Nature, 467, 951–954, 2010.

Kalma, J., McVicar, T., and McCabe, M.: Estimating land surface evaporation: a review of methods using remotely sensed surface temperature data, Surv. Geophys., 29, 421–469, doi:10.1007/s10712-008-9037-z, 2008.

Kelly, R., Chang, A., Tsang, L., and Foster, J.: A prototype AMSR-E global snow area and snow depth algorithm, IEEE T. Geosci. Remote, 41, 230–242, 2003.

Kustas, W. and Norman, J.: A two-source energy balance approach using directional radiometric temperature observations for sparse canopy covered surfaces, Agron. J., 92, 847–854, 2000.

Liu, Y. Y., de Jeu, R. A. M., McCabe, M., Evans, J., and van Dijk, A. I. J. M.: Global long-term passive microwave satellite-based retrievals of vegetation optical depth, Geophys. Res. Lett., 38, L18402, doi:10.1029/2011GL048684, 2011a.

Liu, Y. Y., Parinussa, R. M., Dorigo, W. A., De Jeu, R. A. M., Wagner, W., van Dijk, A. I. J. M., McCabe, M. F., and Evans, J. P.: Developing an improved soil moisture dataset by blending passive and active microwave satellite-based retrievals, Hydrol. Earth Syst. Sci., 15, 425–436, doi:10.5194/hess-15-425-2011, 2011b.

Liu, Y. Y., van Dijk, A. I. J. M., de Jeu, R. A. M., Canadell, J. G., McCabe, M. F., Evans, J. P., and Wang, G.: Recent reversal in loss of global terrestrial biomass, Nat. Clim. Change, 5, 470–474, doi:10.1038/nclimate2581, 2015.

Luojus, K. and Pulliainen, J.: Global Snow Monitoring for Climatic Research: Snow Water Equivalent (SWE) Product Guide, Finnish Meteorological Institute, Helsinki, Finland, 2010.

Mao, J., Fu, W., Shi, X., Ricciuto, D. M., Fisher, J. B., Dickinson, R. E., Wei, Y., Shem, W., Piao, S., Wang, K., Schwalm, C. R., Tian, H., Mu, M., Arain, A., Ciais, P., Cook, R., Dai, Y., Hayes, D., Hoffman, F. M., Huang, M., Huang, S., Huntzinger, D. N., Ito, A., Jain, A., King, A. W., Lei, H., Lu, C., Michalak, A. M., Parazoo, N., Peng, C., Peng, S., Poulter, B., Schaefer, K., Jafarov, E., Thornton, P. E., Wang, W., Zeng, N., Zeng, Z., Zhao, F., Zh, Q., and Zhu, Z: Disentangling climatic and anthropogenic controls on global terrestrial evapotranspiration trends, Environ. Res. Lett., 10, 094008, doi:10.1088/1748-9326/10/9/094008, 2015.

Martens, B., Miralles, D. G., Lievens, H., Fernández-Prieto, D., and Verhoest, N.: Improving terrestrial evaporation estimates over continental Australia through assimilation of SMOS soil moisture, Int. J. Appl. Earth Obs., doi:10.1016/j.jag.2015.09.012, in press, 2016.

McCabe, M. F. and Wood, E.: Scale influences on the remote estimation of evapotranspiration using multiple satellite sensors, Remote Sens. Environ., 105, 271–285, doi:10.1016/j.rse.2006.07.006, 2006.

McCabe, M. F., Ershadi, A., Jimenez, C., Miralles, D. G., Michel, D., and Wood, E. F.: The GEWEX LandFlux project: evaluation of model evaporation using tower-based and globally-gridded forcing data, Geosci. Model Dev. Discuss., 8, 6809–6866, doi:10.5194/gmdd-8-6809-2015, 2015.

Miralles, D. G., Gash, J. H., Holmes, T. R. H., de Jeu, R. A. M., and Dolman, A. J.: Global canopy interception from satellite observations, J. Geophys. Res., 115, D16122, doi:10.1029/2009JD013530, 2010.

Miralles, D. G., Holmes, T. R. H., De Jeu, R. A. M., Gash, J. H., Meesters, A. G. C. A., and Dolman, A. J.: Global land-surface evaporation estimated from satellite-based observations, Hydrol. Earth Syst. Sci., 15, 453–469, doi:10.5194/hess-15-453-2011, 2011a.

Miralles, D. G., De Jeu, R. A. M., Gash, J. H., Holmes, T. R. H., and Dolman, A. J.: Magnitude and variability of land evaporation and its components at the global scale, Hydrol. Earth Syst. Sci., 15, 967–981, doi:10.5194/hess-15-967-2011, 2011b.

Miralles, D. G., Jiménez, C., Jung, M., Michel, D., Ershadi, A., McCabe, M. F., Hirschi, M., Martens, B., Dolman, A. J., Fisher, J. B., Mu, Q., Seneviratne, S. I., Wood, E. F., and Fernandez-Prieto, D.: The WACMOS-ET project – Part 2: Evaluation of global terrestrial evaporation data sets, Hydrol. Earth Syst. Sci., 20, 823–842, doi:10.5194/hess-20-823-2016, 2016.

Monin, A. and Obukhov, A.: Basic laws of turbulent mixing in the surface layer of the atmosphere, Tr. Akad. Nauk SSSR Geophiz. Inst., 24, 163–187, 1954.

Monteith, J.: Evaporation and environment, Sym. Soc. Exp. Biol., 19, 205–234, 1965.

Mooney, P. A., Mulligan, F. J., and Fealya, R.: Comparison of ERA-40, ERA-Interim and NCEP/NCAR reanalysis data with observed surface air temperatures over Ireland, Int. J. Cancer, 31, 545–557, 2011.

Mu, Q., Heinsch, F. A., Zhao, M., and Running, S. W.: Development of a global evapotranspiration algorithm based on MODIS and global meteorology data, Remote Sens. Environ., 111, 519–536, 2007.

Mu, Q., Zhao, M., and Running, S.: Improvements to a MODIS global terrestrial evapotranspiration algorithm, Remote Sens. Environ., 115, 1781–1800, 2011.

Mueller, B., Seneviratne, S. I., Jimenez, C., Corti, T., Hirschi, M., Balsamo, G., Ciais, P., Dirmeyer, P., Fisher, J. B., Guo, Z., Jung, M., Maignan, F., McCabe, M. F., Reichle, R., Reichstein, M., Rodell, M., Sheffield, J., Teuling, A. J., Wang, K., Wood, E. F., and Zhang, Y.: Evaluation of global observations-based evapotranspiration datasets and IPCC AR4 simulations, Geophys. Res. Lett., 38, L06402, doi:10.1029/2010GL046230, 2011.

Mueller, B., Hirschi, M., Jimenez, C., Ciais, P., Dirmeyer, P. A., Dolman, A. J., Fisher, J. B., Jung, M., Ludwig, F., Maignan, F., Miralles, D. G., McCabe, M. F., Reichstein, M., Sheffield, J., Wang, K., Wood, E. F., Zhang, Y., and Seneviratne, S. I.: Benchmark products for land evapotranspiration: LandFlux-EVAL multi-data set synthesis, Hydrol. Earth Syst. Sci., 17, 3707–3720, doi:10.5194/hess-17-3707-2013, 2013.

Murphy, D. and Koop, T.: Review of the vapour pressures of ice and supercooled water for atmospheric applications, Q. J. Roy. Meteorol. Soc., 131, 1539–1565, 2005.

Owe, M., de Jeu, R., and Walker, J.: A methodology for surface soil moisture and vegetation optical depth retrieval using the microwave polarization difference index, IEEE T. Geosci. Remote, 39, 1643–1654, 2001.

Pinty, B., Lavergne, T., Vossbeck, M., Kaminski, T., Aussedat, O., Giering, R., Gobron, N., Taberner, M., Verstraete, M. M., and Widlowski, J.-L.: Retrieving surface parameters for climate models from Moderate Resolution Imaging Spectroradiometer (MODIS)-Multiangle Imaging Spectroradiometer (MISR) albedo products, J. Geophys. Res.-Atmos., 112, D10116, doi:10.1029/2006JD008105, 2007.

Pinty, B., Jung, M., Kaminski, T., Lavergne, T., Mund, M., Plummer, S., Thomas, E., and Widlowski, J.: Evaluation of the JRC-TIP 0.01° products over a mid-latitude deciduous forest site, Remote Sens. Environ., 115, 3567–3581, 2011a.

Pinty, B., Taberner, M., Haemmerle, V., Paradise, S., Vermote, E., Verstraete, M., Gobron, N., and Widlowski, J.-L.: Global-Scale Comparison of MISR and MODIS Land Surface Albedos, J. Climate, 24, 732–749, 2011b.

Priestley, C. and Taylor, R.: On the Assessment of surface heat flux and evaporation using large-scale parameters, Mon. Weather Rev., 100, 81–92, 1972.

Saha, S., Moorthi, S., Pan, H.-L., et al.: The NCEP climate forecast system reanalysis, B. Am. Meteorol. Soc., 91, 1015–1057, 2010.

Seemann, S. W., Borbas, E. E., Knuteson, R. O., Stephenson, G. R., and Huang, H.-L.: Development of a global infrared surface emissivity database for application to clear sky retrievals from multispectral satellite radiance measurements, J. Appl. Meteorol. Clim., 47, 108–123, 2008.

Simard, M., Pinto, N., Fisher, J., and Baccini, A.: Mapping forest canopy height globally with spaceborne lidar, J. Geophys. Res., 116, G04021, doi:10.1029/2011JG001708, 2011.

Stackhouse, P., Gupta, S., Cox, S., Mikovitz, J., Zhang, T., and Chiacchio, M.: 12-year surface radiation budget data set, GEWEX News, 14, 10–12, 2004.

Su, H., McCabe, M. F., and Wood, E. F.: Modeling evapotranspiration during SMACEX: comparing two approaches for local- and regional-scale prediction, J. Hydrometeorol., 6, 910–922, 2005.

Su, Z.: The Surface Energy Balance System (SEBS) for estimation of turbulent heat fluxes, Hydrol. Earth Syst. Sci., 6, 85–100, doi:10.5194/hess-6-85-2002, 2002.

Szczypta, C., Calvet, J.-C., Albergel, C., Balsamo, G., Boussetta, S., Carrer, D., Lafont, S., and Meurey, C.: Verification of the new ECMWF ERA-Interim reanalysis over France, Hydrol. Earth Syst. Sci., 15, 647–666, doi:10.5194/hess-15-647-2011, 2011.

Vinukollu, R. K., Sheffield, J., Wood, E. F., Bosilovich, M. G., and Mocko, D.: Multimodel analysis of energy and water fluxes: intercomparisons between operational analyses, a land surface model, and remote sensing, J. Hydrometeorol., 13, 3–26, 2011.

Wang, K. and Dickinson, R.: A review of global terrestrial evapotranspiration: Observation, modeling, climatology, and climatic variability, Rev. Geophys., 50, RG2005, doi:10.1029/2011RG000373, 2012.

Wang, K., Wan, Z., Wang, P., Sparrow, M., Liu, J., Zhou, X., and Haginoya, S.: Estimation of surface long wave radiation and broadband emissivity using Moderate Resolution Imaging Spectroradiometer (MODIS) land surface temperature/emissivity products, J. Geophys. Res., 110, D11109, doi:10.1029/2004JD005566, 2005.

Xie, P. and Arkin, P. A.: Global precipitation: A 17-year monthly analysis based on gauge observations, satellite estimates, and numerical model outputs, B. Am. Meteorol. Soc., 78, 2539–2558, 1997.

Zhang, K., Kimball, J. S., Nemani, R. R., and Running, S. W.: A continuous satellite-derived global record of land surface evapotranspiration from 1983 to 2006, Remote Sens. Environ., 46, W09522, doi:10.1029/2009WR008800, 2010.

Zhang, K., Kimball, J. S., Nemani, R. R., Running, S. W., Hong, Y., Gourley, J. J., and Yu, Z.: Vegetation Greening and Climate Change Promote Multidecadal Rises of Global Land Evapotranspiration, Sci. Rep., 5, 15956, doi:10.1038/srep15956, 2015.

7

Probabilistic inference of ecohydrological parameters using observations from point to satellite scales

Maoya Bassiouni[1], **Chad W. Higgins**[1], **Christopher J. Still**[2], **and Stephen P. Good**[1]

[1]Department of Biological and Ecological Engineering, Oregon State University, Corvallis, OR 97331, USA
[2]College of Forestry, Oregon State University, Corvallis, OR 97331, USA

Correspondence: Maoya Bassiouni (bassioum@oregonstate.edu)

Abstract. Vegetation controls on soil moisture dynamics are challenging to measure and translate into scale- and site-specific ecohydrological parameters for simple soil water balance models. We hypothesize that empirical probability density functions (pdfs) of relative soil moisture or soil saturation encode sufficient information to determine these ecohydrological parameters. Further, these parameters can be estimated through inverse modeling of the analytical equation for soil saturation pdfs, derived from the commonly used stochastic soil water balance framework. We developed a generalizable Bayesian inference framework to estimate ecohydrological parameters consistent with empirical soil saturation pdfs derived from observations at point, footprint, and satellite scales. We applied the inference method to four sites with different land cover and climate assuming (i) an annual rainfall pattern and (ii) a wet season rainfall pattern with a dry season of negligible rainfall. The Nash–Sutcliffe efficiencies of the analytical model's fit to soil observations ranged from 0.89 to 0.99. The coefficient of variation of posterior parameter distributions ranged from <1 to 15 %. The parameter identifiability was not significantly improved in the more complex seasonal model; however, small differences in parameter values indicate that the annual model may have absorbed dry season dynamics. Parameter estimates were most constrained for scales and locations at which soil water dynamics are more sensitive to the fitted ecohydrological parameters of interest. In these cases, model inversion converged more slowly but ultimately provided better goodness of fit and lower uncertainty. Results were robust using as few as 100 daily observations randomly sampled from the full records, demonstrating the advantage of analyzing soil saturation pdfs instead of time series to estimate ecohydrological parameters from sparse records. Our work combines modeling and empirical approaches in ecohydrology and provides a simple framework to obtain scale- and site-specific analytical descriptions of soil moisture dynamics consistent with soil moisture observations.

1 Introduction

The movement of water from soils, through plants, and back to the atmosphere via transpiration is a critical component of local and global hydrologic cycles and is the largest surface-to-atmosphere water pathway (Good et al., 2015). A realistic analytical description of soil moisture dynamics is key to understanding ecohydrological processes that regulate the productivity of natural and managed ecosystems. Rodriguez-Iturbe et al. (1999) introduced a simple framework using a bucket model of soil-column hydrology forced with stochastic precipitation inputs where soil water losses are only a function of relative soil moisture or soil saturation. Given this ecohydrological framework, the analytical equation for the probability density function (pdf) of soil saturation depends on simple abiotic characteristics such as average climate and soil texture, and biotic characteristics including soil saturation thresholds at which vegetation can influence soil water losses. However, the shapes of analytical soil saturation pdfs are generally not consistent with observations when literature values for model parameters are used (Miller et al., 2007). Some parameters such as field capacity and wilting point do not correspond to conventional definitions, because of simplifications made to describe soil water loss processes in the model, and need to be calibrated (Dralle and Thomspon,

2016). To our knowledge, parameters of the analytical soil saturation pdfs have not been directly calibrated to empirical pdfs derived from measurements beyond the point scale. Observation networks provide freely available point-scale, spatially integrated soil moisture observations, while remotely sensed soil moisture observations are available through satellite products. These data sources create an opportunity to (i) evaluate whether analytical soil saturation pdfs are consistent with observations across a range of scales, and (ii) determine average ecohydrological parameters relevant to each scale.

Estimates of ecohydrological parameters are used in a large range of applications for which the stochastic soil water balance framework has been used and adapted, including the effects of climate, soil, and vegetation on soil moisture dynamics (Laio et al., 2001a; Rodriguez-Iturbe et al., 2001; Porporato et al., 2004); ecohydrological factors driving spatial and structural characteristics of vegetation (Caylor et al., 2006; Manfreda et al., 2017); soil salinization dynamics (Suweis et al., 2010); biological soil crusts (Whitney et al., 2017); vegetation stress; optimum plant water use strategies and plant hydraulic failure (Laio et al., 2001b; Manzoni et al., 2014; Feng et al., 2017); vertical root distributions (Laio et al., 2006); plant pathogen risk (Thomspon et al., 2013); streamflow persistence in seasonally dry landscapes (Dralle et al., 2016); and soil water balance partitioning (Good et al., 2014, 2017). A survey of nearly 400 ecohydrology publications revealed that 40 % of studies relied heavily on simulation, rarely integrated empirical measurements, and were almost never coupled with experimental studies, suggesting a critical need to combine modeling and empirical approaches in ecohydrology (King and Caylor, 2011). Only a few studies have directly confronted the governing equations of the stochastic soil water balance model with observed soil moisture data, and even fewer studies have attempted to optimize model parameters to best fit soil moisture observations. Miller et al. (2007) calibrated soil saturation pdfs to project vegetation stress in a changing climate. Dralle and Thompson (2016) developed an analytical expression for annually integrated soil saturation pdfs under seasonal climates and then calibrated soil saturation thresholds between which evapotranspiration is maximum and zero to compare the model to soil moisture observations at a savanna site. Chen et al. (2008) related evapotranspiration observations at the stand scale to soil moisture values using a Bayesian inversion approach, and Volo et al. (2014) calibrated the soil moisture loss curve to investigate effects of irrigation scheduling and precipitation on soil moisture dynamics and plant stress. The functional form of the soil moisture losses was approximated using conditionally averaged precipitation (Salvucci, 2001; Saleem and Salvucci, 2002) and remotely sensed data (Tuttle and Salvucci, 2014). The timescale of soil moisture dry-downs, derived from the soil moisture loss equations, was parameterized using evapotranspiration measured at micro-meteorological stations (Teuling et al., 2006) and space-borne near-surface soil moisture observations (McColl et al., 2017). These studies indicate that the ecohydrological soil water balance framework is consistent with ground and larger-scale remotely sensed measurements.

Parameters representative of larger-scale observations are necessary to characterize ecohydrological processes at ecosystem scales and are more relevant to ecohydrological modeling. These larger-scale parameters integrate a range of ecohydrological interactions that are poorly understood and difficult to measure. Abiotic controlling factors of soil water balance including rainfall and soil texture can generally be assessed from readily available data, including site measurements, regionalized maps, and satellite observations, but vegetation controls on soil water dynamics are largely unknown and difficult to measure at hydrologically meaningful scales (Li et al., 2017). Vegetation water-use traits are generally observed at the species level and are not easily translated to the simple parameters necessary in soil water balance models. The rate of soil water losses from the near-surface soil layer, where soil moisture measurements are generally made, do not precisely correspond to evapotranspiration observed or calculated from meteorological stations. We thus focused on estimating parameters that are not directly observable, particularly the soil saturation thresholds at which vegetation controls soil water losses and the maximum rate of evapotranspiration from a near-surface soil layer. We use an inverse modeling approach and data that are commonly collected at environmental monitoring sites or measured from satellites. We present an inference framework that provides a means to quantify and compare the sensitivity of soil moisture dynamics at varying scales through estimates of simple ecohydrological parameters.

A number of studies have combined inverse modeling approaches with ground and remotely sensed soil moisture data to extract meaningful hydrologic information (Xu et al., 2006; Miller et al., 2007; Chen et al., 2008; Volo et al., 2014; Wang et al., 2016; Baldwin et al., 2017). Bayesian inference methods are effective in relating prior pdfs of observations to posterior estimates of model parameters (Xu et al., 2006; Chen et al., 2008; Baldwin et al., 2017). The soil water balance model provides a direct analytical equation for soil saturation pdfs that is convenient to use with the Bayesian paradigm because it is a low parameter model with few data inputs. We selected a Bayesian inversion approach instead of a least-squares or maximum likelihood approach because it quantifies the inference uncertainty and improves upon the work of Miller et al. (2007), which used a least-squares approach to calibrate soil saturation pdfs. Measures of inference uncertainty and parameter convergence diagnostics provided by the Bayesian approach can be used to evaluate the validity of model inversion and develop criteria to generalize the presented framework.

We assume that if a sufficient range of soil moisture values are observed at a site, the shape of the empirical soil saturation pdf is constrained by the ecohydrological factors driving soil moisture dynamics. We hypothesize that key information required to determine these ecohydrological factors is encoded in empirical soil saturation pdfs and that this information can be extracted by calculating the inverse of the commonly used stochastic soil water balance. The analysis of soil saturation pdfs is a more robust and integrated approach to investigate ecohydrological factors of soil water dynamics than is time series analysis. Soil saturation pdfs are less sensitive to the many sources of uncertainty, sensor noise, and common gaps in soil moisture observations and do not require high-quality, co-located, and concurrent hydrologic measurements that are often lacking. We tested three key assumptions embedded in the proposed method. (i) The analytical soil saturation pdfs properly describe empirical soil saturation pdfs observed in annual data. Annual soil moisture records can be affected by transitional dynamics between wet and dry seasons, and the appropriate level of model complexity must be used. We compare parameter identifiability using an annual and a seasonal formulation of the analytical soil saturation pdfs. (ii) Parameter estimates and their uncertainty at point, footprint, and satellite scales are different and reflect variability in soil water dynamics. We determine whether the inference approach can be applied at point, footprint, and satellite scales to provide appropriate scale-specific parameters for ecohydrological modeling. (iii) The range of realizable soil moisture values is captured by the selected time series and the soil saturation pdf determined from these observations is not truncated. We determine whether the inference method based on soil saturation pdfs is robust against reduced data availability by repeating the model inversions on subsets of the soil moisture time series and show that the method can be applied to sparse datasets.

Our goal was to match empirical soil saturation pdfs derived from point-, footprint-, and satellite-scale observations to a commonly used analytical model. We demonstrate the use of a Bayesian inversion framework to calibrate the ecohydrological parameters of a simple stochastic soil water balance model that best fit empirical soil saturation pdfs. We first present data sources, define the analytical model for soil saturation pdfs including parameter assumptions, and detail the algorithm used in the Bayesian inversion. Then, we present a summary of the goodness of fit of optimal analytical soil saturation pdfs and estimated parameter uncertainty. We evaluated results to test key method assumptions including model complexity and data availability. Finally, we discuss the potential of the approach to provide a simple means to investigate variability in ecohydrological controlling factors at varying spatial scales. Our work combines modeling and empirical approaches in ecohydrology to provide more realistic analytical descriptions of soil moisture dynamics. Estimates of ecohydrological parameters consistent with observed soil saturation pdfs, from point to ecosystem scales, are needed to better characterize site-specific ecohydrological processes.

2 Data and methods

2.1 Data

We used daily soil moisture observations from three data products at three spatial scales. We used point-scale soil moisture data at a depth of 10 cm from the FLUXNET2015 data product (http://fluxnet.fluxdata.org/data/fluxnet2015-dataset/, last access: 22 October 2016). We used footprint-scale soil moisture data from the Cosmic-ray Soil Moisture Observing System (COSMOS) (http://cosmos.hwr.arizona.edu/Probes/probelist.html, last access: 4 August 2017). The COSMOS soil moisture footprint measures soil moisture at an average depth of 20 cm with a radius ranging from 130 to 240 m, depending on site characteristics (Köhli et al., 2015). Near-surface soil moisture observations at a spatial resolution of 0.25° were taken from the European Space Agency's (ESA) Climate change Initiative (CCI) project. We used the combined soil moisture product (ECV-SM, version 0.2.2) that merges soil moisture retrievals from four passive (SMMR, SMM/I, TMI, and ASMR-E) and two active (AMI and ASCAT) coarse-resolution microwave sensors (Liu et al., 2011, 2012; Wagner et al., 2012). Although the ECV-SM sensing depth is < 5 cm, it has been shown to have a close relation to ground-based observations of soil moisture in the upper 10 cm (Dorigo et al., 2015). We compiled daily rainfall time series from the FLUXNET2015 dataset for the point- and footprint-scale analysis, and the National Aeronautics and Space Administration's (NASA) Tropical Rainfall Measuring Mission (TRMM) dataset (Huffman et al., 2007) for the satellite-scale analysis.

We selected four sites with soil moisture and rainfall data available for the 2012 calendar year (Fig. 1, Table 1). Selected sites spanned a range of land cover types, including crop and grasslands, oak savanna, deciduous forest and pine forest. We determined the dominant soil texture of the upper soil layer from the Harmonized World Soil Database (HWSD) (version 1.2) (FAO/IIASA/ISRIC/ISS-CAS/JRC, 2012) for each site. We used soil porosity values, derived from the HWSD available as ancillary data through the ESA-CCI data product, for the satellite-scale analysis. We used the maximum soil moisture observation during the year 2012 as a site-specific soil porosity estimate for point- and footprint-scale data products. We used soil porosity for each site to calculate soil saturation s ($0 \leq s \leq 1$) from each observed soil moisture value. We do not expect the differences in data quality between data sources and sites to significantly affect empirical soil saturation pdfs and resulting parameter estimates. All sites had full records of daily point- and footprint-scale observations except for US-Me2, which had 55 missing footprint-scale observations during winter when the ground was saturated and frozen. The number of daily satellite-scale observations in the 2012 records ranged from 202 to 283.

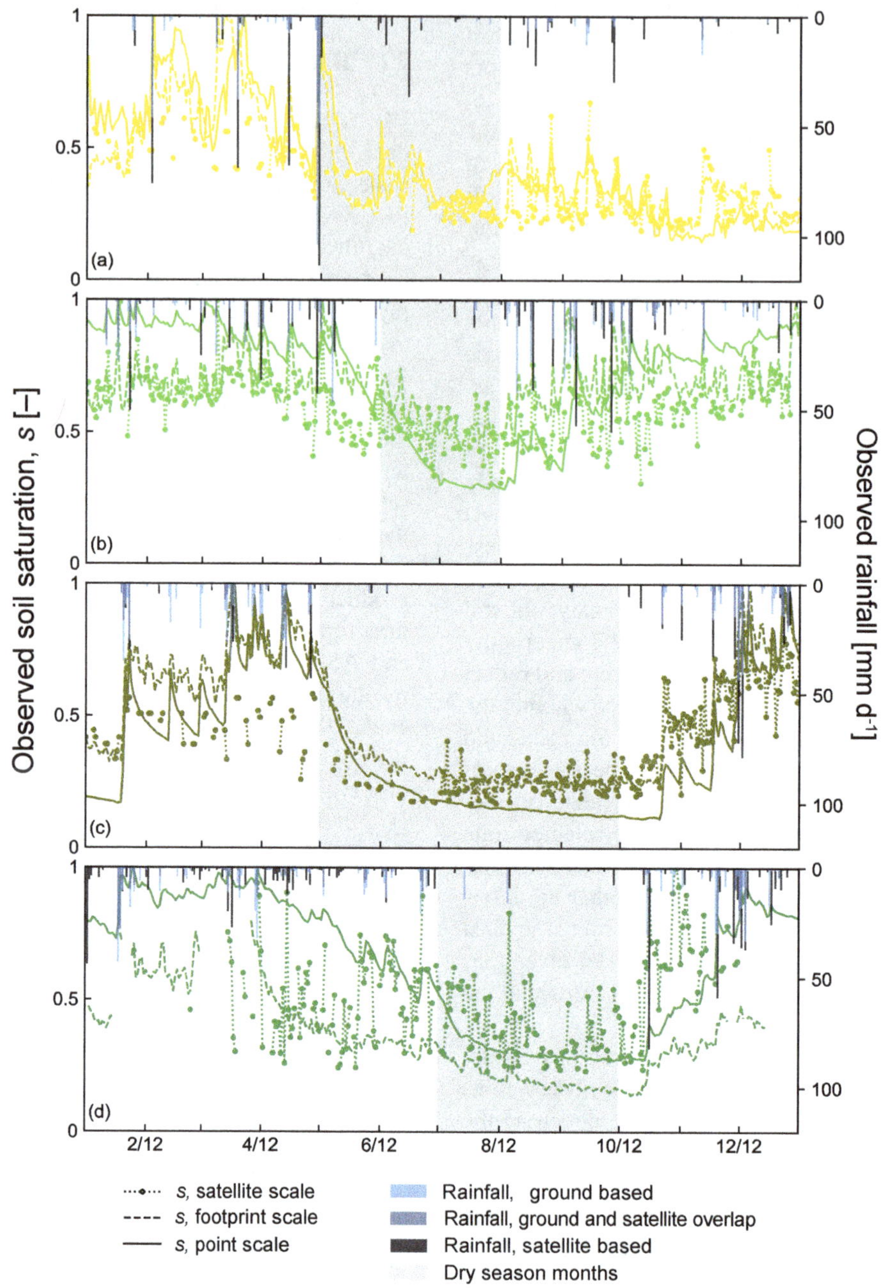

Figure 1. Soil saturation and rainfall time series from **(a)** US-ARM, **(b)** US-MMS, **(c)** US-Ton, and **(d)** US-Me2.

2.2 Analytical model for soil saturation probability density functions (pdfs)

2.2.1 Model definition

Our framework is based on a standard bucket model of soil column hydrology at a point forced with stochastic precipitation inputs and in which soil water losses are a function of soil saturation. We followed the simple formulation of soil water losses in Laio et al. (2001a). We applied two associated analytical formulations for the soil saturation pdf detailed below and derived under the assumption of steady state, wherein parameters are constant for a given period of time. The annual model assumed an annual rainfall pattern and the seasonal model accounted for a wet season rainfall pattern and a dry season of negligible rainfall.

The soil water balance model is defined at a point and a daily time step, for a soil with porosity n, and assuming that soil saturation is uniform in the considered soil column of depth Z. Rainfall, the only input to the soil water balance, is

Table 1. Selected study sites.

Site name	ARM Southern Great Plains	Morgan Monroe State Forest	Tonzi Ranch	Metolius Mature Ponderosa Pine
FLUXNET2015 ID	US-ARM	US-MMS	US-Ton	US-ME2
COSMOS ID	15	27	32	38
Latitude	36.6058 (36.625)	39.3232 (39.375)	38.4316 (38.375)	44.4523 (44.375)
Longitude	−97.4888 (−97.375)	−86.4131 (−86.375)	−120.966 (−120.87)	−97.4888 (−97.375)
Elevation (m)	314	275	177	1253
Vegetation	Crops and grassland	Deciduous forest	Oak savanna	Ponderosa pine forest
Soil texture	Loam	Loam	Loam	Sandy loam
MAT (°C)	14.8	10.9	15.8	6.3
MAP (mm)	843	1032	559	523
α (mm day^{-1})	$21.0^{(p,f)}, 24.4^{(s)}$	$9.04^{(p,f)}, 11.8^{(s)}$	$9.3^{(p,f)}, 16.9^{(s)}$	$8.1^{(p,f)}, 11.6^{(s)}$
α_w (mm day^{-1})	$21.4^{(p,f)}, 26.8^{(s)}$	$9.1^{(p,f)}, 11.9^{(s)}$	$8.7^{(p,f)}, 16.7^{(s)}$	$7.9^{(p,f)}, 11.6^{(s)}$
λ (day^{-1})	$0.05^{(p,f)}, 0.08^{(s)}$	$0.24^{(p,f)}, 0.20^{(s)}$	$0.22^{(p,f)}, 0.10^{(s)}$	$0.24^{(p,f)}, 0.21^{(s)}$
λ_w (day^{1})	$0.07^{(p,f)}, 0.08^{(s)}$	$0.27^{(p,f)}, 0.23^{(s)}$	$0.39^{(p,f)}, 0.17^{(s)}$	$0.31^{(p,f)}, 0.27^{(s)}$
t_d (days)	92	61	153	92
n (–)	$0.35^{(p)}, 0.34^{(f)}, 0.46^{(s)}$	$0.46^{(p)}, 0.66^{(f)}, 0.43^{(s)}$	$0.53^{(p)}, 0.39^{(f)}, 0.43^{(s)}$	$0.36^{(p)}, 0.59^{(f)}, 0.41^{(s)}$
K_s (mm day^{-1})	317	317	317	622
b (–)	4.55	4.55	4.55	3.11
s_h (–)	0.06	0.06	0.06	0.09
s_{fc} (–)	$0.81^{(p)}, 0.75^{(f)}, 0.44^{(s)}$	$0.93^{(p)}, 0.86^{(f)}, 0.69^{(s)}$	$0.75^{(p)}, 0.83^{(f)}, 0.69^{(s)}$	$0.94^{(p)}, 0.60^{(f)}, 0.72^{(s)}$
s_{min} (–)	$0.15^{(p)}, 0.19^{(f)}, 0.19^{(s)}$	$0.28^{(p)}, 0.44^{(f)}, 0.30^{(s)}$	$0.11^{(p)}, 0.22^{(f)}, 0.17^{(s)}$	$0.27^{(p)}, 0.14^{(f)}, 0.23^{(s)}$
s_{max} (–)	$1.0^{(p)}, 1.0^{(f)}, 0.67^{(s)}$	$1.0^{(p)}, 1.0^{(f)}, 1.0^{(s)}$	$1.0^{(p)}, 1.0^{(f)}, 0.80^{(s)}$	$1.0^{(p)}, 1.0^{(f)}, 1.0^{(s)}$
Mean s (–)	$0.44^{(p)}, 0.42^{(f)}, 0.33^{(s)}$	$0.71^{(p)}, 0.68^{(f)}, 0.59^{(s)}$	$0.38^{(p)}, 0.49^{(f)}, 0.38^{(s)}$	$0.64^{(p)}, 0.35^{(f)}, 0.50^{(s)}$
Standard deviation s (–)	$0.21^{(p)}, 0.19^{(f)}, 0.11^{(s)}$	$0.21^{(p)}, 0.11^{(f)}, 0.12^{(s)}$	$0.25^{(p)}, 0.23^{(f)}, 0.17^{(s)}$	$0.25^{(p)}, 0.16^{(f)}, 0.18^{(s)}$

Latitude and longitude in parentheses correspond the centroid of the satellite area associated with the site location; MAT, mean annual temperature from long-term FLUXNET2015 data; MAP, mean annual precipitation from long-term FLUXNET2015 data; soil texture taken from the HWSD; n, porosity; K_s, saturated soil hydraulic conductivity; b, pore size distribution index; s_h, hydroscopic point; s_{fc}, field capacity; α, observed average daily rainfall depth in 2012; the subscript "w" indicates that α was computed for only the wet season months; λ, observed average daily rainfall frequency in 2012; the subscript "w" indicates that λ was computed for only the wet season months; t_d, number of days in the dry season; superscripts $^{(p)}$, $^{(f)}$, and $^{(s)}$ correspond to values used for the point-, footprint-, and satellite-scale analysis. Citations for each FLUXNET2015 site: Biraud (2002), Novick and Phillips (1999), Law (2002), and Baldocchi (2001).

treated as a Poisson distribution characterized by an average event frequency λ and average event intensity α. For simplicity, we assumed that the rainfall applied was equal to the amount that reached the ground surface and that interception by vegetation was negligible. Interception may be a significant component of the soil water balance at forested sites and may need to be considered in future extensions of this work. The daily soil water balance is the difference between the rate of rainfall infiltration φ and the rate of soil moisture losses χ :

$$nZ\frac{\mathrm{d}s(t)}{\mathrm{d}t} = \varphi[s(t);t] - \chi[s(t)]. \tag{1}$$

$\varphi[s(t);\ t]$ is both a stochastic process controlled by rainfall and also a state-dependent process because excess rainfall relative to available soil storage is converted to surface runoff. the soil moisture loss curve, $\chi[s(t)]$, includes leakage losses due to gravity and evapotranspiration and is described in stages determined by five soil saturation thresholds (Laio et al., 2001a). These stages are: (i) the saturation point ($s = 1$), at which all pores are filled with water; (ii) the field capacity (s_{fc}), at which soil-gravity drainage becomes negligible compared to evaporation; (iii) the point of incipient stomata closure (s^*), at which plants begin to reduce transpiration from water stress; (iv) the wilting point (s_w), at which plants cease to transpire; and (v) the hydroscopic point (s_h), at which water is bound to the soil matrix. Soil water losses are controlled by physical soil properties for saturation states above s_{fc}. The rate of leakage due to gravity is assumed maximum when soil is saturated (K_s) and decays exponentially to zero at s_{fc} (Brooks and Corey, 1964). Soil water losses are controlled by micro-meteorological conditions for saturation states between s_{fc} and s^*. The rate of evapotranspiration is assumed to occur at a maximum rate (E_{max}), independent of the saturation state. Soil water losses are controlled primarily by vegetation for saturation states between s^* and s_w. Plants close their stomata in response to soil water deficits that drive leaf water potential gradients, as well as to atmospheric vapor pressure deficits, and evapotranspiration decreases linearly from E_{max} to E_w at s_w. Soil water losses are controlled by soil diffusivity for soil saturation states below s_w, and soil evaporation decreases linearly from E_w to zero at s_h. Soil water losses are negligible for soil saturation states below s_h. For this simplified theoretical description of the soil water loss curve and stochastic rainfall forcing, the analytical solution of the steady state probability distributions of soil saturation, $p(s)$, was given by Laio et al. (2001a):

$$p(s)=\begin{cases} 0, & 0<s\le s_\mathrm{h},\\ \dfrac{C}{\eta_\mathrm{w}}\left(\dfrac{s-s_\mathrm{h}}{s_\mathrm{w}-s_\mathrm{h}}\right)^{\frac{\lambda(s_\mathrm{w}-s_\mathrm{h})}{\eta_\mathrm{w}}-1} e^{-\gamma s}, & s_\mathrm{h}<s\le s_\mathrm{w},\\ \dfrac{C}{\eta_\mathrm{w}}\left[1+\left(\dfrac{\eta}{\eta_\mathrm{w}}-1\right)\left(\dfrac{s-s_\mathrm{w}}{s^*-s_\mathrm{w}}\right)\right]^{\frac{\lambda(s^*-s_\mathrm{w})}{\eta-\eta_\mathrm{w}}-1} e^{-\gamma s}, & s_\mathrm{w}<s\le s^*,\\ \dfrac{C}{\eta}e^{-\gamma s+\frac{\lambda}{\eta}(s-s^*)}\,\dfrac{\eta}{\eta_\mathrm{w}}^{\frac{\lambda(s^*-s_\mathrm{w})}{\eta-\eta_\mathrm{w}}}, & s^*<s\le s_\mathrm{fc},\\ \dfrac{C}{\eta}e^{-(\beta+\gamma)s+\beta s_\mathrm{fc}}\left(\dfrac{\eta e^{\beta s}}{(\eta-m)e^{\beta s_\mathrm{fc}}+me^{\beta s}}\right)^{\frac{\lambda}{\beta(\eta-m)}+1} & \\ \times\dfrac{\eta}{\eta_\mathrm{w}}^{\frac{\lambda(s^*-s_\mathrm{w})}{\eta-\eta_\mathrm{w}}} e^{\frac{\lambda}{\eta}(s_\mathrm{fc}-s^*)}, & s_\mathrm{fc}<s\le 1, \end{cases} \tag{2}$$

where

$$\frac{1}{\gamma}=\frac{\alpha}{nZ},$$

$$\eta_\mathrm{w}=\frac{E_\mathrm{w}}{nZ},$$

$$\eta=\frac{E_\mathrm{max}}{nZ},$$

$$m=\frac{K_\mathrm{s}}{nZ\left(e^{\beta(1-s_\mathrm{fc})}-1\right)},$$

$$\beta=2b+4,$$

where b is an experimentally determined parameter used in the Clapp and Hornberger (1978) soil water retention curve, and the constant C can be obtained numerically to ensure the integral of $p(s)=1$. We used a simplifying relation $E_\mathrm{w}=0.05E_\mathrm{max}$ to reduce the number of parameters.

We adopted Dralle and Thompson's (2016) framework to account for transient dynamics between wet and dry seasons. We defined the dry season as a period of duration t_d in which precipitation was negligible and assumed to not contribute to soil moisture. During that period, we assumed soil saturation decayed from an initial value s_0 to $s(t_\mathrm{d}, s_0)$ given by Laio et al. (2001a). For simplicity, we determined t_d using rainfall records at a monthly step (see Sect. 2.2.2) and s_0 was the soil saturation value on the last day of the wet season. Note that we did not define s_0 as the soil saturation following the last significant storm of the wet season as was done in prior studies (Dralle and Thompson, 2016). We then calculated the annual soil saturation pdf ($p_\mathrm{wd}(s)$) as the weighted sum of the wet and dry season pdfs, $p_\mathrm{w}(s)$ and $p_\mathrm{d}(s)$, respectively.

$$p_\mathrm{wd}(s)=\left(1-\frac{t_\mathrm{d}}{365}\right)p_\mathrm{w}(s)+\frac{t_\mathrm{d}}{365}p_\mathrm{d}(s) \tag{3}$$

The steady-state solution in Eq. (2) was used for the wet season pdf and the dry season pdf is numerically determined by

$$p_\mathrm{d}(s)=\int_0^1 p_{S_\mathrm{d}|S_0}(s,s_0)\,p_0(s_0)\,\mathrm{d}s_0, \tag{4}$$

where $p_0(s_0)$ is the pdf of the initial dry season soil saturation, equal to $p_\mathrm{w}(s)$, and $p_{S_\mathrm{d}|S_0}(s, s_0)$ is the pdf of dry season soil saturation given an initial condition s_0.

$$p_{S_\mathrm{d}|S_0}(s,s_0)=\frac{C_\mathrm{d}}{t_\mathrm{d}}\begin{cases} \dfrac{e^{\beta(s_0-s)}}{\left(\eta^\mathrm{d}-m\right)e^{\beta(s_0-s)}-\eta^\mathrm{d}+m+me^{\beta(s_0-s_\mathrm{fc})}}, & s_\mathrm{fc}<s\le 1,\\ \dfrac{1}{\eta^\mathrm{d}}, & s^*<s\le s_\mathrm{fc},\\ \dfrac{1}{\eta^\mathrm{d}-\eta_\mathrm{w}^\mathrm{d}}\left(\dfrac{s^*-s_\mathrm{w}}{\left(\eta^\mathrm{d}-\eta_\mathrm{w}^\mathrm{d}\right)(s-s_\mathrm{w})+\eta_\mathrm{w}^\mathrm{d}(s^*-s_\mathrm{w})}\right), & s_\mathrm{w}<s\le s^*\\ \dfrac{1}{\eta_\mathrm{w}^\mathrm{d}}\left(\dfrac{s_\mathrm{w}-s_\mathrm{h}}{s-s_\mathrm{h}}\right), & s_\mathrm{h}<s\le s_\mathrm{w}\\ 0, & s_0\le s,\\ 0, & s\le s_\mathrm{h},\\ 0, & s\le s(t_\mathrm{d},s_0) \end{cases} \tag{5}$$

where η^d and $\eta_\mathrm{w}^\mathrm{d}$ are equivalent to η and η_w relative to $E_\mathrm{max}^\mathrm{d}$, the maximum evapotranspiration rate during the dry season, and C_d is a normalization constant. We used the analytical expression for soil saturation decay, $s(t, s_0)$, in the absence of rainfall given by Laio et al. (2001a) to derive $p_{S_\mathrm{d}|S_0}(s, s_0)$.

2.2.2 Climate, soil, and vegetation parameter characterization

We chose readily available data for rainfall characteristics (λ and α), length of the dry period (t_d), and physical soil parameters (s_fc, s_h, K_s, and b) needed in the analytical models of soil saturation pdfs (Eqs. 2 and 3). We focused on estimating the ecohydrological parameters s^*, s_w, and E_max, which describe vegetation control on soil water losses and are not easily observable.

We calculated rainfall characteristics λ and α for the year and wet season months for each site from FLUXNET2015 and TRMM rainfall records following Rodriguez-Iturbe et al. (1984) (Table 1). We used FLUXNET2015 rainfall characteristics for point- and footprint-scale analyses, and we used TRMM rainfall characteristics for the satellite-scale analysis. TRMM rainfall records were generally consistent with ground-based measurements. For each location, we evaluated monthly FLUXNET2015 rainfall depth and categorized consecutive months contributing $<5\,\%$ of the site's annual rainfall as dry season months (Fig. 1). We then calculated the length of the dry period (t_d) as the number of days in those dry months. We used physical soil characteristics for soil textures at each site (s_h, K_s, and b) from Rawls et al. (1982) (Table 1). We estimated s_fc from each soil saturation record (Table 1) to be consistent with the assumption that drainage losses are insignificant compared to evapotranspiration losses the day following a rain event. We identified all days in the 2012 record following an observed decrease in soil saturation and estimated s_fc as the 95th percentile of the soil saturation value of the selected days. Daily soil saturation below s_w and above s_fc is rare (Laio et al., 2001a), so we did not expect the average soil texture values for s_h and K_s to significantly affect the results. Soil depths Z are 10, 20, and 5 cm for the point, footprint, and satellite scales, respectively. E_{max} is only a fraction of the atmospheric moisture demand (or potential evapotranspiration) contributed by that soil depth because we used a soil depth that is shallower than the rooting depth. Consequently our framework includes four (or three if seasonality is ignored) unknown soil water balance parameters, s^*, s_w, E_{max}, and E^d_{max}. We estimated these parameters over the following intervals:

$$\begin{cases} s_h \le s^* \le s_{fc}, \\ s_h \le s_w \le s_{fc}, \\ 0 \le E_{max} \le 10, \\ 0 \le E^d_{max} \le 10, \end{cases} \tag{6}$$

where 10 mm day^{-1} is the pre-defined upper possible boundary for potential evapotranspiration.

2.3 Bayesian inversion approach

2.3.1 Application of the Bayes theorem

We related $p(S)$, the empirical soil saturation pdf of the $j = [1, \ldots, \text{m}]$ soil saturation observations (s_j), and the analytical soil saturation pdfs in Eqs. (2) or (3) derived from the simple soil water balance model in Eq. (1) with up to four unknown soil water balance parameters $\theta = [s^*, s_w, E_{max}, E^d_{max}]$ using the Bayes' theorem defined as

$$p(\theta|S) = \frac{p(S|\theta)p(\theta)}{p(S)}, \tag{7}$$

where the posterior distribution, $p(\theta|S)$, is the solution of the inverse problem and describes the probability of model parameters θ given the set $S = [s_1 s_2, \ldots\ s_m]$ of soil saturation observations. Assuming uninformed prior knowledge, the prior distribution of model parameters θ, $p(\theta)$, were defined by uniform distributions over the intervals (Eq. 6). The conditional probability of observations S given model parameters θ, $p(S|\theta)$, is the likelihood function of model parameters θ.

2.3.2 Parameter estimation

We used the Metropolis–Hastings Markov chain Monte Carlo (MH-MCMC) technique to estimate the posterior distribution of $p(\theta|S)$ by drawing random model samples θ_i from $p(\theta)$ and evaluating $p(S|\theta_i)$ (Metropolis et al., 1953; Hastings, 1970; Xu et al., 2006). We defined the likelihood function of a model i, $p(S|\theta_i)$, as

$$p(S|\theta_i) = \prod_{j=1}^{m} p(s_j|\theta_i), \tag{8}$$

where $p(s_j|\theta_i)$ is the probability of observation s_j given Eqs. (2) or (3) using parameters θ_i.

The MH-MCMC technique converges to a stationary distribution according to the ergodicity theorem in Markov chain theory. The sampling algorithm consisted of repeating two steps: (i) a proposing step, in which the algorithm generates a new model θ'_i using a random function that is symmetric about the previously accepted model θ_i, and (ii) a moving step, to determine whether the model should be accepted or rejected, in which θ'_i is tested against the Metropolis criterion (a) defined as

$$a = \frac{p(S|\theta'_i)}{p(S|\theta_i)}. \tag{9}$$

If $a > 1$, θ_i was accepted and $\theta_{i+1} = \theta'_i$ was used for the next sample. If $a < 1$, a random number $p_* \in [0, 1]$ was drawn from a uniform distribution and compared to a. If $p_* < a$, θ'_i was accepted and $\theta_{i+1} = \theta'_i$ was used for the next sample. If $p_* > a$, θ'_i was rejected and $\theta_{i+1} = \theta_i$ was used for the next sample. If θ'_i was an inconsistent model in which soil saturation thresholds (s_w, s_*) were ranked incorrectly or any of the soil water balance parameters (s^*, s_w, E_{max}, E^d_{max}) were outside of their defined physical bounds, the model likelihood was zero and θ'_i was never accepted. The log-likelihood was more convenient to compute than the likelihood. The symmetric function used in the proposing step was a Gaussian distribution with a mean value equal to the accepted model θ_i and a standard deviation of 1 % of interval range for which each parameter is defined in Eq. (6). We selected this value of the standard deviation of each model parameter after a number of test runs to generally ensure an acceptance rate between 20 and 50 % (Roberts and Rosenthal, 2001). We obtained statistics of the estimated parameters in θ from the union of three run samples of 20 000 simulations each. The burn-in period is the number of simulations after which the running mean and standard deviation are stabilized. We considered a burn-in period of 10,000 simulations, which were discarded for each run sample. If the acceptance rate of a run sample was < 1 or > 90 % after the burn-in period, we discarded the run and concluded that the algorithm was stuck in a local minimum that might be physically impossible. We evaluated convergence by the Gelman–Rubin (GR) diagnostic (Gelman and Rubin, 1992) on the run samples. The GR diagnostic determines that the algorithm reaches convergence when the within-run variability (σ_w) is roughly equal to the between-run variability (σ_b), that is, when σ_w/σ_b approaches one. We verified that the GR diagnostic for each estimated parameter was < 1.1. If the GR diagnostic did not indicate that the three run samples converged, we discarded the run with the lowest likelihood and re-initiated a new run sample until convergence was attained. We counted the number of attempts to quantify how rapidly convergence occurred. We computed mean and standard deviation for each parameter from a total of 30 000 simulations of θ resulting from the three converging run samples. A mean analytical model of soil saturation pdf was determined by applying Eqs. (2) or (3) with the mean values of the 30 000 posterior parameter estimates.

2.4 Model evaluation criteria

We did not have direct measurement to validate the parameters s^*, s_w, and E_{max} estimated through the Bayesian inversion methods. We therefore analysed convergence and uncertainty metrics of the model inversion and goodness of fit between empirical and analytical soil saturation pdfs to evaluate the identifiability of the ecohydrological parameters. We compared the optimum analytical pdf derived from the mean parameter estimates and the empirical pdfs derived from observations. We evaluated the model inversion using the following criteria:

i. Convergence of the Bayesian inversion: a GR diagnostic < 1.1 for all unknown parameters is obtained from the union of three run samples and within ≤ 10 sample runs.

ii. Low uncertainty in parameter estimates: the posterior distributions of parameter estimates are physically plausible and have coefficients of variations $< 20\,\%$.

iii. Goodness of fit: a quantile-level Nash–Sutcliffe efficiency (NSE) (Müller et al., 2014) > 0.85 and a Kolmogorov–Smirnov statistic < 0.2.

2.5 Method assessment

Major assumptions and limitations embedded in the proposed inference framework were tested through the analysis detailed below. We assume, for each scale and location, that the shape of empirical the soil saturation pdfs is controlled by the physical constraints used to parameterize the analytical model of soil saturation pdfs, these parameters can be determined with some certainty and reflect variability in soil water dynamics. We expect that estimated soil saturation thresholds have greater certainty when the empirical soil saturation pdf is defined around those values and greater uncertainty when fewer soil saturation values are observed around the thresholds. We acknowledge that pre-defined rainfall characteristics and physical soil parameters based on observations or literature values may not be exactly representative of the processes at each location or scale and could also create biases and uncertainties in the fitted parameters of interest. We used model evaluation criteria (Sect. 2.4) to investigate the applicability of the inference framework with varying model complexities, scales, locations and data availability.

i. Analytical expressions for soil saturation pdfs were derived under the assumption of steady state. Annual soil moisture records can be affected by transitional dynamics between wet and dry seasons, and the appropriate level of model complexity must be used. We applied the inversion framework to annual soil saturation using variations of the analytical model for soil saturation pdfs of increasing complexity: (i) the annual model in Eq. (2) and (ii) the seasonal model in Eq. (3). We determined whether the added complexity of the dry season pdf increases the identifiability of ecohydrological parameters or if the simpler annual model is sufficiently consistent with annual empirical soil saturation pdfs.

ii. We compared co-located parameter estimates and their uncertainty at point, footprint, and satellite scales for each site. We determine whether the inference approach can provide appropriate scale-specific parameters for ecohydrological modeling at each location.

iii. We assumed that the whole range of realizable soil saturation values was captured within the selected time series at each scale and that the resulting soil saturation pdf was not truncated. If the range of observed values is not representative of the soil saturation pdf because it is truncated or affected by noise in the data, parameter estimates may be biased. Minimum and maximum observed soil saturation values during 2012 (Table 1) indicate the range of observed soil saturation values we used to estimate ecohydrological parameters. We determine whether the inference method based on soil saturation pdfs is robust against reduced data availability by repeating the model inversions on subsets of the soil saturation time series and show that the method can be applied to sparse datasets. We performed the model inversion using subsets of each soil saturation record by randomly resampling fractions of the data down to $10\,\%$ of the annual timeseries and computed goodness of fit statistics between the resulting analytical models and the empirical models based on the full annual record. We determined the number of data points necessary to infer converging model parameters that best match observations and whether the proposed inference method based on soil saturation pdf can be reliably used to identify ecohydrological parameters from sparse datasets.

3 Results and discussion

3.1 Level of model complexity

For each of the four locations (Table 1), we obtained optimal analytical soil saturation pdfs consistent with the empirical pdfs derived from soil saturation observations using the Bayesian inversion framework and a MH-MCMC algorithm. Model inversions for each site and scale and for both annual and seasonal models met the evaluation criteria (see Sect. 2.4). Our results indicated that the framework of Dralle and Thompson (2016) can be applied to sites with low (US-MMS) and high (US-Ton) seasonality in rainfall patterns. Posterior probability distributions of soil water balance parameters (s_w, s^*, E_{max}) were well constrained overall. The parameter estimates and their coefficient of variation as well as the model goodness of fit statistics are summarized in Table 2. Figures 2 through 5 present a comparison between

Table 2. Estimated ecohydrological parameters and goodness of fit of analytical soil saturation pdfs.

Site name	Scale	N		NSE		KS		E_{max}		s^*		s_w	
		p	p_{wd}	p	p_{wd}	p	p_{wd}	p	p_{wd}	p	p_{wd}	p	p_{wd}
US-ARM	point	4	4	0.96	0.96	0.07	0.07	1.1 (11)	1.3 (14)	0.7 (8)	0.74 (5)	0.19 (4)	0.27 (7)
	footprint	3	3	0.94	0.94	0.08	0.06	1.7 (11)	2 (12)	0.62 (7)	0.61 (9)	0.24 (3)	0.29 (2)
	satellite	3	3	0.96	0.97	0.08	0.09	0.7 (13)	0.5 (13)	0.42 (4)	0.42 (4)	0.24 (3)	0.25 (2)
US-Ton	point	3	4	0.95	0.97	0.09	0.08	2.3 (4)	1.9 (10)	0.24 (6)	0.33 (7)	0.12 (1)	0.18 (6)
	footprint	3	3	0.94	0.98	0.13	0.08	2.2 (3)	1.8 (8)	0.29 (2)	0.4 (10)	0.25 (0)	0.26 (1)
	satellite	3	9	0.99	0.99	0.06	0.07	1.2 (15)	1 (13)	0.53 (12)	0.62 (6)	0.22 (3)	0.26 (3)
US-MMS	point	3	4	0.96	0.98	0.12	0.08	1.3 (3)	1.1 (6)	0.34 (3)	0.5 (8)	0.29 (0)	0.31 (2)
	footprint	3	3	0.95	0.95	0.13	0.08	2.7 (6)	4.5 (10)	0.82 (2)	0.79 (3)	0.38 (5)	0.59 (1)
	satellite	3	6	0.95	0.88	0.1	0.14	0.7 (8)	0.9 (10)	0.65 (4)	0.66 (3)	0.28 (9)	0.43 (2)
US-Me2	point	3	8	0.95	0.97	0.16	0.1	1.4 (3)	1.1 (7)	0.33 (3)	0.37 (8)	0.29 (0)	0.29 (1)
	footprint	3	6	0.94	0.94	0.09	0.1	2.1 (2)	2.9 (10)	0.23 (4)	0.45 (5)	0.15 (2)	0.2 (6)
	satellite	3	4	0.89	0.89	0.12	0.1	1.6 (12)	1.4 (15)	0.64 (8)	0.66 (8)	0.25 (3)	0.31 (4)

Values in parentheses correspond to the coefficient of variation of the posterior parameter estimates in percentage. p, analytical model for the soil saturation pdf without seasons, p_{wd}, analytical model for the soil saturation pdf including wet and dry seasons; N, number of 20 000 simulation runs needed to obtain three converging results (see Sect. 2.3.2); NSE, quantile-level Nash–Sutcliffe efficiency; KS, Kolmogorov–Smirnov statistic; E_{max}, maximum evapotranspiration in mm day^{-1} (the weighted average wet and dry season E_{max} is reported for the p_{wd} model); s^*, point of incipient stomatal closure; s_w, wilting point.

empirical as well as analytical pdfs and associated quantile–quantile plots for point, footprint, and satellite scales at the four study sites and for both annual and seasonal models. The goodness of fit between empirical pdfs and analytical models was only slightly better for the seasonal model than for the annual model. However, the coefficient of variation of the posterior parameter distributions was smaller for the annual model and it converged more rapidly. The Bayesian inversion of the annual model is therefore more computationally efficient. The parameter identifiability was not greatly improved by the more complex seasonal model. The estimated soil saturation threshold s_w was consistently smaller for the annual model than for the seasonal model and s^* was often higher, which may indicate that s_w and s^* in the annual model could be biased and may have absorbed dry season dynamics. Previous studies calibrating soil saturation pdf models found ecohydrological parameter values comparable to ours (Table 2). For example, using point-scale observations at US-Ton, best-fit values of s_w and s_{fc} were 0.26 and 0.82, respectively (Dralle and Thompson, 2016), and best-fit values of s^* and E_{max} were 0.3 and 1.9 mm day^{-1}, respectively (Miller et al., 2007). We did not compare soil saturation thresholds s^* and s_w with literature values of soil water potential at which stomata are fully open or closed because the conversion of soil saturation to soil matrix potential is non-linear (Clapp and Hornberger, 1978) and site- and scale-specific soil water retention parameters were unknown. Average parameters derived from soil texture (Rawls et al., 1982) were not compatible with soil moisture data from each scale and site.

3.2 Site and scale considerations

Parameter estimates were most constrained for scales and locations at which soil water dynamics are more sensitive to the fitted ecohydrological parameters of interest. In these cases, convergence of the model inversion was attained less rapidly, but ultimately provided better goodness of fit. Soil saturation states at drier sites may be more controlled by soil water loss parameters, while soil saturation states at wetter sites may also be controlled by rainfall characteristics. Estimated soil saturation thresholds had greater certainty if the empirical soil saturation pdfs were defined around those values and had greater uncertainty if there were fewer soil saturation values observed around the thresholds. For example, uncertainty of s_w was greater for the humid subtropical deciduous forest site (US-MMS) than for the Mediterranean savanna sites (US-Ton), and uncertainty of s^* was greater for US-Ton than US-MMS. Similarly, soil saturation states representing larger spatial scales were less sensitive to specific site characteristics.

Parameter uncertainty for satellite and footprint scales was greater than for the point scale. Estimates of larger-scale soil water balance parameters are more relevant to regional ecohydrological dynamics. Differences in parameter estimates among scales within a site may be associated with differences in soil texture properties, such as porosity and field capacity, that were determined separately for each record. Co-located and concurrent soil saturation pdfs are different at each scale (Figs. 2–5) and suggest variability in observed soil water dynamics at each scale. Differences in driving processes among scales were specifically determined from the model inversion for each scale and provided robust scale-specific parameters for ecohydrological modeling.

3.3 Data availability

For each spatial scale and site, the annual model was inversed, using random subsamples of 100 to 10 % of the 2012 time series (Fig. 6). For all sites and scales the number of observations did not significantly impact model inference.

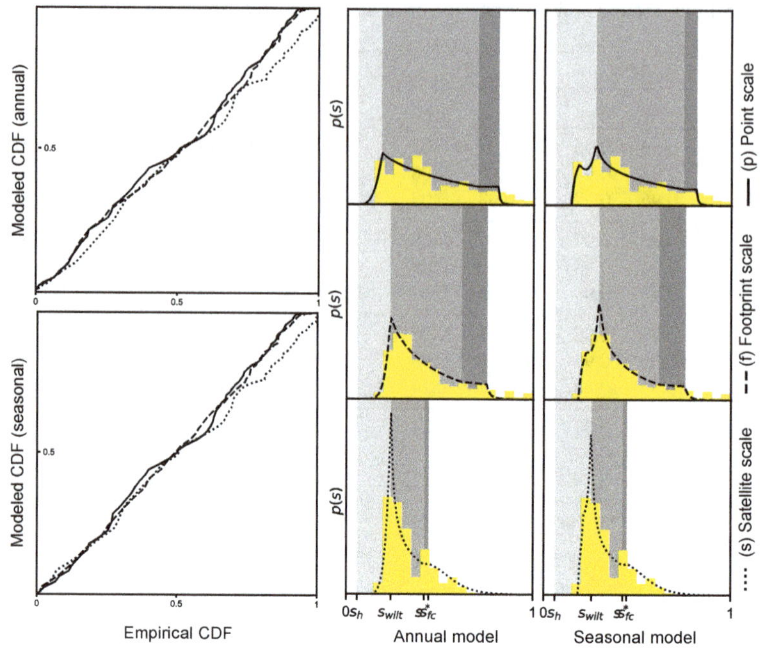

Figure 2. Empirical versus modeled cumulative density functions (CDFs) and soil saturation probability distribution ($p(s)$) for US-ARM. The mean values of the posterior parameter distributions were used with Eq. (2) in the annual model and Eq. (3) in the seasonal model. The grey shaded areas correspond to the soil saturation thresholds (s_{h}, s_{w}, s^*, s_{fc}) in the water balance model.

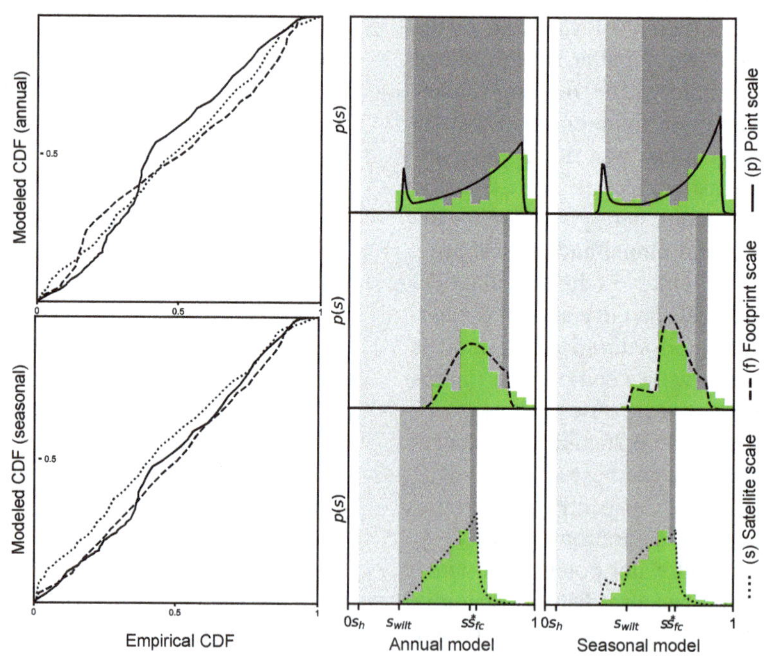

Figure 3. Empirical versus modeled CDFs and soil saturation probability distribution ($p(s)$) for US-MMS. The mean values of the posterior parameter distributions were used with Eq. (2) in the annual model and Eq. (3) in the seasonal model. The grey shaded areas correspond to the soil saturation thresholds (s_{h}, s_{w}, s^*, s_{fc}) in the water balance model.

The NSE, Kolmogorov–Smirnov statistic, and parameter estimates were stable down to about 100 observations. Fitted model parameter values and the variability of parameter estimates among the 10 repetitions in each subsample category were not sensitive to the number of observations used. Results indicate the identifiability of ecohydrological parameters through the inversion of the analytical model of soil saturation pdfs was robust because the mean and standard de-

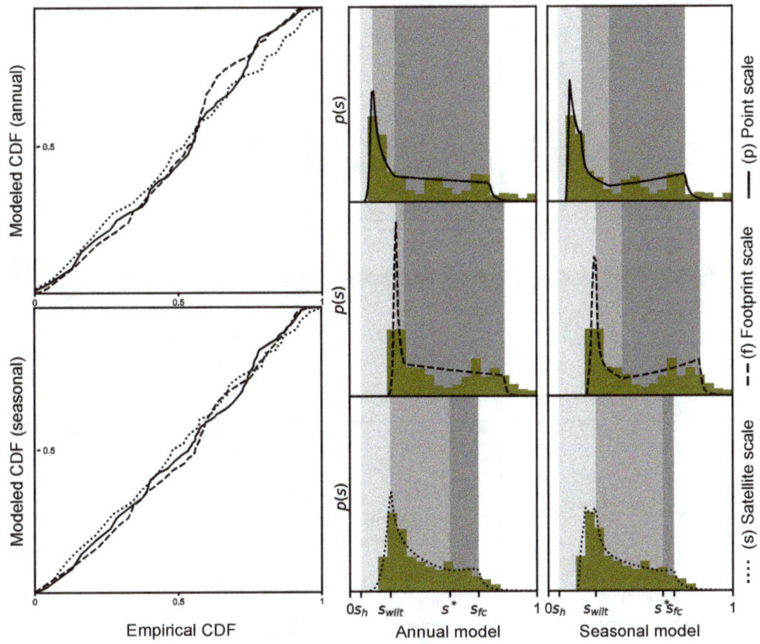

Figure 4. Empirical versus modeled CDFs and soil saturation probability distribution ($p(s)$) for US-Ton. The mean values of the posterior parameter distributions were used with Eq. (2) in the annual model and Eq. (3) in the seasonal model. The grey shaded areas correspond to the soil saturation thresholds (s_h, s_w, s^*, s_fc) in the water balance model.

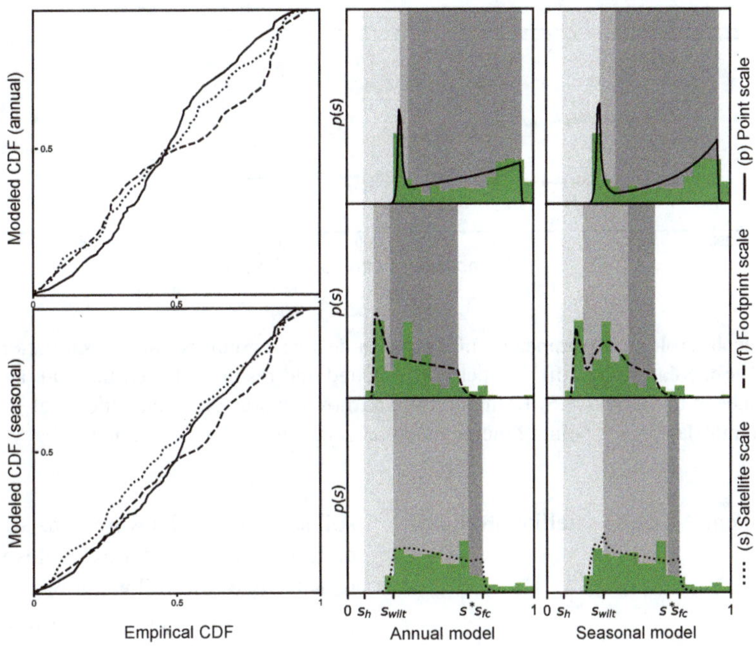

Figure 5. Empirical versus modeled CDFs and soil saturation probability distribution ($p(s)$) for US-Me2. The mean values of the posterior parameter distributions were used with Eq. (2) in the annual model and Eq. (3) in the seasonal model. The grey shaded areas correspond to the soil saturation thresholds (s_h, s_w, s^*, s_fc) in the water balance model.

viation of the randomly selected subsets of annual data were representative of the full record. There was no correlation between the small differences in the mean and standard deviations of the subsamples and the model goodness of fit. The proposed inference method based on soil saturation pdfs can therefore reliably be used to identify ecohydrological parameters from sparse datasets. Inference methods, which do not require continuous data, are particularly relevant to large-

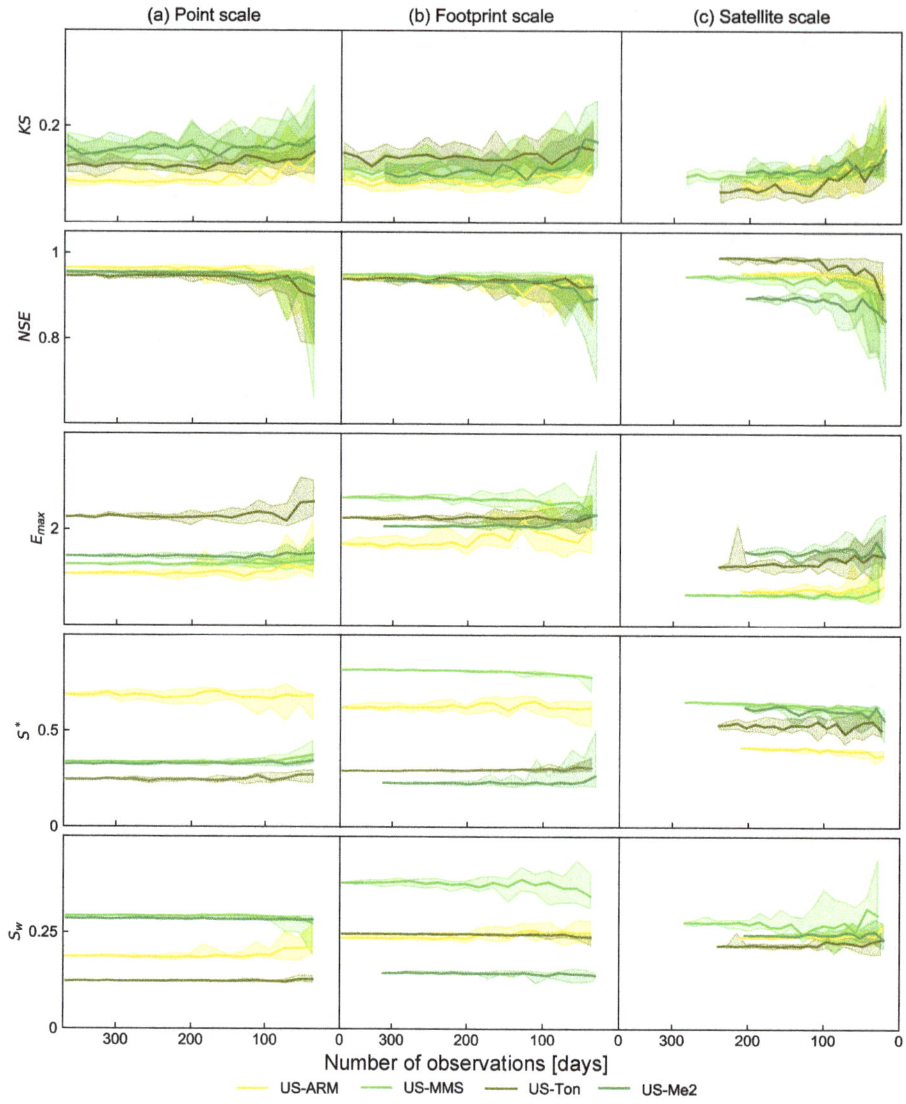

Figure 6. Goodness of fit and ecohydrological parameters inferred with decreasing number of soil saturation observations (annual model). For each subsample category, the median results of 10 repeats are plotted and results between the 90th and 10th percentiles are shaded. Colors correspond to the four sites in the legend. KS, Kolmogorov–Smirnov statistic; NSE, quantile-level Nash–Sutcliffe efficiency; $E_{\max}$, maximum evapotranspiration in mm day^{-1}; s^*, point of incipient stomatal closure; s_{w}, wilting point.

scale soil moisture measurements, such as satellite products, that are not continuous.

4 Conclusions

We document a generalizable Bayesian inversion framework to infer parameter values of the stochastic soil water balance model and their associated uncertainty using freely available rainfall and soil moisture observations at point-, footprint- and satellite-scales. Empirical pdfs derived from soil saturation observations provided key information to determine unknown ecohydrological parameters s^*, s_{w}, and $E_{\max}$. Model assumptions were appropriately met, and optimal analytical soil saturation pdfs were consistent with empirical pdfs. Uncertainty in parameter estimates were small and reflected the sensitivity of the soil water balance model to ecohydrological parameters at varying scales and locations. We demonstrate that the form of the simple ecohydrological model for soil saturation pdfs was consistent with observations from point-, footprint-, and satellite-scales. However, optimal parameters were different at each scale because co-located and concurrent soil saturation pdfs are different at each scale, which may result from spatial heterogeneity in soil water dynamics. We demonstrate the advantage of analyzing soil saturation pdfs instead of time series. We obtained stable results using sparse subsets of the datasets, indicating that the proposed framework is robust and can be used with non-continuous data. Although the seasonal model was conceptually more consistent with our physical understanding of annual soil water dynamics, the annual model provided satisfactory results matching annual empirical pdf sites we analysed. We were not able to determine if some differences in parameters estimated using the seasonal model are physically meaningful because wet and dry season dynamics were better characterized in this more complex model. Our methodology can be customized to characterize site-specific parameters and to test consistency between observed and analytical soil saturation pdfs for any other adaptation of the stochastic ecohydrological framework with more or less complexity depending on the study objectives.

We provide a method based on a parsimonious soil water balance model, requiring a minimum level of data inputs to estimate ecohydrological characteristics that are not directly observable and for which established estimation methods are not available. Our methods can be applied in future studies to better understand differences in soil water dynamics at different scales and to improve scaling of ecohydrological processes. Results demonstrate the value of large-scale near-surface soil moisture observations to improve characterization of soil water dynamics at ecosystem scales. Relations between the soil saturation threshold values inferred from the near-surface soil moisture data and dynamics in the full active rooting zone are unknown. The datasets we used are freely available from sensor networks and global satellite products, and methods can therefore be applied to a large range of sites or to global analyses to improve understanding of spatial patterns in ecohydrological parameters relevant for local and global water cycle analyses.

Code and data availability. We downloaded all datasets from publicly available sources. Point-scale soil moisture and rainfall data are available through FLUXNET2015 (http://fluxnet.fluxdata.org/data/fluxnet2015-dataset/); footprint-scale soil moisture data are available through COSMOS (http://cosmos.hwr.arizona.edu/Probes/probelist.html); remotely sensed soil moisture data are available through ESA CCI (http://www.esa-soilmoisture-cci.org/node/145); remotely sensed rainfall data are available through NASA TRMM (https://pmm.nasa.gov/data-access/downloads/trmm); global soil texture data are available through FAO HWSD (http://www.fao.org/soils-portal/soil-survey/soil-maps-and-databases/harmonized-world-soil-database-v12/en/). Custom scripts in the Python computing language associated with our analysis are open source (Bassiouni, 2018, https://doi.org/10.5281/zenodo.1283371).

Competing interests. The authors declare that they have no conflict of interest.

Acknowledgements. We thank Minghui Zhang, Marc Müller, David Dralle, Xue Feng, and editor Sally Thomspon for their thoughtful reviews and useful feedback on an earlier draft of this paper. This material is based upon work supported by the National Science Foundation Graduate Research Fellowship under grant no. 1314109-DGE. Stephen P. Good acknowledges the financial support of the US National Aeronautics and Space Administration (NNX16AN13G). This work used the Extreme Science and Engineering Discovery Environment (XSEDE) via allocation DEB160018, supported by National Science Foundation grant number ACI-1548562. This work used data acquired and shared by the FLUXNET community, including these networks: AmeriFlux, AfriFlux, AsiaFlux, CarboAfrica, CarboEuropeIP, CarboItaly, CarboMont, ChinaFlux, Fluxnet-Canada, GreenGrass, ICOS, KoFlux, LBA, NECC, OzFlux-TERN, TCOS-Siberia, and USCCC. The FLUXNET eddy covariance data processing and harmonization were carried out by the European Fluxes Database Cluster, AmeriFlux Management Project, and Fluxdata project of FLUXNET, with the support of CDIAC and the ICOS Ecosystem Thematic Center and the OzFlux, ChinaFlux, and AsiaFlux offices.

Edited by: Sally Thompson

References

Baldocchi, D.: AmeriFlux US-Ton Tonzi Ranch, https://doi.org/10.17190/AMF/1245971, 2001.

Baldwin, D., Manfreda, S., Keller, K., and Smithwick, E. A. H.: Predicting root zone soil moisture with soil properties and satellite near-surface moisture data across the conterminous United States, J. Hydrol., 546, 393–404, https://doi.org/10.1016/j.jhydrol.2017.01.020, 2017.

Bassiouni, M.: PIEP, https://doi.org/10.5281/zenodo.1283371, 2018.

Biraud, S.: Ameriflux US-ARM ARM Southern Great Plains site-Lamont, https://doi.org/10.17190/AMF/1246027, 2002.

Brooks, R. and Corey, T.: Hydraulic properties of porous media, hydrology papers, Colorado State University, 1964.

Caylor, K. K., D'Odorico, P., and Rodriguez-Iturbe, I.: On the ecohydrology of structurally heterogeneous semi-arid landscapes, Water Resour. Res., 42, W07424, https://doi.org/10.1029/2005WR004683, 2006.

Chen, X., Rubin, Y., Ma, S., and Baldocchi, D.: Observations and stochastic modeling of soil moisture control on evapotranspiration in a Californian oak savanna: soil moisture control on ET, Water Resour. Res., 44, W08409, https://doi.org/10.1029/2007WR006646, 2008.

Clapp, R. B. and Hornberger, G. M.: Empirical equations for some soil hydraulic properties, Water Resour. Res., 14, 601–604, 1978.

COSMOS Dataset, http://cosmos.hwr.arizona.edu/Probes/probelist.html, last access: 4 August 2016.

Dorigo, W. A., Gruber, A., De Jeu, R. A. M., Wagner, W., Stacke, T., Loew, A., Albergel, C., Brocca, L., Chung, D., Parinussa, R. M., and Kidd, R.: Evaluation of the ESA CCI soil moisture product using ground-based observations, Remote Sens. Environ., 162, 380–395, https://doi.org/10.1016/j.rse.2014.07.023, 2015.

Dralle, D. N. and Thompson, S. E.: A minimal probabilistic model for soil moisture in seasonally dry climates, Water Resour. Res., 52, 1507–1517, https://doi.org/10.1002/2015WR017813, 2016.

Dralle, D. N., Karst, N. J., and Thompson, S. E.: Dry season streamflow persistence in seasonal climates, Water Resour. Res., 52, 90–107, https://doi.org/10.1002/2015WR017752, 2016.

FAO/IIASA/ISRIC/ISS-CAS/JRC: Harmonized World Soil Database (version 1.1), FAO, Rome, Italy and IIASA, Laxenburg, Austria, 2009.

Feng, X., Dawson, T. E., Ackerly, D. D., Santiago, L. S., and Thompson, S. E.: Reconciling seasonal hydraulic risk and plant water use through probabilistic soil-plant dynamics, Global Change Biol., 23, 3758–3769, https://doi.org/10.1111/gcb.13640, 2017.

FLUXNET2015 Dataset: http://fluxnet.fluxdata.org/data/fluxnet2015-dataset/, last access: 22 October 2016.

Gelman, A. and Rubin, D. B.: Inference from iterative simulation using multiple sequences, Stat. Sci., 7, 457–472, 1992.

Global Soil Moisture Dataset (ESA CCI SM v04.2): http://www.esa-soilmoisture-cci.org/node/145, last access: 21 February 2016.

Good, S. P., Soderberg, K., Guan, K., King, E. G., Scanlon, T. M., and Caylor, K. K.: δ^2H isotopic flux partitioning of evapotranspiration over a grass field following a water pulse and subsequent dry down, Water Resour. Res., 50, 1410–1432, https://doi.org/10.1002/2013WR014333, 2014.

Good, S. P., Noone, D., and Bowen, G.: Hydrologic connectivity constrains partitioning of global terrestrial water fluxes, Science, 349, 175–177, 2015.

Good, S. P., Moore, G. W., and Miralles, D. G.: A mesic maximum in biological water use demarcates biome sensitivity to aridity shifts, Nat. Ecol. Evol., 1, 1883–1888, https://doi.org/10.1038/s41559-017-0371-8, 2017.

Harmonized World Soil Database v 1.2: http://www.fao.org/soils-portal/soil-survey/soil-maps-and-databases/harmonized-world-soil-database-v12/en/, last access: 21 February 2016.

Hastings, W. K.: Monte Carlo Sampling Methods Using Markov Chains and Their Applications, Biometrika, 57, 97–109, https://doi.org/10.2307/2334940, 1970.

Huffman, G. J., Bolvin, D. T., Nelkin, E. J., Wolff, D. B., Adler, R. F., Gu, G., Hong, Y., Bowman, K. P., and Stocker, E. F.: The TRMM Multisatellite Precipitation Analysis (TMPA): Quasi-Global, Multiyear, Combined-Sensor Precipitation Estimates at Fine Scales, J. Hydrometeorol., 8, 38–55, https://doi.org/10.1175/JHM560.1, 2007.

King, E. G. and Caylor, K. K.: Ecohydrology in practice: strengths, conveniences, and opportunities, Ecohydrology, 4, 608–612, https://doi.org/10.1002/eco.248, 2011.

Köhli, M., Schrön, M., Zreda, M., Schmidt, U., Dietrich, P., and Zacharias, S.: Footprint characteristics revised for field-scale soil moisture monitoring with cosmic-ray neutrons, Water Resour. Res., 51, 5772–5790, https://doi.org/10.1002/2015WR017169, 2015.

Laio, F., Porporato, A., Ridolfi, L., and Rodriguez-Iturbe, I.: Plants in water-controlled ecosystems: active role in hydrologic processes and response to water stress: II. Probabilistic soil moisture dynamics, Adv. Water Resour., 24, 707–723, 2001a.

Laio, F., Porporato, A., Fernandez-Illescas, C. P., and Rodriguez-Iturbe, I.: Plants in water-controlled ecosystems: active role in hydrologic processes and response to water stress: IV. Discussion of real cases, Adv. Water Resour., 24, 745–762, 2001b.

Laio, F., D'Odorico, P., and Ridolfi, L.: An analytical model to relate the vertical root distribution to climate and soil properties: vertical root distribution, Geophys. Res. Lett., 33, L18401, https://doi.org/10.1029/2006GL027331, 2006.

Law, B.: AmeriFlux US-Me2 Metolius mature ponderosa pine, https://doi.org/10.17190/AMF/1246076, 2002.

Li, Y., Guan, K., Gentine, P., Konings, A. G., Meinzer, F. C., Kimball, J. S., Xu, X., Anderegg, W. R. L., McDowell, N. G., Martínez-Vilalta, J., Long, D. G., and Good, S. P.: Estimating global ecosystem iso/anisohydry using active and passive microwave satellite data, J. Geophys. Res.-Biogeo., 122, 3306–3321, https://doi.org/10.1002/2017JG003958, 2017.

Liu, Y. Y., Parinussa, R. M., Dorigo, W. A., De Jeu, R. A. M., Wagner, W., van Dijk, A. I. J. M., McCabe, M. F., and Evans, J. P.: Developing an improved soil moisture dataset by blending passive and active microwave satellite-based retrievals, Hydrol. Earth Syst. Sci., 15, 425–436, https://doi.org/10.5194/hess-15-425-2011, 2011.

Liu, Y. Y., Dorigo, W. A., Parinussa, R. M., de Jeu, R. A. M., Wagner, W., McCabe, M. F., Evans, J. P., and van Dijk, A. I. J. M.: Trend-preserving blending of passive and active microwave soil moisture retrievals, Remote Sens. Environ., 123, 280–297, https://doi.org/10.1016/j.rse.2012.03.014, 2012.

Manfreda, S., Caylor, K. K., and Good, S. P.: An ecohydrological framework to explain shifts in vegetation organization across climatological gradients: Vegetation pattern in dry environments, Ecohydrology, 10, e1809, https://doi.org/10.1002/eco.1809, 2017.

Manzoni, S., Vico, G., Katul, G., Palmroth, S., and Porporato, A.: Optimal plant water-use strategies under stochastic rainfall, Water Resour. Res., 50, 5379–5394, https://doi.org/10.1002/2014WR015375, 2014.

McColl, K. A., Wang, W., Peng, B., Akbar, R., Short Gianotti, D. J., Lu, H., Pan, M., and Entekhabi, D.: Global characterization of surface soil moisture drydowns:, Geophys. Res. Lett., 44, 3682–3690, https://doi.org/10.1002/2017GL072819, 2017.

Metropolis, N., Rosenbluth, A. W., Rosenbluth, M. N., Teller, A. H., and Teller, E.: Equation of State Calculations by Fast Computing Machines, J. Chem. Phys., 21, 1087–1092, https://doi.org/10.1063/1.1699114, 1953.

Miller, G. R., Baldocchi, D. D., Law, B. E., and Meyers, T.: An analysis of soil moisture dynamics using multi-year data from a network of micrometeorological observation sites, Adv. Water Resour., 30, 1065–1081, https://doi.org/10.1016/j.advwatres.2006.10.002, 2007.

Müller, M. F., Dralle, D. N., and Thompson, S. E.: Analytical model for flow duration curves in seasonally dry climates, Water Resour. Res., 50, 5510–5531, https://doi.org/10.1002/2014WR015301, 2014.

Novick, K. and Phillips, R.: AmeriFlux US-MMS Morgan Monroe State Forest, https://doi.org/10.17190/AMF/1246080, 1999.

Porporato, A., Daly, E., and Rodriguez-Iturbe, I.: Soil water balance and ecosystem response to climate change, Am. Nat., 164, 625–632, 2004.

Rawls, W. J., Brakensiek, D. L., and Saxtonn, K. E.: Estimation of soil water properties, T. ASAE, 25, 1316–1320, 1982.

Roberts, G. O. and Rosenthal, J. S.: Optimal Scaling for Various Metropolis-Hastings Algorithms, Stat. Sci., 16, 351–367, 2001.

Rodriguez-Iturbe, I., Gupta, V. K., and Waymire, E.: Scale considerations in the modeling of temporal rainfall, Water Resour. Res., 20, 1611–1619, 1984.

Rodriguez-Iturbe, I., Porporato, A., Ridolfi, L., Isham, V., and Coxi, D. R.: Probabilistic modelling of water balance at a point: the role of climate, soil and vegetation, P. Roy. Soc. Lond. A, 455, 3789–3805, 1999.

Rodriguez-Iturbe, I., Porporato, A., Laio, F., and Ridolfi, L.: Intensive or extensive use of soil moisture: plant strategies to cope with stochastic water availability, Geophys. Res. Lett., 28, 4495–4497, 2001.

Saleem, J. A. and Salvucci, G. D.: Comparison of soil wetness indices for inducing functional similarity of hydrologic response across sites in Illinois, J. Hydrometeorol., 3, 80–91, 2002.

Salvucci, G. D.: Estimating the moisture dependence of root zone water loss using conditionally averaged precipitation, Water Resour. Res., 37, 1357–1365, 2001.

Suweis, S., Rinaldo, A., Van der Zee, S. E. A. T. M., Daly, E., Maritan, A., and Porporato, A.: Stochastic modeling of soil salinity, Geophys. Res. Lett., 37, L07404, https://doi.org/10.1029/2010GL042495, 2010.

Teuling, A. J., Seneviratne, S. I., Williams, C., and Troch, P. A.: Observed timescales of evapotranspiration response to soil moisture, Geophys. Res. Lett., 33, L23403, https://doi.org/10.1029/2006GL028178, 2006.

Thompson, S., Levin, S., and Rodriguez-Iturbe, I.: Linking Plant Disease Risk and Precipitation Drivers: A Dynamical Systems Framework, Am. Nat., 181, E1–E16, https://doi.org/10.1086/668572, 2013.

TRMM – Tropical Rainfall Measuring Mission Dataset: https://pmm.nasa.gov/data-access/downloads/trmm, last access: 21 February 2016.

Tuttle, S. E. and Salvucci, G. D.: A new approach for validating satellite estimates of soil moisture using large-scale precipitation: Comparing AMSR-E products, Remote Sens. Environ., 142, 207–222, https://doi.org/10.1016/j.rse.2013.12.002, 2014.

Volo, T. J., Vivoni, E. R., Martin, C. A., Earl, S., and Ruddell, B. L.: Modelling soil moisture, water partitioning, and plant water stress under irrigated conditions in desert urban areas, Ecohydrology, 7, 1297–1313, https://doi.org/10.1002/eco.1457, 2014.

Wagner, W., Dorigo, W., de Jeu, R., Fernandez, D., Benveniste, J., Haas, E., and Ertl, M.: Fusion of active and passive microwave observations to create an essential climate variable data record on soil moisture, ISPRS Annals of the Photogrammetry, Remote Sensing and Spatial Information Sciences (ISPRS Annals), 7, 315–321, 2012.

Wang, T., Franz, T. E., Yue, W., Szilagyi, J., Zlotnik, V. A., You, J., Chen, X., Shulski, M. D., and Young, A.: Feasibility analysis of using inverse modeling for estimating natural groundwater recharge from a large-scale soil moisture monitoring network, J. Hydrol., 533, 250–265, https://doi.org/10.1016/j.jhydrol.2015.12.019, 2016.

Whitney, K. M., Vivoni, E. R., Duniway, M. C., Bradford, J. B., Reed, S. C., and Belnap, J.: Ecohydrological role of biological soil crusts across a gradient in levels of development, Ecohydrology, 10, e1875, https://doi.org/10.1002/eco.1875, 2017.

Xu, T., White, L., Hui, D., and Luo, Y.: Probabilistic inversion of a terrestrial ecosystem model: Analysis of uncertainty in parameter estimation and model prediction, Global Biogeochem. Cy., 20, GB2007, https://doi.org/10.1029/2005GB002468, 2006.

8

Quantification of surface water volume changes in the Mackenzie Delta using satellite multi-mission data

Cassandra Normandin[1]**, Frédéric Frappart**[2,3]**, Bertrand Lubac**[1]**, Simon Bélanger**[4]**, Vincent Marieu**[1]**, Fabien Blarel**[3]**, Arthur Robinet**[1]**, and Léa Guiastrennec-Faugas**[1]

[1]EPOC, UMR 5805, Université de Bordeaux, Allée Geoffroy Saint-Hilaire, 33615 Pessac, France
[2]GET-GRGS, UMR 5563, CNRS/IRD/UPS, Observatoire Midi-Pyrénées, 31400 Toulouse, France
[3]LEGOS-GRGS, UMR 5566, CNRS/IRD/UPS, Observatoire Midi-Pyrénées, 31400 Toulouse, France
[4]Dép. Biologie, Chimie et Géographie, groupe BOREAS and Québec-Océan, Université du Québec à Rimouski, 300 allée des ursulines, Rimouski, Qc, G5L 3A1, Canada

Correspondence: Cassandra Normandin (cassandra.normandin@u-bordeaux.fr)

Abstract. Quantification of surface water storage in extensive floodplains and their dynamics are crucial for a better understanding of global hydrological and biogeochemical cycles. In this study, we present estimates of both surface water extent and storage combining multi-mission remotely sensed observations and their temporal evolution over more than 15 years in the Mackenzie Delta. The Mackenzie Delta is located in the northwest of Canada and is the second largest delta in the Arctic Ocean. The delta is frozen from October to May and the recurrent ice break-up provokes an increase in the river's flows. Thus, this phenomenon causes intensive floods along the delta every year, with dramatic environmental impacts. In this study, the dynamics of surface water extent and volume are analysed from 2000 to 2015 by combining multi-satellite information from MODIS multispectral images at 500 m spatial resolution and river stages derived from ERS-2 (1995–2003), ENVISAT (2002–2010) and SARAL (since 2013) altimetry data. The surface water extent (permanent water and flooded area) peaked in June with an area of 9600 km^2 ($\pm$200 km^2) on average, representing approximately 70 % of the delta's total surface. Altimetry-based water levels exhibit annual amplitudes ranging from 4 m in the downstream part to more than 10 m in the upstream part of the Mackenzie Delta. A high overall correlation between the satellite-derived and in situ water heights ($R > 0.84$) is found for the three altimetry missions. Finally, using altimetry-based water levels and MODIS-derived surface water extents, maps of interpolated water heights over the surface water extents are produced. Results indicate a high variability of the water height magnitude that can reach 10 m compared to the lowest water height in the upstream part of the delta during the flood peak in June. Furthermore, the total surface water volume is estimated and shows an annual variation of approximately 8.5 km^3 during the whole study period, with a maximum of 14.4 km^3 observed in 2006. The good agreement between the total surface water volume retrievals and in situ river discharges ($R = 0.66$) allows for validation of this innovative multi-mission approach and highlights the high potential to study the surface water extent dynamics.

1 Introduction

Deltas are vulnerable to both anthropogenic and natural forcing such as socio-economic infrastructure development and global warming. In the Arctic, the latter is particularly severe due to the polar amplification processes and complex positive feedback loops (Holmes et al., 2012). This system is undergoing important changes, as the increase in precipitation at high latitudes increases river discharge and melting of stock ices on land and sea (Stocker and Raible, 2005). These changes may induce an acceleration of the hydrologic cycle (Stocker and Raible, 2005). River discharge may increase from 18 to 70 % from now to the end of the century (Peterson et al., 2002). Improving our knowledge on the dynamics of the surface water reservoir in circumpolar areas is crucial for a better understanding of their role in flood hazard, carbon production, greenhouse gases emission, sediment transport, exchange of nutrients and land–atmosphere interactions.

Mapping surface water extent on the scale of the Mackenzie Delta is an important issue. However, it is nearly impossible to provide long-term monitoring with traditional methods using in situ measurements in such a large and heterogeneous environment. Satellite remote sensing method offers a unique opportunity for the continuous observation of wetlands and floodplains. Remote sensing has been proven to have strong potential to detect and monitor floods during the last 2 decades (Alsdorf et al., 2007; Smith, 1997). Typically, two kinds of sensor are used to map flooded areas at high and moderate resolutions: passive multispectral imagery and active synthetic aperture radar (SAR). The spectral signature of the surface reflectance is used to discriminate between water and land (Rees, 2013). The SAR images provide valuable information on the nature of the observed surface through the backscattering coefficient (Ulaby et al., 1981).

If space missions of radar altimetry were mainly dedicated to estimate ocean surface topography (Fu and Cazenave, 2001), it is now commonly used for monitoring inland water levels (Birkett, 1995; Cazenave et al., 1997; Frappart et al., 2006a, 2015b; Santos da Silva et al., 2010; Crétaux et al., 2011a, 2017). Several studies have shown the possibility to measure water levels variations in lakes, rivers and flooding plains (Frappart et al., 2006b, 2015a; Santos da Silva et al., 2010). In the present study, satellite multispectral imagery and altimetry are used in synergy to quantify surface water extents and the surface water volumes of the Mackenzie Delta and analyse their temporal variations. In the past, this approach has been applied in tropical (e.g. the Amazon, Frappart et al., 2012; Mekong, Frappart et al., 2006b) and peri-Arctic (e.g. the Lower Ob' basin, Frappart et al., 2010) major river basins, allowing direct observations of the spatio-temporal dynamics of surface water storage. Several limitations prevent their use over estuaries and deltas. The first is the too-coarse spatial resolution of the datasets used for retrieving the flood extent that ranges from 1 km with SPOT-VGT images used in the lower Mekong Basin to $\sim 0.25°$ with the Global Inundation Extent from Multi-Satellite (GIEMS, Papa et al., 2010) dataset for the Lower Ob' and the Amazon basins. The second is inherent to the datasets used in these studies. For the Mekong Basin, due to the limited number of spectral bands present in the VGT sensor, a mere threshold on the normalized difference vegetation index (NDVI) was applied. For the Amazon and the Lower Ob', as the GIEMS dataset is using surface temperatures from the Special Sensor Microwave Imager (SSM/I), no valid data are available at less than 50 km from the coast. The originality and novelty of the study stem from the use of multi-space mission data at better spatial, temporal and spectral resolutions than the previous studies to monitor surface water storage changes in a deltaic environment over a 15-year time period.

Earlier studies pointed out (i) the lack of continuous information in the Mackenzie delta to study the spatial distribution of water levels during the flood events and to analyse the relationship between flood severity and the timing and duration of break-up in the delta (Goulding et al., 2009b; Beltaos et al., 2012) and (ii) the importance of the tributaries to the Mackenzie River (i.e. Peel and Arctic Red rivers) on break-up and ice-jam flooding in the delta (Goulding et al., 2009a). As the goal of this study is to characterize the spatio-temporal surface and storage dynamics of surface water in the Mackenzie delta, Northwest Territories of Canada, in response to spring ice break-up and snow melt, over the period 2000–2015, it will provide important new information for a better understanding of the hydro-climatology of the region.

2 Study region

The Mackenzie Delta, a floodplain system, is located in the northern part of Canada (Fig. 1a) and covers an area of 13 135 km^2 (Emmerton et al., 2008), making it the second biggest delta of Arctic with a length of 200 km and a width of 80 km (Emmerton et al., 2008). It is mainly drained by the Mackenzie River (90 % of the delta's water supply) and Peel River (8 % of the delta's water supply, Emmerton et al., 2007). The Mackenzie Delta channels have very mild slopes ($-0.02\ m\ km^{-1}$; Hill et al., 2001), and are ice-covered during 7–8 months per year (Emmerton et al., 2007).

The Mackenzie River begins in the Great Slave Lake and then, flows through the Northwest Territories before reaching the Beaufort Sea. It has a strong seasonality in term of discharge due to spring ice break-up and snowmelt, from about 5000 $m^3\ s^{-1}$ in winter up to 40 000 $m^3\ s^{-1}$ in June during the ice break-up for wet years (Fig. 1b, Macdonald and Yu, 2006; Goulding et al., 2009a, b; Beltaos et al., 2012). The Stamukhi (ground accumulation of sea ice) is responsible for recurrent floods in the Mackenzie Delta. At the flood peak, 95 % of the delta surface is likely to be covered with water (Macdonald and Yu, 2006). Water level peaks are mainly controlled by ice break-up effects and secondarily by the amount of water contained in snowpack (Lesack and Marsh, 2010). This is one of the most important annual hydrologic events in cold regions (Muhammad et al., 2016).

The delta is a complex of multiple channels and numerous shallow and small lakes (over 49 000 lakes), covering nearly $\sim 50\ \%$ of the delta area (Emmerton et al., 2007), which are ecologically sensitive environments largely controlled by river water (Squires et al., 2009). This environment is also one of the most productive ecosystems in northern Canada, with large populations of birds, fish and mammals, which are critical resources for local population (Squires et al., 2009).

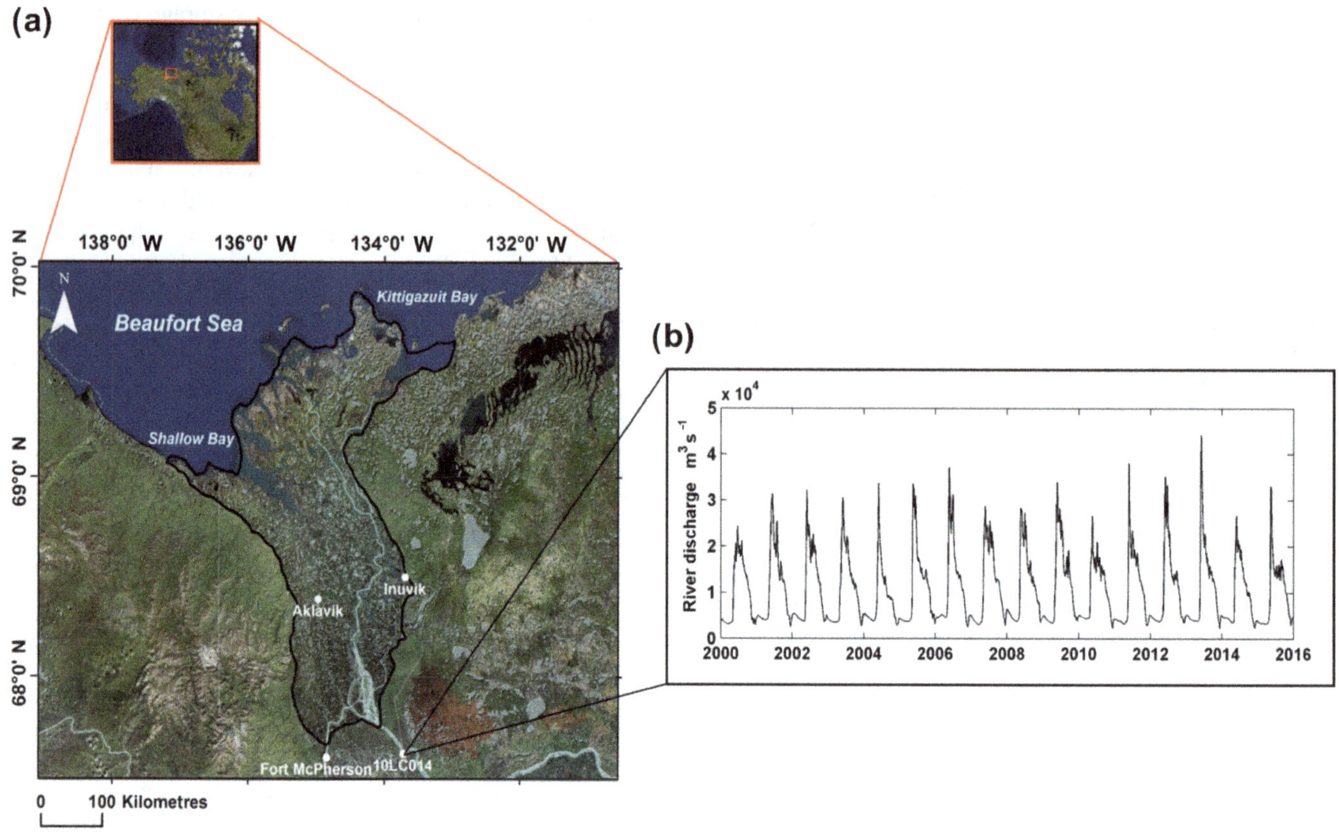

Figure 1. **(a)** Location of the Mackenzie Delta at the mouth of the Mackenzie River in the Northwest Territories of Canada. **(b)** River discharges of the Mackenzie River at 10LC014 station from 2000 to 2015 (133° W, 67° N), 30 km upstream the Mackenzie Delta.

3 Datasets

3.1 Multispectral imagery

3.1.1 MODIS

The Moderate Resolution Imaging Sensor (MODIS) is a spectroradiometer, part of the payload of the Aqua (since 2002) and Terra (since 1999) satellites. The MODIS sensor measures radiances in 36 spectral bands. In this study, the MOD09A1 product (8-day binned level 3, version 6) derived from Terra satellite surface reflectance measurements were downloaded from the United States Geological Survey (USGS) EarthExplorer website (https://ladsweb.modaps.eosdis.nasa.gov/). It consists of gridded, atmospherically corrected surface reflectance acquired in seven bands from visible to shortwave infrared (SWIR) (2155 nm) at a 500 m spatial resolution. This product is obtained by combining for each wavelength the best surface reflectance data of every pixel acquired during an 8-day period. Each MODIS tile covers an area of 1200 km by 1200 km. Two tiles (h12v02 and h13v02) are used to cover the whole study area. In this study, 223 composites, acquired during the ice-free period from June to September over the 2000–2015 time span, are used.

3.1.2 OLI

The Landsat-8 satellite is composed of two Earth-observing sensors, the Operational Land Imager (OLI) and Thermal InfraRed Sensor (TIRS). This satellite was launched in February 2013 and orbits at an altitude of 705 km. The swath is 185 km and the whole Earth surface is covered every 16 days.

The OLI–TIRS sensors measure in 11 spectral bands in the visible (450–680 nm), near-infrared (845–885 nm) and shortwave infrared (1560–2300 nm) portions of the electromagnetic spectrum. In this study, the Landsat-8 OLI surface reflectance products were downloaded from the Landsat-8 USGS portal (http://earthexplorer.usgs.gov/). The multispectral spatial resolutions are 30 and 15 m for panchromatic bands. Two images are necessary to cover the Mackenzie Delta.

Landsat-8 mission is characterized by a lower revisit time than the Terra and Aqua missions. Thus, associated with a high occurrence of clouds over the study area, Landsat-8 yields a small amount of high-quality data. Consequently, OLI images cannot be used in this study to monitor water surface area temporal changes. In this context, MODIS represents a relevant alternative to OLI despite a lower spatial resolution. However, available high-quality OLI data have been used to compare and validate MODIS water surface areas.

3.2 Radar altimetry data

3.2.1 ERS-2

The ERS-2 satellite (European Remote Sensing) was launched in 1995 by the European Space Agency (ESA). Its payload is composed of several sensors, including a radar altimeter (RA), operating at the K_u band (13.8 GHz). It was orbiting sun-synchronously at an altitude of 790 km with an inclination of 98.54° with a 35-day repeat cycle. This orbit was ERS-1's orbit with a ground-track spacing about 85 km at the Equator. ERS-2 provides observations of the topography of the Earth from 82.4° latitude north to 82.4° latitude south. ERS-2 data are disposable from 17 May 1995 to 9 August 2010 but after 22 June 2003, the coverage is limited.

3.2.2 ENVISAT

Envisat mission was launched on 1 March 2002 by ESA. This satellite carried 10 different instruments including the advanced radar altimeter (RA-2). It was based on the heritage of ERS-1 and 2 satellites. RA-2 was a nadir-looking pulse-limited radar altimeter operating at two frequencies at K_u (13.575 GHz) and S (3.2 GHz) bands. Its goal was to collect radar altimetry over ocean, land and ice caps (Zelli, 1999). Envisat remained on its nominal orbit until October 2010 but RA-2 stopped operating correctly at the S band in January 2008. Its initial orbital characteristics are the same as for ERS-2.

3.2.3 SARAL

SARAL mission was launched on 25 February 2013 by a partnership between CNES (Centre National d'Etudes Spatiales) and ISRO (Indian Space Research Organization). Its payload comprised the AltiKa radar altimeter and bi-frequency radiometer, and a triple system for precise orbit determination: the real-time tracking system DIODE of the DORIS instrument, a laser retroflector array (LRA), and the Advance Research and Global Observation Satellite (ARGOS-3). AltiKa is the first radar altimeter to operate at the K_a band (35.75 GHz). It is a solid-state mono-frequency altimeter that provides precise range estimates (Verron et al., 2015). SARAL orbit was earlier utilized by ERS-1 & 2 and ENVSAT missions with a track spacing of 85 km at the Equator (Verron et al., 2015). It has been put on a drifting orbit since 4 July 2016.

Altimetry data used here are contained in the Interim Geophysical Data Records (GDRs) and are the following:

- cycle 001 (17 May 1995) to cycle 085 (7 August 2003) for ERS-2 from the reprocessing of the ERS-2 mission raw waveform performed at Centre de Topographie de l'Océan et de l'Hydrosphère (CTOH) (Frappart et al., 2016)
- GDR v2.1 for ENVISAT from cycle 006 (14 May 2002) to cycle 094 (21 October 2010)
- GDR E for SARAL from cycle 001 (15 March 2015) to cycle 027 (14 October 2015).

These data were made available by CTOH (http://ctoh.legos.obs-mip.fr/). Data were acquired along the altimeter track at 18, 20 and 40 Hz for ENVISAT, ERS −2 and SARAL respectively (high-frequency mode commonly used over land and coastal areas where the surface properties are changing more rapidly than over the open ocean). They consist of the satellite locations and acquisition times and all the parameters necessary to compute the altimeter heights (see Sect. 4.3).

3.3 In situ water levels and discharges

The altimetry-based water level time series derived from radar altimetry were compared to gauge records from in situ stations for validation purpose. Data from 10 gauge stations were found in close vicinity to altimetry virtual stations (VSs; at a distance of less than 20 km along the streams). Virtual stations are built at intersections between an orbit ground rack and a water body (lake, river and floodplain) (Crétaux et al., 2017). Besides, surface water storage variations were compared to the river flow entering the delta, summing the records from three gauge stations located in upstream part of the delta. Daily data of water level and discharge were downloaded for free from the Canadian government website (http://wateroffice.ec.gc.ca).

4 Methods

4.1 Quantification of surface water extent

Multispectral imagery is commonly used for delineating flood extent using spectral indices (e.g. Frappart et al., 2006b; Sakamoto et al., 2007; Crétaux et al., 2011b; Verpoorter et al., 2014; Ogilvie et al., 2015; Pekel et al., 2016). As we do not have any external information to perform a supervised classification using the current state-of-the-art machine learning techniques (Pekel et al., 2016; Tulbure et al., 2016; Klein et al., 2017), we used the approach proposed by Sakamoto et al. (2007) to monitor the water surface area extent in the Mackenzie Delta (Fig. 2). This approach is based on the application of thresholds on the enhanced vegetation index (EVI), the land surface water index (LSWI) and the difference value between EVI and LSWI (DVEL = EVI − LSWI) to determine the status (non-flooded, mixed, flooded and permanent water body) of any pixel in an 8-day MODIS composite image of surface reflectance. As the spectral response of the near infrared (NIR) and shortwave infrared bands is highly dependent on the Earth surface nature, in particular water versus soil–vegetation surfaces, their complement was used to define LSWI. For instance, the surface reflectance presents low values (a few percentage points) over non-turbid water bodies and high values (a few tens of percentage points) over vegetation feature in the NIR spectral bands. The spectral response in the SWIR is mainly dominated by strong water absorption bands, which is directly sensitive to moisture content in the soil and the vegetation. For water surface area, the signal in the SWIR is assumed to be zero even in turbid waters (Wang and Shi, 2005). Thus, LSWI is expected to get values close to 1 for water surface areas and lower values for non-water surface areas.

The two indices, used in this approach, are defined as follows (Huete et al., 1997; Xiao et al., 2005):

$$\mathrm{EVI} = a \times \frac{\rho_{\mathrm{NIR}} - \rho_{\mathrm{red}}}{\rho_{\mathrm{NIR}} + b \times \rho_{\mathrm{red}} - c \times \rho_{\mathrm{blue}} + d}, \quad (1)$$

$$\mathrm{LSWI} = \frac{\rho_{\mathrm{NIR}} - \rho_{\mathrm{SWIR}}}{\rho_{\mathrm{NIR}} + \rho_{\mathrm{SWIR}}}, \quad (2)$$

where for MODIS, ρ_{blue} is the surface reflectance value in the blue (459–479 nm, band 3), ρ_{red} is the surface reflectance value in the red (621–670 nm, band 1), ρ_{NIR} is the surface reflectance value in the NIR (841–875 nm, band 2), and ρ_{SWIR} is the surface reflectance in the SWIR (1628–1652 nm, band 6). For OLI, ρ_{blue}, ρ_{red}, ρ_{NIR}, and ρ_{SWIR} are associated with channel 2 (452–512 nm), channel 4 (636–673 nm), channel 5 (851–879 nm), and channel 6 (1570–1650 nm), respectively. The constants a, b, c and d are equal to 2.5, 6, 7.5 and 1, respectively, for both MODIS and OLI (USGS, product guide).

To process multispectral images, the first step consists of removing the cloud-contaminated pixels by applying a cloud masking based on a threshold of the surface reflectance in the blue band ($\rho_{\mathrm{blue}} \geq 0.2$). Then, spectral indices are computed. Note that, contrary to Sakamoto et al. (2007), no smoothing was applied on spectral index time series. In a second step, the identification of the status of each pixel is performed by applying thresholds on EVI, LSWI and their differences (Fig. 2), which reduce the noise component. Thresholds determined by Sakamoto et al. (2007) were validated for our study site using OLI images acquired on 1 July and 2 August 2013 and compared to MODIS (Fig. S1 in the Supplement). Histograms show a similar bi-modal distribution for both EVI, LSWI and EVI-LSWI between MODIS and OLI 500 m (Figs. S1 and S2). For EVI, pixels with a value lower than 0.1 are clearly associated with water land surfaces, while pixels with a value higher than 0.3 are associated with soil and vegetation features. Other pixels, with an EVI value between 0.1 and 0.3, are identified as mixed surface types. For LSWI, pixels with a value higher than 0.5 are clearly associated with water land surfaces, while pixels with a value lower than 0.3 are associated with vegetation features or soil land surfaces when LSWI values are negative. Other pixels, with an LSWI value between 0.3 and 0.5, are identified as mixed

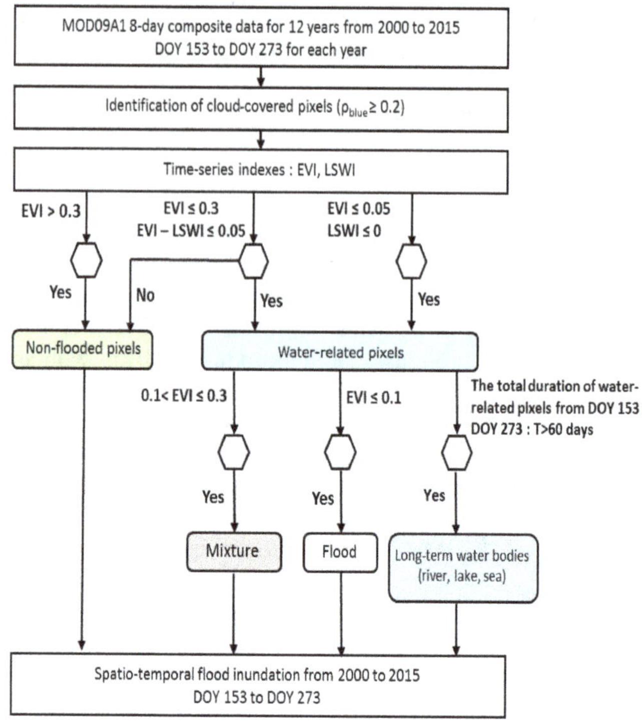

Figure 2. Flow chart of the method (adapted from Sakamoto et al., 2007) used to classify each pixel of the multispectral images acquired over the Mackenzie Delta in four categories (non-flooded, mixed, flooded and permanent water bodies) for each year from 2000 to 2015 using MODIS 8-day composite data from the day of the year (DOY) 169 to 257.

surface types. Contrary to what was found by Sakamoto et al. (2007) in the Mekong Basin, no negative values of LSWI were observed over our study area. This threshold was not applied in this study. For EVI–LSWI, pixels with a value lower than −0.05 represent water land surface and values between −0.05 and 0.1 are associated with mixed pixels. Other pixels, with values higher than 0.1 are represented vegetation features or soil land surfaces (Fig. S2). Each pixel was then classified in two main categories: non-flooded ($\mathrm{EVI} > 0.3$ or $\mathrm{EVI} \leq 0.3$ but $\mathrm{EVI} - \mathrm{LSWI} > 0.05$) and water-influenced ($\mathrm{EVI} \leq 0.3$ and $\mathrm{EVI} - \mathrm{LSWI} \leq 0.05$ or $\mathrm{EVI} \leq 0.05$) (Fig. 2). The second category was divided into three sub-classes: mixed pixels ($0.1 < \mathrm{EVI} \leq 0.3$), flooded pixels ($\mathrm{EVI} \leq 0.1$) and permanent water bodies (e.g. lake, river and sea), the latter denoting when the total duration of a pixel classified as flooded is longer than 70 days out of 105 days for the study period. This annual duration for our study corresponds roughly to two-thirds of the study period, as proposed by Sakamoto et al. (2007). The spatio-temporal variations of floods have been characterized for the months between June and September over the 2000–2015 period.

Hereafter, in this paper we define water surface area as permanent water bodies with flooded areas, although inundated surfaces include only inundated areas.

4.2 Validation of MODIS retrievals using OLI

Evaluation of the performance of the water surface area detection from MODIS is based on the comparison between land surface water estimated from MODIS at a 500 m resolution, OLI at a 30 m resolution, and OLI re-sampled at 500 m resolution. For validation purposes, MODIS and OLI images are selected when (1) the time difference between the acquisitions of two satellite images is lower than 3 days and (2) the presence of cloud over the area is lower than 5 %. Following these criteria, only two cloud-free OLI composites were selected between 1 July and 2 August 2013.

4.3 Satellite-derived water level time series in the Mackenzie Delta

The concept of radar altimetry is explained below. The radar emits an electromagnetic (EM) wave towards the surface and measures the round-trip time (Δt) of the EM wave. Taking into account propagation corrections caused by delays due to the interactions of electromagnetic waves in the atmosphere, and geophysical corrections, the height of the reflecting surface (h) with reference to an ellipsoid can be estimated as follows (Crétaux et al., 2017):

$$h = H - (R + \sum \Delta R_{\text{propagation}} + \Delta R_{\text{geophysical}}), \tag{3}$$

where H is the satellite centre of mass height above the ellipsoid, R is the nadir altimeter range from the satellite centre of mass to the surface (taking into account instrument corrections; $R = c\Delta t/2$, where c is the light velocity in the vacuum), and $\sum \Delta R_{\text{propagation}}$ is the sum of the geophysical and environmental corrections applied to the range.

$$\sum \Delta R_{\text{propagation}} = \Delta R_{\text{ion}} + \Delta R_{\text{dry}} + \Delta R_{\text{wet}}, \tag{4}$$

where ΔR_{ion} is the atmospheric refraction range delay due to the free electron content associated with the dielectric properties of the ionosphere, ΔR_{dry} is the atmospheric refraction range delay due to the dry gas component of the troposphere, and ΔR_{wet} is the atmospheric refraction range delay due to the water vapour and the cloud liquid water content of the troposphere.

$$\sum \Delta R_{\text{geophysical}} = \Delta R_{\text{solid Earth}} + \Delta R_{\text{pole}}, \tag{5}$$

where $\Delta R_{\text{solid Earth}}$ and ΔR pole are the corrections respectively accounting for crustal vertical motions due to the solid Earth and pole tides. The propagation corrections applied to the range are derived from model outputs: the global ionospheric maps (GIMs) and Era-Interim datasets from the European Centre for Medium-Range Weather Forecasts (ECMWF) for the ionosphere and the dry and wet troposphere range delays respectively. The changes in the altimeter height h over the hydrological cycles are related to variations in water level. Here, the Multi-mission Altimetry Processing Software (MAPS) was used to precisely select valid altimetry data at every virtual station location (see Sect. 3.3) series in the Mackenzie Delta. Data processing consisted of four steps (Frappart et al., 2015b):

- the rough delineation of the river–lake cross sections with overlaying altimeter tracks using Google Earth (distances of ± 5 km from the river banks are generally considered);
- the loading of the altimetry over the study area and the computation of the altimeter heights from the raw data contained in the GDRs;
- the selection of valid altimetry data through a refined process that consists of eliminating outliers and measurements over non-water surface areas based on visual inspection (the shape of the altimeter along-track profiles permit identification of the river that is generally materialized as a shape of "V" or "U", with the lower elevations corresponding to the water surface area; see Santos da Silva et al., 2010 and Baup et al., 2014 for more details);
- the computation of the time series of water level.

4.4 Surface water volume storage

The approach used to estimate the anomalies of surface water volume is based on the combination of the surface water extent derived from MODIS images with altimetry-based water levels estimated at virtual stations distributed all over the delta (Fig. 5). Surface water level maps were computed from the interpolation of water levels over the water surface areas using an inverse-distance weighting spatial interpolation technique following Frappart et al. (2012). Hence, water level maps were produced every 8 days from 2000 to 2015. For each water pixel, the minimal height of water during 2000–2015 is estimated. As ERS-2, ENVISAT and SARAL had a repeat cycle of 35 days, water levels are linearly interpolated every 8 days to be combined with the MODIS composite images.

Surface water volume time series are estimated over the Mackenzie Delta following Frappart et al. (2012):

$$V = \sum_{j \in S} [h(\lambda_j, \varphi_j) - h_{\min}(\lambda_j, \varphi_j)] \cdot \delta_j \cdot \Delta S, \tag{6}$$

where V is the anomaly of surface water volume (km^3), S is the surface of the Mackenzie Delta (km^2), h (λj, φj) the water level, $h_{\min}$ (λ_j, φ_j) the minimal water level for the pixel of coordinates (λ_j, φ_j) inside the Mackenzie Delta, δ_j equals 1 if the jth pixel is associated with permanent water body/inundated and 0 if not and ΔS the pixel surface ($0.25\ \text{km}^2$).

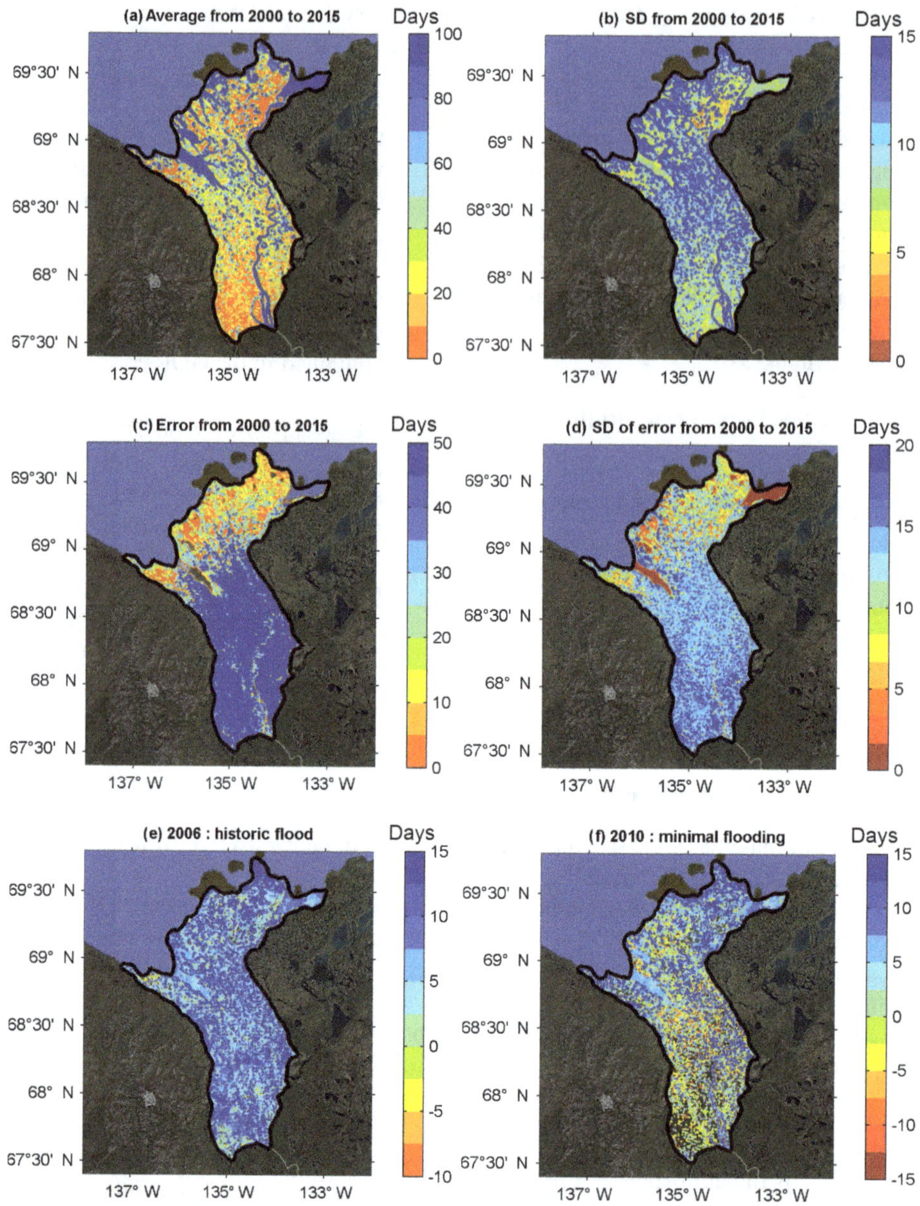

Figure 3. Maps of surface water extent duration for **(a)** annual average from 2000 to 2015, **(b)** annual standard deviation from 2000 to 2015, **(c)** error average from 2000 to 2015, **(d)** standard deviation from 2000 to 2015, difference between annual average water surface area duration from 2000 to 2015 and water surface area duration during **(e)** 2006, the period associated with the highest flood event, and **(f)** 2010, the period associated with the lowest flood event recorded over the period.

5 Results

5.1 MODIS-based land water extent and their validation

Following the method of Sakamoto et al. (2007), all pixels of 8-day image have been classified into four groups: class 0 corresponding to vegetation, class 1 to permanent water, class 2 to inundation, and class 3 to a mixture of land and water. A map of annual average of water surface area, composed of inundated and permanent water bodies (classes 1 and 2), was obtained at spatial and temporal resolutions of 500 m and 8 days respectively from June to September over the 2000–2015 period (Fig. 3a). A map of annual average of water surface area duration along with associated standard deviation over 2000–2015 during an ice-free period of 3.5 months (105 days) is presented in Fig. 3b. Permanent water bodies (i.e. identified as water surface area more than 70 days annually) are located along the Mackenzie River main channel, its tributaries (Reindeer, Peel, Middle and East channels) and major lakes of the Delta. The longer water areas (i.e. identified as flooded between 30 and 70 days annually) are surrounding permanent water bodies. Other areas of the delta are annually inundated up to 30 days (Fig. 3a). The map of standard deviation of the annual flood duration shows ranges from a few days over the areas affected by floods during a short time span to 15 days close to permanent water bodies (Fig. 3b).

Maps of errors made on water surface area duration with associated standard deviation are shown in Fig. 3c and d over 2000–2015. Mixed pixels have been used to calculate the error for each pixel on water surface area duration, corresponding to the class 3 "mixed" of Sakamoto et al. (2007) classification. Standard deviation is presented in Fig. 3d. Maximal error and standard deviation is obtained for pixels of potential flooding area in the delta. If short differences – lower than 20 ± 12 days – can be observed in the downstream part of the delta (over 69° N), longer differences (30 to 50 ± 15 to 20 days) are present in the upstream part. They can be attributed to the presence of small permanent lakes in this area. Important inter-annual differences can be observed between wetter (Fig. 3e) and dryer (Fig. 3f) years.

Surface water extent (the sum of permanent bodies and inundated areas) were also estimated by applying the approach described in Sect. 3.1 for OLI images at 30 m of spatial resolution, and resampled at 500 m of spatial resolution. They were compared to MODIS-based surface water extent for the closest date (Table 1). Figure S3a, b and c present the maps of the surface water extent determined using MODIS, OLI 500 m and OLI 30 m respectively, acquired in July 2013. Medium and large-scale (with a minimal size of 300 m) land water features are well detected, as displayed in the enlarged part of the images. Figure S3c presents an enlarged image of surface water extent using OLI 30 m with permanent and inundated bodies. Surface water extent from OLI 500 m and MODIS are similar for both dates, with differences lower than 20 % (Table 1). For example in July 2013, water surface area is about 4499 km^2 for OLI 500 and 3798 km^2 for MODIS (Table 1). Percentages of common detection of surface water were estimated for the pixels detected as water surface area in the pair of satellite images. These percentages are 73 and 74 % for July and August 2013, respectively. Areas detected as water by both sensors correspond to the main channels and connected floodplains. Differences appear on the boundaries of areas commonly detected as inundated and on small scales and can be attributed to the difference of acquisition dates between MODIS and OLI (Fig. S4). These results highlight the robustness of the method of Sakamoto et al. (2007) for accurate water surface area retrievals. These surface water extents have been compared with surface water extents (channels and wetlands) determined by Emmerton et al. (2007) in Table 1. For MODIS, differences are lower than 15 % and for OLI 500 differences are about 25 % (Table 1).

However, the comparison between surface water extent estimated from OLI 30 m and MODIS 500 m shows important differences. In July 2013, surface water extent is about 3798 km^2 from MODIS and 7685 km^2 from OLI 30. The surface extents are higher for OLI 30 by a factor of 2 (Table 1). According to Emmerton et al. (2007), the Mackenzie Delta is composed of 49 000 lakes with a mean area of 0.0068 km^2 and 40 % of the total number of lakes have an area inferior to 0.25 km^2. The pixel sizes of OLI 30 m and MODIS 500 m are approximately 0.0009 and 0.25 km^2, respectively. Thus, the high difference between the water surface areas detected using OLI 30 m and MODIS is probably associated with a spatial sample bias. Small-scale water features detected from OLI cannot be detected from MODIS due to a lower spatial resolution.

Surface water extents determined using OLI 30 have been compared to Emmerton et al. (2007) surface water extents (including channels, wetlands and lakes). Emmerton et al. (2007) classified the Mackenzie Delta habitat in lakes, channels, wetlands and dry floodplains using information from topographic maps derived from aerial photographs taken during the 1950s for low-water periods. Differences between surface water extent of OLI 30 and Emmerton et al. (2007) are lower than 15 % (Table 1).

In order to investigate the assumption of spatial sample bias associated with MODIS 500 m, a satellite validation of surface water extent is performed (Table 2). Permanent water and inundated surfaces have been calculated for MODIS, OLI 500 and OLI 30. For OLI 30 and OLI 500, pixels identified as surface water for the two dates are considered as permanent waters (Table 2). In July 2013, inundated surfaces are nearly equal, about 577 km^2 for MODIS, 690 km^2 for OLI 500 and 627 km^2 for OLI 30 (Table 2). In August, inundated surfaces are equal to 250 km^2 and are 2.5 times more important than OLI 30 (98 km^2), if we consider OLI 30 as truth.

Time series of surface water extent in the Mackenzie Delta were derived from the 8-day maps of surface water extent (Fig. 4). Surface extent water varies from 1500 to 14 284 km^2 between 2000 and 2015 along the hydrological cycle. Each year, water surface area extent is at its maximum in June in response to the spring ice break-up and snow melt that occurred in May (between day of year, DOY, 110 and 130 on average) in the Delta and decreases to reach a minimum in September, as previously observed by Goulding et al. (2009a, b). On average, during the study period, maximum surface water extent is $\sim$9600 km^2 . The largest water surface area extent was reached in June 2006 with an inundated area of 14 284 km^2, which represents $\sim$85 % of the delta total surface (Fig. 4). Large surface water extents ($\sim$12 500 km^2) were also detected in 2011 and 2013 in accordance with high discharge peaks reported for these years (http://wateroffice.ec.gc.ca/) and the historic inundation that occurred in Aklavik in 2006 (Beltaos and Carter, 2009).

5.2 Altimetry-based water levels and their validation

The Mackenzie Delta is densely covered with altimetry tracks from the ERS-2, ENVISAT and SARAL missions that were all on the same nominal orbit. A total of 22, 27 and

Table 1. Validation of surface water extents (km^2) determined using OLI 30 m, OLI 500 m, and MODIS 500 m images with the results of Emmerton et al. (2007).

	MODIS: 4 July 2013 OLI: 1 July 2013	MODIS: 5 August 2013 OLI: 2 August 2013
MODIS 500 m	3798	3298
OLI 500 m	4499	3859
Emmerton et al. (2007) (channels + wetlands, km^2)	3358	3358
Difference between MODIS 500 and Emmerton et al. (2007)	440 km^2 (13 %)	60 km^2 (2 %)
Difference between OLI 500 and Emmerton et al. (2007)	1141 (34 %)	500 (15 %)
OLI 30 m	7685	7156
Emmerton et al. (2007) (channels + lakes + wetlands, km^2)	6689	6689
Difference between OLI 30 and Emmerton et al. (2007)	996 km^2 (13 %)	467 km^2 (7 %)

Table 2. Satellite validation of surface water extent using OLI 30, OLI 500 and MODIS 500 m.

Date	Permanent water MODIS (km^2)	Permanent water OLI 500 (km^2)	Permanent water OLI 30 (km^2)	Inundated surfaces MODIS (km^2)	Inundated surfaces OLI 500 (km^2)	Inundated surfaces OLI 30 (km^2)
MODIS : 4 July 2013 OLI : 1 July 2013	3167	3809	7058	577	690	627
MODIS : 5 August 2013 OLI : 2 August 2013	2885	3809	7058	250	50	98

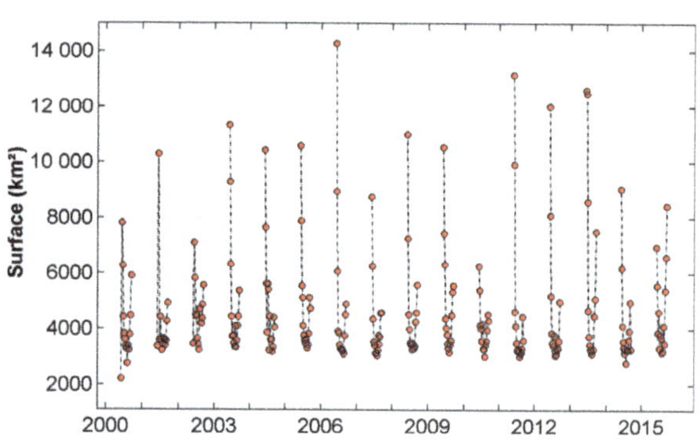

Figure 4. Time series of surface water extent from 2000 to 2015, between June and September, derived from the MODIS images.

24 altimetry virtual stations were built at the cross section of an altimetry track with a water body for these three missions respectively (see Fig. 5 for their locations). A water level temporal series is obtained for each virtual station.

The quality of altimetry-based water levels was evaluated using in situ gauge records. Only 6 virtual stations are located near in situ stations (with a distance lower than 20 km) for ERS-2 data, with 10 for ENVISAT and 8 for SARAL data. Characteristics of these virtual stations are given in Table 3. For ERS-2 and SARAL comparisons, the correlation r is low at the station 0114-c, i.e. −0.38 and 0.15 respectively (Table 3).

For ERS-2, quite high correlation coefficients are obtained for four virtual stations out of six, with $r \geq 0.69$ and RMS ≤ 1 m (Table 3). For the two other stations, no correlation is observed (−0.38 and 0.08 for ERS-2-0114c and ERS-2-0200-d respectively with a RMS ≥ 1 m) (Table 3).

For ENVISAT, 8 out of 10 stations have a correlation coefficient ranging between 0.66 and 0.93 (Table 3). Except for ENV-0572-a, which is located 22 km away from the nearest in situ station, higher correlations were found when the river is larger at the VS (Table 3). For example, ENV-0114-b exhibits a negative correlation ($r = -0.27$) where the cross

Table 3. Statistic parameters obtained between altimetry-based water levels from altimetry multi-mission and in situ water levels.

Virtual station (SV)	In situ station	Altimetry mission	Distance (km)	River width at VS (m)	N	r	RMS (m)	R^2	Bias (m)	Bias ICESat (m)
0439-a	10MC008	ERS-2	11.44	1950	5	0.76	0.5	0.58	0.55	1.36
		ENVISAT			24	0.89	0.5	0.81	0.15	0.65
		SARAL			8	0.96	0.35	0.93	−0.95	−0.15
0983-c	10MC003	ERS-2	3.1	360	20	0.69	0.7	0.47	–	–
		ENVISAT			26	0.66	0.89	0.44	–	–
		SARAL			6	0.9	0.4	0.8	–	–
0114-c	10MC022	ERS-2	1.9	430	14	−0.38	2.82	0.14	–	–
		ENVISAT			23	0.8	1.17	0.64	–	–
		SARAL			7	0.14	0.73	0.02	–	–
0200-d	10MC023	ERS-2	4.11	630	17	0.08	4.3	0	–	–
		ENVISAT			22	0.87	0.33	0.75	–	–
		SARAL			6	0.76	0.3	0.57	–	–
0744-a	10MC010	ERS-2	5.16	850	5	0.88	0.1	0.77	–	−1.28
		ENVISAT			24	0.93	0.15	0.87	–	−1.17
		SARAL			2	0.99	0.15	0.99	–	−2.19
0439-d	10LC015	ERS-2	7.2	380	20	0.92	0.83	0.86	–	–
		ENVISAT			28	0.65	1.75	0.43	–	–
		SARAL			5	0.95	1.3	0.9	–	–
0525-a	10MC002	ENV	16.31	500	29	0.77	1.45	0.6	–	–
0028-a	10LC014	ENV	16.05	1360	17	0.83	1.84	0.7	–	2.35

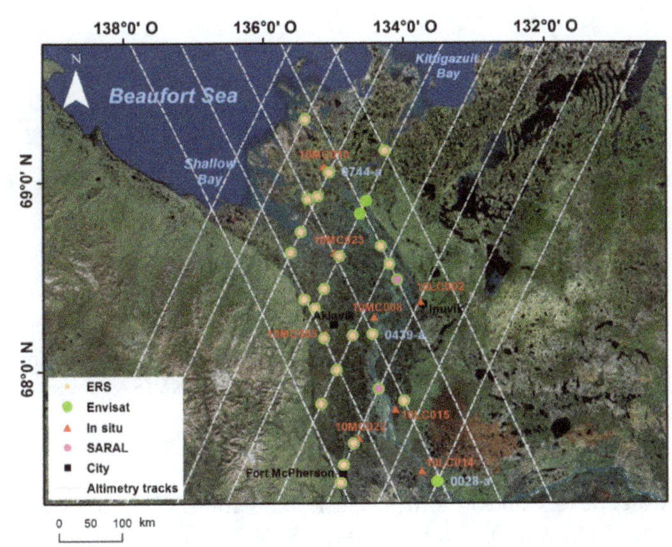

Figure 5. Locations of virtual stations (VSs) in the Mackenzie Delta for ERS-2 (yellow dots), Envisat (green dots) and SARAL (purple dots) altimetry missions. Altimetry tracks appear in grey. In situ stations are represented using red triangles.

section was only 150 m wide (Table 3). This station is also located near the city of Inuvik. The presence of the town in the altimeter footprint could exert a strong impact on the radar echo and explain this low correlation.

For SARAL, five out of six virtual stations have a good correlation r coefficient higher than 0.76 with a low RMS (Table 3) due to its narrower footprint with an increase in the along-track sampling.

Comparisons between water levels derived from altimetry and in situ are shown for two stations for ERS-2 (called ERS-2-0744-a and ERS-2-0439-a; Figs. 6a and 7a), three for ENVISAT (ENV-0744-a, ENV-0439-a and ENV-0028-a; in Figs. 6b, 7b and 8) and two for SARAL (SARAL-0744-a and SARAL-0439-a; Figs. 6c and 7c). Virtual station 0744-a is located in the downstream part of the delta, 0439-a in the centre and 0028-a in the upstream part (Fig. 5). For each station, water levels obtained by altimetry and water levels obtained by in situ gauges are superposed (Figs. 6, 7 and 8). Then, water level anomalies, which are computed as the average water level minus the water level, have been calculated for altimetry and in situ data.

The virtual station 0744-a is located in the north of the Mackenzie Delta (Fig. 5). Water level time series have been processed between 1995 and 2015 and compared to in situ

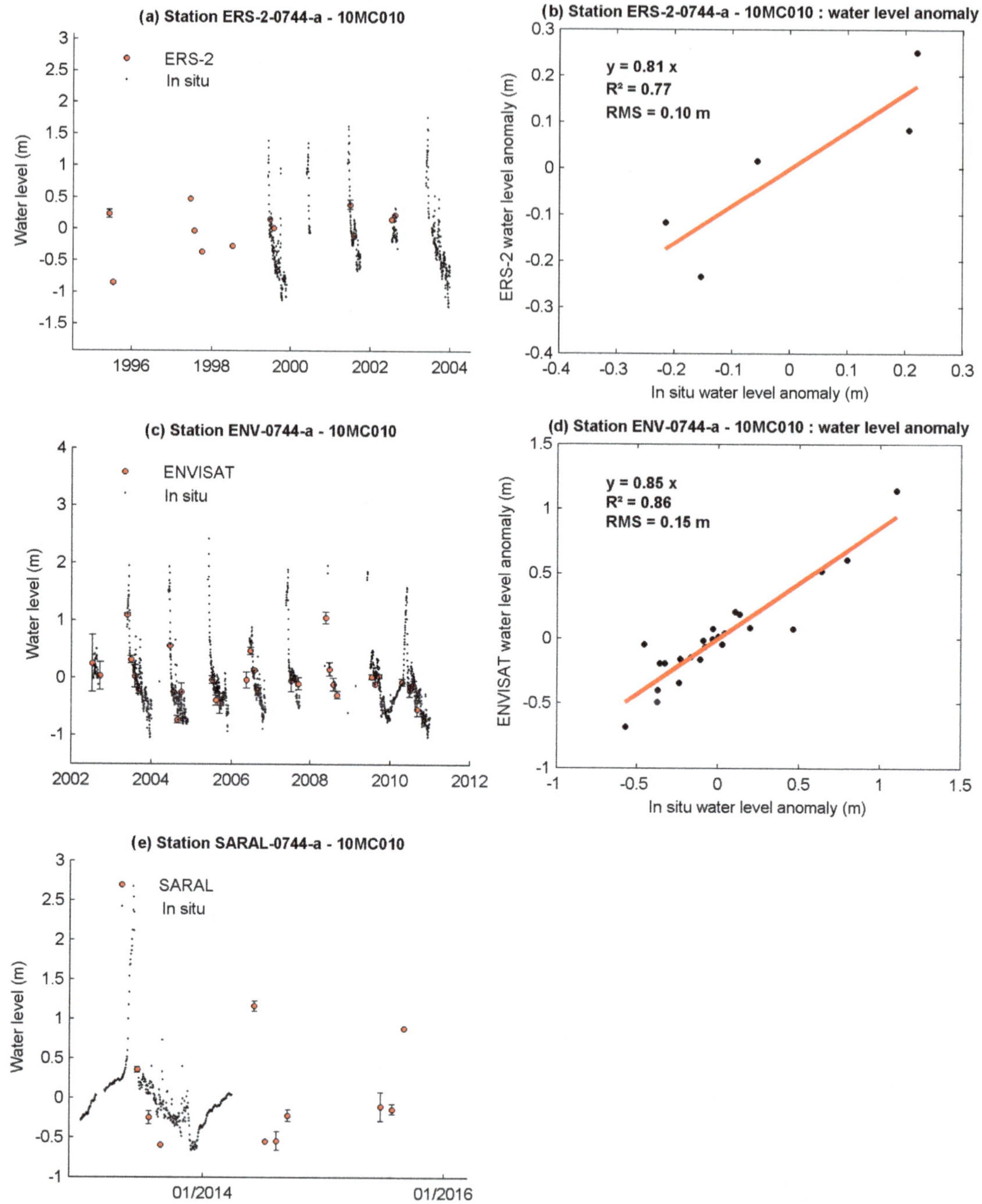

Figure 6. Altimetry-based water levels from 1995 to 2015 compared with in situ water levels for the station 0744-a, located in the downstream part in the Mackenzie Delta, using **(a)** the ERS-2 mission and **(b)** a water level anomaly with statistic parameters, **(c)** the ENVISAT mission and **(d)** a water level anomaly with statistic parameters, and **(e)** using SARAL mission.

data of the station 10MC010 for each mission ERS-2, ENVISAT and SARAL (Fig. 6). In situ data are not continuous since the river is frozen from October to April. With regard to altimetry, data have been acquired throughout the year, but during frozen periods water levels are unrealistic due to the presence of river ice. Thus, the processing is done only from the beginning of June to the end of September for multispectral imagery treatment. The correlation r between altimetry water levels and in situ levels is 0.88 for ERS-2, 0.93 for ENVISAT and 0.99 for SARAL (Table 3). For the three missions, RMS is weak, lower than 0.15 m (Table 3). At this station, the variation of water level is about 2 m on average with an important water level in June that decreases to September (Fig. 7a, c and e).

The virtual station ERS-2-0439-a is in the centre of the Mackenzie Delta and water level time series have been

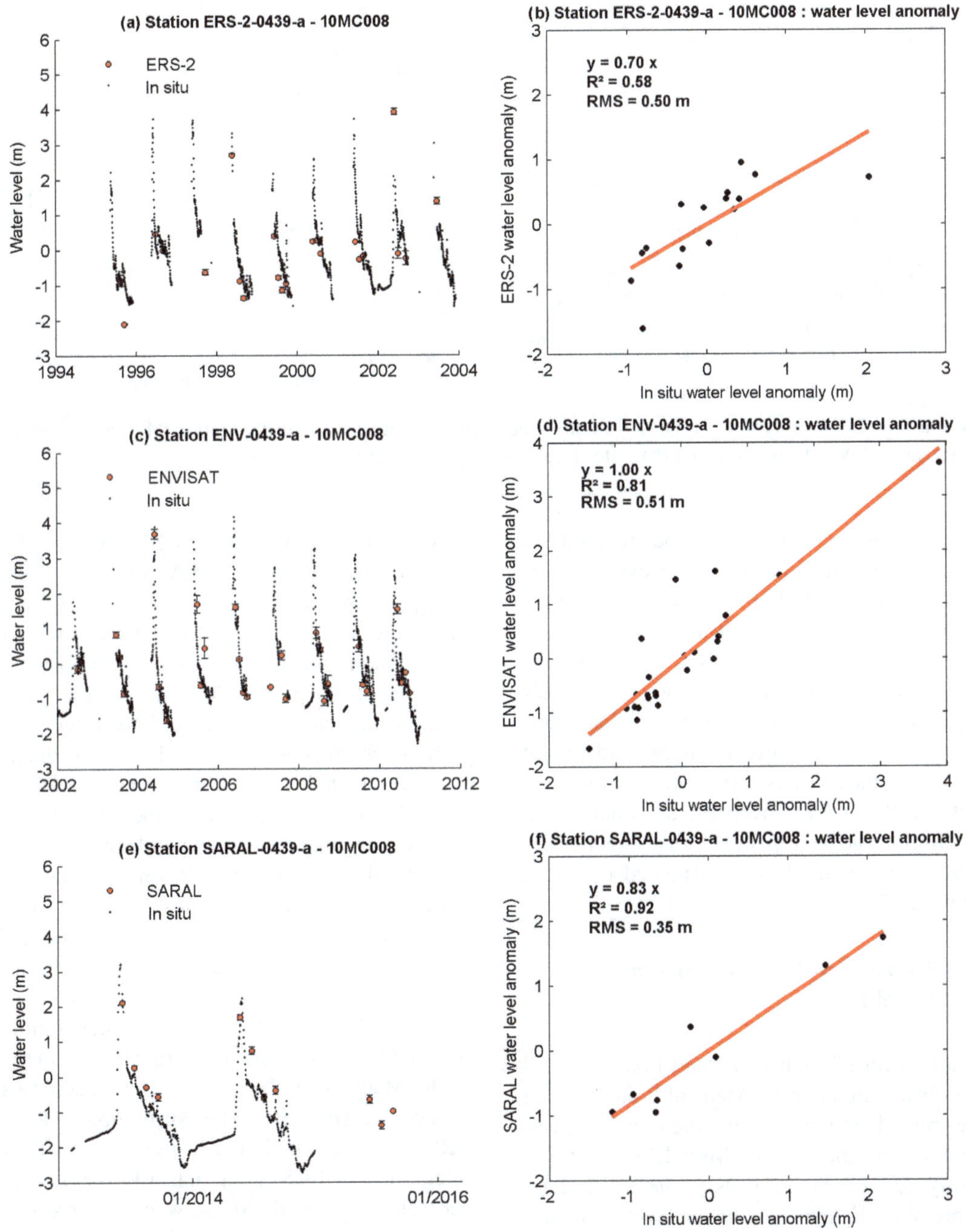

Figure 7. Altimetry-based water levels from 1995 to 2015 compared with in situ water levels for the station 0439-a located in the centre of the Mackenzie Delta using **(a)** the ERS-2 mission and **(b)** a water level anomaly with statistic parameters, **(c)** the ENVISAT mission and **(d)** a water level anomaly with statistic parameters, and **(e)** the SARAL mission and **(f)** a water level anomaly with statistic parameters.

recorded between 1995 and 2015 and compared to in situ data of the station 10MC008 for the three missions ERS-2, ENVISAT and SARAL (Fig. 7). The correlation between altimetry water levels and water levels from in situ gauges is about 0.76 for ERS-2, 0.89 for ENVISAT and 0.96 for SARAL (Table 3). RMS is included between 0.35 and 0.5 m for the three missions. On average at this station, water levels variations are about 4 m, with a maximal water level in June that decreases to reach a minimal value in September (Fig. 7a, c and e).

Water level time series between 2002 and 2010 at the virtual station ENV-0028-a, located upstream of the Mackenzie Delta, have been compared to in situ data of the station 10LC014 (Fig. 8). A good correlation was found for this station too, with a coefficient correlation r of 0.83 and a RMS of 1.84 m (Table 3). For this station, variations of water lev-

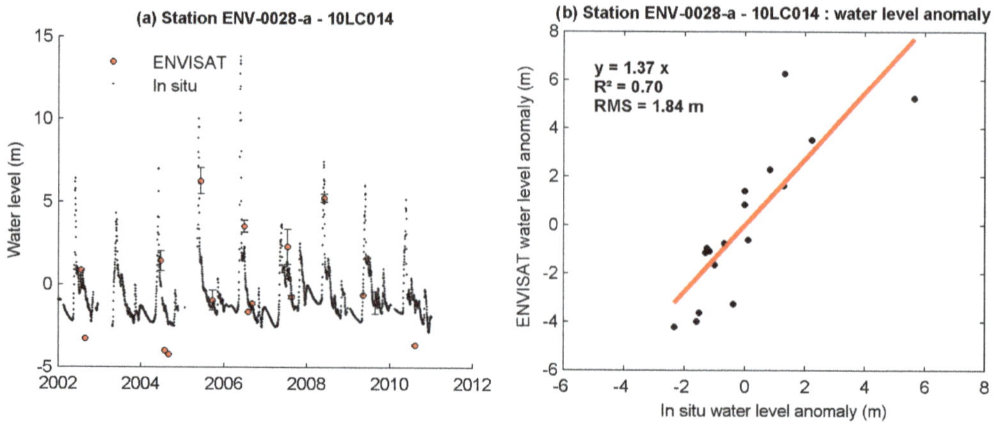

Figure 8. Altimetry-based water levels from 2002 to 2010 compared to in situ water levels for the station 0439-a located in the centre of the Mackenzie Delta **(a)** using ENVISAT mission and **(b)** water level anomaly with statistic parameters.

els are much higher, with 9 m on average, but reach 12 m during the 2006 extreme event (Fig. 8a). Water level time series were constructed only for ENVISAT mission since for the two others (ERS-2 and SARAL), altimetry water levels were not consistent, exhibiting values around 70 m. Therefore, water levels determined by altimetry and water levels from in situ gauges have a difference, which is probably explained by the distance between virtual stations and in situ gauges (16.31 km) since the slope is about $-0.02\,\mathrm{m\,km^{-1}}$ in the Delta (Hill et al., 2001). Moreover, the seasonal cyclic thawing and freezing of the active layer causes cyclic settlement and heave at decimetre levels, estimated to 20 cm (Szostak-Chrzanowski, 2013).

5.3 Time series of surface water storage anomalies in the Mackenzie Delta

The minimum water level of each inundated pixel was determined over the observation period. Maps of 8-day surface water levels were created after subtracting the minimum water level to water level at time t, using MODIS-based flood extent and altimetry-derived water levels in the entire delta from June to September. Example of water level maps are presented for 2006 at 4 different dates (in June, July, August and September), characterized as an historic flood (Fig. 9).

Over the study period, water level maps show a realistic spatial pattern with a gradient of water level from south to north, consistent with flow direction in the delta. In Fig. 9a, in June 2006, for example, water levels are higher (about 5 m) upstream than downstream (about 0.5 m). The surface water storage reaches its maximal extent in June (Fig. 9a) and then decreases during the following months, reaching 1 m in September in the entire delta (Fig. 9b, c and d).

The time series of surface water volume variations was estimated from 2000 to 2010 and then from 2013 to 2015, between June and September, following a similar approach to that in Frappart et al., 2012 (Fig. 10). Surface water storage was estimated from 2000 to 2003 using ERS-2 data, from 2003 to 2010 using ENVISAT data and from 2013 to 2015 using SARAL. Between 2010 and 2013, surface water storage could not be estimated due to lack of RA data over the delta. The impact of the presence of a virtual station located in the upstream part of the delta and the inclusion of ERS-2 data on our satellite-based surface water volume estimation were assessed. For ERS-2 and SARAL data, no virtual station was created in the upstream part due to unreliable water levels in the upstream part of the delta. During the SARAL observation period, in situ water levels from 10LC014 station were used. One curve corresponds to surface water volume with virtual stations in the upstream part of the delta (2002–2015; red) and another one without virtual stations in the upstream part of the delta (2000–2015; green). Correlations between river discharges and surface water volumes with and without (2002–2015) upstream virtual stations are the same (0.66). In the presence of a virtual station in the upstream part of the Mackenzie, the water volume decreases by $\sim 0.3\,\mathrm{km^3}$ on average (Fig. 10). The correlation is lower (0.63) when ERS-2 data are included in the analysis (2000–2015). The integration of ERS-2 data has a lower accuracy and slightly decreases the correlation between water storage and flux.

In terms of temporal variability, a clear seasonal cycle is visible, with a yearly maximum of water surface area volume occurring in June (about $9.7\,\mathrm{km^3}$ on average), followed by a decrease until September (Fig. 10). The peak generally corresponds to the presence of the extensive flood covering the delta in June, and during summer the volume decreases to reach its minimal in September ($\sim 0.2\,\mathrm{km^3}$). The largest surface water volume occurred in 2006 with a volume of $14.4\,\mathrm{km^3}$ (Fig. 10), known as an historic flood (Beltaos and Carter, 2009). These results showed that the satellite-based surface water volume estimation is consistent with the Mackenzie River discharge, which is the main driver of the delta flooding.

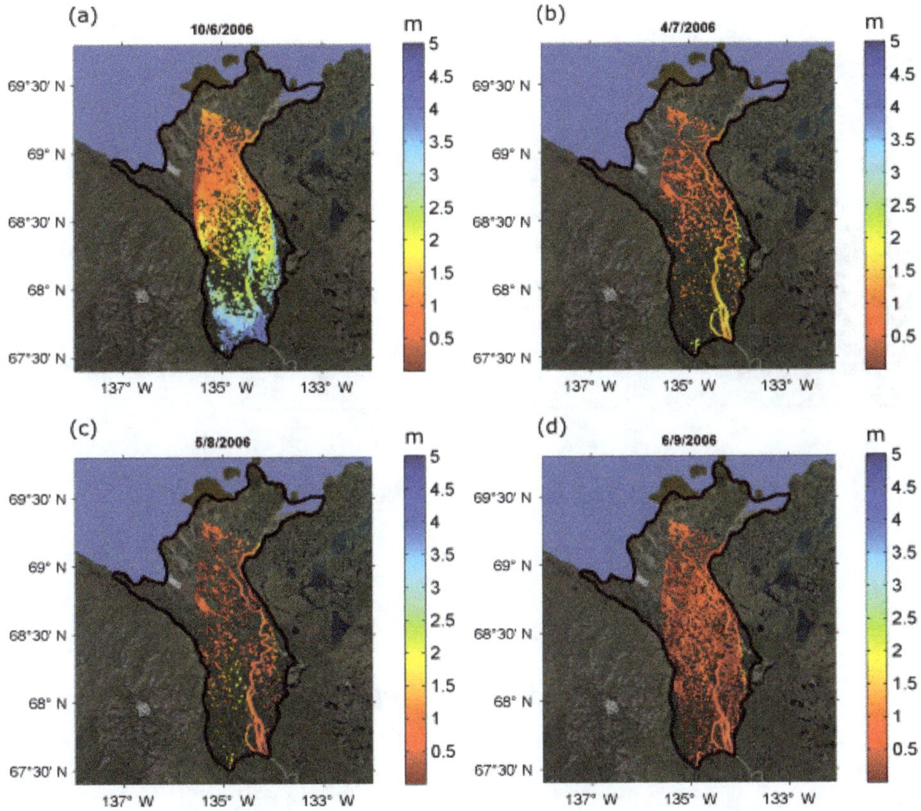

Figure 9. Water level maps in the Mackenzie Delta in 2006 (historic flood) obtained by combining inundated surfaces determined using MODIS images with altimetry-derived water levels **(a)** in June, **(b)** in July, **(c)** in August and **(d)** in September.

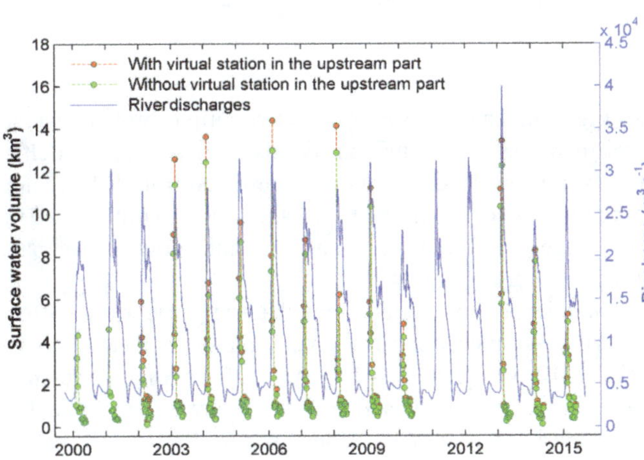

Figure 10. Surface water volume from 2000 to 2015, determined by combining inundated surfaces from MODIS with altimetry data; 167 red points correspond to surface water volume obtained with a virtual station located in the upstream part of the Delta, green points to surface water volume without a virtual station located in the upstream part of the Delta. The Mackenzie River Delta discharges at 10LC014 gauge station appear in blue.

6 Discussion

6.1 Spatio-temporal dynamics of surface water extent

Maps of surface water extent duration for annual average from 2000 to 2015 exhibit important spatio-temporal variations along the Mackenzie Delta (Fig. 3a). Areas with open water present during the whole study period are located along the Mackenzie River and its tributaries. On the contrary, areas covered with open water for a duration lower than 30 days in the study period of 120 days are mostly located in the western upstream and eastern downstream parts of the delta but also in some locations in the western downstream part and along the Mackenzie mainstream (Fig. 3a). They correspond to regions only inundated in June during the floods caused by spring ice break-up and snow melt occurring in May (see Fig. 9 for the temporality of the flood extent). The central part of the Mackenzie is inundated between 40 and 70 days per year (Fig. 3a). As can be seen in Fig. 9, this area is not continuously inundated except for during two flood events in June in response to snowmelt and in August and September in response to an increase in river discharges of the Mackenzie River. This secondary peak ranges from 3000 to 5000 km^2 in comparison with the earlier one that ranges from 4000 to 10 000 km^2 (Fig. 4). Maps of difference between the duration

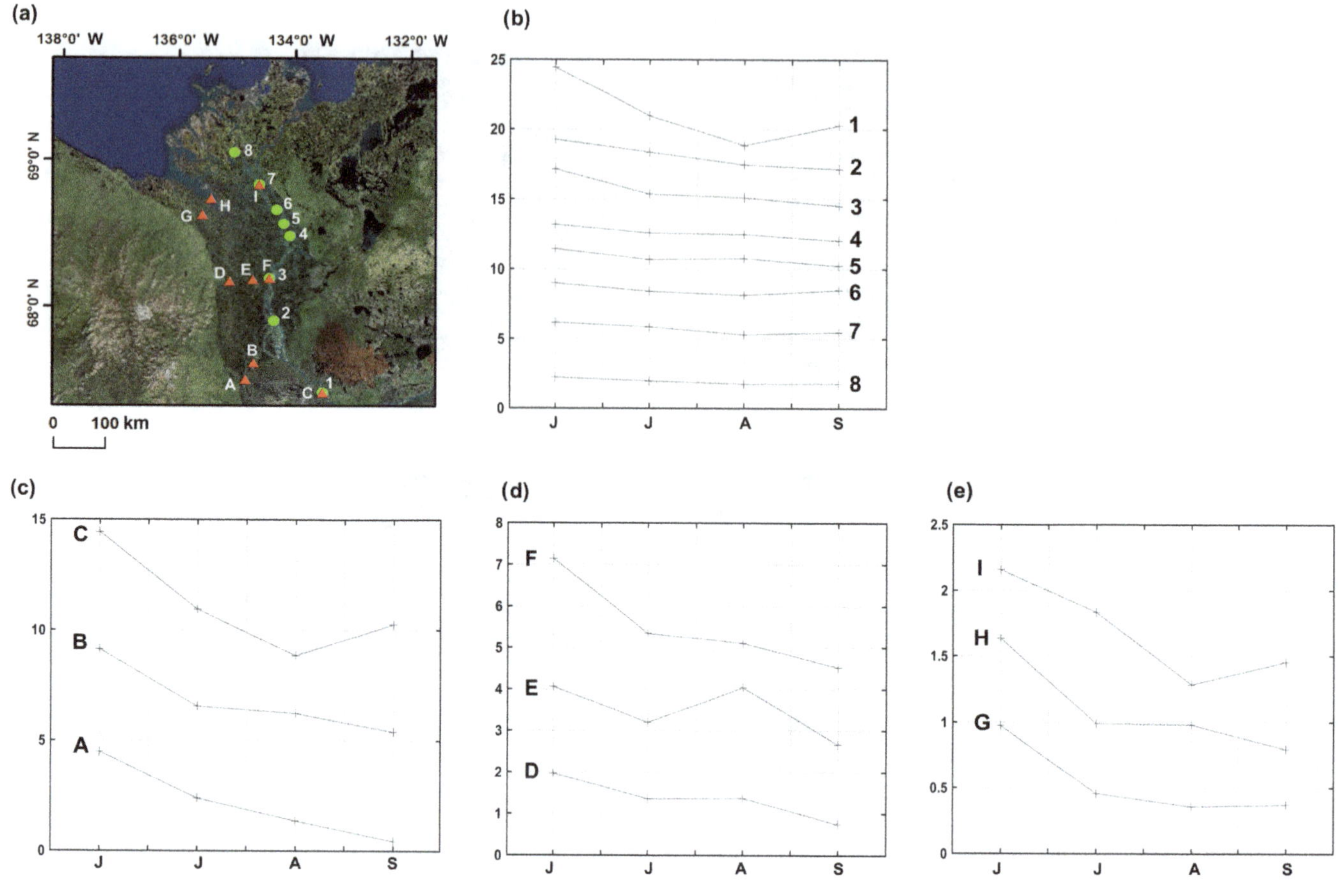

Figure 11. Temporal and spatial variations of surface water levels (in metres) in the Mackenzie Delta. **(a)** Location of virtual stations used to analyse spatial variations; green dots are corresponding to latitudinal variations along the Mackenzie River (from number 1 to 8) and red triangles are corresponding to longitudinal variations at three different latitudes (letters from A to I). **(b)** Surface water level time series along the Mackenzie River at different latitudes. Panels **(c–e)** show surface water levels time series at three different latitudes with three virtual stations at each latitude to analyse longitudinal spatial variations.

of extreme surface water surface area and the average water surface area duration from 2000 to 2015 were estimated for the large historic flood that occurred in 2006 (Fig. 3e) and for the minimal flood that occurred in 2010 (Fig. 3f). The whole Mackenzie Delta was practically covered in water in 2006, whereas large areas, especially in the downstream part of the delta, were not inundated in 2010 (Fig. 3f).

6.2 Spatio-temporal dynamics of surface water levels in the Mackenzie delta

For all stations and RA missions, a strong seasonal cycle can be seen, with a maximum water level reached in June after the spring ice break-up and snow melt that decreases to reach a minimal value in September, in good accordance with the hydrological cycle of the Mackenzie Delta. The delta is frozen from October to May, and during spring–early summer the freshwater meets an ice dam that was formed in winter, which provokes river discharge variations from 5000 m^3 to 25 000 m^3 on average (http://wateroffice.ec.gc.ca/, Fig. 1b). Then, these important variations provoke water level increases and significant floods each year in the delta. However, water level variations as revealed from RA are not equal over the delta. In the upstream part, variations are 9 m on average, 4 m in the centre and 3 m in the downstream part of the Mackenzie Delta.

Water level time series from data acquired by the ENVISAT mission between June and September, averaged over 2002–2010, are presented in Fig. 11. Each time series has been shifted manually and errors are not shown here for clarity purposes. Virtual stations used to discuss the spatio-temporal variations were chosen along the Mackenzie River from upstream to downstream and at similar latitudes on the Mackenzie River and its tributaries. They are represented using green dots for variations along the Mackenzie River and red triangles for latitudinal variations (Fig. 11a). Time series from Fig. 11b are located along the Mackenzie River, number 1 corresponding to the upstream part and number 8 to the downstream part. Logically, a stronger seasonal cycle is observed upstream than downstream. If the primary peak of flooding that occurs in June clearly appears for all the stations, the secondary peak of August–September is not well marked for all the stations. This could be due to either local differences in the hydrodynamics of the river or due to the low temporal frequency of acquisition of the altimeters that is not sufficient to fully capture all specificities of the hydrological cycle (see Biancamaria et al., 2017 for instance). Latitudinal differences can also be noticed (Fig. 11c). Larger annual amplitudes of water levels can be observed in the Mackenzie River than over its tributaries. The second flood event occurs earlier in the central part (August) than in the western and eastern parts (September).

6.3 Spatio-temporal dynamics of surface water storage

The spatio-temporal dynamics of surface water storage is presented in Fig. 9 for 2006. A strong upstream–downstream gradient of water levels can be observed in June with water levels ranging 0 to 5 m from north to south (Fig. 9a). It strongly decreases in July (0 to 1.5 m in Fig. 9b) and does not appear in August (Fig. 9c) and September (Fig. 9d). For these two later months, differences in water levels are more homogeneous of the whole delta (except in a region located around 135° W and between 68° N and 68°30′ N in August). Our results were compared to the ones estimated by Emmerton et al. (2007) under the assumption of a storage change as a rectangular water layer added to the average low-water volume for a stage variation from 1.231 m above sea level during a low-water period and 5.636 m above sea level during peak flooding. Using this approach, Emmerton et al. (2007) found an increase in water volume of 14.14 km^3 over the floodplains and 7.68 km^3 over the channels. With our method, maximal water volume is around 9.6 km^3 on average and can reach 14 km^3. As can be seen in Fig. 11, water levels present a strong decreasing gradient of amplitude over the delta towards the mouth and are, on average, lower than 5.636 m from Emmerton et al. (2007). The difference of approaches is likely to account for such discrepancy. The comparison between storage and flux (discharge) exhibits quite a good correlation ($R = 0.66$ with no time-lag) between these two quantities. Several studies demonstrated that there are no linear relationships among surface water extent, surface water volume and river discharge due to the presence of floodplains non-connected to the river (e.g. Frappart et al., 2005; Heimhuber et al., 2017). Due to the small area of the non-connected lakes present in the Mackenzie delta, they are detected in our approach based on the use of MODIS images at 500 m of spatial resolution, as mixture areas (except during the June flood event where almost all the delta is inundated and all the flooded areas are connected to the river). Only the floodplains connected to river are considered in this study.

7 Conclusion

This study provides surface water estimates (permanent water of rivers, lakes and inundated surfaces connected to the rivers) dynamics both in extent and storage in the Mackenzie Delta from 2000 to 2015 using MODIS images at 500 m of spatial resolution and altimetry-based water levels. Surface water exhibits a maximal extent in the beginning of June and decreases to reach a minimal value in September. In June, the extent of water surface area is on average about 9600 km^2. The highest value was observed in 2006 ($\sim$ 14 284 km^2), during the historic flood described by Beltaos and Carter (2009). Despite the lower resolution of MODIS images in comparison with Landsat-8 images, surface water extent estimates are quite similar when using both sensors over the river channels and the floodplains, with an underestimation of 20 % found for MODIS. But the numerous small lakes present in the Mackenzie Delta are not detected using MODIS. Nevertheless, the MODIS-based inundation product provides important information on flooding patterns along the hydrological cycle (flood events of June and August–September).

Virtual stations, or river–lake cross sections, have been created across the Mackenzie Delta for the three radar altimetry missions (ERS-2, 1993–2003; ENVISAT, 2002–2010; SARAL, since 2013). Due to the lack of valid data acquired in interferometry SAR mode by Cryosat-2, no information on surface water levels is available in 2011 and 2012. The water levels determined by altimetry at those stations have been validated with in situ river levels, with good correlation coefficients (> 0.8) for the three missions. The dense network of altimetry virtual stations composed of 22 stations for ERS-2, 27 for ENVISAT and 24 for SARAL allowed the analysis of the spatio-temporal variations of water levels across the delta.

The combination between land water extent determined by MODIS imagery and the water levels derived from altimetry permitted estimation of surface water storage variations in the Mackenzie Delta at 8-day temporal resolution. Maps of surface water levels showed a clear upstream–downstream gradient in June that decreases with time. Temporal variations in surface water volume calculated from 2000 to 2010 and from 2012 to 2015 showed a maximal volume in June (on average 9.6 km^3) and a minimal volume in September (about 0.1 km^3). A relatively strong correlation was found between surface water volume and the Mackenzie River discharges ($R = 0.66$).

These products provide a unique long-term dataset that allows continuous monitoring of the changes affecting the surface water reservoir before the launch of the NASA–CNES Surface Water and Ocean Topography (SWOT) mission in 2021. This approach can be applied to any other deltaic and estuarine environments as MODIS and altimetry data are available globally. The major limitations are (i) the presence of clouds and dense vegetation cover that prevent the use of

MODIS images, (ii) the relatively coarse spatial resolution of MODIS images, and (iii) the coarse coverage of altimetry tracks. They can be overcome (i) using SAR images for flood extent monitoring as in Frappart et al. (2005), (ii) using images with a higher spatial resolution, and (iii) combing information on the different altimetry missions orbiting simultaneously. The recent launches of Sentinel-1, Sentinel-2 and Sentinel-3 offer new opportunities for flood extent monitoring at higher spatial (from 10 to 300 m) and temporal (a few days) resolutions. Associated with aquatic colour radiometry (Mouw et al., 2015), the approach developed here should provide useful information for the study of fluvial particle transport along the river-to-coastal ocean continuum and its potential impacts on ecosystems.

Data availability. Surface water extent, level and storage were estimated in the Mackenzie Delta, Canadian Northern Territories, using multi-sensor satellite observations. Surface water area extents were estimated from 2000 to 2015 using MODIS Terra reflectances (8-day mosaic) at 500 m of spatial resolutions based on the Sakamoto et al. (2007) approach. Water levels were estimated using radar altimetry data from ERS-2 (1995–2003), ENVISAT (2003–2010) and SARAL (2013–2016). In total, 22, 27 and 24 virtual stations were built under the ground-tracks of these missions on their nominal orbit using the Multi-mission Altimetry Processing Software (MAPS) (Frappart et al., 2015b). Monthly surface water storage changes were obtained combining the two former datasets following the approach proposed in Frappart et al. (2006, 2012) on the common period of availability of the datasets. These datasets will be soon available on Hydroweb (http://hydroweb.theia-land.fr/). If you would like to access them before they are published here, you can contact Cassandra Normandin (cassandra.normandin@u-bordeaux.fr) and Frédéric Frappart (frederic.frappart@legos.obs-mip.fr).

Competing interests. The authors declare that they have no conflict of interest.

Acknowledgements. This study was supported by an internship grant from LabEX Côte (Université de Bordeaux) and a PhD grant from Ministère de l'Enseignement Supérieur et de la Recherche and also by the CNES TOSCA CTOH grant. The authors also thank David Doxaran for fruitful discussion.

Edited by: Florian Pappenberger

References

Alsdorf, D. E., Rodríguez, E., and Lettenmaier, D. P.: Measuring surface water from space, Rev. Geophys., 45, RG2002, https://doi.org/10.1029/2006RG000197, 2007.

Baup, F., Frappart, F., and Maubant, J.: Combining high-resolution satellite images and altimetry to estimate the volume of small lakes, Hydrol. Earth Syst. Sci., 18, 2007–2020, https://doi.org/10.5194/hess-18-2007-2014, 2014.

Beltaos, S. and Carter, T.: Field studies of ice breakup and jamming in the Mackenzie Delta, St John's, Newfoundland and Labrador, CGU HS Committee on River Ice Processes and the Environment 15th Workshop on River Ice St. John's, Newfoundland and Labrador, 15–17 June 2009.

Beltaos, S., Carter, T., and Rowsell, R.: Measurements and analysis of ice breakup and jamming characteristics in the Mackenzie Delta, Canada, Cold Reg. Sci. Technol., 82, 110–123, https://doi.org/10.1016/j.coldregions.2012.05.013, 2012.

Biancamaria, S., Frappart, F., Leleu, A.-S., Marieu, V., Blumstein, D., Desjonquères, J.-D., Boy, F., Sottolichio, A., and Valle-Levinson, A.: Satellite radar altimetry elevations performance over a 200 m wide river: Evaluation over the Garonne River, Adv. Space Res., 59, 128–146, 2017.

Birkett, C. M.: The contribution of TOPEX/POSEIDON to the global monitoring of climatically sensitive lakes, J. Geophys. Res., 100, 179–204, 1995.

Cazenave, A., Bonnefond, P., Dominh, K., and Schaeffer, P.: Caspian Sea level from Topex-Poseidon altimetry: Level now falling, Geophys. Res. Lett., 24, 881–884, https://doi.org/10.1029/97GL00809, 1997.

Crétaux, J.-F., Jelinski, W., Calmant, S., Kouraev, A., Vuglinski, V., Bergé-Nguyen, M., Gennero, M.-C., Nino, F., Abarca Del Rio, R., Cazenave, A., and Maisongrande, P.: SOLS: A lake database to monitor in the Near Real Time water level and storage variations from remote sensing data, Adv. Space Res., 47, 1497–1507, https://doi.org/10.1016/j.asr.2011.01.004, 2011a.

Crétaux, J.-F., Bergé-Nguyen, M., Leblanc, M., and Abarca Del Rio, R.: Flood Mapping Inferred From Remote Sensing Data, Fifteenth International Water Technology Conference, IWTC-15 2011, Alexandria, Egypt, 2011b.

Crétaux, J. F., Nielsen, K., Frappart, F., Papa, F., Calmant, S., and Benveniste, J.: Hydrological applications of satellite altimetry: rivers, lakes, man-made reservoirs, inundated areas, in: Satellite Altimetry Over Oceans and Land Surfaces, edited by: Stammer, D. and Cazenave, A., 644 pp., CRC press, 2017.

Emmerton, C. A., Lesack, L. F. W., and Marsh, P.: Lake Abundance, potential water storage, and habitat distribution in the Mackenzie River Delta, western Canadian Arctic, Water Resour. Res., 43, W05419, https://doi.org/10.1029/2006WR005139, 2007.

Emmerton, C. A., Lesack, L. F. W., and Vincent, W. F.: Nutrient and organic matter patterns across the Mackenzie River, estuary and shelf during the seasonal recession of sea-ice, J. Mar. Syst., 74, 741–755, https://doi.org/10.1016/j.jmarsys.2007.10.001, 2008.

Frappart, F., Seyler, F., Martinez, J. M., Léon, J. G., and Cazenave, A.: Floodplain water storage in the Negro River basin estimated from microwave remote sensing of inundation area and water levels, Remote Sens. Environ., 99, 387–399, https://doi.org/10.1016/j.rse.2005.08.016, 2005.

Frappart, F., Calmant, S., Cauhope, M., Seyler, F., and Cazenave, A.: Preliminary results of ENVISAT RA-2-derived water levels validation over the Amazon basin, Remote Sens. Environ., 100, 252–264, https://doi.org/10.1016/j.rse.2005.10.027, 2006a.

Frappart, F., Minh, K. D., L'Hermitte, J., Cazenave, A., Ramillien, G., Le Toan, T., and Mognard-Campbell, N.: Water volume change in the lower Mekong from satellite altimetry and imagery data, Geophys. J. Int., 167, 570–584, https://doi.org/10.1111/j.1365-246X.2006.03184.x, 2006b.

Frappart, F., Papa, F., Güntner, A., Werth, S., Ramillien, G., Prigent, C., Rossow, W. B., and Bonnet, M.-P.: Interannual variations of the terrestrial water storage in the Lower Ob' Basin from a multisatellite approach, Hydrol. Earth Syst. Sci., 14, 2443–2453, https://doi.org/10.5194/hess-14-2443-2010, 2010.

Frappart, F., Papa, F., Santos da Silva, J., Ramillien, G., Prigent, C., Seyler, F., and Calmant, S.: Surface freshwater storage and dynamics in the Amazon basin during the 2005 exceptional drought, Environ. Res. Lett., 7, 044010, https://doi.org/10.1088/1748-9326/7/4/044010, 2012.

Frappart, F., Papa, F., Malbeteau, Y., León, J., Ramillien, G., Prigent, C., Seoane, L., Seyler, F., and Calmant, S.: Surface Freshwater Storage Variations in the Orinoco Floodplains Using Multi-Satellite Observations, Remote Sens., 7, 89–110, https://doi.org/10.3390/rs70100089, 2015a.

Frappart, F., Papa, F., Marieu, V., Malbeteau, Y., Jordy, F., Calmant, S., Durand, F., and Bala, S.: Preliminary Assessment of SARAL/AltiKa Observations over the Ganges-Brahmaputra and Irrawaddy Rivers, Mar. Geod., 38, 568–580, https://doi.org/10.1080/01490419.2014.990591, 2015b.

Frappart, F., Legrésy, B., Niño, F., Blarel, F., Fuller, N., Fleury, S., Birol, F., and Calmant, S.: An ERS-2 altimetry reprocessing compatible with ENVISAT for long-term land and ice sheets studies, Remote Sens. Environ., 184, 558–581, https://doi.org/10.1016/j.rse.2016.07.037, 2016.

Fu, L.-L. and Cazenave, A. (Eds.): Satellite altimetry and earth sciences: a handbook of techniques and applications, Academic Press, San Diego, 2001.

Goulding, H. L., Prowse, T. D., and Bonsal, B.: Hydroclimatic controls on the occurrence of break-up and ice-jam flooding in the Mackenzie Delta, NWT, Canada, J. Hydrol., 379, 251–267, https://doi.org/10.1016/j.jhydrol.2009.10.006, 2009a.

Goulding, H. L., Prowse, T. D., and Beltaos, S.: Spatial and temporal patterns of break-up and ice-jam flooding in the Mackenzie Delta, NWT, Hydrol. Process., 23, 2654–2670, https://doi.org/10.1002/hyp.7251, 2009b.

Heimhuber, V., Tulbure, M. G., and Broich, M.: Modeling multidecadal surface water inundation dynamics and key drivers on large river basin scale using multiple time series of Earth-observation and river flow data, Water Resour. Res., 53, 1251–1269, https://doi.org/10.1002/2016WR019858, 2017.

Hill, P. R., Lewis, C. P., Desmarais, S., Kauppaymuthoo, V., and Rais, H.: The Mackenzie Delta: sedimentary processes and facies of a high-latitude, fine-grained delta, Sedimentology, 48, 1047–1078, https://doi.org/10.1046/j.1365-3091.2001.00408.x, 2001.

Holmes, R. M., McClelland, J. W., Peterson, B. J., Tank, S. E., Bulygina, E., Eglinton, T. I., Gordeev, V. V., Gurtovaya, T. Y., Raymond, P. A., Repeta, D. J., Staples, R., Striegl, R. G., Zhulidov, A. V., and Zimov, S. A.: Seasonal and Annual Fluxes of Nutrients and Organic Matter from Large Rivers to the Arctic Ocean and Surrounding Seas, Estuaries Coasts, 35, 369–382, https://doi.org/10.1007/s12237-011-9386-6, 2012.

Huete, A. R., Liu, H. Q., Batchily, K., and van Leeuwen, W.: A comparison of vegetation indices over a global set of TM images for EOS-MODIS, Remote Sens. Environ., 59, 440–451, 1997.

Klein, I., Gessner, U., Dietz, A. J., and Kuenzer, C.: Global WaterPack – A 250 m resolution dataset revealing the daily dynamics of global inland water bodies, Remote Sens. Environ., 198, 345–362, https://doi.org/10.1016/j.rse.2017.06.045, 2017.

Lesack, L. F. W. and Marsh, P.: River-to-lake connectivities, water renewal, and aquatic habitat diversity in the Mackenzie River Delta, Water Resour. Res., 46, W12504, https://doi.org/10.1029/2010WR009607, 2010.

Macdonald, R. W. and Yu, Y.: The Mackenzie Estuary of the Arctic Ocean, in Estuaries, vol. 5H, edited by: P. J. Wangersky, 91–120, Springer-Verlag, Berlin/Heidelberg, available at: http://link.springer.com/10.1007/698_5_027 (last access: 14 December 2016), 2006.

Mouw, C. B., Greb, S., Aurin, D., DiGiacomo, P. M., Lee, Z., Twardowski, M., Binding, C., Hu, C., Ma, R., Moore, T., Moses, W., and Craig, S. E.: Aquatic color radiometry remote sensing of coastal and inland waters: Challenges and recommendations for future satellite missions, Remote Sens. Environ., 160, 15–30, https://doi.org/10.1016/j.rse.2015.02.001, 2015.

Muhammad, P., Duguay, C., and Kang, K.-K.: Monitoring ice break-up on the Mackenzie River using MODIS data, The Cryosphere, 10, 569–584, https://doi.org/10.5194/tc-10-569-2016, 2016.

Ogilvie, A., Belaud, G., Delenne, C., Bailly, J.-S., Bader, J.-C., Oleksiak, A., Ferry, L., and Martin, D.: Decadal monitoring of the Niger Inner Delta flood dynamics using MODIS optical data, J. Hydrol., 523, 368–383, https://doi.org/10.1016/j.jhydrol.2015.01.036, 2015.

Papa, F., Prigent, C., Aires, F., Jimenez, C., Rossow, W. B., and Matthews E.: Interannual variability of surface water extent at the global scale, 1993–2004, J. Geophys. Res., 115, D12111, https://doi.org/10.1029/2009JD012674, 2010.

Pekel, J., Cottam, A., Gorelick, N., and Belward, A. S.: High resolution mapping of global surface water and its long-term changes, Nature, 540, 418–422, https://doi.org/10.1038/nature20584, 2016

Peterson, B. J., Holmes, R. M., McClelland, J. W., Vörösmarty, C. J., Lammers, R. B., Shiklomanov, A. I., Shiklomanov, I. A., and Rahmstorf, S.: Increasing River Discharge to the Arctic Ocean, Science, 298, 2171–2173, https://doi.org/10.1126/science.1077445, 2002.

Rees, W. G.: Physical principles of remote sensing, Cambridge University press, 460 pp., 2013.

Sakamoto, T., Van Nguyen, N., Kotera, A., Ohno, H., Ishitsuka, N., and Yokozawa, M.: Detecting temporal changes in the extent of annual flooding within the Cambodia and the Vietnamese Mekong Delta from MODIS time-series imagery, Remote Sens. Environ., 109, 295–313, https://doi.org/10.1016/j.rse.2007.01.011, 2007.

Santos da Silva, J., Calmant, S., Seyler, F., Rotunno Filho, O. C., Cochonneau, G., and Mansur, W. J.: Water levels in the Amazon basin derived from the ERS 2 and ENVISAT radar altimetry missions, Remote Sens. Environ., 114, 2160–2181, https://doi.org/10.1016/j.rse.2010.04.020, 2010.

Smith, L. C.: Satellite remote sensing of river inundation area, stage, and discharge: a review, Hydrol. Process., 11, 1427–1439, https://doi.org/10.1002/(SICI)1099-1085(199708)11:10<1427::AID-HYP473>3.0.CO;2-S, 1997.

Squires, M. M., Lesack, L. F. W., Hecky, R. E., Guildford, S. J., Ramlal, P., and Higgins, S. N.: Primary Production and Carbon Dioxide Metabolic Balance of a Lake-Rich Arctic River Floodplain: Partitioning of Phytoplankton, Epipelon, Macrophyte, and Epiphyton Production Among Lakes on the Mackenzie Delta, Ecosystems, 12, 853–872, https://doi.org/10.1007/s10021-009-9263-3, 2009.

Stocker, T. F. and Raible, C. C.: Water cycles shifts gear, Nature, 434, 830–831, 2005.

Szostak-Chrzanowski, A.: Study of Natural and Man-induced Ground Deformation in MacKenzie Delta Region, Acta Geodyn. Geomater., 11, 117–123, https://doi.org/10.13168/AGG.2013.0060, 2013.

Tulbure, M. G., Broich, M., Stehman, S. V., and Kommareddy, A.: Surface water extent dynamics from the three decades of seasonally continuous Landsat time series at subcontinental scale in a semi-arid region, Remote Sens. Environ., 178, 142–157, https://doi.org/10.1016/j.rse.2016.02.034, 2016.

Ulaby, F. T., Moore, R. K., and Fung, A. K.: Microwave remote sensing: Active and passive, Volume 1 – Microwave remote sensing fundamentals and radiometry, available at: https://ntrs.nasa.gov/search.jsp?R=19820039342 (last access: 15 December 2016), 1981.

Verpoorter, C., Kutser, T., Seekell, D. A., and Tranvik, L. J.: A global inventory of lakes based on high-resolution satellite imagery, Geophys. Res. Lett., 41, 6396–6402, https://doi.org/10.1002/2014GL060641, 2014.

Verron, J., Sengenes, P., Lambin, J., Noubel, J., Steunou, N., Guillot, A., Picot, N., Coutin-Faye, S., Sharma, R., Gairola, R. M., Murthy, D. V. A. R., Richman, J. G., Griffin, D., Pascual, A., Rémy, F., and Gupta, P. K.: The SARAL/AltiKa Altimetry Satellite Mission, Mar. Geod., 38, 2–21, https://doi.org/10.1080/01490419.2014.1000471, 2015.

Wang, M. and Shi, W. Estimation of ocean contribution at the MODIS near-infrared wavelengths along the east coast of the U.S.: Two case studies, Geophys. Res. Lett., 32, L13606, https://doi.org/10.1029/2005GL022917, 2005.

Xiao, X., Boles, S., Liu, J., Zhuang, D., Frolking, S., Li, C., Salas, W., and Moore, B.: Mapping paddy rice agriculture in southern China using multi-temporal MODIS images, Remote Sens. Environ., 95, 480–492, https://doi.org/10.1016/j.rse.2004.12.009, 2005

Zelli, C.: ENVISAT RA-2 Advances Radar Altimeter: instrument design and pre-launch performance assessment review, Acta Astronaut., 44, 323–333, 1999.

9

Evaluation of the ability of the Weather Research and Forecasting model to reproduce a sub-daily extreme rainfall event in Beijing, China using different domain configurations and spin-up times

Qi Chu[1,2,3], **Zongxue Xu**[1,2], **Yiheng Chen**[3], **and Dawei Han**[3]

[1]College of Water Sciences, Beijing Normal University, Beijing, 100875, China
[2]Beijing Key Laboratory of Urban Hydrological Cycle and Sponge City, Beijing, 100875, China
[3]Department of Civil Engineering, University of Bristol, Bristol, BS8 1TR, UK

Correspondence: Zongxue Xu (zongxuexu@vip.sina.com)

Abstract. The rainfall outputs from the latest convection-scale Weather Research and Forecasting (WRF) model are shown to provide an effective means of extending prediction lead times in flood forecasting. In this study, the performance of the WRF model in simulating a regional sub-daily extreme rainfall event centred over Beijing, China is evaluated at high temporal (sub-daily) and spatial (convective-resolving) scales using different domain configurations and spin-up times. Seven objective verification metrics that are calculated against the gridded ground observations and the ERA-Interim reanalysis are analysed jointly using subjective verification methods to identify the likely best WRF configurations. The rainfall simulations are found to be highly sensitive to the choice of domain size and spin-up time at the convective scale. A model run covering northern China with a 1 : 5 : 5 horizontal downscaling ratio (1.62 km), 57 vertical layers (less than 0.5 km), and a 60 h spin-up time exhibits the best performance in terms of the accuracy of rainfall intensity and the spatial correlation coefficient (R'). A comparison of the optimal run and the initial run performed using the most common settings reveals clear improvements in the verification metrics. Specifically, R' increases from 0.226 to 0.67, the relative error of the maximum precipitation at a point rises from −56 to −11.7 %, and the root mean squared error decreases by 33.65 %. In summary, re-evaluation of the domain configuration options and spin-up times used in WRF is crucial for improving the accuracy and reliability of rainfall outputs used in applications related to regional sub-daily heavy rainfall (SDHR).

1 Introduction

The possibility that sub-daily heavy rainfall (SDHR) will increase with climate change is of significant societal concern. SDHR-driven flash floods (FFs) are among the most destructive natural hazards that threaten many urban areas in northern and central China and many other parts of the world. In these regions, SDHR is mainly triggered by regional mesoscale circulation systems (MCSs) and occurs with increased intensity and frequency in warm seasons (Yu et al., 2007; Chen et al., 2012). Records from the Emergency Events Database (EM-DAT) indicate that the damages and losses caused by FF events in China have increased significantly over the past several decades. The risks are expected to continue to grow, given the increase in the magnitude of SDHR predicted by most general circulation models (Chen et al., 2012; Willems et al., 2012; Westra et al., 2014). The accelerating pace of urbanization also contributes to the increase in risk; urbanization has already changed the hydrologic characteristics of the land surface considerably, resulting in higher peak flows and shorter flow concentration times (Xu and Zhao, 2016; Gao et al., 2017). In such cases, very short-term (< 6 h) rainfall predictions are not sufficient to provide adequate warning and mobilize emergency response activities. Recently developed statistically based rainfall generation methods and remote sensing data have been shown to enable the extension of the lead time to 24 h (Yu et al., 2016). However, this lead time is still insufficient to provide effective flood mitigation for medium or large urban areas with very short hydrologic response times (Shih et al., 2014; Li et al., 2017). Therefore, numerical weather prediction (NWP), which represents a means of forecasting heavy rainfall with lead times exceeding 24 h, has come into wide use in flood-related studies and applications (Cuo et al., 2011).

Precipitation uncertainty accounts for a large proportion of the uncertainty in flood forecasts. Hence, given the large uncertainties in NWP, its use in flood forecasting has long been questioned (Castelli, 1995; Bartholmes and Todini, 2005). Its usefulness was not realized until the end of the 20th century; substantial improvements in the predictive skill of NWP were made that resulted from the increases in computational power and storage capacity, which enable parallel processing of high-resolution forcing data and the resolution of convective-scale physical processes (Done et al., 2004; Clark et al., 2016). The NWP models developed during and after this period can perform regional and convective-scale modelling and display good performance in simulating heavy rainfall. Experimental studies have shown that NWP models of this kind, such as the WRF model (Skamarock et al., 2008), tend to capture greater numbers of small-scale processes and the triggers of convective storms (Klemp, 2006; Prein et al., 2015). Increasing numbers of meteorological operational centres and research groups are adopting these new NWP models to carry out simulations of heavy rainfall events or real-time forecasting. The resolutions of the rainfall products have improved from tens of kilometres to less than a kilometre, and the lead times have increased from less than a day to more than a week (WMO, 2013). Meanwhile, case studies have been carried out using regional convective-resolving models to evaluate the local rainfall predictions generated by sophisticated regional nesting techniques or the global smooth grid transition approach on unstructured grids (Swinbank and James Purser, 2006; Hong and Lee, 2009; Soares et al., 2012; Sikder and Hossain, 2016; Heinzeller et al., 2016). The results of these studies demonstrate that, over relatively short periods of time, regional modelling is often superior to large-scale modelling because it better resolves surface heterogeneities, topography, and small-scale features in air flow such as growing instabilities (Miguez-Macho et al., 2004; Yu et al., 2010; Prein et al., 2015; Brömmel et al., 2018).

Despite the great potential of NWP models to predict heavy rainfall, a number of uncertainties remain that must be considered. The errors induced by the initial and boundary conditions represent one source of these uncertainties; others stem from cognitive errors and the scale effect in the solution of physical models, both of which may be exacerbated by the chaotic nature of NWP. In regional simulations, these uncertainties are expected to be further magnified by downscaling or the use of mesh transition procedures, so re-evaluation and calibration of the related model configurations are commonly required (Warner, 2011; Vrac et al., 2012; Liu et al., 2012). As an example, running the WRF model at convective scales means that convective processes are more likely to be resolved by explicit physical schemes than when sub-grid parameterizations are used, which may incorporate new structural uncertainties related to the model physics (Done et al., 2004; Ruiz et al., 2010; Crétat et al., 2012). In addition to model physics, several other aspects of model configuration, such as the domain size, the spatial resolution, and the spin-up time, may also have a substantial impact on the uncertainty in rainfall forecasts through their effects on the initial and boundary conditions (Aligo et al., 2009; Fierro et al., 2009; Cuo et al., 2011). However, these aspects of model configuration have received less attention in regional case studies because of their insignificant effects on rainfall forecasts in coarse-resolution and long-term model simulations when compared to the physics of the WRF model. Generally, these model configuration aspects are left at the common settings recommended by the official website of the WRF model and by some experimental regional heavy rainfall studies.

Precipitation is one of the most sensitive variables to NWP model uncertainties. In this study, a re-evaluation of WRF is performed to explore whether the recommended configuration of WRF represents the best choice in reproducing a regional SDHR event that happened in Beijing. The WRF model is assessed here because of its superior scalability and computational efficiency; these traits are valued in interdisciplinary studies (Klemp, 2006; Foley et al., 2012; Coen et al., 2013; Yucel et al., 2015). As the latest NWP community model, WRF incorporates up-to-date developments in physics, numerical methods, and data assimilation and is thus widely used in theoretical studies and practical applications (Powers et al., 2017). The selected regional SDHR event occurred on 21 July 2012 and was centred over Beijing, China. Beijing is among the most vulnerable cities to SDHR-induced floods in central China (Yu et al., 2007). The precipitation in this area is caused mainly by monsoon weather systems and enhanced by local orographic effects, and 60–80 % of the total annual precipitation occurs during just a few SDHR events (Xu and Chu, 2015). The SDHR event that occurred on 21 July 2012 caused the most disastrous urban flood in Beijing since 1950. The national operational NWP system failed to predict this event, which resulted in 79 deaths and more than USD 1.6 billion in damage (Brömmel et al., 2018; Wang et al., 2013; Zhou et al., 2014). Thus, several convective-scale studies have been carried out to re-evaluate the optimal combination of the physics options used in the WRF model, such as Di et al. (2015) and Wang et al. (2015). These studies represent the background information that stimulates this research.

The second question we attempt to explore is to what extent rainfall simulations could be improved through the use of the likely best set of settings if the recommended model configurations are not the best choices. The aspects of the model configuration that are evaluated in this study are the domain size, vertical resolution, horizontal resolution, and spin-up time. These options have been found to have substantial impacts on daily-scale extreme rainfall outputs (Leduc and Laprise, 2009; Aligo et al., 2009; Goswami et al., 2012).

A comparative test with four scenarios is designed. Each scenario evaluates one model configuration option to ensure that the simulated disparities can be attributed solely to a single factor each time. In addition, the test is conceived as a progressive process: the optimal setting identified in each scenario will be adopted as the primary choice for the next scenario to help quantify the overall improvement in the accuracy of rainfall outputs. The "ground truth" datasets are gridded observations obtained from Beijing Normal University and the China Meteorological Administration. A coarser-resolution reanalysis called ERA-Interim (Dee et al., 2011) is also employed in identifying departures of the WRF simulations from the driving weather fields as the model setup is varied. Seven objective verification metrics that reflect different features of the model performance are adopted and considered jointly as part of a subjective verification process because no single verification approach has been shown to provide comprehensive information about the quality of rainfall simulations (Sikder and Hossain, 2016). Most of the metrics adopted here are those used to assess the performance of WRF over daily or longer time periods (Liu et al., 2012; Tian et al., 2017). In this research, these metrics are calculated on an hourly basis and averaged over different sub-daily time spans to evaluate the performance of the WRF model using different configurations from a sub-daily and convective-scale perspective.

2 Numerical model used to forecast heavy rainfall

The advanced WRF (ARW-WRF) model, version 3.7.1, is used as the dynamical downscaling tool. ARW-WRF is a compressible non-hydrostatic and convection-permitting regional NWP model that employs the conservative form of the dynamic Euler equations. As the latest regional NWP community system, WRF is composed of two dynamic cores, a data assimilation system and a platform that facilitates parallel computation and function portability. Observations, model output or assimilated reanalysis output can be used to initialize WRF. In terms of discretization, WRF uses a third-order Runge–Kutta method for temporal separation and an Arakawa C-grid staggering scheme for spatial discretization. The model is capable of conducting either one-way or two-way nested runs for regional downscaling. A detailed introduction to the physics and numerical properties of ARW-WRF can be found in Skamarock et al. (2008). Given its emphasis on efficiency, portability, and updates to reflect the state of the art, WRF has been employed in settings ranging from research to applications and has been incorporated into various operational systems, such as the Hurricane-WRF system for hurricane forecasting and the WRF-Hydro system for hydrologic prediction.

In WRF, the domain size implicitly determines the large-scale dynamics and terrain effects, whereas the vertical and horizontal grid spacings determine the smallest resolvable scale (Goswami et al., 2012). Together, these domain configuration options affect the spectrum of the resolved scales and the nature of scale interactions in the model dynamics (Leduc and Laprise, 2009). Thus, they are responsible for the generation and distribution of precipitation. In regional simulations, small domain sizes are commonly preferred for computational efficiency. Seth and Rojas (2003) demonstrated that simulations with small domain sizes are more likely to benefit from the lateral boundary conditions (LBCs) by dampening the feedback from local perturbations on the large-scale general circulation. However, insufficiently large domains have been shown to prevent the full development of small-scale features over areas of interest. To solve this issue, the official website of WRF provides general guidance (Warner, 2011). This guidance recommends that the ranges of domains should include the major features of the leading MCSs and local surface perturbations, and more than five grid points should exist between adjacent nested domains to allow for sufficient relaxation.

As for grid spacing, it appears plausible that WRF model runs performed with relatively small grid spacings would provide more accurate outputs because such runs would resolve more small-scale phenomena of interest that are not present in the LBCs. This statement is generally accepted as true when a relatively coarse-resolution run (> 10 km horizontally or > 1 km vertically) is compared with a relatively finely resolved run at the convective scale (1–5 km horizontally or < 1 km vertically) in representing a convective storm. However, this conclusion is controversial when the comparison is conducted among convective-scale model runs. Taking the horizontal resolution as an example, although there is evidence to show that WRF runs performed at relatively high resolution capture more convective-scale features, the accuracy of rainfall outputs either shows considerable or no statistical improvement (Roberts and Lean, 2008; Kain et al., 2008; Schwartz et al., 2009). In one study, Fierro et al. (2009) suggested that some features detected in convective-scale runs with too small horizontal grid spacings tend to weaken the kinetic structures that favour torrential rainfall. A similar conclusion was drawn by Aligo et al. (2009) in evaluating the impact of the vertical grid spacing on simulations of summer rainfall performed using WRF. Thus, horizontal and vertical grid spacings of approximately 4 and 1 km, respectively, have been employed as a reasonable compromise between accuracy and computational efficiency in several regional studies.

In regional modelling, a spin-up period is often required to balance the inconsistencies between the results simulated by the model physics and the initial and boundary conditions provided by the forcing data (Luna et al., 2013). The proper spin-up time depends on the time needed for initialization, which can be affected by the size of the domain and the local boundary perturbations (Warner et al., 1997; Kleczek et al., 2014). Moreover, the presence of chaotic behaviour, which causes reductions in the predictive skill of models over time, imposes an upper bound on the spin-up time. Therefore, in

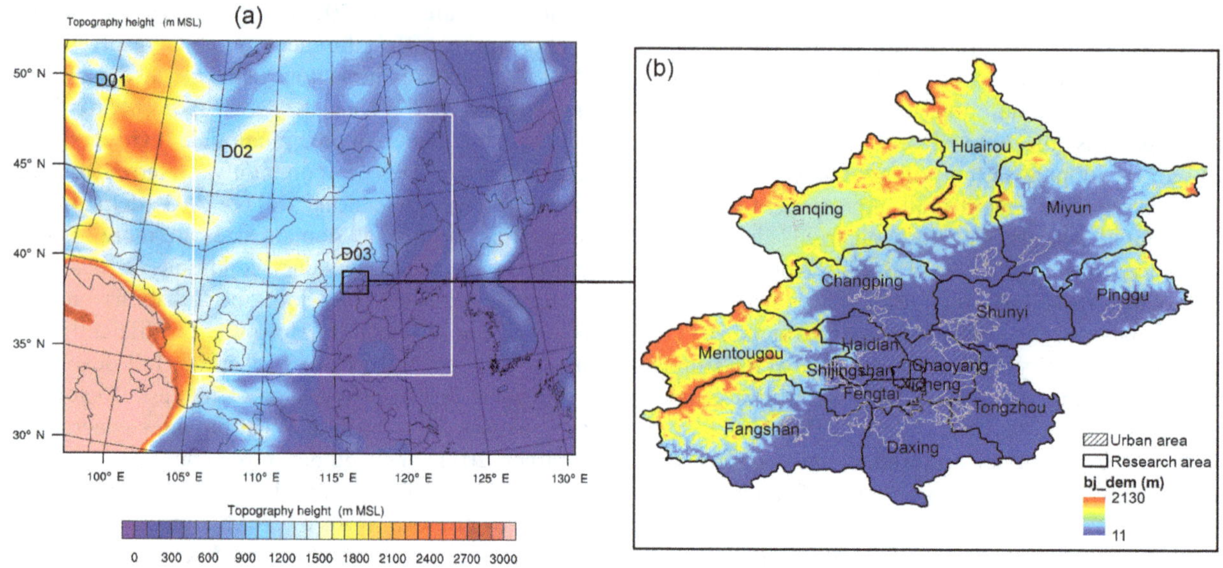

Figure 1. Relative location of the study area. **(a)** Shows the three nested domains adopted in most of the experiments, of which domain three (D03) covers the entire Beijing area; **(b)** depicts the geographic features of the Beijing area.

cases where short forecast lead times are expected, e.g. real-time rainfall forecasting, the spin-up time is mainly determined by the domain size and the regional initial and boundary conditions. However, in cases where long forecast lead times are needed, e.g. warnings of extreme rainfall, the effects of chaotic behaviour should be relatively evident. In practice, this issue is commonly addressed by regularly updating the lateral boundary information derived from the latest forecasts or analyses to maintain consistency between the regional model solutions and the atmospheric forcing conditions. In such cases, the best-fit performance may occur for model runs with long spin-up times. Based on most previous studies, a spin-up time of 12 h is recommended to obtain an initial state; however, this spin-up time is often regarded as the suitable choice in many regional case studies without further verification.

3 Studied event and experimental design

As mentioned above, one aim of this study is to re-evaluate whether the recommended WRF domain configuration options and spin-up time represent the optimal model configuration for reproducing a regional SDHR event when evaluated at a sub-daily timescale. Here, the SDHR event that occurred on 21 July 2012 and was centred on Beijing, China is selected as a case study. The reasons why this event is selected, the synoptic and physical features that drove this event, and the model physics adopted in this study are presented before the entire procedure of the experimental design is introduced.

3.1 Study event selection and WRF physical schemes

Beijing is selected as the study area because it is one of the most vulnerable cities to SDHR-induced FF hazards in China. Beijing is located in central China. It has an area of 16 411 km^2, and its weather is mainly affected by the semi-humid warm continental monsoon climate. The flows of air that favour local precipitation are cold, dry flows of air from high-latitude areas to the north and hot, wet flows of air from the ocean to the south. The interactions between these two flows of air lead to clear divergence in the temporal distribution of rainfall amount; 60–80 % of the annual precipitation occurs during just a few heavy rainfall events during the warm season. Of all of the heavy rainfall events, the intensity and frequency of SDHR events have been shown to display the greatest increasing tendencies over the past several decades. Meanwhile, Beijing, as the capital of China, has experienced a significant expansion of its urban area and rapid increases in its population and economic development. The negative effects of this expansion, such as losses of natural water bodies, increases in land cover with low permeability, and increases in urban drainage pipe networks, have led to continuous decreases in the hydrologic response time. In addition, most of the population lives in the southwestern plain area. This region is downstream of mountainous areas with steep terrain that varies in elevation from 60 to 2300 m (Fig. 1). All of these factors contribute to the continuing increase in the exposure of this city to the high risks of flooding and waterlogging caused by SDHR events (Xu and Chu, 2015).

The case study examines the largest heavy rainfall event that has occurred in Beijing in the past 65 years. The rainfall event lasted for 16 h (from 02:00 to 18:00 UTC) on 21 July 2012, and the highest hourly rainfall intensity ($100\,\text{mm}\,\text{h}^{-1}$) was experienced in the southwestern part of the plain area. The associated FF hazard led to 79 deaths and damages totalling USD 1.6 billion, and more than 1.6 million people were affected. In addition to Beijing, the adjacent provinces, including Hubei and Liaoning, were all significantly affected by this event and experienced severe FF hazards. The synoptic features that triggered the rainfall were an eastward-moving vortex in the middle to high troposphere, a northward-moving zone of subtropical high pressure, and sharp vertical wind shear (Sun et al., 2013). The rainfall event as a whole can be divided into two phases. From 02:00 to 14:00 UTC, the convective rain was dominated and enhanced by the orographic effect. The frontal rain was then followed by the arrival of a cold front moving from the northwest until 18:00 UTC (Guo et al., 2015). The rainfall intensity in the second phase was relatively low compared to that in the first phase, due to the lack of strong kinetic forcing to maintain the occurrence of precipitation.

The ERA-Interim reanalysis and 30-second static geographical data are employed to initialize the surface and meteorological fields of the WRF. ERA-Interim is produced by an integrated forecasting system (IFS) used by the European Centre for Medium-Range Weather Forecasts (ECMWF). The IFS is an Earth system model that incorporates a data assimilation system and an atmospheric model that is fully coupled with land-surface and oceanic processes. The atmospheric model provides output every 30 min at a spectral resolution of T255 (approximately 81 km over Beijing). This output is then employed as prior information and combined with available observations twice a day to produce the reanalysis output using the four-dimensional variation (4D-Var) assimilation system. The final reanalysis product, ERA-Interim, is a global gridded dataset that is available at a spectral resolution of T255 and at both the 60 levels used in the model and 38 interpolated pressure levels for all dates beginning on 1 January 1979 (Berrisford et al., 2009; Dee et al., 2011). Here, the ERA-Interim pressure-level data are selected as the initial forcing. One reason is that, as is necessary, the vertical grid spacing between the adjacent pressure layers is less than 1 km in the free troposphere, where the convective processes mainly occurred during the Beijing SDHR event. In addition, the NWP models used by the China Meteorological Administration mainly employ 31 vertical levels in regional forecasting (WMO, 2013).

As shown in Fig. 2, ERA-Interim captures the vortex and the subtropical high pressure well that occurred at the beginning of the rainfall event. In addition, the patterns of the leading MCSs and the primary synoptic features shown in this figure also correspond well to those described in previous studies (Zhou et al., 2014). The setup of the model physics is based mainly on the results of sensitive, high-resolution studies on the physics of the WRF model in simulating the same event (Wang et al., 2015; Di et al., 2015). The "resolved rain" is driven by the single-moment 6-class microphysics scheme (Hong and Lim, 2006), whereas the "convective rain" is resolved using the Grell–Dévényi cumulus parameterization scheme (Grell and Dévényi, 2002). The Noah land-surface model (Chen and Dudhia, 2001) is used and coupled with the Monin–Obukhov surface layer model (Ek et al., 2003). The radiation processes are represented by the RRTMG shortwave radiation and the RRTMG longwave radiation schemes (Iacono et al., 2008). For the planetary boundary layer scheme, the Yonsei University method (Hong et al., 2006) is adopted.

3.2 Experimental design: domain configuration options and spin-up time

The comparative test is designed as a progressive process to help quantify the overall improvement in the performance of WRF after re-evaluating the WRF experiments performed using different domain configuration options and spin-up times. The test is classified into four successive scenarios. The first three scenarios investigate the domain configuration options, including the domain size, vertical resolution, and horizontal resolution; the fourth scenario concerns the spin-up time. During the entire procedure, the optimum configuration identified in each scenario is then adopted as the primary choice for the corresponding configuration in the following scenario. The initial datasets and the model physics are the same for all of the domains throughout the entire comparative procedure. Because the area of interest is located in the middle latitudes, the Lambert conformal projection is employed in all of the experiments, which is centred on the same latitude (42.25° N) and longitude (114.0° E). Moreover, sigma vertical coordinates with a top level of 50 hPa are used in all of the experiments.

Initially, the WRF domain configuration options and the spin-up time are set to the recommended values described in Sect. 2. Three levels of two-way nested domains are adopted so that the horizontal resolution in the smallest domain is sufficiently high to explicitly resolve convective-scale processes (Fig. 1). An odd downscaling ratio (1 : 3 : 3) is selected to reduce the initial error introduced by interpolating the initial fields to the assigned Arakawa grid. For the same reason, the boundaries of each domain are set along specific grid lines of the ERA-Interim dataset. Of the three nested domains, the outermost domain (D01) has the largest horizontal grid spacing of 40.5 km over north-central China, where the main perturbed synoptic features occur. The innermost domain (D03) has the smallest horizontal grid spacing of nearly 4.5 km over the area of interest, Beijing. The second domain (D02) is the child of D01 and the parent of D03 and has a horizontal grid spacing of 13.5 km. The distance between D01 and D02 is similar to that between D02 and D03, both of which exceed five grid points. The grid numbers of D01, D02, and D03 are 40×40, 72×72, and 90×90, respectively. The eta values utilized in the initial run are set based on the pressure values

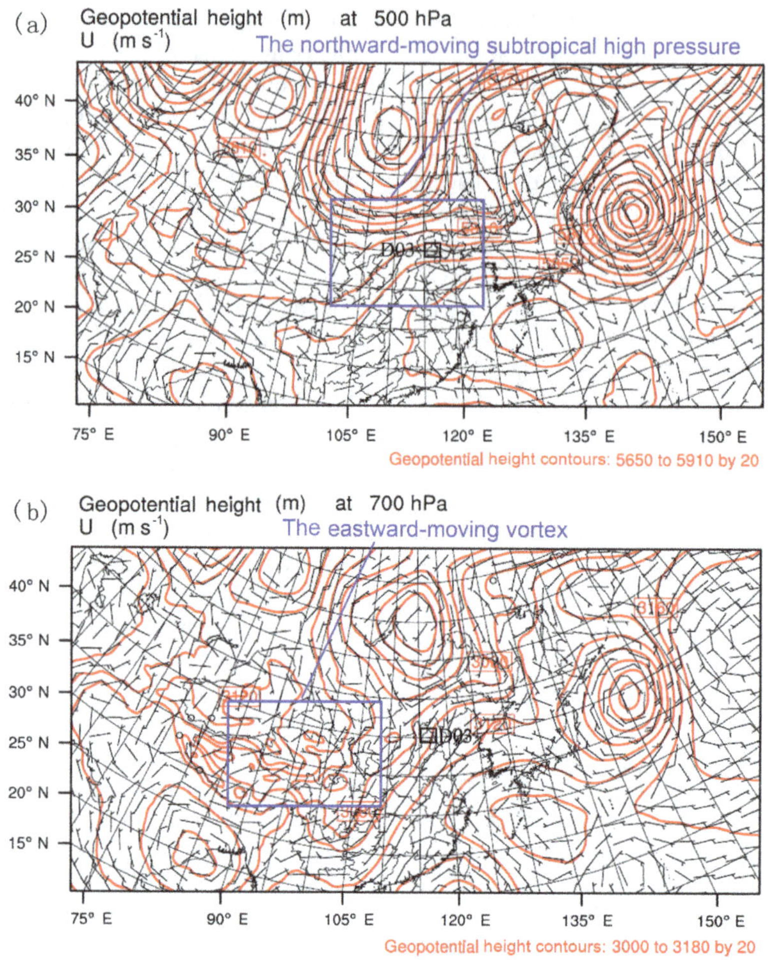

Figure 2. Initial wind field and geopotential height field at 00:00 UTC on 20 July 2012 over the Northern Hemisphere obtained from the ERA-Interim reanalysis. **(a)** The fields at 500 hPa; **(b)** the fields at 700 hPa.

at the 29 vertical layers of the ERA-Interim pressure-level data. A spin-up time of 12 hours (12 h) is selected; the outputs are saved every 3 h in D03 and every hour in D02. The LBCs are updated every 6 h using ERA-Interim.

As shown in Table 1, the first experiment (C0) adopts the model configuration options mentioned above. To determine whether the domain configuration options and the spin-up time used in C0 are the likely best set, four scenarios are designed. The first scenario (S1) focuses on evaluating the effect of the WRF domain size. For computational efficiency, the MCS systems that drive the local synoptic features are not completely contained within the outermost domain of C0, the information of which is compensated by the updated LBCs from ERA-Interim. Two comparative experiments, C1 and C2, are devised to verify that the domain size assigned to C0 is large enough to enable the full development of small-scale features. Of the three experiments, C2 has the largest outermost domain size, which incorporates the leading MCS systems over the entire Northern Hemisphere. The intermediate domain, which is centred between the outermost and innermost domains, is then adopted as the outermost domain of C1. The purpose of scenario two (S2) is to evaluate whether the use of a higher vertical resolution in a WRF model run results in better performance. In this scenario, the starting experiment is the optimal experiment identified in S1 (OS1), forced by the ERA-Interim pressure-level data with 29 vertical levels. This starting experiment is then followed by two experiments, C3 and C4, which incorporate 1 and 2 times more vertical levels than OS1 (57 and 85 vertical levels), respectively. In the Beijing SDHR event, the pressure-level data meet the requirement of a grid spacing of less than 1 km in the troposphere; however, this condition is not necessarily satisfied in other regions. Thus, an experiment forced by the ERA-Interim model-level data with 38 vertical levels (C5) is also designed for comparison. The three experiments (OS2, C6, and C7) in scenario three (S3) differ in terms of their horizontal resolutions and nesting ratios, with increased nesting ratio of 1 : 3 : 3 (4.5 km grid spacing in D03), 1 : 5 : 5 (1.62 km in D03), and 1 : 7 : 7 (0.826 km in D03). The last scenario (S4) is designed to identify a reasonable optimal

Table 1. Categories of experiments with different domain sizes, vertical resolutions, horizontal resolutions and spin-up times.

Scenario	Experiment number	Domain size	Vertical levels	Horizontal resolution (nesting ratio)	Spin-up time
Domain size (S1)	Case 0 (C0)	D01 40 × 40; D02 72 × 72; D03 90 × 90	29 (pressure level)	D01 40.5 km; D02 13.5 km; D03 4.5 km 1 : 3 : 3	12 h
	Case 1 (C1)	D01 80 × 64; D02 120 × 120	as C0	as C0	as C0
	Case 2 (C2)	D01 160 × 128; D02 240 × 192	as C0	as C0	as C0
Vertical resolution (S2)	Optimal case in S1 (OS1)	as OS1	29	as C0	as C0
	Case 3 (C3)	as OS1	57	as C0	as C0
	Case 4 (C4)	as OS1	85	as C0	as C0
	Case 5 (C5)	as OS1	38 (model level)	as C0	as C0
Horizontal resolution (S3)	Optimal case in S2 (OS2)	as OS1	as OS2	1 : 3 : 3	as C0
	Case 6 (C6)	as OS1	as OS2	D01 40.5 km; D02 8.1 km; D03 1.62 km 1 : 5 : 5	as C0
	Case 7 (C7)	as OS1	as OS2	D01 40.5 km; D02 5.785 km; D03 0.826 km 1 : 7 : 7	as C0
Spin-up time (S4)	Optimal case in S3 (OS3)	as OS1	as OS2	as OS3	12 h
	Case 8 (C8)	as OS1	as OS2	as OS3	0 h
	Case 9–Case 19 (C9–C19)	as OS1	as OS2	as OS3	24–144 h per 12 h

model run with the maximum spin-up time after minimizing the uncertainties introduced by inappropriate domain configuration options. It contains 1 starting experiment (OS3) and 12 comparative experiments (C8–C19). Except for C8, which includes no spin-up time, the remaining experiments (C9–C19) include spin-up times that increase from 24 to 144 h by every 12 h.

4 Verification schemes

Both objective and subjective verification methods are applied to the innermost domain (D03) at a sub-daily scale. D03 is selected because it covers the area of interest, Beijing, and the convective processes in this domain can be explicitly resolved in all of the experiments. The rainfall data used for comparison in D03 are 3-hourly 0.05° data that were produced by fusing rain gauge observations and the CMORPH data (Huang et al., 2013). The ERA-Interim reanalysis is utilized as well to monitor the possible departures of the model simulations from the driving fields. Because the sub-daily rainfall is not available from the reanalysis, the atmospheric precipitable water vapour (PW), which determines the possible maximum precipitation, is instead compared with the model outputs every 6 h. In addition, the model outputs that cover a larger domain (D02) are compared with an hourly 0.1° gridded dataset obtained from the China Meteorological Administration (http://data.cma.cn/data/cdcdetail/dataCode/SEVP_CLI_CHN_MERGE_CMP_PRE_HOUR_GRID_0.10.html, last access: 17 June 2018). The comparison over domain two is used only as an auxiliary method for subjective verification, based on the assumption that an experiment with good performance in the inner domain should also capture the large-scale features in the outer domain, as the appropriate representation of these large-scale features will result in more accurate boundary conditions.

Seven error metrics that describe different features of precipitation are selected for use as objective verification metrics. Five are rainfall-related and compared by bilinear interpolation of the output of the simulations to the grid of the ground truth data. The accumulated areal rainfall is assessed using the relative error of the total precipitation (RE_{TP}). The percentage of correct rainfall hits is measured using the probability of detection (POD) with a threshold of 0.1 mm. The root mean squared error (RMSE) represents the amount of continuous error in the predicted precipitation. Detailed illustrations of these three metrics can be found in Liu et al. (2012) and Tian et al. (2017). The other two rainfall-

Table 2. Correlations between the original and rescaled objective verification metrics.

Original metrics	Representative meaning	Rescaled metrics	Threshold value
POD	probability of detection	$\text{POD}' = \text{POD}$	n/a
RMSE	root mean squared error	$\text{RMSE}' = 1 - \text{RMSE}/\text{RMSE}_{\text{max}}$	+62.5 max
R	Pearson correlation coefficients	$R' = R$	n/a
WRMSE	RMSE of the precipitable water	$\text{WRMSE}' = 1 - \text{WRMSE}/\text{WRMSE}_{\text{max}}$	+8.3 max
WR	R of the precipitable water	$\text{WR}' = \text{WR}$	n/a
RE_{PMAX}	relative error of the maximum precipitation	$\text{PMAX}' = \text{RE}_{\text{PMAX}} + 1$	n/a
RE_{TP}	relative error of the total precipitation	$\text{TP}' = \text{RE}_{\text{TP}} + 1$	n/a

n/a = not applicable.

related metrics are the relative error of the maximum grid precipitation (RE_{PMAX}) and the Pearson correlation coefficient (R), which describe the spatial association between the simulations and the ground truth data (Eqs. 1 and 2). The two metrics selected for the verification of PW (PW-related metrics) are the root mean squared error (WRMSE) and the Pearson correlation coefficient (WR). For comparison, the PW fields of the reanalysis are remapped to the grids of the model outputs using the WRF Preprocessing System (WPS). In this study, all of the metrics are calculated between the simulations and the reference data on the same grid at each time step (3 h in D03). The values of these metrics are then averaged over four different time periods (6, 12, 18, and 24 h) counted from 00:00 UTC on 21 July 2012. Different time periods are selected with the purpose of determining whether the performance of WRF differs when the evaluation is conducted using different durations.

$$\text{RE} = \frac{1}{N}\sum_{i=1}^{N}\left[\frac{f-r}{r} \times 100\,\%\right], \tag{1}$$

$$R = \frac{1}{N}\sum_{i=1}^{N}\left(\frac{\sum_{j=1}^{M}\left(f_j - \overline{f}\right)\left(r_j - \overline{r}\right)}{\sqrt{\sum_{j=1}^{M}\left(f_j - \overline{f}\right)^2 \sum_{j=1}^{M}\left(r_j - \overline{r}\right)^2}}\right). \tag{2}$$

Here, R is the empirical spatial correlation coefficient; M is the total number of grid points within the evaluated domain of the starting experiment; f_j is the value of the jth grid point in the tested field at time step i; r_j is the value of the reference field; N is the total number of time steps, depending on the time period considered; and RE is the relative error. For the maximum precipitation, f is the tested value of the maximum gridded precipitation over the area of interest, and r is the reference value of the maximum gridded precipitation over the same area.

To facilitate evaluation, the metrics are further adjusted to ensure that the ideal value of all of the metrics is 1. In this study, RMSE and WRMSE are first divided by a rescaling factor to fall into the range of 0–1 and then subtracted from 1 to provide an indication of good performance. The rescaled metrics, RMSE′ and WRMSE′, have the value 1 representing the lowest accumulated error (highest accuracy). The factor used for rescaling is determined by the largest values of each error metric in all of the experiments and is kept at the same value for all of the evaluated time periods (Sikder and Hossain, 2016). RE_{PMAX} and RE_{TP} are added by 1 to have the ideal value of 1. The rescaled metrics are PMAX′ and TP′, respectively. The other metrics are not rescaled because they already have ideal values of 1, but they are assigned a new set of symbols to distinguish them from the original metrics used before rescaling. For example, POD is replaced with POD′, and R is replaced with R'. Table 2 shows the correlations between the original metrics and the rescaled metrics. Given that the metrics describe different features of the rainfall simulations, the values of these metrics are checked and considered together in subjective verification to determine the likely best set of domain configuration options and to search for the longest reasonable spin-up time.

5 Results and analyses

In each scenario, the metrics are compared among the experiments that consider different durations and cover the same domain (D03). The results are presented in four sub-graphs; each sub-graph shows the values of the metrics calculated for individual evaluated time periods. The spatial distribution of rainfall is also presented over domain two (D02) when evident discrepancies are noted in the results obtained for the inner domain (D03) and the outer domain (D02). Table 1 shows the categories of the scenarios and the model configurations adopted in each experiment. In the following section, the domain size scenario (S1) is evaluated first, followed by the vertical resolution scenario (S2) and the horizontal resolution scenario (S3).

5.1 Results of the domain size scenario

Figure 3 shows the spatial values of the verification metrics for the WRF domain size experiments. The performance of

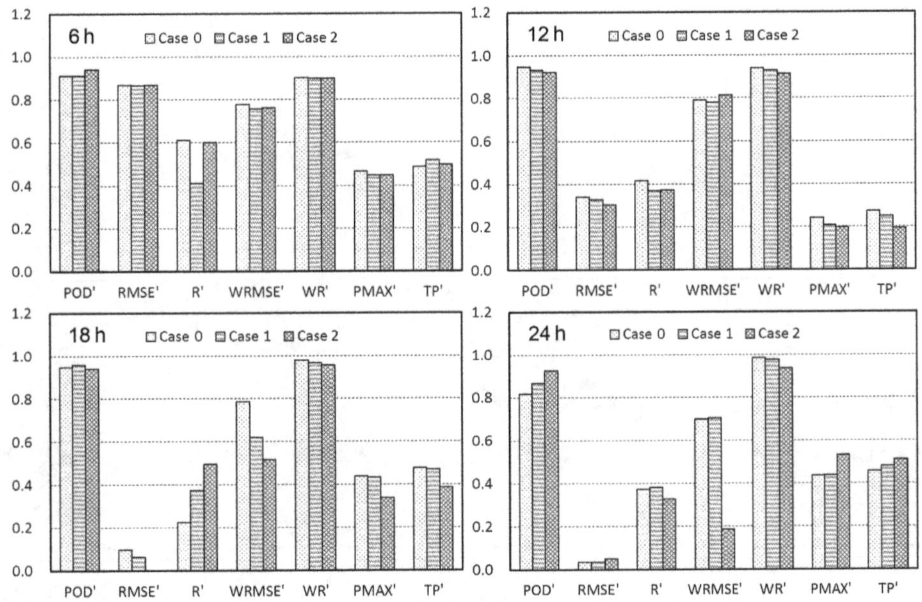

Figure 3. Spatial values of the verification metrics for the WRF domain size experiments, calculated over different temporal durations and over domain three. Case 0 (C0) incorporates the smallest domain, which covers north-central China; Case 1 (C1) incorporates a domain of intermediate size that covers northern China and part of Mongolia; and Case 2 (C2) incorporates the largest domain, which covers the Northern Hemisphere. The metrics are calculated over time periods of 6, 12, 18, and 24 h that begin at 12:00 UTC on 21 July 2012.

the experiments clearly worsens as the evaluated temporal duration increases from 6 to 24 h. The most evident deteriorations are detected in the point-to-point accuracy of the rainfall; the reversed root mean squared error (RMSE′) decreases by 0.8, which represents a 6-fold increase in the cumulative spatial error. The spatial association between the simulations and the gridded observations also declines; the Pearson correlation coefficient (R') decreases by 0.3 on average. Although a slight increase is observed in the percentage of correct hits (POD′) during the first 18 h, this increase is followed by a rapid decrease of nearly 14 % during the last stage of the rainfall event. The relative bias in the accumulated areal rainfall (TP′) indicates that the total rainfall amount is underestimated throughout the entire evaluated temporal period. The maximum gridded precipitation (PMAX′) is also underestimated; the largest negative bias occurs during the heavy convective rainfall stage. For PW, a slight decrease is found in the reversed accumulated error (WRMSE′), whereas an increase of 5–9 % is detected in the spatial correlation coefficient (WR′). Such variations may be attributable to the role of the updated boundary conditions in adjusting the local model solutions to approach the large-scale atmospheric circulation conditions.

Comparison of the four sub-graphs shows that the values of the metrics do not point to a single perfect experiment in a given period, and their ranked predictive skills determined using a given metric differ when evaluated over different time periods. During the early stage of the rainfall event (6 h), C0 yields better performance than C1 and C2 in terms of RMSE′, R', and PMAX′; it simultaneously displays the lowest value of POD′ and the largest bias in estimating the total precipitation. Although the superiority of C0 is more evident in the second period, a sharp deterioration is then observed in capturing the point-to-point accuracy of precipitation for the 18 h duration, where the lowest R' is obtained. Meanwhile, C1, which employs a domain of moderate size, displays greater skill than C0 in capturing the correct hits and the spatial pattern of the simulated rainfall. C2 employs the largest domain. Although it shows the best fit to the rainfall observations on the daily scale (24 h), it displays the worst performance over the three shorter time periods. For the PW fields, the highest similarity with the ERA-Interim reanalysis is found for C0, whereas the lowest similarity is found for C2. These results demonstrate indirectly that small domains are more likely to be influenced by updated boundary conditions.

In this scenario, if the experiments are merely evaluated in D03, the conclusion that C0 displays the best performance during most of the evaluated time periods may be reached. However, when evaluated in D02, clear differences between C0 and the ground truth in both the spatial characteristics of the rainfall and the magnitude of the maximum precipitation are detected. Figure 4 shows the spatial distribution of the accumulated 6 h precipitation over the domain two area of C0. Note that the speed of movement of the belt of heavy rain simulated in C0 is a few kilometres per hour faster than those in C1 and C2, leading to an early end of the heavy rainfall event. This difference may explain why the modelling skill of C0 declines significantly as the end of the rainfall event

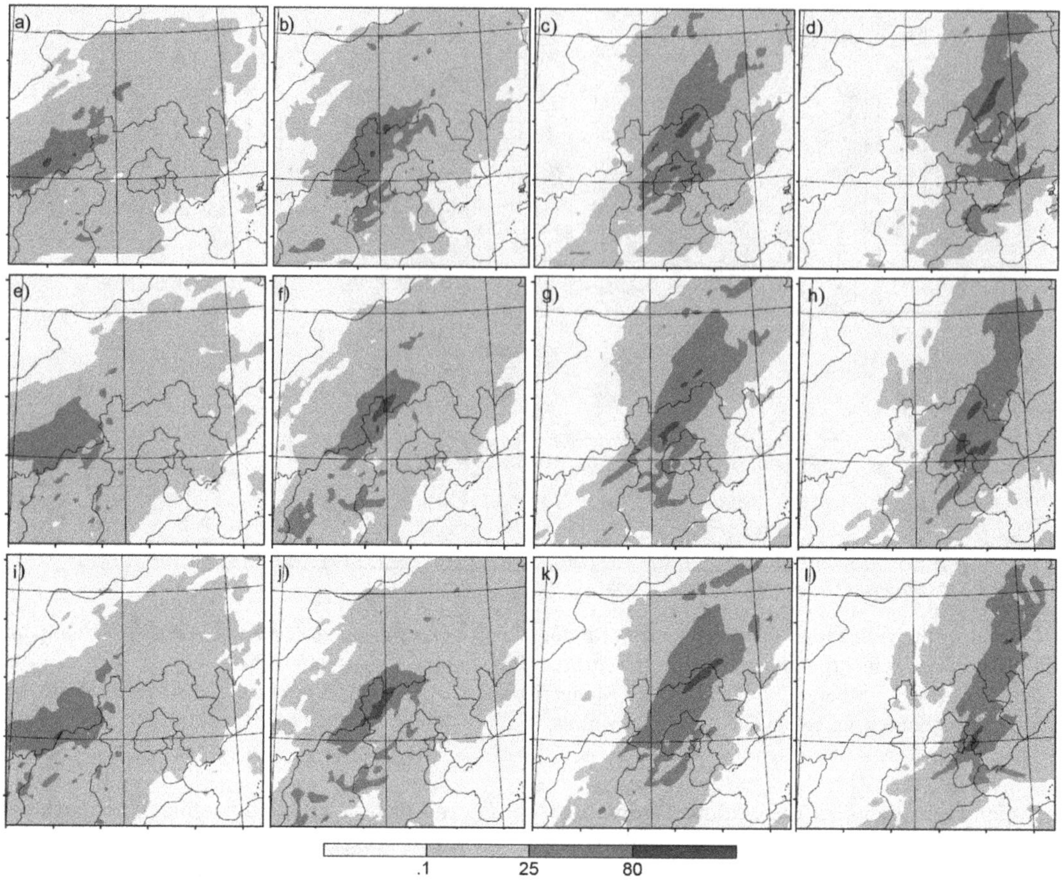

Figure 4. Spatial distribution of 6-h accumulated precipitation for the domain size experiments over the domain two area of C0 during the Beijing heavy rainfall event beginning at 12:00 UTC on 21 July 2012. **(a)** Accumulated precipitation (AP) in C0 during the first 6 h period (00:00–06:00 UTC); **(b)** AP in C0 during the second 6 h period (06:00–12:00 UTC); **(c)** AP in C0 during the third 6 h period (12:00–18:00 UTC); **(d)** AP in C0 during the fourth 6 h period (18:00–00:00 UTC); **(e)** AP in C1 during the first 6 h period; **(f)** AP in C1 during the second 6 h period; **(g)** AP in C1 during the third 6 h period; **(h)** AP in C1 during the fourth 6 h period; **(i)** AP in C2 during the first 6 h period; **(j)** AP in C2 during the second 6 h period; **(k)** AP in C2 during the third 6 h period; and **(l)** AP in C2 during the fourth 6 h period.

approaches. The belt of heavy rain in C0 displays an orientation that is shifted nearly 10 degrees northward from those simulated in C1 and C2 during the first 6 h, and the storm centre in C0 displays the smallest range; it is nearly half of the area in C2. The results indicate that the domain size of C0 is not broad enough to allow the model physics to fully develop the small-scale features that favour heavy rainfall. The spatial characteristics of precipitation are relatively similar in the other two experiments, but C1 outperforms C2 in both the rainfall-related and the PW-related features over domain two. It may be that C2 does not yield better performance than C1 because of its inefficient use of boundary conditions to adjust the false perturbations generated by the local model run. Therefore, C1 is verified as reasonable from both statistical and physical perspectives and is chosen as the optimal experiment in the domain size scenario (OS1).

5.2 Results of the vertical resolution scenario

Based on the analysed results, C1 is selected as the starting experiment in the vertical resolution scenario. As mentioned above, C1 is forced with the ERA-Interim pressure-level data with 29 vertical levels. C3 and C4 are forced with the same pressure-level data with 57 and 85 vertical levels, respectively, whereas C5 is forced with the model-level data with 38 vertical levels. As shown in Fig. 5, a decline in model performance is also obtained for all of the vertical resolution experiments as the evaluated time period increases in length. Moreover, the largest deterioration in RMSE′ is also observed; it decreases by 0.82 on average. The values of TP′ and PMAX′ derived from the simulations are slightly higher than those predicted in S1 but are still less than those calculated for the actual precipitation over the entire rainfall event. POD′ displays an evident decrease during the end stage of the rainfall event, and its magnitude decreases 50 % less relative to that shown in C1. The most obvious difference from the

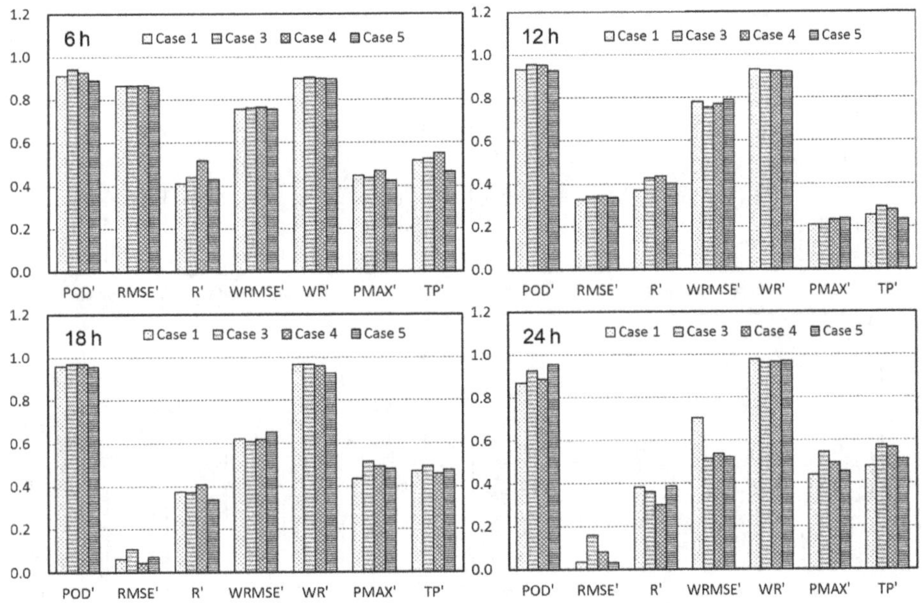

Figure 5. As in Fig. 3, but for the experiments in scenario two with different vertical resolutions. Case 1 is forced by the ERA-Interim pressure-level data with 29 vertical levels; cases 3 and 4 are forced by the same data but include double and triple the number of vertical levels, respectively; Case 5 is forced by the ERA-Interim model-level data with 38 vertical levels.

domain size scenario is that the values of R' calculated between the simulations and the ground truth vary slightly and remain almost the same between the different time periods. In addition, the performance of the vertical resolution experiments seems to be less sensitive to the boundary conditions because they result in relatively small variations in WRMSE′ and WR′.

Unlike the apparent discrepancies noted in the metrics obtained for the domain size experiments, the differences in the rainfall-related metrics among the experiments with different numbers of vertical levels are not evident, especially during the less rainy period (6 h) and the period when convective rainfall dominates (12 h). During the first 12 h, C4 displays better agreement with the gridded observations than the other three experiments in terms of the accuracy and spatial correlation of the rainfall amount. However, over the longer time periods, C3 displays the greatest skill, according to most of the verification metrics. Comparing C3 and C1 shows that increases in the vertical resolution may increase WRF's ability to explicitly resolve small-scale physical processes and improve the accuracy of the amount and distribution of the simulated rainfall. Comparing C3 and C4 shows that, although C4 includes further refinement of the vertical resolution, the C4 performance is worse than that of C3 when the evaluated time period increases from 6 h to more than 12 h. This result may occur because progressive reductions in the vertical grid spacing magnify the propagation of surface perturbations through the vertical grid columns, potentially weakening the kinetic energy that favours precipitation. Examining the values of WRMSE′ and WR′ shows that the differences between the simulations and the reanalysis are more distinct in C3 and C4 than in C1. This discrepancy may occur due to the exaggeration of the initial errors introduced by the interpolation process and the incorporation of false surface perturbations introduced by the limited accuracy and resolution of the initial forcing data. C5 shows either better or worse performance than C1 in each period but produces less accurate rainfall simulations than C3 over most of the evaluated durations. As such, C3 is identified as yielding the best performance in the vertical resolution scenario.

5.3 Results of the horizontal resolution scenario

Based on the results obtained for scenario S2, C3 is selected as the starting experiment in the horizontal resolution scenario. The modelling skill of the S3 experiments shows similar temporal trends as that of the S2 experiments (Figs. 5 and 6). However, the sensitivity of the metrics to the variation in the horizontal resolution is more evident than that with different vertical resolutions. Over most of the evaluated time periods, C6, which has a grid spacing of 1.62 km, displays better performance than C3 and C7 having grid spacings of 4.5 and 0.826 km, respectively. Comparison of C3 and C6 shows that C6 tends to produce more accurate spatial patterns of rainfall throughout the heavy rainfall event in Beijing. Higher values of PMAX′ and TP′ are also detected in C6 when compared to C3. This result stems in part from the explicit resolution of the convective processes by the WRF microphysics scheme, which may explain why the PMAX′ of C7 is higher than C6 over most of the tested durations. Note that the modelling skill of C7 deteriorates

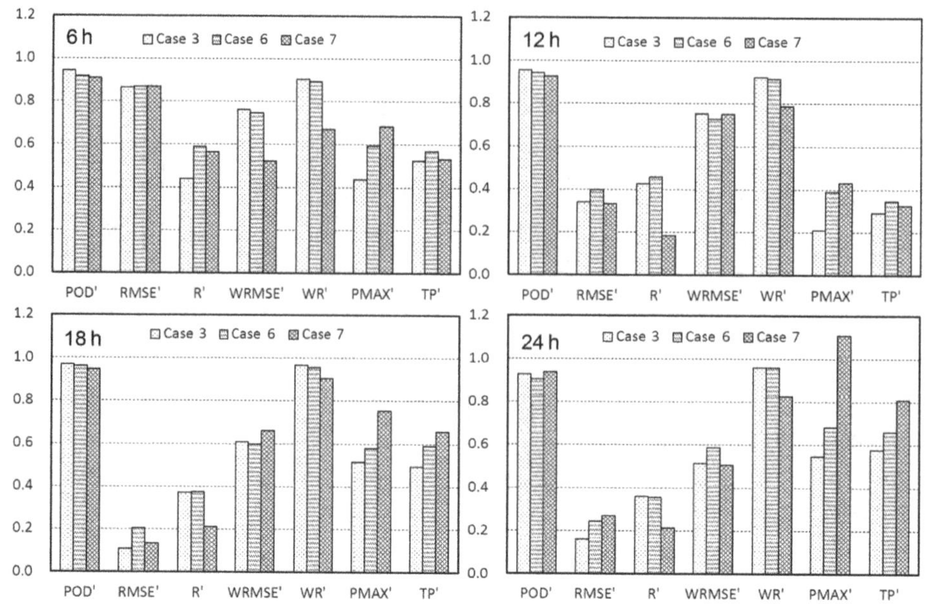

Figure 6. As in Fig. 3, but for the experiments in scenario three with different horizontal resolutions. Case 3 has an initial downscaling ratio of 1 : 3 : 3 with horizontal grid spacing of 40.5, 13.5, and 4.5 km, whereas cases 6 and 7 have the same large horizontal grid spacing with nesting ratios of 1 : 5 : 5 and 1 : 7 : 7, respectively. The innermost grid spacing is 1.62 km in Case 6 and 0.826 km in Case 7.

rapidly after the heavy rain begins (12 h); the lowest POD′ and R' values of the three experiments are obtained for this simulation and time period. Analysis of the WRMSE′ values suggests that simulation C7 displays significant departures from the coarser-scale PW fields that are used to force the model. Thus, model simulations with excessively high horizontal resolutions may also display poor performance. Theoretically, this deterioration may be attributed to the accumulated errors introduced by the imperfect model physics or biases in the initial and boundary conditions, which can be exaggerated by the chaotic nature of NWP systems. According to the above analysis, C6 yields the best agreement with the ground truth data among the horizontal resolution experiments.

5.4 Searching for the likely ideal spin-up time

To limit the effects of the chaotic nature of NWP on the model simulations and extend the lead time, the scenario in which the spin-up time used in WRF is varied is placed at the end of the experimental design, after the possible errors introduced by inappropriate domain configuration options have been reduced. In S4, C6 is adopted as the starting experiment (OS3). Unlike the previous scenarios, the ranks of the spin-up time experiments, as sorted by the metrics, are nearly the same across the different time periods. Hence, Fig. 7 presents only the modelling skill of the spin-up time experiments over the time period of 18 h. The model performance of WRF in simulating heavy rainfall clearly varies with the spin-up time. For most of the metrics, an obvious diurnal tendency is found from 0 to 60 h, followed by a short-term

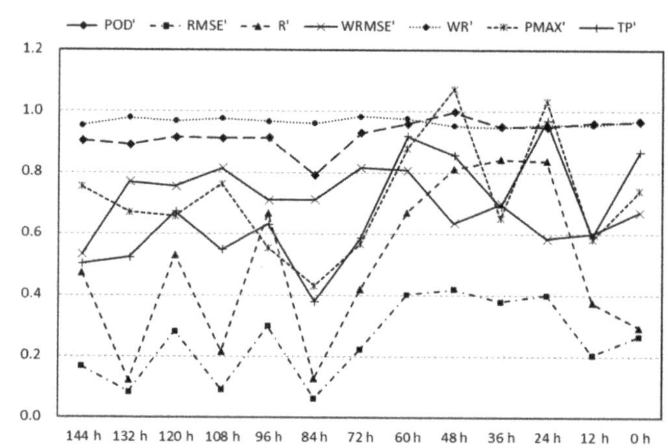

Figure 7. Spatial values of the verification metrics for the WRF spin-up experiments, calculated over 18 h periods and over domain three. Case 6 employs an initial spin-up time of 12 h; Case 8 employs a spin-up time of 0 h; and from Case 9 to Case 19, the spin-up time is increased from 24 to 144 h by every 12 h.

decrease until 72 h; random fluctuations occur after 72 h. Before 72 h, the variations in the rainfall and PW metrics are almost consistent; thus, the good fits of the simulations produced by the model runs with longer spin-up times are also physically reasonable within this period. The discrepancies among these experiments may be due to differences in the initial conditions (e.g. the water vapour amounts and the times of day when the simulations begin).

From TP′, it is found that all of the spin-up time experiments underestimate the total rainfall amount during the

Table 3. Comparison of the values of the error metrics in the initial experiment and the optimum experiments identified for each scenario.

Experiment number	POD′	RMSE′	R'	WRMSE′	WR′	PMAX′	TP′
Case 0 (C0)	0.950	0.098	0.226	0.789	0.980	0.440	0.478
Case 1 (C1)	0.960	0.064	0.376	0.622	0.967	0.436	0.471
Case 3 (C3)	0.969	0.110	0.373	0.610	0.967	0.515	0.496
Case 6 (C6)	0.963	0.205	0.375	0.600	0.956	0.582	0.592
Case 12 (C12)	0.959	0.402	0.670	0.807	0.977	0.883	0.920

heavy rainfall event. Of all of the rainfall-related metrics, POD′ is found to display the least sensitivity to the spin-up time; however, it displays similar variations over time as PMAX′, R', and RMSE′ before 72 h, with the highest values shown in the experiment with a spin-up time of 48 h (C11). Positive biases are detected in PMAX′ in C9 (which is run 24 h ahead) and C11, in which the largest positive biases are detected in the simulated amount of water vapour across the analysed periods and earlier (during the initialization period). This result may occur because the atmospheric water vapour content determines the maximum possible rainfall amount. C12, which includes a spin-up time of 60 h, is ranked third in terms of PMAX′, whereas it displays better performance than C9 and C11 in terms of TP′, WR′, and WRMSE′. As seen in Fig. 8, C9, C11 and C12 also rank in the top three, based on the values of the rainfall-related metrics calculated over domain two. However, larger departures from the forcing PW fields are seen in C9 and C11 than in C12. The difference is that C12 shows the best agreement with the ground truth data in terms of both the rainfall- and PW-related fields. Overall, C12 is regarded as the experiment that best reproduces the Beijing SDHR event with the optimal set of domain configuration options and the longest spin-up time.

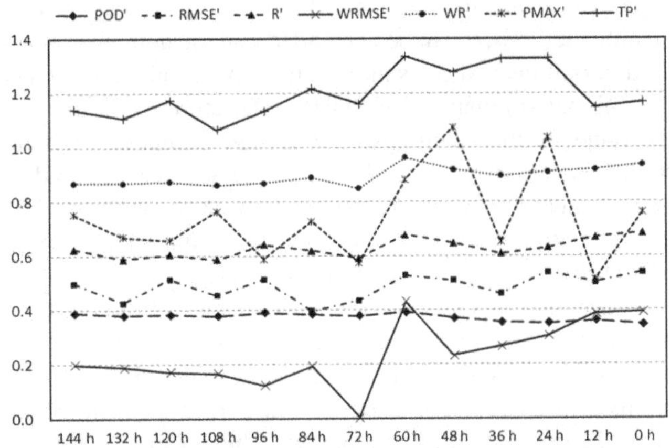

Figure 8. As in Fig. 7, but the metrics are calculated over 18 h periods and over domain two in Case 6.

6 Discussion

The results reveal that the initial experiment with the most commonly employed WRF domain settings does not yield the best performance in reproducing the temporal and spatial characteristics of SDHR on the convective scale. In S1, the assigned domain size of C0 is not sufficiently broad to allow the model physics to fully develop local small-scale features, resulting in obvious reductions in modelling skill as the evaluated time duration increases from 12 to 24 h. Further refinement of the grid spacing of C0 in S2 and S3 is shown to enable more explicit resolution of convective processes, leading to more accurate rainfall simulations. The comparison made in S4 suggests that the proper spin-up time is determined by both the time needed for model initialization and the accuracy of the initial conditions fed into the model run. Moreover, experiments with too large domains, too high spatial resolutions, or too long spin-up times also yield poor performance in rainfall simulations. Therefore, the reasonableness of these WRF settings should be checked before the model is used in regional NWP systems for flood forecasting or as a reference for the design of flood mitigation strategies.

In addition to exploring whether the recommended WRF domain configuration options and spin-up time are optimal for application in SDHR-prone urban areas, the performance of the model is quantified, and its total improvement is evaluated by comparing the values of the verification metrics yielded by the experiments. Table 3 compares the values of the verification metrics obtained for the optimal experiments in each scenario with the values obtained for the initial experiment. Here, the 18 h time duration is selected for evaluation because it covers most of the heavy rainfall event, and the metrics calculated over this period display a greater range and thus greater ability in identifying the simulation with the best performance. One exception is the domain size scenario, in which C0 presents the most obvious reduction in performance during the last stage of the rainfall event (24 h). Therefore, the improvement in C1 relative to C0 is mainly represented by R' and POD′ across D03 over the 18 h time period. The improvement produced by refining the vertical resolution is indicated by all of the rainfall-related metrics but is accompanied by a decrease in WRMSE′ that stems in part from the reduction in kinetic energy, which promotes rainfall. C6 yields higher values of POD′, RMSE′, R', and PMAX′ when compared with C3, indicating that appropriate increases in the horizontal resolution can increase the accuracy of rainfall simulations. The largest differences in the metrics between C6 and C12 occurs for PMAX′, which may relate to the different initial weather conditions at the different starting times of the model runs.

Overall, although the magnitudes of the increases in the rainfall metrics differ, they all reflect an increase in model skill after the re-evaluation process has been conducted. Specifically, R' increases from 0.226 in C0 to 0.67 in C12, RMSE′ increases from 0.098 to 0.402, and PMAX′ increases from 0.44 to 0.883. As the complete assessment is based on objective verification metrics and checked by subjective verification methods, it can be concluded that the domain configuration options and the spin-up time have significant effects on regional simulations of SDHR. Therefore, re-evaluating the values of those settings used in high-resolution regional studies is certainly worthwhile, and the accuracy of predictions of heavy rain clearly benefit from these analyses. For the evaluated metrics, evaluations based on a single type of metric or a single time period may clearly result in partially accurate conclusions. The use of datasets from multiple sources in verification can help increase the comprehensiveness of the analyses, such as the use of WRMSE′ and WR′ in this study. The use of different time periods helps to determine the optimal configurations with higher physical rationality, such as the selection of the proper domain size. In addition, the verification results may also depend on the fields and temporal–spatial scales of interest. To further understand the effects of WRF model configuration options on regional simulations of sub-daily heavy rainfall, more objective verification metrics for SDHR should be developed, and more case studies of SDHR events are also needed. Given that the uncertainties in the regional NWP studies result mainly from the inaccurate boundary conditions associated with grid nesting techniques, methods that can serve as alternate schemes to reduce these uncertainties are also worth studying. One example includes the mesh transitions approach used on irregular grids. In addition, more accurate simulations are expected when the model is driven with forcing data with higher temporal or spatial resolutions than those of the ERA-Interim reanalysis because the uncertainties and errors introduced by the input data could be further reduced.

7 Conclusions

In this study, a comparative test is designed to evaluate the effects of WRF domain configuration options and the spin-up time on simulations of the precipitation during the SDHR event that occurred on 21 July 2012 in Beijing, China. Three nested domains are established: D01 is the largest, has the coarsest resolution, and covers the leading synoptic features; and D03 is the smallest and covers the area of interest, Beijing. The initial conditions of the three domains are provided by the ERA-Interim reanalysis and the 30-second static geographical datasets. For the LBCs, D01 is forced by the ERA-Interim reanalysis, whereas D02 is forced by D01, and D03 is forced by D02. The reference ground truth data used for verification is 3-hourly 0.05 gridded rainfall observations and the coarser-scale ERA-Interim reanalysis. Five rainfall-related error metrics and two PW-related indices that monitor the departure of the model simulations from the driving fields are calculated at the convective-resolving scale over different sub-daily time spans. These metrics are then checked and considered together as part of a subjective verification process that is intended to pinpoint the likely best combination of the domain configuration options and spin-up time and to help quantify the possible improvements in the model performance of WRF in reproducing severe SDHR events after carrying out the entire re-evaluation process.

Precipitation simulations are sensitive to changes in domain size, vertical resolution, horizontal resolution, and spin-up time. Of all of the configurations, the most obvious variations are found when adjusting the domain size and the spin-up time. This analysis shows that domains that cover only the area of interest may be insufficiently broad to permit full development of small-scale features, resulting in poor performance in capturing the spatial pattern of heavy rainfall, especially in the early stages of rainfall events. Despite the dominant role of chaotic processes, it is still possible that model runs with longer spin-up times may result in better rainfall simulations, given favourable initial weather conditions. The effects of the vertical and horizontal resolutions are smaller, but the accuracy of the rainfall amount and the correct hits exhibit evident increases in runs with slightly higher spatial resolutions. A comparison of C12, which uses the evaluated optimum configurations, and C0, which uses the recommended settings, shows that the metrics clearly increase. Specifically, R' increases from 0.226 to 0.67, $\mathrm{RE}_{\mathrm{PMAX}}$ rises from -56 to $-11.7\,\%$, and RMSE decreases by $33.65\,\%$. Thus, substantial benefits may result from re-evaluating the WRF domain configuration options and spin-up times used in regional studies of SDHR.

Given the intensification of SDHR and the increased risks posed by SDHR-induced hazards, the demands of the operational flood management community for more accurate rainfall predictions with longer lead times, especially over highly affected areas with very short hydrologic response times, are increasing. One method that has now been proven to be effective is to dynamically downscale freely available global NWP products to areas of interest using high-resolution regional NWP models (e.g. WRF). Therefore, the uncertainties associated with the downscaling process, such as errors in boundary conditions and the issues associated with grid nesting, should be carefully evaluated to ensure that the rainfall simulations produced are both statistically accurate and physically reasonable before they are employed in flood forecasting systems. This study illustrates the importance of re-evaluating the domain configuration options and spin-up times used in WRF for improving regional rainfall simulations. Comparisons of the metrics indicate that evaluations based on just one category of metrics or values of metrics calculated over only one time period (e.g. 24 h) do not result in comprehensive comparisons and may lead to partially accurate conclusions. The use of PW fields calculated against reanalysis output is verified to be helpful in determining the optimal set of model configurations when analyses of rainfall-related metrics do not yield uniform conclusions. In addition, evaluations conducted over larger-scale domains are demonstrated to be useful in establishing the reasonableness of the evaluated results. Overall, the evaluation process is partly subjective. To simplify the assessment process, verification methods that can replace this subjective verification procedure should be developed. More regional case studies are also needed to further investigate the effects of configuration options in simulations of regional SDHR and to explore methods of reducing the uncertainties in regional NWP modelling associated with the scale-variation procedures. In addition, the use of more accurate forcing data with higher temporal and spatial resolutions is also expected to reduce the errors in the initial and boundary conditions and could thus be helpful in further improving the accuracy of rainfall simulations and extending the lead times of forecasts.

Competing interests. The authors declare that they have no conflict of interest.

Acknowledgements. This study is supported by the key research projects "Sponge city construction and urban flooding/waterlogging disaster in the sub-centre of Beijing City" (Z171100002217080), Beijing Municipal Science and Technology Commission, and "Urban storm flooding/waterlogging disasters under changing environment" (2017YFC1502701), national key research and development plan, Ministry of Science and Technology, China. Support is also received from the the Resilient Economy and Society by Integrated SysTems modelling (RESIST), the Newton Fund via Natural Environment Research Council (NERC) and the Economic and Social Research Council (ESRC; NE/N012143/1), and the National Natural Science Foundation of China (no. 4151101234). The China Scholarship Council supports the first author for her academic visit to the University of Bristol, UK.

Edited by: Uwe Ehret

References

Aligo, E. A., Gallus Jr., W. A., and Segal, M.: On the impact of WRF model vertical grid resolution on Midwest summer rainfall forecasts, Weather Forecast., 24, 575–594, 2009.

Bartholmes and Todini: Coupling meteorological and hydrological models for flood forecasting, Hydrol. Earth Syst. Sci., 9, 333–346, https://doi.org/10.5194/hess-9-333-2005, 2005.

Berrisford, P., Dee, D. P., Fielding, K., Fuentes, M., Kallberg, P., Kobayashi, S., and Uppala, S. M.: The ERA-Interim Archive, ERA Report Series, 1, 1–16, 2009.

Brömmel, D., Frings, W., and Wylie, B.: Technical Report Juqueen Extreme Scaling Workshop 2015, Tech. rep., Jülich, Germany, available at: http://hdl.handle.net/2128/8435, last access: 17 June 2018.

Castelli, F.: Atmosphere modeling and hydrologic-prediction uncertainty, U.S. – Italy Research Workshop on the Hydrometeorology, impacts and management of extreme floods, Perugia, 1995.

Chen, F. and Dudhia, J.: Coupling an advanced land surface-hydrology model with the Penn State-NCAR MM5 modeling system. Part I: Model implementation and sensitivity, Mon. Weather Rev., 129, 569–585, 2001.

Chen, H., Sun, J., Chen, X., and Zhou, W.: CGCM projections of heavy rainfall events in China, Int. J. Climatol., 32, 441–450, 2012.

Clark, P., Roberts, N., Lean, H., Ballard, S. P., and Charlton-Perez, C.: Convection-permitting models: a step-change in rainfall forecasting, Meteorol. Appl., 23, 165–181, 2016.

Coen, J. L., Cameron, M., Michalakes, J., Patton, E. G., Riggan, P. J., and Yedinak, K. M.: WRF-Fire: coupled weather–wildland fire modeling with the weather research and forecasting model, J. Appl. Meteorol. Clim., 52, 16–38, 2013.

Crétat, J., Pohl, B., Richard, Y., and Drobinski, P.: Uncertainties in simulating regional climate of Southern Africa: sensitivity to physical parameterizations using WRF, Clim. Dynam., 38, 613–634, 2012.

Cuo, L., Pagano, T. C., and Wang, Q. J.: A review of quantitative precipitation forecasts and their use in short-to medium-range streamflow forecasting, J. Hydrometeorol., 12, 713–728, 2011.

Dee, D. P., Uppala, S. M., Simmons, A. J., Berrisford, P., Poli, P., Kobayashi, S., Andrae, U., Balmaseda, M. A., Balsamo, G., Bauer, P., Bechtold, P., Beljaars, A. C. M., van de Berg, L., Bidlot J., Bormann, N., Delsol, C., Dragani, R., Fuentes, M., Geer, A. J., Haimberger, L., Healy, S. B., Hersbach, H., Hólm, E. V., Isaksen, L., Kållberg, P., Köhler, M., Matricardi, M., McNally, A. P., Monge-Sanz, B. M., Morcrette, J.-J., Park, B.-K., Peubey, C., de Rosnay, P., Tavolato, C., Thépaut, J.-N., and Vitart, F.: The ERA-Interim reanalysis: configuration and performance of the data assimilation system, Q. J. Roy. Meteorol. Soc., 137, 553–597, https://doi.org/10.1002/qj.828, 2011.

Di, Z. H., Duan, Q. Y., Gong, W., Wang, C., Gan, Y. J., Quan, J. P., Li, J. D., Miao, C. Y., Ye, A. Z., and Tong, C.: Assessing WRF model parameter sensitivity: A case study with five-day summer precipitation forecasting in the Greater Beijing Area, Geophys. Res. Lett., 42, 579–587, 2015.

Done, J., Davis, C. A., and Weisman, M.: The next generation of NWP: Explicit forecasts of convection using the Weather Research and Forecasting (WRF) model, Atmos. Sci. Lett., 5, 110–117, 2004.

Ek, M. B., Mitchell, K. E., Lin, Y., Rogers, E., Grunmann, P., Koren, V., Gayno, G., and Tarpley, J. D.: Implementation of Noah land surface model advances in the National Centers for Environmental Prediction operational mesoscale Eta model, J. Geophys. Res.-Atmos., 108, 8851, https://doi.org/10.1029/2002JD003296, 2003.

Fierro, A. O., Rogers, R. F., Marks, F. D., and Nolan, D. S.: The impact of horizontal grid spacing on the microphysical and kinematic structures of strong tropical cyclones simulated with the WRF-ARW model, Mon. Weather Rev., 137, 3717–3743, 2009.

Foley, A. M., Leahy, P. G., Marvuglia, A., and McKeogh, E. J.: Current methods and advances in forecasting of wind power generation, Renewable Energy, 37, 1–8, 2012.

Gao, Y., Yuan, Y., Wang, H., Schmidt, A. R., Wang, K., and Ye, L.: Examining the effects of urban agglomeration polders on flood events in Qinhuai River basin, China with HEC-HMS model, Water Sci. Technol., 75, 2130–2138, 2017.

Goswami, P., Shivappa, H., and Goud, S.: Comparative analysis of the role of domain size, horizontal resolution and initial conditions in the simulation of tropical heavy rainfall events, Meteorol. Appl., 19, 170–178, 2012.

Grell, G. A. and Dévényi, D.: A generalized approach to parameterizing convection combining ensemble and data assimilation techniques, Geophys. Res. Lett., 29, 1693, https://doi.org/10.1029/2002GL015311, 2002.

Guo, C., Xiao, H., Yang, H., and Tang, Q.: Observation and modeling analyses of the macro-and microphysical characteristics of a heavy rain storm in Beijing, Atmos. Res., 156, 125–141, 2015.

Heinzeller, D., Duda, M. G., and Kunstmann, H.: Towards convection-resolving, global atmospheric simulations with the Model for Prediction Across Scales (MPAS) v3.1: an extreme scaling experiment, Geosci. Model Dev., 9, 77–110, https://doi.org/10.5194/gmd-9-77-2016, 2016.

Hong, S. Y. and Lee, J. W.: Assessment of the WRF model in reproducing a flash-flood heavy rainfall event over Korea, Atmos. Res., 93, 818–831, 2009.

Hong, S. Y. and Lim, J. O. J.: The WRF single-moment 6-class microphysics scheme (WSM6), J. Korean Meteorol. Soc., 42, 129–151, 2006.

Hong, S. Y., Noh, Y., and Dudhia, J.: A new vertical diffusion package with an explicit treatment of entrainment processes, Mon. Weather Rev., 134, 2318–2341, 2006.

Huang, C., Zheng, X., Tait, A., Dai, Y., Yang, C., Chen, Z., Li, T., and Wang, Z.: On using smoothing spline and residual correction to fuse rain gauge observations and remote sensing data, J. Hydrol., 508, 410–417, 2013.

Iacono, M. J., Delamere, J. S., Mlawer, E. J., Shephard, M. W., Clough, S. A., and Collins, W. D.: Radiative forcing by long-lived greenhouse gases: calculations with the AER radiative transfer models, J. Geophys. Res.-Atmos., 113, D13103, https://doi.org/10.1029/2008JD009944, 2008.

Kain, J. S., Weiss, S. J., Bright, D. R., Baldwin, M. E., Levit, J. J., Carbin, G. W., Schwartz, C. S., Weisman, M. L., Droegemeeier, K. K., Weber, D. B., and Thomas, K. W.: Some practical considerations regarding horizontal resolution in the first generation of operational convection-allowing NWP, Weather Forecast., 23, 931–952, 2008.

Kleczek, M. A., Steeneveld, G. J., and Holtslag, A. A.: Evaluation of the weather research and forecasting mesoscale model for GABLS3: impact of boundary-layer schemes, boundary conditions and spin-up, Bound.-Lay. Meteorol., 152, 213–243, 2014.

Klemp, J. B.: Advances in the WRF model for convection-resolving forecasting, Adv. Geosci., 7, 25–29, https://doi.org/10.5194/adgeo-7-25-2006, 2006.

Leduc, M. and Laprise, R.: Regional climate model sensitivity to domain size, Clim. Dynam., 32, 833–854, 2009.

Li, J., Chen, Y., Wang, H., Qin, J., Li, J., and Chiao, S.: Extending flood forecasting lead time in a large watershed by coupling WRF QPF with a distributed hydrological model, Hydrol. Earth Syst. Sci., 21, 1279–1294, https://doi.org/10.5194/hess-21-1279-2017, 2017.

Liu, J., Bray, M., and Han, D.: Sensitivity of the Weather Research and Forecasting (WRF) model to downscaling ratios and storm types in rainfall simulation, Hydrol. Process., 26, 3012–3031, 2012.

Luna, T., Castanheira, M., and Rocha, A.: Assessment of WRF-ARW forecasts using warm initializations, available at: http://climetua.fis.ua.pt/publicacoes/APMG_extended_abstract_2013_Luna_et_al.pdf (last access: 17 June 2018), 2013.

Miguez-Macho, G., Stenchikov, G. L., and Robock, A.: Spectral nudging to eliminate the effects of domain position and geometry in regional climate model simulations, J. Geophys. Res.-Atmos., 109, D13104, https://doi.org/10.1029/2003JD004495, 2004.

Powers, J. G., Klemp, J. B., Skamarock, W. C., Davis, C. A., Dudhia, J., Gill, D. O., Coen, J. L., and Grell, G. A.: The Weather Research and Forecasting Model: Overview, System Efforts, and Future Directions, B. Am. Meteorol. Soc., 98, 1717–1737, https://doi.org/10.1175/BAMS-D-15-00308.1, 2017.

Prein, A. F., Langhans, W., Fosser, G., Ferrone, A., Ban, N., Goergen, K., Keller, M., Tölle, M., Gitjahr, O., Feser, F., Brisson, E., Kollet, S., Schmidli, J., van Lipzig, N. P. M., and Leung, R.: A review on regional convection-permitting climate modeling: Demonstrations, prospects, and challenges, Rev. Geophys., 53, 323–361, 2015.

Roberts, N. M. and Lean, H. W.: Scale-selective verification of rainfall accumulations from high-resolution forecasts of convective events, Mon. Weather Rev., 136, 78–97, 2008.

Ruiz, J. J., Saulo, C., and Nogués-Paegle, J.: WRF model sensitivity to choice of parameterization over South America: validation against surface variables, Mon. Weather Rev., 138, 3342–3355, 2010.

Schwartz, C. S., Kain, J. S., Weiss, S. J., Xue, M., Bright, D. R., Kong, F. Y., Thomas, K. W., Levit, J. J., and Coniglio, M. C.: Next-day convection-allowing WRF model guidance: A second look at 2-km versus 4-km grid spacing, Mon. Weather Rev., 137, 3351–3372, 2009.

Seth, A. and Rojas, M.: Simulation and sensitivity in a nested modeling system for South America. Part I: Reanalyses boundary forcing, J. Climate, 16, 2437–2453, 2003.

Shih, D. S., Chen, C. H., and Yeh, G. T.: Improving our understanding of flood forecasting using earlier hydro-meteorological intelligence, J. Hydrol., 512, 470–481, 2014.

Sikder, S. and Hossain, F.: Assessment of the weather research and forecasting model generalized parameterization schemes for advancement of precipitation forecasting in monsoon-driven river basins, J. Adv. Model. Earth Syst., 8, 1210–1228, 2016.

Skamarock, W. C., Klemp, J. B., Dudhia, J., Gill, D. O., Barker, D. M., Duda, M. G., Huang, X. Y., Wang, W., and Powers, J. G.: A Description of the Advanced Research WRF Version 3, NCAR Note NCAR/TN-475, NCAR, Boulder, Colorado, USA, 2008.

Soares, P. M., Cardoso, R. M., Miranda, P. M., de Medeiros, J., Belo-Pereira, M., and Espirito-Santo, F.: WRF high resolution dynamical downscaling of ERA-Interim for Portugal, Clim. Dynam., 39, 2497–2522, 2012.

Sun, M. S., Yang, L. Q., Yin, Q., Niu, Z. Y., and Gao, L. M.: Analysis of the cause of a torrential rain occurring in Beijing on 21 July 2012, Torrent. Rain Disast., 32, 218–223, 2013.

Swinbank, R. and James Purser, R.: Fibonacci grids: A novel approach to global modeling, Q. J. Roy. Meteorol. Soc., 132, 1769–1793, 2006.

Tian, J., Liu, J., Yan, D., Li, C., and Yu, F.: Numerical rainfall simulation with different spatial and temporal evenness by using a WRF multiphysics ensemble, Nat. Hazards Earth Syst. Sci., 17, 563–579, https://doi.org/10.5194/nhess-17-563-2017, 2017.

Vrac, M., Drobinski, P., Merlo, A., Herrmann, M., Lavaysse, C., Li, L., and Somot, S.: Dynamical and statistical downscaling of the French Mediterranean climate: uncertainty assessment, Nat. Hazards Earth Syst. Sci., 12, 2769–2784, https://doi.org/10.5194/nhess-12-2769-2012, 2012.

Wang, K., Wang, L., Wei, Y. M., and Ye, M.: Beijing storm of July 21, 2012: observations and reflections, Nat. Hazards, 67, 969–974, 2013.

Wang, S. L., Kang, H. W., Gu, X. Q., and Ni, Y. Q.: Numerical Simulation of Mesoscale Convective System in the Warm Sector of Beijing '7.21' Severe Rainstorm, Meteorol. Mon., 41, 544–553, 2015.

Warner, T. T.: Quality assurance in atmospheric modeling, B. Am. Meteorol. Soc., 92, 1601–1610, 2011.

Warner, T. T., Peterson, R. A., and Treadon, R. E.: A tutorial on lateral boundary conditions as a basic and potentially serious limitation to regional numerical weather prediction, B. Am. Meteorol. Soc., 78, 2599–2617, 1997.

Westra, S., Fowler, H. J., Evans, J. P., Alexander, L. V., Berg, P., Johnson, F., Kendon, E. J., Lenderink, G., and Roberts, N. M.: Future changes to the intensity and frequency of short-duration extreme rainfall, Rev. Geophys., 52, 522–555, 2014.

Willems, P., Arnbjerg-Nielsen, K., Olsson, J., and Nguyen, V. T. V.: Climate change impact assessment on urban rainfall extremes and urban drainage: methods and shortcomings, Atmos. Res., 103, 106–118, 2012.

WMO: Anticipated advances in numerical weather prediction, and the growing technology gap in weather forecast, available at: https://www.wmo.int/pages/prog/www/swfdp/Meetings/documents/Advances_NWP.pdf (last access: 17 June 2018), 2013.

Xu, Z. X. and Chu, Q.: Climatological features and trends of extreme precipitation during 1979–2012 in Beijing, China, P. Int. Assoc. Hydrolog. Sci., 369, 97–102, 2015.

Xu, Z. X. and Zhao, G.: Impact of urbanization on rainfall-runoff processes: case study in the Liangshui River Basin in Beijing, China, P. Int. Assoc. Hydrolog. Sci., 373, 7–12, 2016.

Yu, E.-T., Wang, H.-J., and Sun, J.-Q.: A quick report on a dynamical downscaling simulation over China using the nested model, Atmos. Ocean. Sc. Lett., 3, 325–329, 2010.

Yu, R., Xu, Y., Zhou, T., and Li, J.: Relation between rainfall duration and diurnal variation in the warm season precipitation over central eastern China, Geophys. Res. Lett., 34, L13703, https://doi.org/10.1029/2007GL030315, 2007.

Yu, W., Nakakita, E., Kim, S., and Yamaguchi, K.: Impact Assessment of Uncertainty Propagation of Ensemble NWP Rainfall to Flood Forecasting with Catchment Scale, Adv. Meteorol., 4, 1–17, https://doi.org/10.1155/2016/1384302, 2016.

Yucel, I., Onen, A., Yilmaz, K. K., and Gochis, D. J.: Calibration and evaluation of a flood forecasting system: Utility of numerical weather prediction model, data assimilation and satellite-based rainfall, J. Hydrol., 523, 49–66, 2015.

Zhou, Y. S., Liu, L., Zhu, K. F., and Li, J. T.: Simulation and evolution characteristics of mesoscale systems occurring in Beijing on 21 July 2012, Chinese J. Atmos. Sci., 38, 885–896, 2014.

10

Use of reflected GNSS SNR data to retrieve either soil moisture or vegetation height from a wheat crop

Sibo Zhang[1,2], **Nicolas Roussel**[3], **Karen Boniface**[2,3,4,a], **Minh Cuong Ha**[3], **Frédéric Frappart**[3,5], **José Darrozes**[3], **Frédéric Baup**[4], **and Jean-Christophe Calvet**[1]

[1]CNRM – UMR3589 (Météo-France, CNRS), Toulouse, France
[2]Fondation STAE, Toulouse, France
[3]GET – UMR5563, CNRS/Université Paul Sabatier, UR254 IRD, Toulouse, France
[4]CESBIO, Université de Toulouse, CNES/CNRS/IRD/UPS, Toulouse, France
[5]LEGOS – UMR566 (CNES, CNRS, IRD, UPS), Toulouse, France
[a]now at: Joint Research Centre/European Commission, Ispra, Italy

Correspondence to: Jean-Christophe Calvet (jean-christophe.calvet@meteo.fr)

Abstract. This work aims to estimate soil moisture and vegetation height from Global Navigation Satellite System (GNSS) Signal to Noise Ratio (SNR) data using direct and reflected signals by the land surface surrounding a ground-based antenna. Observations are collected from a rainfed wheat field in southwestern France. Surface soil moisture is retrieved based on SNR phases estimated by the Least Square Estimation method, assuming the relative antenna height is constant. It is found that vegetation growth breaks up the constant relative antenna height assumption. A vegetation-height retrieval algorithm is proposed using the SNR-dominant period (the peak period in the average power spectrum derived from a wavelet analysis of SNR). Soil moisture and vegetation height are retrieved at different time periods (before and after vegetation's significant growth in March). The retrievals are compared with two independent reference data sets: in situ observations of soil moisture and vegetation height, and numerical simulations of soil moisture, vegetation height and above-ground dry biomass from the ISBA (interactions between soil, biosphere and atmosphere) land surface model. Results show that changes in soil moisture mainly affect the multipath phase of the SNR data (assuming the relative antenna height is constant) with little change in the dominant period of the SNR data, whereas changes in vegetation height are more likely to modulate the SNR-dominant period. Surface volumetric soil moisture can be estimated ($R^2 = 0.74$, RMSE $= 0.009\,\mathrm{m^3\,m^{-3}}$) when the wheat is smaller than one wavelength (~ 19 cm). The quality of the estimates markedly decreases when the vegetation height increases. This is because the reflected GNSS signal is less affected by the soil. When vegetation replaces soil as the dominant reflecting surface, a wavelet analysis provides an accurate estimation of the wheat crop height ($R^2 = 0.98$, RMSE $= 6.2$ cm). The latter correlates with modeled above-ground dry biomass of the wheat from stem elongation to ripening. It is found that the vegetation height retrievals are sensitive to changes in plant height of at least one wavelength. A simple smoothing of the retrieved plant height allows an excellent matching to in situ observations, and to modeled above-ground dry biomass.

1 Introduction

In situ observations of soil moisture and vegetation variables are key to validate land surface models and satellite-derived products. Recent international initiatives, such as the International Soil Moisture Network (Dorigo et al., 2013) or the Committee on Earth Observation Satellites (CEOS) Land Product Validation group (Morisette et al., 2006), have improved the access to such observations. However, they remain very sparse and there is a need to develop new automatic techniques to monitor land surface variables on a local scale. Global Navigation Satellite System (GNSS) reflectometry could be a solution. A number of studies demonstrated that GNSS multipath signals can be used to retrieve various geophysical parameters of the surface surrounding a GNSS receiving antenna (Motte et al., 2016). Over land, variables such as soil moisture, snow depth and vegetation status can be observed (Larson et al., 2008; Small et al., 2010; Larson and Nievinski, 2013; Wan et al., 2015; Boniface et al., 2015; Larson, 2016; Roussel et al., 2016). GNSS satellites operate at the L-band microwave frequency domain (between 1.2 and 1.6 GHz). At these relatively low frequencies, the microwave signal is less perturbed by atmospheric effects and can better penetrate clouds and heavy rains than higher-frequency signals. This ensures continuous operation, in all weather conditions, either daytime or nighttime. The L-band signal emitted or reflected by terrestrial surfaces is related to surface parameters like surface soil moisture, roughness or vegetation characteristics. These properties have been exploited by, for example, the Soil Moisture and Ocean Salinity (SMOS) satellite and the Soil Moisture Active Passive (SMAP) missions (Kerr et al., 2001; Chan et al., 2016) for Earth's surface remote sensing applications. While SMOS is a radiometer and measures the Earth's surface microwave emission (passive microwaves), GNSS satellites emit a radar signal (active microwaves). Active microwaves can present improved temporal and spatial resolutions, but the signal may be more sensitive to the structure of the surface such as soil roughness or vegetation effects than for passive microwaves (Wigneron et al., 1999; Njoku et al., 2002).

Existing geodetic-quality GNSS networks have the potential to provide a large number of in situ observations, depending on the receiver technology: (1) waveform acquisition with a specific receiver using two antennas (one zenith-oriented antenna and one surface-oriented antenna), called the GNSS reflectometry (GNSS-R) technique (Zavarotny et al., 2014) or (2) GNSS signal strength represented by the signal-to-noise ratio (SNR) acquired with a classical geodetic receiver using one antenna, called the SNR GNSS interferometric reflectometry (GNSS-IR) technique (Larson, 2016). GNSS networks can be used to monitor small or large areas depending on the antenna height and satellite elevation (Roussel et al., 2014). Continuous monitoring of surface soil moisture can be made over a long period at spatial scales ranging from $100\,m^2$ (antenna height of about 2 m) to $8000\,m^2$ (antenna height of about 150 m) for a classical geodetic receiver but can reach a few thousand square kilometers with waveform receivers embedded on satellites (e.g., the TechDemoSat-1 mission; Foti et al., 2015).

Using the SNR GNSS-IR technique, Larson et al. (2008) showed that SNR data obtained from existing networks of single ground-based geodetic antennas can be used to infer soil moisture. Other GNSS methods (besides reflectometry) can be used. For example, Koch et al. (2016) used three geodetic GNSS antennas (one was installed above the soil, the other two were buried at a depth of 10 cm) to measure the GNSS signal strength attenuation and to retrieve soil moisture over bare soil.

A network called PBO H_2O based on single GNSS antennas at Plate Boundary Observatory (PBO) sites is currently used in western regions of the USA to monitor surface soil moisture (Larson et al., 2013; Chew et al., 2016) and snow depth (Larson and Nievinski, 2013; Boniface et al., 2015). It must be noted that most of the 161 GNSS stations of this network are located in mountainous areas or in areas of California characterized by a relatively arid climate. They are surrounded by sparse vegetation and are therefore not adapted to vegetation growth studies.

With the SNR GNSS-IR technique, the interference between the direct and the reflected signals is observed through temporal variations of the SNR data (Bilich and Larson, 2007; Zavorotny et al., 2010; Chew et al., 2014). Changes in geophysical or biophysical parameters affect the phase, amplitude and frequency of the SNR modulation pattern. The SNR is also influenced by surface roughness, and by the position of the antenna with respect to the surface and to the satellite (Larson and Nievinski, 2013; Chew et al., 2016). The SNR modulation primarily depends on the following:

- the relative height of the GNSS antenna above the reflecting surface (ground or vegetation surface),
- satellite elevation,
- the superposition of the direct signal and of the reflected signal, which varies along with changes in the satellite track positions,
- the right hand circular polarization (RHCP) and left hand circular polarization (LHCP) gain pattern of the receiving antenna, (RHCP usually increases the SNR when the satellite elevation angle increases, LHCP is related to imperfections of the antenna and is greater than RHCP for the reflected signal),
- reflection coefficients for the reflecting surface, related to the water content and to the ground mineralogical content of the reflecting surface,
- surface topography and roughness and
- the satellite transmitted power.

A soil moisture retrieval algorithm from SNR data was derived by Chew et al. (2014) for bare soil. In subsequent modeling studies Chew et al. (2015) showed that the vegetation canopies affected the SNR modulation pattern. They showed that vegetation growth tended to trigger a decrease in the SNR amplitude. Because the vegetation effects tended to perturb the soil moisture retrieval, Chew et al. (2016) proposed an improved algorithm for soil moisture retrieval in vegetated environments, which used the amplitude decrease extent to decide when vegetation influence was too large. They used a model database for the SNR of the L2C signal to remove the most significant vegetation effects for the sites they considered in western USA. Small et al. (2016) further compared different algorithms of GNSS-IR soil moisture retrieval in the presence of vegetation. Roussel et al. (2016) integrated both GPS and GLONASS SNR data to retrieve soil moisture over bare soil. Using data from a field study, Wan et al. (2015) showed that the amplitude of the SNR data presented a good linear relationship with the vegetation water content (VWC), but it was restricted to VWC values of less than $\sim 1\ \mathrm{kg\,m^{-2}}$. In addition to the amplitude of the SNR data, it was also possible to infer VWC by the $MP1_{rms}$ index, which is a linear combination of L1 and L2 carrier phase data and L1 pseudorange data (Small et al., 2010), and by the NMRI (Normalized Microwave Reflection Index) which is derived from the $MP1_{rms}$ (Small et al., 2014; Larson and Small, 2014).

In this study, the SNR GNSS-IR technique was used to analyze GNSS SNR data obtained with a single classical geodetic antenna receiver over an intensively cultivated wheat field in southwestern France. The data were used to retrieve either soil moisture or relative vegetation height during the growing period of the wheat crop. The method proposed by Chew et al. (2016) (hereafter referred to as CH16) was used to retrieve soil moisture. Moreover, we performed a wavelet analysis in order to extract the dominant period of the SNR. We investigated to what extent vegetation height influenced the dominant period resulting from the wavelet analysis. The main justification for investigating the impact of vegetation height was that it impacted the relative antenna height (the distance from the antenna to the reflecting surface). Vegetation growth tended to decrease the relative antenna height and broke up the constant height assumption used in soil moisture retrieval algorithms. In this context, key objectives of this study were to (1) assess the soil moisture retrieval technique in either short or tall vegetation conditions and (2) retrieve vegetation height along the wheat growth cycle.

2 Materials and methods

2.1 SNR data and preprocessing

The GNSS SNR data were acquired from an antenna at 2.51 m above the soil surface over an experimental field covered by rainfed winter wheat in Lamasquère, France (43°29′10″ N, 1°13′57″ E; see Fig. S1 in the Supplement). These GNSS data were collected by GET (Géosciences Environnement Toulouse) for a whole growing season, from January to July 2015. A Leica GR25 receiver equipped with an AS10 antenna was used and data were acquired at a sample frequency of 1 Hz. Only the S1C SNR signal strength on the civilian L1 C/A channel of the GPS constellation was used in this study because the used receiver could not track the L2C signal. The latter is only transmitted by the recent Block IIR-M ("Replenishment Modernized") and IIF ("Follow-on") GPS satellites. Vey et al. (2016) showed that soil moisture root mean square difference between L2C and L1 was $0.03\ \mathrm{m^3\,m^{-3}}$. The quality of the more recently available L2C signal (used by PBO H_2O; CH16) is higher than either L1 C/A or L2P from non-code tracking receivers. However, a number of studies (e.g., Vey et al., 2016) showed that the SNR of the L1 C/A signal can be used to provide reliable soil moisture estimates over sparse vegetation and bare soil surface, although it is less precise than the L2C signal. Although data from other constellations were also acquired (e.g., GLONASS, GALILEO), their orbital parameters such as satellite track positions or satellite altitude were not the same. In order to be consistent with the GPS-only studies of Larson et al. (2008), CH16, and Small et al. (2016), we only used GPS SNR data. For our site, four GPS satellites out of 32 were excluded from the analysis because their data were incomplete (GPS03, 20, 26; these numbers corresponding to their pseudo-random noise – PRN – numbers) or not received (GPS08). Finally, GPS SNR data were missing for only nine days: 8 and 9 February, 3 April and from 13 to 18 May 2015.

Following the method proposed by Larson et al. (2010), a low-order polynomial was fit to the SNR data, and the modulation pattern was then derived from the SNR by subtracting this polynomial from the SNR data. The logarithmic dB-Hz units were converted to a linear scale in $\mathrm{V\,V^{-1}}$ using the following conversion equation: $SNR_{linear} = 10^{\frac{SNR}{20}}$ (Vey et al., 2016). Figure 1a shows an example of the detrended multipath SNR data for the ascending track of GPS01 on 21 January 2015. The periodic signature of the multipath SNR data is visible. We only analyzed the modulation patterns in a valid segment for satellite configurations corresponding to low elevation angles, ranging from 5 to 20°. This corresponded to a valid segment data recording of less than one hour (40 to 50 min). We excluded very low elevation angles (less than 5°) in order to avoid spurious effects from trees and artificial surfaces surrounding the field. Because the SNR signal amplitude was much reduced and the wave pattern was not visible at high elevations for our field observations, we excluded elevation angles larger than 20°.

2.2 Soil moisture and vegetation characteristics

The field campaign was part of a coordinated effort led by CESBIO (Centre d'Etudes Spatiales de la BIOsphère) to monitor crops in southwestern France using both in situ and satellite Earth observation data. Independent in situ observations of soil moisture and vegetation height were made together with model simulations of these quantities. Both observations and simulations were used to validate soil moisture and vegetation height retrievals.

Since the whole wheat growing cycle was examined, both soil moisture and vegetation modulated the multipath SNR pattern. Soil roughness was considered as stable in time from sowing to harvest. Soil in the close vicinity of the antenna

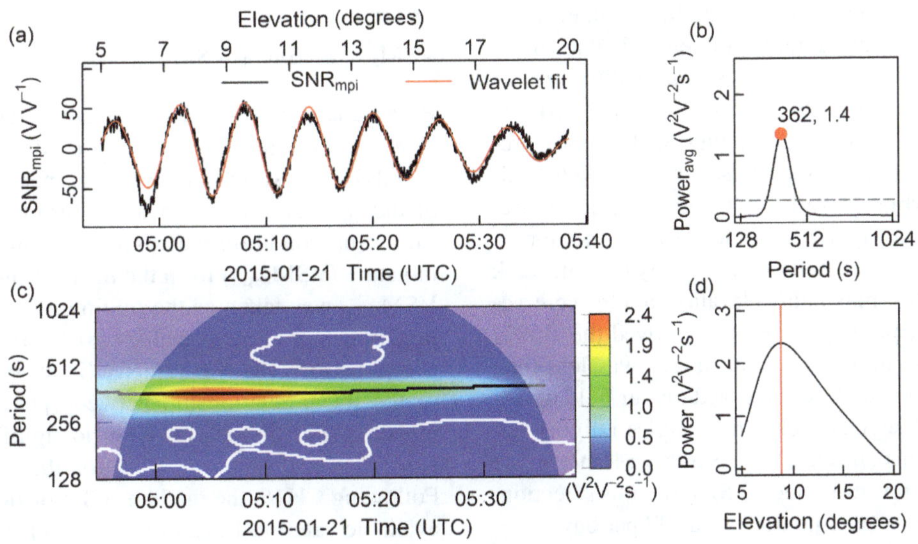

Figure 1. Example of a usable GPS01 ascending track SNR data set from 04:50 to 05:38 UTC on 21 January 2015: **(a)** multipath SNR data (in V V^{-1}), **(b)** average power spectrum with its maximum value (red dot) and **(c)** power spectrum for periods from 128 to 1024 s. The red line in **(a)** is the reconstructed SNR data by the daughter wavelet corresponding to the peak period (362 s) indicated in **(b)**. The power at the peak period across elevation angles **(d)** presents a maximum value at an elevation angle of about 9°.

consisted of 18 % sand, 41 % clay and 41 % silt. The row spacing of the wheat crop was 15 cm.

The wheat was sown during the autumn, on 1 October 2014 and was harvested from 26 to 30 June 2015. Volumetric soil moisture (VSM) was measured by FDR (frequency domain reflectometry) ML2 Thetaprobes and was continuously monitored at a depth of 5 cm from 16 January to 10 March 2015 and from 30 March to 26 May 2015. Measurements of crop height were performed at seven dates during the plant growing cycle. The canopy height did not exceed 0.1 m in wintertime and rapidly increased in springtime: it reached 0.2 m on 10 March 2015 and 1 m on 29 May. It dropped to 0.39 m on 18 June because of a lodging event. The exact date of lodging could not be precisely determined. It could be inferred that lodging happened between 29 May and 18 June.

In addition to in situ observations, simulations of surface soil moisture (0–10 cm top-soil layer), plant height and above-ground dry biomass were performed for this site by CNRM (Centre National de Recherches Météorologiques) using the ISBA (Interactions between Soil, Biosphere, and Atmosphere) land surface model within the SURFEX (version 8.0) modeling platform (Masson et al., 2013). The ISBA configuration and the atmospheric analysis used to force the model are described in Lafont et al. (2012). The C3 crop plant functioning type and a multilayer representation of the soil hydrology are considered. The model soil depth is 12 m, with 15 layers and the layer thickness increases from the top surface layer to the deepest layers (Decharme et al., 2011). These simulations were used as an independent benchmark for soil moisture and vegetation variables.

2.3 Multipath SNR characteristics

Due to the motion of the GPS satellites, the path delay between the direct and reflected signals causes an interference pattern in the signal power of SNR data. The distance from the antenna to the dominant reflecting surface directly affects the SNR frequency/period.

As noted by Georgiadou and Kleusberg (1988) and Bilich and Larson (2007), assuming the ground surface is horizontal, the additional distance (δ) traveled by a reflected signal relative to the direct signal is

$$\delta = 2h\sin(\theta), \tag{1}$$

where h is the relative antenna height, and θ is the satellite elevation angle. This path delay δ can also be expressed in terms of the multipath relative phase ψ:

$$\psi = 2\pi\frac{\delta}{\lambda}, \tag{2}$$

where λ represents the L1 wavelength (0.1903 m).

Thus the multipath frequency (f) and period (T) can be written as the following:

$$\omega = 2\pi f = \frac{\mathrm{d}\psi}{\mathrm{d}t} = \frac{4\pi}{\lambda}h\cos(\theta)\frac{\mathrm{d}\theta}{\mathrm{d}t}, \tag{3}$$

$$\frac{1}{T} = f = \frac{2h\cos(\theta)}{\lambda}\frac{\mathrm{d}\theta}{\mathrm{d}t}. \tag{4}$$

This means that the relative antenna height (h) directly affects the multipath frequency f and period T. Antennas far above the reflecting surface have higher multipath frequencies (smaller multipath periods) than antennas closer to the reflecting surface. Furthermore, satellite geometric information and motion substantially influence T due to the $\cos(\theta)$ and $d\theta/dt$ terms in Eq. (4). When satellite passes reach high elevation angles $d\theta/dt$ becomes larger (Bilich and Larson, 2007). Conversely, satellites with passes presenting small maximum elevations present smaller $d\theta/dt$ values than satellites orbiting overhead. Contrasting configurations are illustrated in Fig. S2. In order to limit the impact of these differences from satellite motion, only the full-track data with at least 40° maximum elevation angle were selected. Among the remaining tracks we removed the slowly moving tracks whose maximum $\cos\theta \cdot d\theta/dt$ was less than $9.5 \times 10^{-5}\ \mathrm{rad\,s^{-1}}$ (threshold value based on our field observations) of the valid segment (elevation angles ranging between 5 and 20°). This specific data sorting was only made for vegetation height retrieval (Sect. 2.5). After this selection, the number of available satellite tracks was 37 per day.

Provided the reflecting surface is stable, the a priori antenna height can be used to estimate the SNR frequency. The SNR frequency is used to calculate the multipath SNR phase, and then the SNR phase is used to estimate VSM (Sect. 2.4). If the reflecting surface is changing in response to vegetation growth, relative vegetation height can be retrieved instead of VSM by directly estimating the dynamic SNR frequency/period with a wavelet analysis (Sect. 2.5).

2.4 Soil moisture retrieval

As the SNR frequency is known (Eq. 4), it is possible to estimate the SNR amplitude and phase. Larson et al. (2008, 2010) showed that phase varies linearly with VSM in $\mathrm{m^3\,m^{-3}}$ ($R^2 = 0.76$ to 0.90). Retrieving absolute VSM values in $\mathrm{m^3\,m^{-3}}$ is possible after a calibration phase. This result was used by Chew et al. (2014) to develop an algorithm to estimate surface soil moisture (top 5 cm) from bare soil.

For bare soil, changes in surface soil moisture affect the signal penetration depth. The latter can be very small in wet conditions and tends to increase in dry conditions, up to a few centimeters (Chew et al., 2014; Roussel et al., 2016). This is a small change with respect to the antenna height (2.51 m in this study). Consequently, the relative antenna height (h) is considered as a constant ($h_c = 2.51$ m) in this section. Using sine of the elevation angle ($\sin(\theta)$) as the independent variable, the modulation frequency becomes proportional to h_c. Then the multipath SNR ($\mathrm{SNR_{mpi}}$) can be expressed as the following (Larson et al., 2008):

$$\mathrm{SNR_{mpi}} = A\cos\left(\frac{4\pi h_c}{\lambda}\sin(\theta) + \phi_{\mathrm{mpi}}\right). \quad (5)$$

The least square estimation method proposed by Larson et al. (2008) is used to estimate the multipath amplitude (A) and multipath phase (ϕ_{mpi}) from the multipath SNR data. Then, ϕ_{mpi} can be used to estimate the soil moisture changes (CH16),

$$\mathrm{VSM}_t = S \cdot \Delta\phi_t + \mathrm{VSM_{resid}}. \quad (6)$$

Phase changes $\Delta\phi_t = \phi_t - \phi_0$ are calculated with respect to ϕ_0, the reference phase. We used the method proposed by CH16 by estimating ϕ_0 as the mean of the lowest 15 % of the ϕ_{mpi} data for each track during the retrieval period. The same condition was used to estimate the residual soil moisture ($\mathrm{VSM_{resid}}$) from the in situ VSM observations. The $\mathrm{VSM_{resid}}$ was taken as the minimum soil moisture observation, which presented a value of $0.252\ \mathrm{m^3\,m^{-3}}$ during the retrieval period. The S parameter (in $\mathrm{m^3\,m^{-3}\,{}^{\circ -1}}$) is the slope of the linear relationship between phase changes and soil moisture. For time series with no significant vegetation effects, $S = 0.0148\ \mathrm{m^3\,m^{-3}\,{}^{\circ -1}}$ for the L2C signal (CH16). Following CH16, the median soil moisture estimate from all available satellite tracks (66 per day) that passed at different times during the day was used as the final soil moisture estimate.

We also used the in situ VSM_t, $\Delta\phi_t$ and $\mathrm{VSM_{resid}}$ to fit a locally adjusted slope. The retrieval of the S parameter requires at least 1 or 2 months of VSM in situ observations because soil moisture conditions ranging from dry to wet need to be sampled. However, if a scaled soil wetness index is used instead of soil moisture, no in situ VSM observations are needed.

Alternatively, the phase time series can be normalized for each satellite track, and using S is not needed. We considered the median value of the normalized phases from all available satellite tracks (66 per day) as the final scaled soil wetness index (ϕ_{index}) for each day:

$$\phi_{\mathrm{index}} = \frac{\phi - \phi_{\min}}{\phi_{\max} - \phi_{\min}}. \quad (7)$$

VSM could then be estimated from ϕ_{index}:

$$\mathrm{VSM} = \mathrm{VSM_{obs\_min}} + \phi_{\mathrm{index}} \cdot \left(\mathrm{VSM_{obs\_max}} - \mathrm{VSM_{obs\_min}}\right). \quad (8)$$

$\mathrm{VSM_{obs\_min}}$ and $\mathrm{VSM_{obs\_max}}$ are the minimum and maximum in situ VSM observations during the experimental time period, respectively.

CH16 defined the normalized amplitude (A_{norm}) as the ratio of amplitude to the average of the top 20 % amplitude values. The A_{norm} time series can be used to assess whether or not vegetation effects are significant. Values of A_{norm} above 0.78 (dimensionless) indicate that vegetation effects are small (CH16). In conditions of significant vegetation effects CH16 used an algorithm able to correct the phase for vegetation effects. This algorithm is based on an unpublished lookup table. Since we were not able to correct for vegetation effects, we retrieved surface soil moisture during a period with rather sparse vegetation, from 16 January to 5 March. During this time span, A_{norm} was above 0.78 as shown in Fig. 2 (black dots).

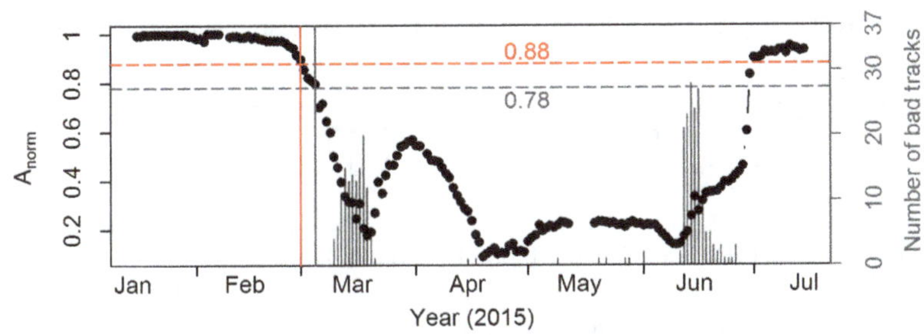

Figure 2. Normalized amplitude (A_{norm}) time series (black dots) and probability distribution (grey bars) of low quality tracks among all available satellite tracks on a daily basis from 16 January to 15 July 2015. The empirical A_{norm} threshold (0.78) is shown by the grey dashed line, and the soil moisture can be retrieved from 16 January to 5 March 2015 depending on it. Our field intuitive estimated A_{norm} threshold (0.88) depending on the A_{norm} in post-harvest (after 30 June) is shown by the red dashed line, and it indicates the soil moisture can be retrieved from 16 January to 1 March 2015.

2.5 Vegetation height retrieval using a wavelet analysis

While vegetation grows, the vegetation surface gradually replaces the bare soil surface as the dominant reflecting surface. As a consequence, the height (h) of the antenna above the reflecting surface decreases. Equation (4) shows that changes in h impact T. This property allows the use of changes in T values to infer changes in h, and further estimate relative vegetation height. To retrieve relative vegetation height we propose a new approach based on wavelet analysis. Wavelets have been used for many years in signal processing studies in geosciences (Ouillon et al., 1995; Darrozes et al., 1997; Gaillot et al., 1999), astrophysics (Escalera and MacGillivray, 1995), meteorology (e.g., Hagelberg and Helland, 1995; Torrence and Compo, 1998), hydrology (Labat, 2005) and in many other fields. The wavelet analysis is well suited for analyzing time series with non-stationary power and frequency changes across time as illustrated by Fig. 1. Our wavelet analysis methodology is based on the WaveletComp R-package (Roesch et al., 2014). To analyze the period structure, we used a well-known Morlet mother function which comes from a combination of a Gaussian and a sinusoidal function (Fig. S3). Due to its shape, Morlet daughters allow detection of singularities on all scales/periods of the spectrum. Morlet wavelets are also well suited for environmental analysis (Grinsted et al., 2004). We calculate the Morlet wavelet transform of the multipath SNR and evaluate the power spectrum of the multipath SNR signal (see Eqs. S1–S4).

Vegetation height can be retrieved using the dominant SNR period (T_d), which is the peak period of the average power spectrum derived from a wavelet analysis of SNR, from the multipath SNR segment at elevation angles from 5 to 20°. After obtaining T_d time series, the relative antenna height (h) can be derived from Eq. (4) as the following:

$$h = \frac{\lambda}{2\cos\theta_{E9}\cdot\frac{d\theta_{E9}}{dt}\cdot T_d}. \tag{9}$$

The T_d value is used to represent the multipath SNR data in order to estimate h. Also, changes in the elevation angle (θ) and in $d\theta/dt$ have to be accounted for. In this study, changes in h were surveyed across dates at an elevation angle of 9° (e.g., $\theta_{E9} = 9°$; see Sect. 3.2).

Changes in relative antenna height (h) during vegetation growth are directly related to vegetation height increase by

$$\Delta H = h_0 - h. \tag{10}$$

Similarly to the phase change estimates ($\Delta\phi_t$ in Sect. 2.4), h_0 is the median value of the top 15 % h data during the whole wheat growth cycle for each track.

The final retrieved vegetation height (H) is based on the mean relative antenna height change from all available satellite tracks ($N = 37$), plus one wavelength

$$H = \frac{\sum_N \Delta H}{N} + \lambda. \tag{11}$$

The minimum value of H is one wavelength. Therefore Eq. (11) can only be applied when the wheat height is higher than one wavelength (0.19 m for L1).

It must be noted that it is not necessary to retrieve soil moisture before retrieving vegetation height.

2.6 GDD (growing degree days) model

Because of the lack of in situ records for the wheat field growth stages, we built a reference GDD model based on the wheat growth stage dates observed at the same location in 2010 (Duveiller et al., 2011; Fieuzal et al., 2013; Betbeder et al., 2016). The GDD model is described in Eqs. (S5–S6) and Fig. S4.

3 Results

3.1 Soil moisture retrieval

Figure 3 presents the surface soil moisture retrievals from 16 January to 5 March 2015, together with independent

Table 1. Soil moisture scores from 16 January to 5 March 2015. MAE is the mean absolute error, RMSE is the root mean square error and SDD is the standard deviation of differences.

	GPS vs. in situ	GPS vs. ISBA	GPS vs. in situ	GPS vs. ISBA	GPS (ϕ_{index}) vs. in situ	GPS (ϕ_{index}) vs. ISBA	ISBA vs. in situ
S ($m^3\,m^{-3}\,{}^{\circ-1}$)	0.0148	0.0148	0.0033	0.0033	–	–	–
N	47	43	47	43	47	43	43
MAE ($m^3\,m^{-3}$)	0.036	0.034	0.011	0.018	0.007	0.009	0.009
RMSE ($m^3\,m^{-3}$)	0.046	0.041	0.014	0.022	0.009	0.012	0.010
SDD ($m^3\,m^{-3}$)	0.036	0.037	0.009	0.012	0.008	0.011	0.006
Mean bias ($m^3\,m^{-3}$)	0.029	0.019	−0.010	−0.018	0.003	−0.005	0.008
R^2	0.73	0.63	0.73	0.63	0.74	0.65	0.88

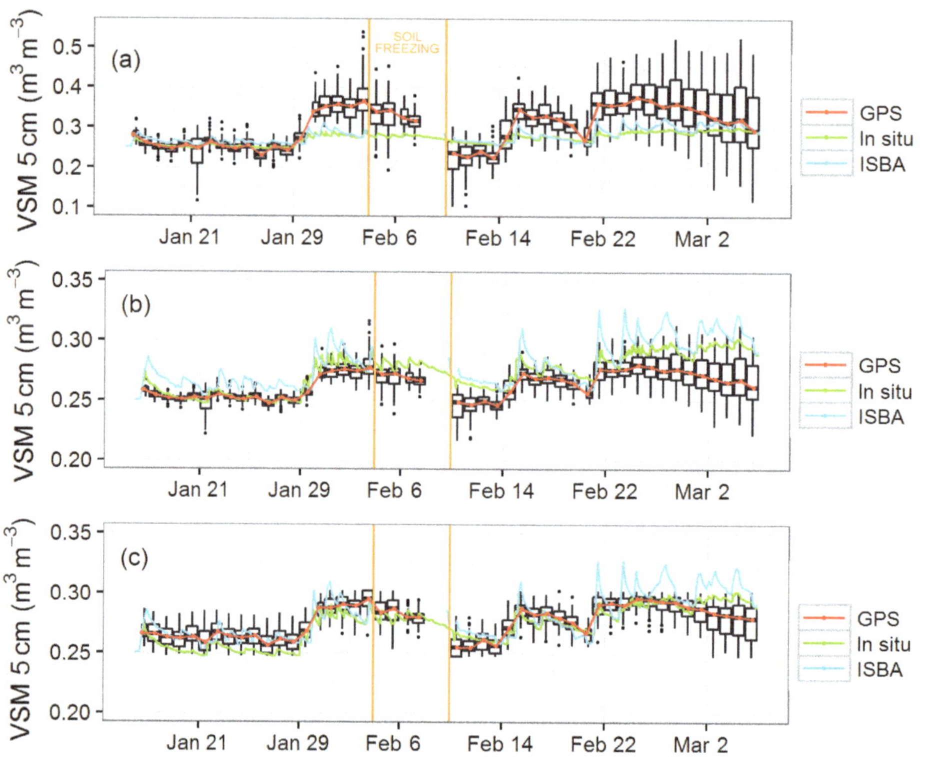

Figure 3. In situ surface volumetric soil moisture (VSM) observations at 5 cm depth (green line), ISBA simulations (blue line) and median of the daily GPS retrievals **(a)** with the a priori slope ($S = 0.0148\,m^3\,m^{-3}\,{}^{\circ-1}$) (red line), **(b)** with a locally adjusted slope ($S = 0.0033\,m^3\,m^{-3}\,{}^{\circ-1}$) (red line) and **(c)** from scaled soil wetness index (red line), and their daily statistical distribution (black box plots) for all available satellite tracks from 16 January to 5 March 2015. Boxes: 25–75 % percentiles; bars: maximum (minimum) values below (above) 1.5 IQR (Inter Quartile Range, corresponding to the 25–75 % percentile interval); dots: data outside the 1.5 IQR interval. The ISBA simulations indicate soil freezing (i.e., the presence of ice in the top soil layer) from 4 to 9 February.

in situ VSM observations and ISBA simulations. The VSM retrievals are derived from GPS SNR observations using Eq. (6) in sparse vegetation conditions, when A_{norm} is above 0.78, with the a priori S value of $0.0148\,m^3\,m^{-3}\,{}^{\circ-1}$ (Fig. 3a) and the adjusted local slope $S = 0.0033\,m^3\,m^{-3}\,{}^{\circ-1}$ (Fig. 3b). This adjusted S value is the mean of slope values obtained for satellite tracks whose phase presented a linear correlation with in situ soil moisture higher than 0.9. This occurred for the ascending tracks of GPS 13, 21, 24 and 30 and for the descending tracks of GPS 05, 09, 10, 15 and 23. Figure 3c shows the VSM retrievals from the scaled soil wetness index (Eq. 8).

The GPS and ISBA scores are given in Table 1. The mean soil moisture values during the experimental period are 0.27, 0.28, 0.31, 0.26, and $0.28\,m^3\,m^{-3}$ for in situ VSM measurements, ISBA simulations, GPS retrievals with $S = 0.0148\,m^3\,m^{-3}\,{}^{\circ-1}$, GPS retrievals with

Table 2. Sub-daily variability (standard deviation, in $m^3\,m^{-3}$) of VSM estimates.

	Minimum	Maximum	Average value
In situ observations	0.000	0.009	0.002
ISBA simulations	0.000	0.021	0.005
GPS retrievals with $S=0.0148\,m^3\,m^{-3\,\circ-1}$	0.012	0.090	0.036
GPS retrievals with $S=0.0033\,m^3\,m^{-3\,\circ-1}$	0.003	0.020	0.008
GPS retrievals from scaled soil wetness indexes	0.005	0.017	0.009

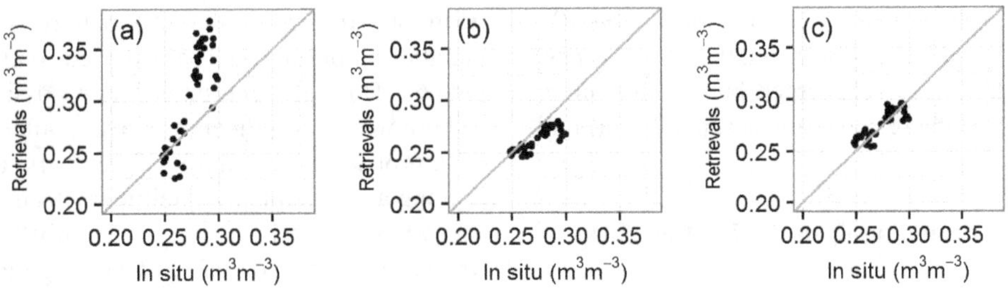

Figure 4. VSM GPS retrievals ($N=47$) vs. daily mean in situ VSM observations ($m^3\,m^{-3}$) at 5 cm from 16 January to 5 March 2015, **(a)** with the a priori slope $S=0.0148\,m^3\,m^{-3\,\circ-1}$, $\mathrm{VSM}=0.0148\Delta\phi+0.252$, **(b)** with the locally adjusted slope $S=0.0033\,m^3\,m^{-3\,\circ-1}$, $\mathrm{VSM}=0.0033\Delta\phi+0.252$ and **(c)** from scaled soil wetness indexes, $\mathrm{VSM}=0.055\phi_{\mathrm{index}}+0.247$. More scores can be referred from Table 1.

$S=0.0033\,m^3\,m^{-3\,\circ-1}$, and GPS retrievals from the scaled soil wetness index, respectively.

In Fig. 3, the sub-daily statistical distribution of the VSM retrievals is indicated by box plots. The range of daily standard deviation value of the various VSM estimates is shown in Table 2. The in situ VSM measurements present the smallest sub-daily variability, with a mean standard deviation value of $0.002\,m^3\,m^{-3}$. The largest variability is obtained for the GPS retrievals based on the a priori slope value $S=0.0148\,m^3\,m^{-3\,\circ-1}$, with a mean standard deviation value of $0.036\,m^3\,m^{-3}$. GPS retrievals based on the adjusted slope value $S=0.0033\,m^3\,m^{-3\,\circ-1}$ presents intermediate values ($0.008\,m^3\,m^{-3}$), together with those based on the scaled soil wetness index ($0.009\,m^3\,m^{-3}$) and with the ISBA simulations ($0.005\,m^3\,m^{-3}$). Figure 3 shows that the sub-daily variability of GPS VSM retrievals tends to increase during the last 10 days of the retrieval period.

It must be noted that GPS data are missing on 8 and 9 February, and that the ISBA simulations indicate soil freezing (i.e., the presence of ice in the top soil layer) from 4 to 9 February. This period was excluded from the comparison. In the end, there were 47 valid observation days for the statistical analysis of the retrieved surface VSM, among which 43 days could be compared with model simulations.

The GPS VSM daily mean retrievals based on the CH16 method present a good agreement with both in situ observations and ISBA simulations: MAE (mean absolute error) and RMSE (root mean square error) are lower than $0.05\,m^3\,m^{-3}$, and SDD (standard deviation of differences) does not exceed $0.04\,m^3\,m^{-3}$ (Table 1). The errors are reduced by at least 50 % when the local adjusted slope is used. When the scaled soil wetness index is used, the errors are further reduced.

Figure 4a and b show the retrieved soil moisture as a function of the in situ observations for a priori and adjusted slopes ($S=0.0148$ and $S=0.0033\,m^3\,m^{-3\,\circ-1}$, respectively) from all available satellite tracks (66 per day), not only those tracks used for fitting the slope (see Fig. S5). The corresponding improvements in score values are given in Table 1: the MAE decreases from 0.036 to $0.011\,m^3\,m^{-3}$, the RMSE decreases from 0.046 to $0.014\,m^3\,m^{-3}$ and the SDD decreases from 0.036 to $0.009\,m^3\,m^{-3}$. The retrievals based on the a priori slope markedly overestimate VSM in wet conditions. On the other hand, the retrievals based on the adjusted slope only slightly underestimate VSM. This shows that adjusting the slope is critical and has a major impact on the retrieval accuracy. Furthermore, Fig. 4c gives the retrievals based on the scaled soil wetness index. Scores are further improved: the MAE decreases to $0.007\,m^3\,m^{-3}$, RMSE to $0.009\,m^3\,m^{-3}$ and SDD to $0.008\,m^3\,m^{-3}$.

We also compared the retrievals with the independent ISBA simulations. The ISBA model VSM simulations present a better agreement with the in situ VSM observations than the GPS retrievals, for all the scores, as shown by Table 1 (last column) and Fig. 3. In particular, $R^2=0.88$ for ISBA simulations, against $R^2=0.74$ for GPS retrievals. This shows that the ISBA simulations can be used as a reference to assess local GPS retrievals for this site. The statistical scores resulting from the comparison between the GPS retrievals and the simulations are similar to those based on in situ observations.

After 5 March, A_{norm} drops below 0.78 (Fig. 2), and the VSM retrievals are not valid. We made an attempt to retrieve VSM from 6 to 15 March. We obtained 10 VSM retrieved values and we compared them with ISBA VSM simulations, because in situ observations were lacking. The R^2 score decreased from 0.63 before 6 March (Table 1) to only 0.21 from 6 to 15 March. This result confirms that the empirical A_{norm} threshold (0.78) is a good way to assess the VSM retrieval feasibility from vegetated areas. Additionally, we found that adjusting the A_{norm} threshold from 0.78 to 0.88 permitted making a distinction between harvest and post-harvest (after 30 June), A_{norm} values in Fig. 2. Four more days (2–5 March) are excluded. Figure 3 shows that the 25–75 % percentile intervals for these days are larger, but the maximum retrieval differences for these days are acceptable, around $0.03\,\mathrm{m}^3\,\mathrm{m}^{-3}$.

3.2 Dominant SNR period analysis during the wheat growth cycle

Figure 1 shows an example of the multipath SNR data from the ascending track of GPS01 on 21 January 2015. Its average power spectrum (Fig. 1b) derived from a wavelet analysis is also shown, together with the power spectrum (Fig. 1c) for periods ranging from 128 to 1024 s. The average power spectrum presents a single peak and the corresponding peak period is 362 s. The SNR data is reconstructed well (red line in Fig. 1a), using this peak period. Both phases and amplitudes match very well. This shows that the peak period from the average power spectrum can be used to represent the multipath SNR data. Limiting elevation angle values from 5 to 20° (Sect. 2.1) ensures a relatively stable value of the peak period. The peak period is considered as the dominant period (T_{d}) of the multipath SNR data.

Additionally, the major part of the signal power is concentrated on elevation angles ranging from 7 to 11° (see Fig. 1). A preliminary analysis for the entire wheat growing cycle showed that, more often than not, the best elevation angle corresponding to the peak power was around 9°. In this study, elevation and its change rate at 9° are used to represent the SNR data for all available satellite tracks (37 per day). It must be noted that this reference elevation angle is specific to the gain pattern and height of the antenna encountered in this experiment. It could present different values in other antenna configurations.

During the wheat growth cycle, preliminary tests showed that the average power spectrum could present multiple peaks together with a reduced maximum average power. This made T_{d} unsuitable for the representation of the multipath SNR data. Under this situation the quality of the T_{d} value was considered as poor and the data were not used. An example of T_{d} time series is shown in Fig. 5 for GPS01 ascending tracks. Poor quality data (e.g., on 17–20 March and 12–16 June) are indicated. We sorted out the data acquired in two situations: (1) track data presenting more than one peak in the highest 80 % percentile of the power spectrum and (2) T_{d} value smaller by 10 s than the mean value of the lowest 10 % of the dominant periods (e.g., $T_{\mathrm{d}} < 352$ s for GPS01). This is further illustrated in Fig. 6, comparing a usable track and an unusable track. On 1 May, there is one peak in the average power spectrum (Fig. 6b), and the dominant period (456 s) obtained can be used to fit the SNR data in Fig. 6a. While on 15 June, there are two peaks in the average power spectrum as shown in Fig. 6d. Furthermore, the maximum average power is only 0.54, which is significantly smaller than the maximum average power of 1.0 observed on 1 May 2015 (Fig. 6b). In Fig. 6c, the SNR pattern is clearly noisier, with smaller amplitudes and a less clear pattern than in Figs. 1a and 6a. This data set is unusable. A possible cause is the more inhomogeneous reflecting surface after the lodging event. The probability distribution (grey bars) of bad quality tracks among all available 37 satellite tracks is shown in Fig. 2 on a daily basis from 16 January to 15 July 2015. Most unsuitable tracks are observed during two time periods: (1) at the beginning of spring, from 10 to 20 March, and (2) at the beginning of summer, from 12 to 26 June. The latter corresponded to lodging of vegetation, which occurred during a strong wind event and affected the reflecting surface height. The in situ observation of wheat height was only 39 cm on 18 June.

As shown in Sect. 2.4, vegetation effects on the SNR signal became significant after 5 March. After this date, A_{norm} (black dots in Fig. 2) decreased drastically, in relation to plant growth. After 10 March, wheat height exceeded one wavelength (> 0.19 m). In addition to lower A_{norm} values, an increasing number of unsuitable tracks were observed until 20 March, together with low values of the peak power (Fig. 5). During this time period, the vegetation gradually decreased the strength of the signal reflected from the soil surface and more signal was reflected by the vegetation. This triggered multiple peaks for some tracks. Such tracks were not used. When the vegetation surface completely replaced the soil surface as the dominant reflecting surface of the GNSS signal, a single peak period was observed again and its value increased in response to the rise of the reflecting surface. For example, T_{d} increased from 362 s (7 March) to 397 s (22 March) for GPS01 ascending tracks. Figure 5 shows that T_{d} is not sensitive to vegetation height when vegetation height is smaller than one wavelength. Therefore, it can be concluded that this relative vegetation height (at satellite elevation of 9°) retrieval technique does not work for vegetation height below one λ (~ 0.19 m for L1) and when multiple peaks are observed in the average power spectrum.

3.3 Vegetation height retrieval

Figure 7 shows the retrieved vegetation height from 16 January to 15 July 2015, together with seven in situ vegetation height measurements and daily vegetation height simulations by ISBA. Since the original H retrievals present a marked leveling effect, the moving average of the GPS

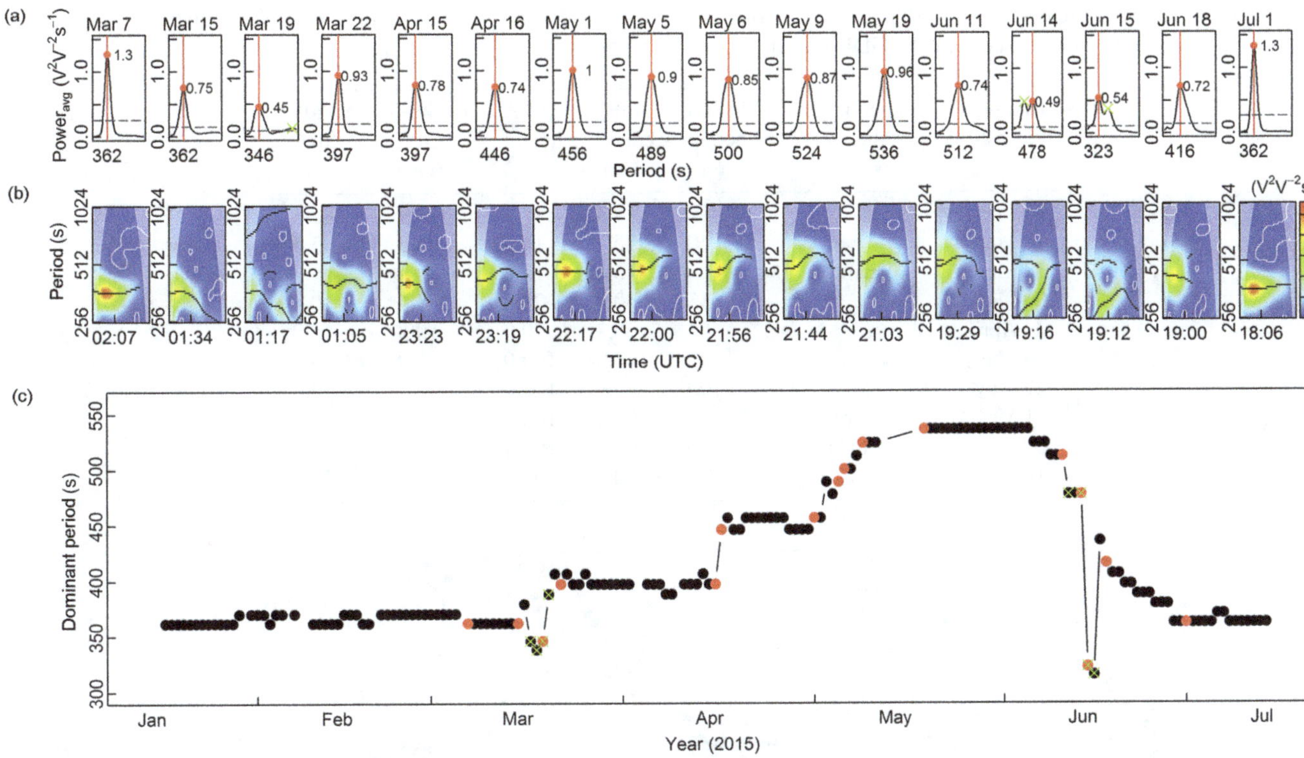

Figure 5. SNR-dominant period (T_d) time series (black dots in **c**) derived from the GPS01 ascending tracks, with the green crosses indicating that more than one peak are recognized as bad quality data, from 16 January to 15 July 2015. **(a)** the average power spectrums with their maximum values (red dots) and **(b)** power spectrums on the selected days (red dots in **c**) are also shown.

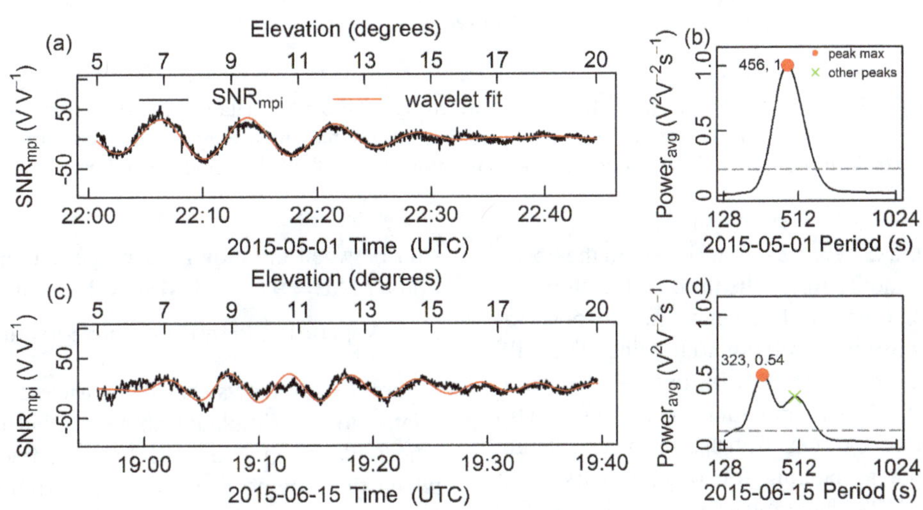

Figure 6. Examples of **(a)** usable and **(c)** unusable track data sets from the ascending tracks of GPS01 on 1 May and 15 June 2015, respectively: **(a, c)** multipath SNR data and **(b, d)** average power spectrums. The red lines in **(a, c)** are the reconstructed SNR data by the daughter wavelet corresponding to the maximum peak periods in **(b, d)**, respectively. The green cross in **(d)** shows there is more than one peak in this track data, indicating bad quality, unusable data.

height retrievals computed using a centered gliding window of 21 days is shown. The relative vegetation height retrievals are compared with ISBA height simulations and in situ height observations in Table 3. The differences between the seven in situ observations and the original H retrievals are −8, +4, −5, −10, −6, −2 and −2 cm. Most of them exhibit a negative bias. In comparison with the errors between the in situ observations and the ISBA simulations (−5, +6, +10, −15,

Table 3. Vegetation height retrievals from GPS and simulations from ISBA, and their relative deviations for each in situ height observation. The phenological statuses are derived from the GDD model.

Dates (year 2015)	Phenological status	In situ height (cm)	GPS height (cm)	ISBA height (cm)	In situ–GPS (cm)	In situ–ISBA (cm)
20 January	–	10	18.4	15.4	−8.4	−5.4
10 March	–	20	15.7	14.5	4.3	5.5
12 March	tillering	–	15.5	15.6	–	–
30 March	–	35	40.4	24.6	−5.4	10.4
24 April	–	55	65.3	70.0	−10.3	−15.0
19 May	–	97	102.9	100.0	−5.9	−3.0
29 May	–	100	101.7	100.0	−1.7	0.0
31 May	flowering	–	102.4	100.0	–	–
3 June	ripening	–	101.9	100.0	–	–
18 June	–	39	40.5	100.0	−1.5	−61.0

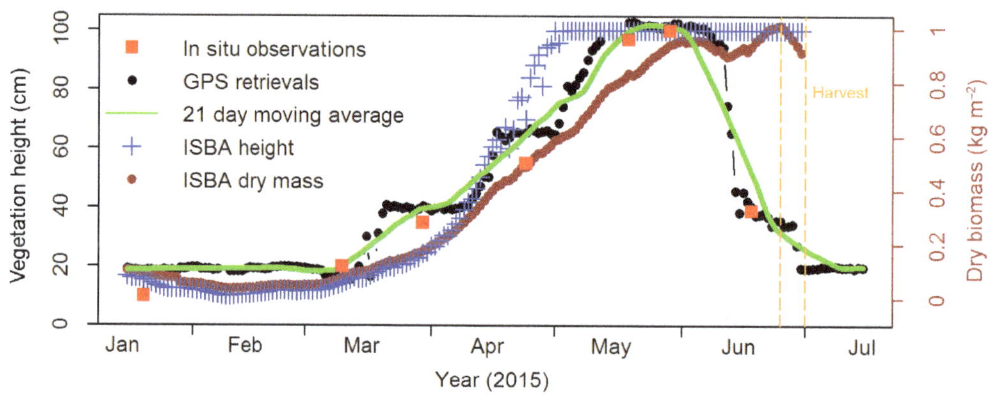

Figure 7. Wheat canopy height from 16 January to 15 July 2015 derived from GPS SNR data (black dots), from in situ observations (red squares), and from ISBA simulations (blue crosses). The green line represents the moving average of the GPS retrievals, computed using a centered gliding window of 21 days. Wheat above-ground dry biomass simulated by the ISBA model is indicated by dark red dots.

−3, 0 and −61 cm), the GPS retrievals are closer to the observations on 30 March and 24 April (the third and forth in situ observations). On 18 June, the last height in situ observation before harvest is 39 cm, in relation to lodging. The GPS retrieval is very close to this value with only −2 cm error. On the other hand, the ISBA simulation on 18 June is still at 1 m with an error of −61 cm, because the wheat height was simulated without accounting for lodging. This result shows that the in situ GPS height retrievals are able to detect local changes in vegetation height. Figure 7 and the scores given in Table 4 show that the GPS retrievals are closer to the observed growing trend than the ISBA simulations. Additionally, the moving average height presents a much better fit to the in situ measurements than the raw GPS retrievals. We also compared the GPS retrievals with the ISBA model simulations. We obtained the following score values from 10 March to 11 June 2015: MAE = 8.9 cm, RMSE = 12.4 cm and $R^2 = 0.89$. Similar values were obtained for the comparison between the moving average height and ISBA simulations: MAE = 9.0 cm, RMSE = 11.6 cm and $R^2 = 0.91$.

3.4 Vegetation height vs. above-ground dry biomass

Figure 7 also shows that the retrieved vegetation height is related to the simulated above-ground dry biomass of the wheat (dark red line). We found a linear relationship between the moving average height from GPS retrievals and the above-ground dry biomass simulated by the ISBA model from 10 March to 29 May 2015 (when the maximum vegetation height, 1 m, was measured), during the time period from tillering to flowering. The correlation coefficient between the moving average height and the above-ground dry biomass, with 81 observations, was 0.996.

A similar result was obtained using the in situ height and above-ground dry biomass measurements in Wigneron et al. (2002) from another wheat crop site (*Triticum durum,* cultivar *prinqual*) in spring 1993 (see Eqs. S7–S8 and Fig. S6).

Table 4. Vegetation height scores from 10 March to 11 June 2015. MAE is the mean absolute error, RMSE is the root mean square error and SDD is the standard deviation of differences

	GPS vs. in situ	Moving average (21 days) GPS vs. in situ	GPS vs. ISBA	Moving average (21 days) GPS vs. ISBA	ISBA vs. in situ
N	5	5	87	94	5
MAE (cm)	5.5	3.7	8.9	9.0	6.8
RMSE (cm)	6.2	5.0	12.4	11.6	8.6
SDD (cm)	5.5	3.8	12.5	11.6	9.6
Mean bias (cm)	3.8	3.7	−0.6	−0.8	0.4
R^2	0.98	0.99	0.89	0.91	0.95

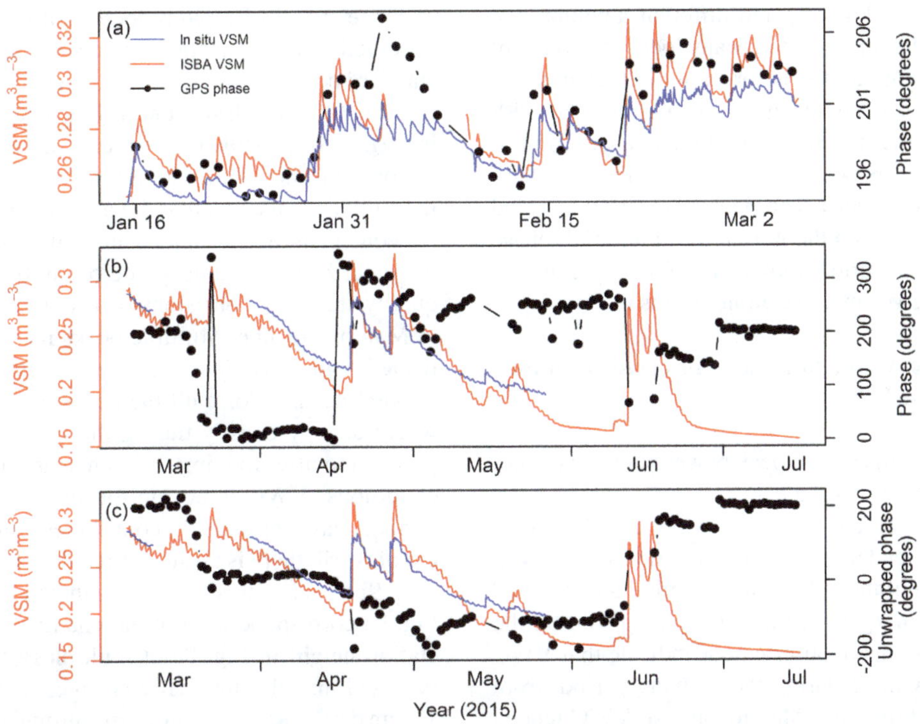

Figure 8. Example of a track data set (descending tracks from GPS10): **(a)** from 16 January to 5 March, with no significant vegetation effects and **(b, c)** from 6 March to 15 July, with significant vegetation effects. In **(a, b)**, multipath phases (black dots) are compared with in situ VSM measurements at 5 cm (blue line) and ISBA simulations (red line). In **(c)**, unwrapped multipath phases (black dots) are used to compare with in situ and simulated VSM.

4 Discussion

4.1 Can soil moisture be retrieved under significant vegetation effects?

Our results show that over a wheat field the vegetation gradually replaces the soil as the dominant reflecting surface when plant height becomes comparable to, or larger than, one wavelength.

We tested the relationship between the multipath phase in Eq. (5) and soil moisture for the whole wheat growing cycle (Fig. 8). We found that when the vegetation effects are not significant ($A_{norm} > 0.78$), the multipath phase correlates well ($R = 0.92$, $N = 47$ for the GPS10 descending tracks) with the in situ soil moisture observations (Fig. 8a). During this time period, the variation of multipath phase is about 12°, for in situ VSM values ranging from 0.25 to $0.30\,m^3\,m^{-3}$. But when the vegetation effects are significant ($A_{norm} < 0.78$), the multipath phase (without or with unwrapping, Fig. 8b and c) is no longer linearly related to soil moisture. For example, when vegetation height starts exceeding one wavelength, multipath phase rapidly decreases from

207 to 43° (between 10 and 20 March). Changes in multipath phase are disconnected from ISBA VSM simulations. This is consistent with CH16, who showed that soil moisture cannot be retrieved unless vegetation effects are corrected for.

4.2 Why does the locally adjusted *S* parameter differ from CH16?

In our experiment, the possible VSM retrieval duration was less than two months, in relatively wet conditions and VSM varied little: $0.25\,m^3\,m^{-3} < VSM < 0.30\,m^3\,m^{-3}$. This is probably not enough to represent the full yearly range of soil moisture. This might affect the representativeness of the S parameter (Sect. 2.4) we derived from our field observations. Furthermore, different signal wavelengths (L1 = 19.03 cm, L2 = 24.45 cm) and different antenna gain patterns also affect the S parameter. Many local environment factors such as vegetation effects, precipitation, changes in soil roughness and soil composition can perturb the GPS VSM estimates. All these factors contribute to changes in S, and further affect the retrieval accuracy and the sub-daily variability of VSM estimates. That is why we used a scaled soil wetness index based on the normalized multipath phase for each track, without a priori knowledge of the S parameter. This approach also gives more accurate results.

4.3 Can vegetation water content be inferred from the wavelet analysis?

We found that VWC impacts the peak power but we were not able to retrieve VWC at this stage.

Figure 7 shows that the retrieved vegetation height is consistent with independent height measurements. However, vegetation height is not the only factor affecting the reflected GPS signal. VWC (in $kg\,m^{-2}$) may also play a role on the reflected GPS signal. In situ observations indicate that VWC increased together with H during the growing period, from March to mid-May. From mid-May to harvest, VWC tended to decrease but H also decreased in relation to lodging. Can this specific behavior of VWC be detected from the results of the wavelet analysis? The latter provides three quantities: the dominant period (Sect. 2.5), A_{norm} and the peak power.

The amplitude (A_{norm}) is related, to some extent, to VWC (see Sect. 1). However, A_{norm} is calculated assuming the relative antenna height is constant. Because the wheat height increased from 10 to 100 cm, the relative antenna height was reduced, and this assumption was not satisfied. This affected the estimates of the amplitude of the multipath SNR data, especially when the wheat was tall. Comparing Fig. 6a and c, it can be observed that the signal amplitude is larger on 1 May than that on 15 June. But A_{norm} (0.15) on 1 May is even smaller than the A_{norm} (0.33) on 15 June (Fig. 2). It is likely that A_{norm} was underestimated on 1 May. Therefore, it is difficult to unequivocally relate A_{norm} to vegetation characteristics, as illustrated in Fig. 2. However, the drop in A_{norm} observed at the beginning of June (Fig. 2) could be related to the drop in VWC.

From the wavelet analysis, we also obtained the peak power when we searched for the peak period from the average power spectrum. Peak power can represent changes in the multipath SNR strength. Figure 9 shows daily box plots of the peak power for all available satellite tracks from 16 January to 15 July 2015, together with the distribution of bad quality tracks (as in Fig. 2), and rainfall. There are two major possible causes for a sudden reduction of the strength of the SNR signal: (1) the attenuation of the signal by the rain intercepted by vegetation or in the troposphere and (2) the occurrence of more than one dominant reflecting surface at different heights, and these two causes can occur at the same time.

Three events of rapid reduction of the peak power can be observed in Fig. 9a. These events are related to larger daily standard deviation (SD) values of vegetation height retrievals (see Fig. 9b). The last event in June could be related to lodging. However, whether maximum SD is an indicator of lodging or not is unclear. It seems that these events are not related to rainfall events, and that the attenuation by intercepted water content is not a major cause of peak power drops. On the other hand, the emergence of multiple peaks and of bad quality tracks is consistent with the rapid power reduction in March and June. Multiple peaks may indicate that the reflected signal originates from surfaces at different heights. A possible cause of multiple peaks is a more heterogeneous wheat canopy density during the first stage of the growing period and after lodging. In such sparse or mixed vegetation conditions, VWC is not uniformly distributed and the soil surface may significantly contribute to the SNR. In the middle of April, there is no such effect but the SD score increases (Fig. 9b). It is interesting to note that the peak power drops in Fig. 9a corresponding to rapid changes in the retrieved vegetation height in Fig. 9c at multiples of λ or 0.5λ. It must be noted that absolute daily changes in H (and h), of about $1.1\,cm\,day^{-1}$ are fairly uniform throughout the growing period. Since h decreases when plants grow, relative changes in h tend to increase. According to Eq. (4), T behaves similarly. This means that the sensitivity of the retrieval method to changes in H is larger at the end of the growing period. This is probably why leveling is more pronounced between mid-March and mid-April than at the end of April (see Fig. 9c). Leveling is less noticeable in May.

4.4 Can the unwrapped multipath phase be used to retrieve vegetation height?

Our results indicate that using the dominant period to retrieve vegetation height is more relevant than using the multipath phase.

The relationship between the multipath phase (Fig. 8) in Eq. (5) and vegetation height was investigated. Because changes in relative antenna height exceeded λ during vegeta-

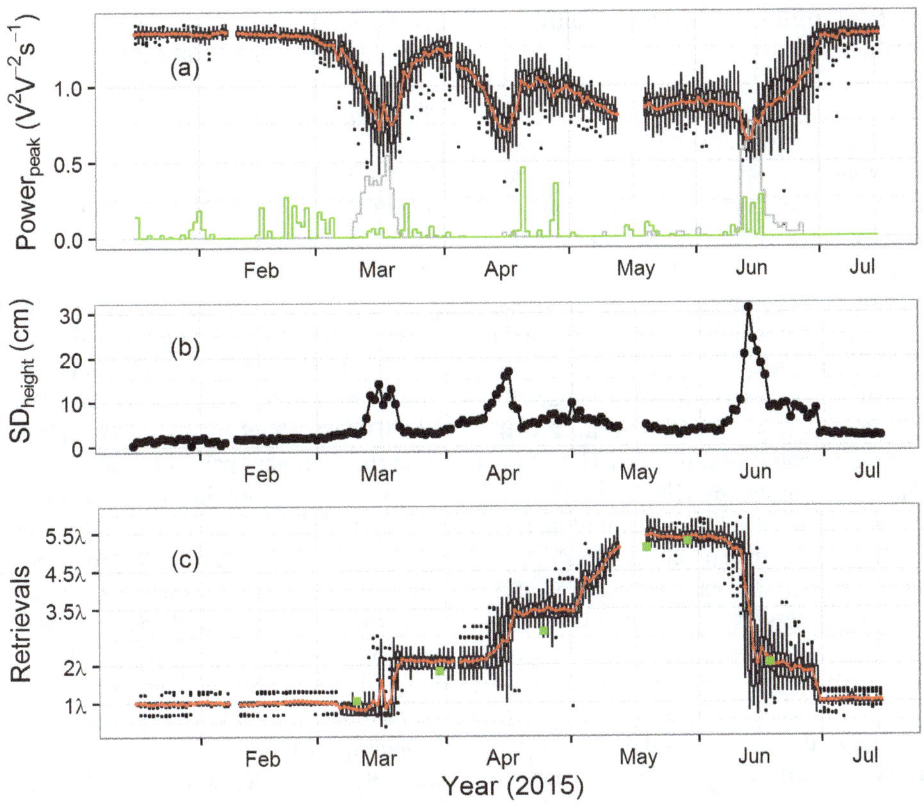

Figure 9. The box plots of **(a)** the peak power from a wavelet analysis, **(b)** standard deviation (SD) score of the retrieved vegetation height and **(c)** the retrieved vegetation height (rescaled in λ units) for all available satellite tracks from 16 January to 15 July 2015. The mean value of the peak power in **(a)** and of the retrievals in **(c)** are shown by red lines. In **(a)**, the grey line shows the statistical distribution of bad quality tracks (the number of the bad quality tracks can be obtained multiplying by 37) and the green line represents the rainfall (daily precipitation in mm day^{-1} can be obtained multiplying by 50). In **(c)**, the rescaled in situ observations are shown by green squares.

tion growth, the multipath phase had to be unwrapped. When the vegetation height was smaller than λ (before 10 March), the multipath phase (around 200°) presented little changes (about 12°). From 21 March to 18 April, the multipath phase was much smaller (around 10°) and relatively stable. On the other hand, the variability increased from 19 April to 11 June (Fig. 8c), and no relationship with plant growth could be found. It can be noted that the multipath phase and dominant period are relatively stable when the vegetation height is smaller than λ. Both tend to aggregate at several value levels.

4.5 Can wheat phenological stages be inferred?

Figure 9 shows that the occurrence of multiple peaks together with a drop of the peak power can be used as an indicator of the start of the most active part of the growing season, and of the end of the senescence period preceding the harvest.

We applied the GDD model (see Sect. 2.6) to year 2015 and we obtained the following dates for tillering, flowering and ripening: 12 March, 31 May, and 3 June, respectively (see Fig. S3). The obtained tillering date (12 March) is close to the start date (10 March) of the multiple peaks (see Sect. 3.2). Tillering in wheat triggers nitrogen uptake and the accumulation of biomass (Gastal and Lemaire, 2002). This is consistent with the rapid changes in the indicators derived from the wavelet analysis: drop in A_{norm} values and high rate of multiple peaks (Fig. 2), rise in the retrieved H (Fig. 7), and drop in peak power (Fig. 9). For our site, the tillering date also corresponded to the period when H reached a value of about 0.2 m. This was the case in 2015 and also in 2010 at the same site (Betbeder et al., 2016).

Flowering and ripening did not trigger abrupt changes in the GPS retrievals. However, these stages corresponded to a change in H trend. This is illustrated in Fig. S7, which shows the difference between retrieved vegetation height at a given date and retrieved vegetation height 15 days before. Flowering and ripening occur towards the end of the growing period when the vegetation height is no longer increased compared with 15 days before but slightly declines due to wheat heads tipping down (Wigneron et al., 2002). In order to confirm these findings, it could be recommended to perform GNSS-IR measurements over other wheat fields and other crops, together with phenological stage observations combined with in situ height measurements.

4.6 Potential future applicability and transferability of the retrieval method

In situ VSM observations are not widespread in France and in situ vegetation height observations are generally not available. Therefore, ISBA simulations are key for water resource monitoring on the country scale. It must be noted that the ISBA model is forced by the SAFRAN atmospheric analysis (Durand et al., 1993, 1999) and that SAFRAN is able to integrate thousands of in situ rain gauge observations. ISBA is also able to simulate vegetation characteristics such as vegetation height, leaf area index and above-ground dry biomass. However, in situ VSM observations are needed to validate the model simulations (e.g., Albergel et al., 2010). From this point of view, the spatial resolution of GNSS retrievals is an asset. The area sampled by GNSS retrievals is much larger than what can be achieved using individual soil moisture probes and much smaller than pixel size of satellite-derived products. Longer continuous time periods of GNSS retrievals should be envisaged to serve as independent validation data sources in statistical methods such as Triple Collocation (Dorigo et al., 2010).

We successfully assessed the surface soil moisture retrieval technique for a wheat crop field, during the start of the growing period. However, the rather narrow range of surface soil moisture values during the corresponding experimental time period limited the representativeness of the obtained retrieval accuracy. Furthermore, our data set did not include GNSS data and in situ VSM measurements for periods of bare soil. Longer periods presenting a bare soil surface should be investigated in future studies. At the same time, more in situ vegetation measurements should be carried out.

The retrieved vegetation height was based on the dominant period of the average power spectrum. The latter was derived from GPS multipath SNR data for elevation angles between 5 and 20°. We only considered the dominant period variations, without accounting for instantaneous phase changes. The accuracy of the retrieved vegetation height could probably be improved considering changes in both period and phase of the multipath SNR oscillations.

In this study, only the SNR data of the L1 C/A signal was used, SNR data from different wavelengths (e.g., L1 C/A, L2C and L5) should also be compared or combined to survey canopy characteristics.

A linear relationship between wheat height and dry biomass was observed during the period from wheat tillering to ripening. Retrieving dry biomass is a motivation for further research because most current satellite vegetation products focus on retrieving vegetation indexes or leaf area index. The dry biomass is directly related to the wheat yield, and retrieving wheat height could have applications in crop monitoring. In this study, only wheat is considered. Other crops should be investigated in the future.

5 Conclusions

GNSS SNR data were obtained using the SNR GNSS-IR technique over an intensively cultivated wheat field in southwestern France. The data were used to retrieve either soil moisture or relative vegetation height during the growing period of wheat. Vegetation growth tended to decrease the relative antenna height and broke up the constant height assumption used in soil moisture retrieval algorithms. Soil moisture could not be retrieved after wheat tillering. A new algorithm based on a wavelet analysis was implemented and used to extract the dominant period of the SNR and to retrieve vegetation height. The dominant period was derived from the peak period of the average power spectrum derived from a wavelet analysis of SNR. The method proposed by CH16 was used to retrieve soil moisture under sparse vegetation conditions, before wheat tillering. Soil moisture was retrieved on a daily basis with a precision (SDD) of $0.008\,m^3\,m^{-3}$. Before tillering, only one stable peak was observed in the average power spectrum, because the soil surface was the dominant GNSS reflecting surface. During and after tillering (10–20 March), the reflected GNSS signal included contributions from both soil and vegetation. More than one peak was observed in the average power spectrum together with low values of peak power, showing that there was no clear dominant reflecting surface. Wheat growth gradually raised the reflecting surface of the GNSS signal, from the soil surface to the vegetation surface, which significantly modulated the dominant period of the multipath SNR data. In these conditions, vegetation effects could not be ignored and soil moisture could not be retrieved. The retrieved vegetation height was in good agreement with the in situ observations, and was consistent with a lodging event. However, the retrieved height consisted of several levels. Using a moving average on the retrieved height permitted a better match with the in situ height measurements: a precision of 3.8 cm could be achieved, against 5.5 cm for the original retrievals. Furthermore, several indicators derived from the wavelet analysis could be used to detect tillering. We also found that VWC impacts the peak power but the latter cannot be used to retrieve VWC at this stage.

Competing interests. The authors declare that they have no conflict of interest.

Acknowledgements. The work of Sibo Zhang and Karen Boniface was supported by the STAE (Sciences et Technologies pour l'Aéronautique et l'Espace) foundation, in the framework of the PRISM (Potentialités de la Réflectométrie GNSS In-Situ et Mobile) project. Authors would also like to thank the farmer, A. Blanquet, for his time and the person who helped for collecting ground data.

Edited by: Alberto Guadagnini

References

Albergel, C., Calvet, J. C., De Rosnay, P., Balsamo, G., Wagner, W., Hasenauer, S., Naeimi, V., Martin, E., Bazile, E., Bouyssel, F., and Mahfouf, J. F.: Cross-evaluation of modelled and remotely sensed surface soil moisture with in situ data in southwestern France, Hydrol. Earth Syst. Sci., 14, 2177–2191, https://doi.org/10.5194/hess-14-2177-2010, 2010.

Betbeder, J., Fieuzal, R., Philippets, Y., Ferro-Famil, L., and Baup, F.: Contribution of multitemporal polarimetric synthetic aperture radar data for monitoring winter wheat and rapeseed crops, J. Appl. Remote Sens., 10, 026020, https://doi.org/10.1117/1.JRS.10.026020, 2016.

Bilich, A. and Larson, K. M.: Mapping the GPS multipath environment using the signal-to-noise ratio (SNR), Radio Sci., 42, RS6003, https://doi.org/10.1029/2007RS003652, 2007.

Boniface, K., Braun, J. J., McCreight, J. L., and Nievinski, F. G.: Comparison of snow data assimilation system with GPS reflectometry snow depth in the western United States, Hydrol. Processe., 29, 2425–2437, https://doi.org/10.1002/hyp.10346, 2015.

Chan, S. K., Bindlish, R., O'Neill, P. E., Njoku, E., Jackson, T., Colliander, A., Chen, F., Burgin, M., Dunbar, S., Piepmeier, J., Yueh, S., Entekhabi, D., Cosh, M. H., Caldwell, T., Walker, J., Wu, X., Berg, A., Rowlandson, T., Pacheco, A., McNairn, H., Thibeault, M., Martínez-Fernández, J., González-Zamora, A., Seyfried, M., Bosch, D., Starks, P., Goodrich, D., Prueger, J., Palecki, M., Small, E. E., Zreda, M., Calvet, J.-C., Crow, W., and Kerr, Y.: Assessment of the SMAP passive soil moisture product, IEEE T. Geosci. Remote, 54, 4994–5007, https://doi.org/10.1109/TGRS.2016.2561938, 2016.

Chew, C. C., Small, E. E., Larson, K. M., and Zavorotny, V. U.: Effects of near-surface soil moisture on GPS SNR data: development of a retrieval algorithm for soil moisture, IEEE T. Geosci. Remote, 52, 537–543, https://doi.org/10.1109/TGRS.2013.2242332, 2014.

Chew, C. C., Small, E. E., Larson, K. M., and Zavorotny, V. U.: Vegetation sensing using GPS-interferometric reflectometry: theoretical effects of canopy parameters on signal-to-noise ratio data, IEEE T. Geosci. Remote, 53, 2755–2764, https://doi.org/10.1109/TGRS.2014.2364513, 2015.

Chew, C. C., Small, E. E., and Larson, K. M.: An algorithm for soil moisture estimation using GPS-interferometric reflectometry for bare and vegetated soil, GPS Solutions, 20, 525–537, https://doi.org/10.1007/s10291-015-0462-4, 2016.

Darrozes, J., Gaillot, P., De Saint-Blanquat, M., and Bouchez, J. L.: Software for multi-scale image analysis: The normalized optimized Anisotropic Wavelet Coefficient method, Comput. Geosci., 23, 889–895, https://doi.org/10.1016/S0098-3004(97)00063-0, 1997.

Decharme, B., Boone, A., Delire, C., and Noilhan, J.: Local evaluation of the Interaction between Soil Biosphere Atmosphere soil multilayer diffusion scheme using four pedotransfer functions, J. Geophys. Res., 116, D20126, https://doi.org/10.1029/2011JD016002, 2011.

Dorigo, W. A., Scipal, K., Parinussa, R. M., Liu, Y. Y., Wagner, W., de Jeu, R. A. M., and Naeimi, V.: Error characterisation of global active and passive microwave soil moisture datasets, Hydrol. Earth Syst. Sci., 14, 2605–2616, https://doi.org/10.5194/hess-14-2605-2010, 2010.

Dorigo, W. A., Xaver, A., Vreugdenhil, M., Gruber, A., Hegyiová, A., Sanchis-Dufau, A. D., Zamojski, D., Cordes, C., Wagner, W., and Drusch, M.: Global automated quality control of *in situ* soil moisture data from the International Soil Moisture Network, Vadose Zone J., 12, 21, https://doi.org/10.2136/vzj2012.0097, 2013.

Durand, Y., Brun, E., Merindol, L., Guyomarc'h, G., Lesaffre, B., and Martin, E.: A meteorological estimation of relevant parameters for snow models, Ann. Geophys., 18, 65–71, https://doi.org/10.1017/S0260305500011277, 1993.

Durand, Y., Giraud, G., Brun, E., Merindol, L., and Martin, E.: A computer-based system simulating snow-pack structures as a tool for regional avalanche forecasting, Ann. Glaciol., 45, 469–484, https://doi.org/10.1017/S0022143000001337, 1999.

Duveiller, G., Weiss, M., Baret, F., and Defourny, P.: Retrieving wheat green area index during the growing season from optical time series measurements based on neural network radiative transfer inversion, Remote Sens. Environ., 115, 887–896, https://doi.org/10.1016/j.rse.2010.11.016, 2011.

Escalera, E. and MacGillivray, H. T.: Topology in galaxy distributions: method for a multi-scale analysis. A use of the wavelet transform, Astron. Astrophys., 298, 1–21, 1995.

Fieuzal, R., Baup, F., and Marais-Sicre, C.: Monitoring wheat and rapeseed by using synchronous optical and radar satellite data – From temporal signatures to crop parameters estimation, Adv. Remote Sens., 2, 33222, https://doi.org/10.4236/ars.2013.22020, 2013.

Foti, G., Gommenginger, C., Jales, P., Unwin, M., Shaw, A., Robertson, C., and Rosello, J.: Spaceborne GNSS reflectometry for ocean winds: First results from the UK TechDemoSat-1 mission, Geophys. Res. Lett., 42, 5435–5441, https://doi.org/10.1002/2015GL064204, 2015.

Gaillot, P., Darrozes, J., and Bouchez, J. L.: Wavelet transform: a future of rock fabric analysis?, J. Struct. Geol., 21, 1615–1621, https://doi.org/10.1016/S0191-8141(99)00073-5, 1999.

Gastal, F. and Lemaire, G.: N uptake and distribution in crops: an agronomical and ecophysiological perspective, J. Exp. Bot., 53, 789–799, https://doi.org/10.1093/jexbot/53.370.789, 2002.

Georgiadou, Y. and Kleusberg, A.: On carrier signal multipath effects in relative GPS positioning, Manuscripta Geodaetica, 13, 172–179, 1988.

Grinsted, A., Moore, J. C., and Jevrejeva, S.: Application of the cross wavelet transform and wavelet coherence to geophysical time series, Nonlin. Processes Geophys., 11, 561–566, https://doi.org/10.5194/npg-11-561-2004, 2004.

Hagelberg, C. and Helland, J.: Thin-line detection in meteorological radar images using wavelet transforms, J. Atmos. Ocean. Tech., 12, 633–642, https://doi.org/10.1175/1520-0426(1995)012<0633:TLDIMR>2.0.CO;2, 1995.

Kerr, Y., Waldteufel, P., Wigneron, J., Martinuzzi, J., Font, J., and Berger, M.: Soil moisture retrieval from space: The Soil Moisture and Ocean Salinity (SMOS) mission, IEEE T. Geosci. Remote,

39, 1729–1735, https://doi.org/10.1109/36.942551, 2001.

Koch, F., Schlenz, F., Prasch, M., Appel, F., Ruf, T., and Mauser, W.: Soil Moisture Retrieval Based on GPS Signal Strength Attenuation, Water, 8, 276, https://doi.org/10.3390/w8070276, 2016.

Labat, D.: Recent advances in wavelet analyses: Part 1. A review of concepts, J. Hydrol., 314, 275–288, 2005.

Lafont, S., Zhao, Y., Calvet, J.-C., Peylin, P., Ciais, P., Maignan, F., and Weiss, M.: Modelling LAI, surface water and carbon fluxes at high-resolution over France: comparison of ISBA-A-gs and ORCHIDEE, Biogeosciences, 9, 439–456, https://doi.org/10.5194/bg-9-439-2012, 2012.

Larson, K. M.: GPS interferometric reflectometry: applications to surface soil moisture, snow depth, and vegetation water content in the western United States, Wiley Interdisciplin. Rev.: Water, 3, 775–787, https://doi.org/10.1002/wat2.1167, 2016.

Larson, K. M. and Nievinski, F. G.: GPS snow sensing: results from the EarthScope Plate Boundary Observatory, GPS Solutions, 17, 41–52, https://doi.org/10.1007/s10291-012-0259-7, 2013.

Larson, K. M. and Small, E. E.: Normalized microwave reflection index: A vegetation measurement derived from GPS networks, IEEE J. Select. Top. Appl. Earth Obs. Rem. Sens., 7, 1501–1511, https://doi.org/10.1109/JSTARS.2014.2300116, 2014.

Larson, K. M., Small, E. E., Gutmann, E. D., Bilich, A. L., Braun, J. J., and Zavorotny, V. U.: Use of GPS receivers as a soil moisture network for water cycle studies, Geophys. Res. Lett., 35, L24405, https://doi.org/10.1029/2008GL036013, 2008.

Larson, K. M., Braun, J. J., Small, E. E., Zavorotny, V. U., Gutmann, E. D., and Bilich, A. L.: GPS multipath and its relation to near-surface soil moisture content, IEEE J. Select. Top. Appl. Earth Obs. Rem. Sens., 3, 91–99, https://doi.org/10.1109/JSTARS.2009.2033612, 2010.

Larson, K. M., Small, E. E., Chew, C. C., Nievinski, F. G., Pratt, J., McCreight, J. L., Braun, J., Boniface, K., and Evans, S. G.: PBO H_2O: Plate Boundary Observatory Studies of the Water Cycle, in: Fall Meeting, American Geophysical Union, 9–13 December 2013, San Francisco, 2013.

Masson, V., Le Moigne, P., Martin, E., Faroux, S., Alias, A., Alkama, R., Belamari, S., Barbu, A., Boone, A., Bouyssel, F., Brousseau, P., Brun, E., Calvet, J.-C., Carrer, D., Decharme, B., Delire, C., Donier, S., Essaouini, K., Gibelin, A.-L., Giordani, H., Habets, F., Jidane, M., Kerdraon, G., Kourzeneva, E., Lafaysse, M., Lafont, S., Lebeaupin Brossier, C., Lemonsu, A., Mahfouf, J.-F., Marguinaud, P., Mokhtari, M., Morin, S., Pigeon, G., Salgado, R., Seity, Y., Taillefer, F., Tanguy, G., Tulet, P., Vincendon, B., Vionnet, V., and Voldoire, A.: The SURFEXv7.2 land and ocean surface platform for coupled or offline simulation of earth surface variables and fluxes, Geosci. Model Dev., 6, 929–960, https://doi.org/10.5194/gmd-6-929-2013, 2013.

Morisette, J. T., Baret, F., Privette, J. L., Myneni, R. B., Nickeson, J. E., Garrigues, S., Shabanov, N., Weiss, M., Fernandes, R., Leblanc, S., Kalacska, M., Sánchez-Azofeifa, G. A., Chubey, M., Rivard, B., Stenberg, P., Rautiainen, M., Voipio, P., Manninen, T., Pilant, A. N., Lewis, T. E., Iiames, J. S., Colombo, R., Meroni, M., Busetto, L., Cohen, W., Turner, D.P ., Warner, E. D., Petersen, G. W., Seufert, G., and Cook, R.: Validation of global moderate-resolution LAI products: A framework proposed within the CEOS land product validation subgroup, IEEE T. Geosci. Remote, 44, 1804–1817, 2006.

Motte, E., Egido, A., Roussel, N., Boniface, K., Frappart, F., Baghdadi, N., and Zribi, M.: Applications of GNSS-R in continental hydrology, Land Surface Remote Sensing in Continental Hydrology, Elsevier, Amsterdam, the Netherlands, 281–321, https://doi.org/10.1016/B978-1-78548-104-8.50009-7, 2016.

Njoku, E. G., Wilson, W. J., Yueh, S. H., Dinardo, S. J., Li, F. K., Jackson, T. J., Lakshmi, V., and Bolten, J.: Observations of soil moisture using a passive and active low-frequency microwave airborne sensor during SGP99, IEEE T. Geosci. Remote, 40, 2659–2673, https://doi.org/10.1109/TGRS.2002.807008, 2002.

Ouillon, G., Sornette, D., and Castaing, C.: Organisation of joints and faults from 1-cm to 100-km scales revealed by optimized anisotropic wavelet coefficient method and multifractal analysis, Nonlin. Processes Geophys., 2, 158–177, https://doi.org/10.5194/npg-2-158-1995, 1995.

Roesch, A., Schmidbauer, H., and Roesch, M. A.: Package 'WaveletComp', https://pbil.univ-lyon1.fr/CRAN/web/packages/WaveletComp/WaveletComp.pdf (last access: September 2017), 2014.

Roussel, N., Frappart, F., Ramillien, G., Darrozes, J., Desjardins, C., Gegout, P., Pérosanz, F., and Biancale, R.: Simulations of direct and reflected wave trajectories for ground-based GNSS-R experiments, Geosci. Model Dev., 7, 2261–2279, https://doi.org/10.5194/gmd-7-2261-2014, 2014.

Roussel, N., Frappart, F., Ramillien, G., Darrozes, J., Baup, F., Lestarquit, L., and Ha, M. C.: Detection of Soil Moisture Variations Using GPS and GLONASS SNR Data for Elevation Angles Ranging From 2° to 70°, IEEE J. Select. Top. Appl. Earth Obs. Rem. Sens., 9, 4781–4794, https://doi.org/10.1109/JSTARS.2016.2537847, 2016.

Small, E. E., Larson, K. M., and Braun, J. J.: Sensing vegetation growth with reflected GPS signals, Geophys. Res. Lett., 37, L12401, https://doi.org/10.1029/2010GL042951, 2010.

Small, E. E., Larson, K. M., and Smith, W. K.: Normalized microwave reflection index: validation of vegetation water content estimates from Montana grasslands, IEEE J. Select. Top. Appl. Earth Obs. Rem. Sens., 7, 1512–1521, https://doi.org/10.1109/JSTARS.2014.2320597, 2014.

Small, E. E., Larson, K. M., Chew, C. C., Dong, J., and Ochsner, T. E.: Validation of GPS-IR soil moisture retrievals: Comparison of different algorithms to remove vegetation effects, IEEE J. Select. Top. Appl. Earth Obs. Rem. Sens., 9, 4759–4770, https://doi.org/10.1109/JSTARS.2015.2504527, 2016.

Torrence, C. and Compo, G. P.: A practical guide to wavelet analysis, B. Am. Meteorol. Soc., 79, 61–78, https://doi.org/10.1175/1520-0477(1998)079<0061:APGTWA>2.0.CO;2, 1998.

Vey, S., Güntner, A., Wickert, J., Blume, T., and Ramatschi, M.: Long-term soil moisture dynamics derived from GNSS interferometric reflectometry: A case study for Sutherland, South Africa, GPS Solutions, 20, 641–654, https://doi.org/10.1007/s10291-015-0474-0, 2016.

Wan, W., Larson, K. M., Small, E. E., Chew, C. C., and Braun, J. J.: Using geodetic GPS receivers to measure vegetation water content, GPS Solutions, 19, 237–248, https://doi.org/10.1007/s10291-014-0383-7, 2015.

Wigneron, J. P., Ferrazzoli, P., Olioso, A., Bertuzzi, P., and Chanzy, A.: A simple approach to monitor crop biomass from C-band radar data, Remote Sens. Environ., 69, 179–188, https://doi.org/10.1016/S0034-4257(99)00011-5, 1999.

Wigneron, J. P., Chanzy, A., Calvet, J. C., Olioso, A., and Kerr, Y.: Modeling approaches to assimilating L band passive microwave observations over land surfaces, J. Geophys. Res.-Atmos., 107, 4219, https://doi.org/10.1029/2001JD000958, 2002.

Zavorotny, V. U., Larson, K. M., Braun, J. J., Small, E. E., Gutmann, E. D., and Bilich, A. L.: A physical model for GPS multipath caused by land reflections: Toward bare soil moisture retrievals, IEEE J. Select. Top. Appl. Earth Obs. Rem. Sens., 3, 100–110, https://doi.org/10.1109/JSTARS.2009.2033608, 2010.

Zavorotny, V. U., Gleason, S., Cardellach, E., and Camps, A.: Tutorial on remote sensing using GNSS bistatic radar of opportunity, IEEE Geosci. Rem. Sens. Mag., 2, 8–45, https://doi.org/10.1109/MGRS.2014.2374220, 2014.

11

High-resolution land surface modeling utilizing remote sensing parameters and the Noah UCM: a case study in the Los Angeles Basin

P. Vahmani[1] **and T. S. Hogue**[1,2]

[1]University of California, Los Angeles, CA, USA
[2]Colorado School of Mines, Golden, CO, USA

Correspondence to: T. S. Hogue (thogue@mines.edu)

Abstract. In the current work we investigate the utility of remote-sensing-based surface parameters in the Noah UCM (urban canopy model) over a highly developed urban area. Landsat and fused Landsat–MODIS data are utilized to generate high-resolution (30 m) monthly spatial maps of green vegetation fraction (GVF), impervious surface area (ISA), albedo, leaf area index (LAI), and emissivity in the Los Angeles metropolitan area. The gridded remotely sensed parameter data sets are directly substituted for the land-use/lookup-table-based values in the Noah-UCM modeling framework. Model performance in reproducing ET (evapotranspiration) and LST (land surface temperature) fields is evaluated utilizing Landsat-based LST and ET estimates from CIMIS (California Irrigation Management Information System) stations as well as in situ measurements. Our assessment shows that the large deviations between the spatial distributions and seasonal fluctuations of the default and measured parameter sets lead to significant errors in the model predictions of monthly ET fields (RMSE = 22.06 mm month^{-1}). Results indicate that implemented satellite-derived parameter maps, particularly GVF, enhance the capability of the Noah UCM to reproduce observed ET patterns over vegetated areas in the urban domains (RMSE= 11.77 mm month^{-1}). GVF plays the most significant role in reproducing the observed ET fields, likely due to the interaction with other parameters in the model. Our analysis also shows that remotely sensed GVF and ISA improve the model's capability to predict the LST differences between fully vegetated pixels and highly developed areas.

1 Introduction

Urbanization introduces significant changes to land surface characteristics that ultimately perturb land–atmosphere fluxes of sensible heat, latent heat, and momentum which, in turn, alter atmospheric properties as well as local weather and climate (Landsberg, 1981; Kalnay and Cai, 2003; Miao et al., 2009; De Ridder et al., 2012). Urban surfaces are covered with variety of materials with distinct thermal, radiative, and moisture properties influencing surface energy and water budgets (Arnfield, 2003). Moreover, contrasting aerodynamic properties of buildings significantly change surface roughness (Cotton and Pielke, 1995). The effects associated with modified urban landscapes extend to air quality (Taha et al., 1997), local temperatures (Bornstein, 1987; Van Wevenberg et al., 2008), local and regional atmospheric circulation (Pielke et al., 2002; Marshall et al., 2004; Niyogi et al., 2006), and regional precipitation patterns (Changnon and Huff, 1986; Changnon, 1992; Lowry, 1998).

Mesoscale meteorological models have been increasingly applied over urban areas to examine the urban–atmosphere exchange of heat, moisture, momentum, or pollutants. Recently updated parameterization in the community Weather Research and Forecasting (WRF) model includes coupling between the Noah LSM (land surface model) and a single-layer urban canopy model (UCM; Kusaka et al. 2001; Kusaka and Kimura, 2004), which has substantially advanced the understanding and modeling of the mesoscale impact of cities. The coupled WRF–Noah UCM has been applied to major metropolitan regions around the world (e.g., Houston, Beijing, Guangzhou/Hong Kong, Salt Lake City, and Athens) to better understand the contribution of urbanization to changes in urban heat island, surface ozone, horizontal convective rolls, boundary layer structure, contaminant transport and dispersion, and heat wave events (Chen et al., 2004; Jiang et al., 2008; Miao and Chen, 2008; Miao et al., 2009; Wang et al., 2009; Tewari et al., 2010; Wei-guang et al., 2011; Giannaros et al., 2013). A common concern with the use of these complex mesoscale models, however, is the high level of uncertainty in the specification of surface cover and geometric parameters (Loridan et al., 2010; Chen et al., 2011). Although realistic representation of surface properties is critical for accurate simulation of the physical processes occurring in urban regions, the majority of previous modeling studies rely on traditional land use data and lookup tables to define surface parameters.

Remotely sensed observations provide important spatial information on urban-induced physical modifications to the Earth's surface (Jin and Shepherd, 2005). Airborne lidar (light detection and ranging) systems and photogrammetric techniques have been utilized to produce morphological parameters over urban areas (Burian et al., 2004, 2006, 2007; Taha, 2008; Ching et al., 2009). Burian et al. (2004) used airborne lidar data, at 1 m resolution, to generate data sets of 20 urban canopy parameters (e.g., building height, height-to-width ratio, and roughness length) for an air quality modeling study over Houston, Texas. Taha (2008) introduced an alternative and low-cost approach for generating urban canopy parameters input for the uMM5 over Sacramento region, California. The study relied on commercially available Google Earth Pro imagery to generate urban geometry parameters (e.g., pavement land-cover fraction, roof cover fraction, and mean building height). Using lidar-based three-dimensional data sets of buildings and vegetation, Ching et al. (2009) presented a high-resolution database of the geometry, density, material, and roughness properties of the morphological features for applications in WRF and other models over Houston, Texas. While promising, the availability of such data sets is currently limited to a few geographical locations and the reproduction of such data sets is extremely challenging due to high collection costs and data management difficulties associated with the extremely large size of lidar data sets (Burian et al., 2006; Ching et al., 2009).

Observations from satellites, on the other hand, have been utilized in model validation processes over urban areas (Miao et al., 2009; Giannaros et al, 2013). In addition to in situ observations, Giannaros et al. (2013) included MODIS (Moderate Resolution Imaging Spectroradiometer)-based land surface temperature (LST) products in their modeling study of the urban heat island (UHI) over Athens, Greece. Similarly, Miao et al. (2009) utilized 1 km-resolution MODIS data to verify the WRF–Noah-UCM-simulated LST distribution in Beijing. Other studies have employed satellite data to replace outdated urban land use maps in atmospheric models with new remote sensing products (Cheng and Byun, 2008; Cheng et al., 2013). Focusing on boundary layer mixing conditions and local wind patterns in the Houston Ship Channel, Cheng and Byun (2008) reported that the Noah LSM and planetary boundary layer (PBL) scheme performances in the MM5 were improved when land-use-type distributions were correctly represented in the model using high-resolution Landsat-based land use data. Cheng et al. (2013) compared WRF simulations in the Taiwan area using US Geological Survey (USGS)-, MODIS-, and SPOT (Système Pour l'Observation de la Terre)-based land use data. Using the new high-resolution land use types obtained from SPOT satellite imagery, the WRF predictions of daytime temperatures and onshore sea breezes had the best agreement with observed data. Furthermore, more accurate surface wind speeds were simulated when MODIS and SPOT data replaced conventional USGS land use maps in the WRF runs due to the more realistic representation of roughness length in the remotely sensed databases. Although these and other previous studies (e.g., Jin and Shepherd, 2005) have recognized the usefulness of satellite imagery (e.g., NASA's Terra, Aqua, and Landsat data) in specifying surface physical characteristics in urban environments, very few have directly incorporated high-resolution gridded satellite-based parameters (e.g., impervious surface area, albedo, and emissivity) into parameter estimation within land surface/atmospheric modeling systems.

In the current work we investigate the utility of remote-sensing-based surface parameters in the Noah-UCM modeling framework over a highly developed urban area. Among parameters that can be related to a measurable physical quantity, we evaluate those routinely and freely obtained from satellite-based platforms. The derived parameter sets are implemented in the Noah UCM with a focus on simulated surface energy and water cycles that are essential feedback to the widely used WRF model. Landsat and fused Landsat–MODIS data are utilized to generate high-resolution (30 m) monthly spatial maps of green vegetation fraction (GVF), impervious surface area (ISA), albedo, leaf area index (LAI), and emissivity in the Los Angeles metropolitan area. The temporal and spatial distributions of newly assigned parameters are compared with those based on the model lookup tables. Next, gridded remotely sensed parameter data sets are directly incorporated into the Noah-UCM modeling framework replacing the land-use/lookup-table-based values. The sensitivity of the simulated energy and water fluxes to the newly developed spatial metrics of parameters is presented. The model's performance in reproducing evapotranspiration (ET) and LST fields is evaluated utilizing Landsat-based land surface temperature and ET estimates from CIMIS (California Irrigation Management Information System) stations as well as in situ measurements. Finally, the influence of each parameter set on the urban energy and water budgets is investigated.

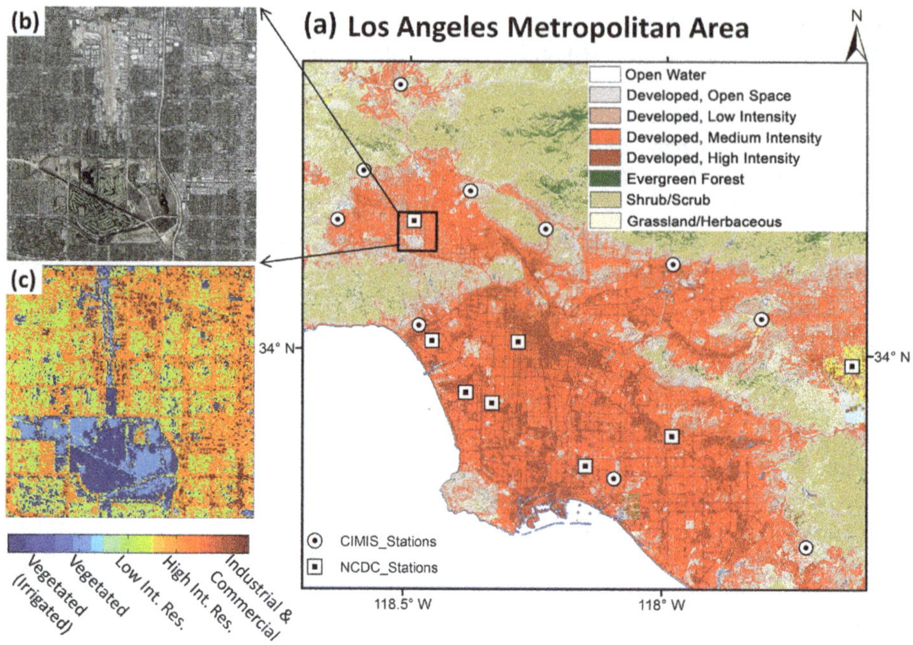

Figure 1. (a) NOAA C-CAP land cover map of the Los Angeles metropolitan area including study domain, 10 CIMIS stations (white circles), and 8 NCDC stations (white squares); **(b)** Google image of the study domain; and **(c)** the Noah-UCM urban land cover classification of the study domain.

2 Study area

The study domain is a 49 km^2 highly developed neighborhood in the city of Los Angeles (Fig. 1). Los Angeles is the second most populous city in the United States with a population of 3.8 million (US Census, 2011), covering an area of 1215 km^2 in southern California. The city has a Mediterranean climate and receives 381 mm of annual precipitation, mostly over the winter months (NOAA-CSC, 2003; SCDWR, 2009). Due to the semiarid nature of the region, the city's water supply is heavily dependent on imported water (52 % from the Colorado River and 36 % from the Los Angeles Aqueduct (Los Angeles Department of Water and Power (LADWP), 2010). Regional water demands and the extensive dependence on external sources make accurate spatial representation of the metropolitan area in regional land surface/atmospheric models imperative for predicting current and future water budgets. The study domain includes commercial/industrial as well as low- and high-intensity residential land cover types and a large park with both irrigated and nonirrigated landscapes (Fig. 1b and c).

3 Remotely sensed parameters

Remote sensing data are retrieved from Landsat ETM+ images with a nominal pixel resolution of 30 m in the shortwave bands and 60 m in the thermal band. The level 1 Gt ETM+ imagery from USGS EROS, spanning years 2010–2011, are calibrated and atmospherically corrected through the Landsat Ecosystem Disturbance Adaptive Processing System (LEDAPS). Study domain data are not affected by the failure of the Landsat-7 ETM+ Scan Line Corrector in 2003 (SLC-off). Using a knowledge-based approach, similar to the one introduced by Song and Civco (2002), several binary masks are applied to the images to detect contaminated areas (cloud and shadow). Images with cloud and/or shadow are distinguished and omitted in the following parameter retrievals. A total of 24 pure images, acquired over 2 years, are utilized in the parameter estimation processes.

In addition to Landsat observations, MODIS products from Terra and Aqua satellite platforms are also utilized. The MODIS MCD43A BRDF (bidirectional reflectance distribution function) products, concurrent with pure Landsat images, are collected for use in the parameter calculations. The 500 m BRDF products are generated by the MODIS Adaptive Processing System (MODAPS) at the Goddard Space Flight Center (GSFC), using a kernel-driven linear model, and distributed through the Land Processes DAAC (Distributed Active Archive Center; Justice et al., 2002; Schaaf et al., 2002; Shuai et al., 2008). The described Landsat and MODIS-based data are used to produce a group of six remotely sensed derivatives, as follows.

Green vegetation fraction (GVF)

GVF spatial maps are derived according to Gutman and Ignatov (1998) utilizing NDVI (normalized difference vegetation index) measurements. First, atmospherically corrected reflectance values from the red (ρ_{ETM3}) and near-infrared (ρ_{ETM4}) bands of Landsat ETM+ are used to derive NDVI maps for each date of imagery based on Eq. (1). Next, assuming the vegetated part of a pixel is covered by dense vegetation (i.e., it has a high LAI), GVF is calculated using Eq. (2).

$$\mathrm{NDVI} = \frac{\rho_{ETM4} - \rho_{ETM3}}{\rho_{ETM4} + \rho_{ETM3}}, \quad (1)$$

$$\mathrm{GVF} = \frac{\mathrm{NDVI} - \mathrm{NDVI_o}}{\mathrm{NDVI_\infty} - \mathrm{NDVI_o}}, \quad (2)$$

where $\mathrm{NDVI_o}$ and $\mathrm{NDVI_\infty}$ are constant values computed using signals from bare soil and densely vegetated pixels in the study domain, respectively.

Impervious surface area (ISA)

ISA is shown to be inversely proportional to vegetation fraction where non-vegetated pervious surfaces are rare (Bauer et al., 2007). Since the majority of pervious surfaces in the studied domain are vegetated and heavily irrigated throughout the year, ISA is assumed to be the complement of the vegetation fraction:

$$\mathrm{ISA} = (1 - \mathrm{GVF_{max}}) \cdot 100, \quad (3)$$

where $\mathrm{GVF_{max}}$ is the maximum GVF detected over the 2-year study period. The produced ISA map shows high accuracy (>95 %) when compared to a previously developed high-resolution land cover map, based on QuickBird remote sensing data, aerial photographs, and geographic information systems over the city of Los Angeles (McPherson et al., 2008). We speculate that one cause that may contribute to the high accuracy of this assumption is that ISA overestimation, induced by non-vegetated pervious surfaces, is offset by tree canopies that cover areas larger than underlying pervious surfaces.

Albedo

With a recent methodology by Shuai et al. (2011) employed, 30 m land surface albedo maps are generated utilizing Landsat surface reflectance and anisotropy information from concurrent 500 m MODIS BRDF products. Landsat data are reprojected from UTM to MODIS sinusoidal projection and aggregated from 30 to 500 m. Using USGS-based land cover types, the percentage of each land cover class within each MODIS pixel is computed, then relatively pure pixels (>85 % purity) are selected for each class. MCD43A2 quality assessment product is used to choose highest-quality MODIS MCD43A1 BRDF parameters for the pure pixels. The concurrent parameters are used to calculate nadir reflectance, white-sky albedo, and black-sky albedo under the solar geometry at Landsat overpass time and MODIS scale. Next, the spectral albedo-to-nadir reflectance ratios, for white-sky and black-sky albedos, are calculated over the pure pixels. The resultant ratios, specific to each land cover class, are applied to Landsat surface reflectance to generate the spectral white-sky and black-sky albedos for each Landsat pixel. A further narrowband-to-broadband conversion based on extensive radiative transfer simulations by Liang (2000) is applied to generate the broadband albedos in a shortwave regime. Finally, albedo (blue sky) is modeled as an interpolation between the black-sky (α_{bs}) and white-sky (α_{ws}) albedos as a function of the fraction of diffuse skylight ($S(\theta,\tau(\lambda))$) which is estimated by the 6S (Second Simulation of the Satellite Signal in the Solar Spectrum) codebase (Eq. 4; Schaaf et al., 2002).

$$\propto(\theta,\lambda) = \{1 - S(\theta,\tau(\lambda))\}\alpha_{bs}(\theta,\lambda) + S(\theta,\tau(\lambda))\alpha_{ws}(\theta\lambda), \quad (4)$$

where τ, θ, and λ are optical depth, solar zenith, and wavelength, respectively.

Leaf area index (LAI)

Stenberg et al. (2004) showed that a reduced simple ratio (RSR) explains 63–75 % of the variations in LAI and that maps of projected LAI, based on RSR, have good agreement with observations. In the current study, LAI values are retrieved based on the LAI–RSR correlations, which are specified utilizing table-based LAI estimates in pure (fully vegetated) pixels and remotely sensed RSR maps. The atmospherically corrected reflectance values of Landsat ETM spectral channels red (ρ_{ETM3}), near infrared (ρ_{ETM4}), and mid-infrared (ρ_{ETM5}), implemented in the following Eq. (5), define RSR:

$$\mathrm{RSR} = \frac{\rho_{ETM4}}{\rho_{ETM3}} \frac{\rho 5_{max} - \rho_{ETM5}}{\rho 5_{max} + \rho 5_{min}}, \quad (5)$$

where $\rho_{ETM5min}$ and $\rho_{ETM5max}$ are the smallest and largest mid-infrared reflectance detected in the Landsat ETM images over the study domain, excluding open water pixels.

Emissivity

Among various methods developed to define land surface emissivity, the NDVI thresholds method ($\mathrm{NDVI^{THM}}$) has been widely applied to urban areas (Stathopoulou and Cartalis, 2007; Stathopoulou et al., 2007; Tan and Li, 2013). $\mathrm{NDVI^{THM}}$ is superior to other methods since the consideration of the internal reflections (cavity effects), caused by heterogeneous surfaces minimizes the overall error in this approach (Sobrino et al., 2001). This methodology, originally introduced by Sobrino and Raissouni (2000) and modified later by Stathopoulou et al. (2007) for urban areas, is selected for land surface emissivity estimation in the current work. Using the Landsat-based NDVI thresholds, the study area is divided into four classes: (1) fully vegetated (NDVI >0.5), (2) built-up areas with sparse vegetation (NDVI $\leq$ 0.2), (3) a mixture of man-made material and vegetation (NDVI >0.2 and ≤ 0.5), and (4) water bodies (NDVI <0). Mean emissivity values of 0.980, 0.920, and 0.995 are then used for fully vegetated, built-up, and water pixels (Similar to Tan and Li, 2013). Emissivity values (ε) for mixed pixels (class 3) are estimated using the following equations (for details see Stathopoulou et al., 2007):

$$\varepsilon = 0.017 P_V + 0.963, \tag{6}$$

$$P_V = \frac{(\mathrm{NDVI} - 0.2)^2}{(0.5 - 0.2)^2}. \tag{7}$$

Land surface temperature (LST)

The emissivity-corrected land surface temperature (LST) is calculated as follows (Artis and Carnahan, 1982):

$$\mathrm{LST} = \frac{\mathrm{BT}}{\left\{1 + \left[\frac{\lambda \cdot \mathrm{BT}}{\rho} \cdot \ln \varepsilon\right]\right\}}, \tag{8}$$

where BT is Landsat at sensor brightness temperature (K), λ and ε are the wavelength of emitted radiance (11.5 μm) and surface emissivity, and $\rho = \mathrm{hc}/\sigma$ (1.438×10^{-2} m K), where σ, h, and c are Boltzmann constant, Planck's constant, and the velocity of light, respectively.

4 Numerical modeling system

4.1 Noah LSM–UCM model

Land surface processes are parameterized using the offline Noah LSM (Chen and Dudhia, 2001) coupled with the single-layer UCM (Kusaka et al. 2001; Kusaka and Kimura, 2004). The Noah LSM is based on a diurnally dependant Penman potential evaporation approach, a multilayer soil parameterization, a canopy resistance model, surface hydrology, and frozen ground physics (Chen et al., 1996, 1997; Chen and Dudhia, 2001; Ek et al., 2003). The UCM parameterization includes urban building geometry, shadowing from buildings, reflections and trapping of radiation in a street canyon, and an exponential wind profile. The Noah LSM provides surface sensible and latent heat fluxes and surface skin temperature for vegetated areas (e.g., parks and trees), and the UCM calculates the fluxes for impervious surfaces. The outputs from the Noah LSM and UCM are coupled through the urban surface fractions.

4.2 Irrigation module

Irrigation is accounted for in the Noah-UCM modeling framework by incorporating an urban irrigation module developed in our previous work (Vahmani and Hogue, 2013, 2014). The developed irrigation scheme mimics the effects of urban irrigation by increasing soil moisture content in vegetated portion of grid pixels at a selected interval. Added anthropogenic soil moisture contribution is a function of the soil moisture deficit, which is the difference between irrigated soil moisture content and actual soil moisture content in the top soil layer. The irrigation module calculates irrigated soil moisture content ($\mathrm{SMC_{IRR}}$; m^3 m^{-3}), soil moisture deficit (DEF; m^3 m^{-3}), and irrigation water (IRR; kg m^{-2} s^{-1}) as

$$\mathrm{SMC_{IRR}} = \alpha \mathrm{SMC_{max}}, \tag{9}$$

$$\mathrm{DEF} = \max\{[\mathrm{SMC_{IRR}} - \mathrm{SMC_1}], 0\}, \tag{10}$$

$$\mathrm{IRR} = \frac{\rho_\mathrm{w}}{\Delta t} \mathrm{DEFD_1}, \tag{11}$$

where the saturation soil moisture content ($\mathrm{SMC_{max}}$; m^3 m^{-3}) and irrigation demand factor (α; unitless) define irrigated soil moisture content (Eq. 9); and $\mathrm{D_1}$ is top soil layer thickness (10 cm); ρ_w (kg m^{-3}) and Δt stand for water density and Noah-UCM time step (3600 s), respectively. The parameter α, ranging from zero to one, regulates the amount of irrigation water added to the soil each time the scheme increases the soil moisture, simulating an irrigation event. Similar to previous studies (Hanasaki et al. 2008a, b; Pokhrel et al. 2012) an irrigation demand factor of 0.75 is utilized in the current work. The irrigation interval is set to three times per week according to the water restrictions implemented by LADWP in 2010.

4.3 Improving the UCM-simulated LST

The calculation of the impervious surface temperature in the UCM version used in this study has been shown to be inaccurate (Li and Bou-Zeid, 2014). This is due to the fact that the turbulent transfer coefficient (C_h) for the whole pixel is calculated using only momentum and thermal roughness lengths of vegetated portion, ignoring the developed surface impact on C_h. Li and Bou-Zeid (2014) showed that this inconsistency could result in large biases in simulated LST values. In the current study, an alternative LST calculation, proposed by Li and Bou-Zeid (2014), is used as follows. First, a revised surface temperature of the impervious part of the pixel ($\mathrm{T_s}$) is calculated based on canyon temperature ($\mathrm{T_c}$) and roof surface temperature (T_r):

$$T_\mathrm{s} = f_\mathrm{r} \times T_\mathrm{r} + (1 - f_\mathrm{r}) \times T_\mathrm{c}, \tag{12}$$

where f_r is the roof fraction of the impervious surface. Note that the T_c calculated by the UCM is an equivalent aerodynamic surface temperature aggregated for canyon surfaces, including walls and roads. Next, the LST for the whole grid cell is computed as a weighted average based on the T_s and surface temperature of pervious part (T_1):

$$\mathrm{LST} = f_\mathrm{urb} \times T_\mathrm{s} + (1 - f_\mathrm{urb}) \times T_1, \tag{13}$$

where f_urb is the urban fraction of the pixel.

4.4 Land cover data and forcing fields

The Noah-UCM modeling system requires static data to describe physical characteristics of the surface, including soil

Table 1. Model scenarios (1–9) and the incorporated remotely sensed parameter sets.

	GVF	ISA	Albedo	LAI	Emissivity
Scenario 1 (def. par.)	–	–	–	–	–
Scenario 2 (R.S. GVF)	X	–	–	–	–
Scenario 3 (R.S. ISA)	–	X	–	–	–
Scenario 4 (R.S. albedo)	–	–	X	–	–
Scenario 5 (R.S. LAI)	–	–	–	X	–
Scenario 6 (R.S. emissivity)	–	–	–	–	X
Scenario 7 (R.S. par. group 1)	X	X	–	–	–
Scenario 8 (R.S. par. group 2)	–	–	X	X	X
Scenario 9 (R.S. par. all)	X	X	X	X	X

type, slope type, vegetation type, and urban type. A combination of the Soil Data Mart (http://sdmdataaccess.nrcs.usda.gov/) and the Los Angeles Department of Public Works (LADPW) databases are used to gather soil classification information. Land use and land cover are parameterized using the 30 m NOAA C-CAP-2006 land cover data which is transformed to urban and vegetation type spatial maps over the study domain. High-, medium-, and low-intensity developed land cover types, recognized by NOAA, are converted to UCM industrial/commercial, high-intensity, and low-intensity residential types, respectively. The developed open space, along with natural land types, is categorized as 1 of the 27 Noah LSM vegetation classes.

The offline Noah LSM–UCM is forced utilizing hourly ground-based observations from CIMIS and National Climatic Data Center (NCDC) stations for the period from 1 January 2010 to 31 December 2011. There are 10 CIMIS and 8 NCDC stations within close proximity of the study domain (Fig. 1a). The NCDC stations, which use Automated Surface Observing Systems (ASOS), are located at smaller local airports (six stations), one major airport (Los Angeles International Airport), and a university campus (University of Southern California, USC) within the Los Angeles metropolitan area. Reporting the meteorological conditions, the NCDC stations are used for wind speed, air temperature, relative humidity, air pressure, and incoming longwave radiation. All NCDC data are gathered at a standard reference height of 2 m. The regional CIMIS stations are utilized for solar radiation (using LI200S pyranometer) and tipping bucket rain gauges in 18 stations (NCDC and CIMIS) are included in collection of precipitation data. Inverse-distance weighting (second power) is employed to create the spatially gridded forcing fields. Linear interpolation and data from the nearest gage are utilized to replace missing data.

5 Numerical experiments and evaluation methods

5.1 Remote-sensing-based parameterization

To investigate the sensitivity of the Noah-UCM model to integration of the developed remotely sensed parameters, nine simulation scenarios are designed (Table 1). A control experiment (scenario 1) is conducted in which all default parameters are utilized in the Noah UCM. Scenarios 2 to 6 explicitly assess each individual parameter effects on urban energy and water budgets using the newly incorporated remote sensing parameters. Scenario 7 analyzes the effects of employing both remotely sensed GVF and ISA, while scenario 8 assesses simultaneous integration of albedo, LAI, and emissivity. We are interested in the comparison of scenarios 7 and 8 as the Noah-UCM parameterizations use GVF and ISA to select albedo, LAI, emissivity, and roughness length values from the predefined ranges in the parameter tables. It is worth mentioning that GVF alters the roughness length values over pervious or natural areas. However, roughness length and building height over the impervious surfaces are kept at the default values listed by Chen et al. (2011). scenarios 7 and 8 help quantify the contribution of each parameter group to the model's ability to reproduce the observed surface states and fluxes. Finally, the last experiment (scenario 9) implements all five remotely sensed parameter sets in the simulations. It should be noted that the GVF and LAI measurements over mixed pixels (vegetated urban areas) are scaled up by multiplying the remotely sensed values by 1/(1–urban fraction) since in the Noah-UCM modeling framework these parameters characterize only the pervious portion (1–urban fraction) of each pixel. However, remotely sensed albedo and emissivity values over each pixel are assigned to both pervious and impervious surfaces for that pixel. Other than the implemented remote-sensing-based parameters, the rest of the model parameters are kept at default values. All experiments incorporate the irrigation module and irrigation rates are kept constant in all scenarios. All scenarios are run at 30 m spatial and 1 h temporal resolutions, spanning 2010 and 2011, with the first 3 months used as model initialization.

5.2 Model evaluation approach

In order to evaluate the performance of the Noah-UCM modeling framework, simulated LSTs are compared with concurrent Landsat observations and simulated latent heat flux time series are assessed against CIMIS-based ET observations. The CIMIS network was established in 1982 by the CDWR (California Department of Water Resources) and the University of California at Davis in order to provide real-time weather conditions and irrigation water need estimates for California's agricultural community. The automated CIMIS stations measure hourly surface solar radiation, temperature, humidity, wind, precipitation, soil temperature, and surface pressure (http://wwwcimis.water.ca.gov). The reference ET (ET_0) is calculated for each site, utilizing observed meteorological fields over a well-watered soil. Utilizing a methodology introduced by CDWR (2000), actual urban landscape ET is estimated using ET_0 and a landscape coefficient, which is a function of species, density, and microclimate factors. Based on the authors' knowledge in the study landscape as well as a report by CDWR (2000), we assume "moderate" (trees and shrubs) and "high" (turf grass) water needs. Following the CDWR (2000) instructions on irrigation zones with mixed water need categories (i.e., low, moderate, and high), a value from the high category is selected (average species factor = 0.80). Assuming the "average" category for vegetation density, a density factor of 1 is used. Furthermore, a "high" category of microclimate condition is used (microclimate factor = 1.25) for the current highly developed study domain. This factor is utilized to take into account the contribution of the developed surfaces to the water loss from vegetated areas, through anthropogenic heating, reflected light, and high temperatures of surrounding heat-absorbing surfaces (e.g., paving and buildings). Using these factors, a landscape coefficient of 1 (landscape coefficient = species factor × density factor × microclimate factor) is prescribed. This coefficient and ET_0 estimations from 10 CIMIS stations within close proximity of the study domain (Fig. 1a) are utilized to compute the urban landscape ET. Inverse-distance weighting (second power) is employed to create spatially gridded ET maps over fully vegetated pixels in the study area which are then used in validation processes of the Noah UCM. ET output of the model is also evaluated against recent ET measurements in the greater Los Angeles area (Moering, 2011). Moering (2011) employed a previously developed chamber approach to measure instantaneous ET in an irrigated and a nonirrigated park in the Los Angeles metropolitan area during WY (water year) 2011 (WY is defined as 1 October of the previous year to 30 September of the designated year). They reported an annual ET of about 1224 mm over the observed irrigated park, which is located within our study domain.

6 Sensitivity study of surface parameter

6.1 Temporal evaluation

The monthly time series of the default Noah UCM and remote-sensing-based GVF, ISA, albedo, and LAI are compared and modeled cumulative monthly sensible and latent heat fluxes, using default and newly estimated parameters, are presented over fully vegetated, low-intensity residential, and industrial/commercial areas (Fig. 2). Fluxes from high-intensity residential areas are not presented as they behave similarly to those from the industrial/commercial areas. Except for the summer months, GVF values are significantly increased throughout the year when remote sensing products are utilized (Fig. 2a). Moreover, the default seasonal variations of GVF values, assumed over all the land cover types, are not detected in Landsat imagery (Fig. 2a). The reason for this is the significant and year-round irrigation in the Los Angeles area, which is not accounted for in the default parameter tables. This is confirmed by previous studies (Johnson and Belitz, 2012) that reported urban vegetation supported by water delivery, in contrast to common seasonal behavior of greening in the winter/spring and browning in the summer, maintains constant greenness, which is reflected in NDVI and GVF estimates. GVF plays a dominant role in the Noah-UCM simulations as it defines the vegetated fraction of the natural areas, and specifies albedo, LAI, emissivity, and roughness length values from the predefined ranges in the model lookup tables. Furthermore, GVF partitions the total ET between soil direct and canopy ET. The simulated latent heat flux is considerably decreased (up to $139\,\mathrm{MJ\,m^{-2}\,month^{-1}}$) in the summertime and increased over the remaining months, when remotely sensed GVF is incorporated in the fully vegetated areas (Fig. 2b). Since any increase of latent heat flux that does not alter the radiative balance leads to a reduction in sensible flux, the newly developed GVF values, in turn, cause enhancements (up to $103\,\mathrm{MJ\,m^{-2}\,month^{-1}}$) in the simulated summer sensible heat fluxes and a reduction in the sensible heat fluxes during the remaining months (Fig. 2b). Latent and sensible heat fluxes from the low-intensity residential pixels show similar but less significant changes (up to 66.1 and $31.0\,\mathrm{MJ\,m^{-2}\,month^{-1}}$, respectively) when the new parameter sets are implemented. Adding remotely sensed GVF causes insignificant changes in the industrial/commercial area fluxes due to the small percentage of vegetated land cover in such areas (Fig. 2d).

There are also large deviations between the lookup-table-based ISAs and the remotely sensed values. Averaged ISA is decreased (10 %) over industrial and commercial pixels and increased (49 %) over low-intensity residential areas when remote sensing products are utilized in the parameter estimation process (Fig. 2e). These changes in the impervious surface area, or urban fraction values, have significant effects on monthly latent and sensible heat fluxes over the developed pixels (Fig. 2g and h), due to the critical role

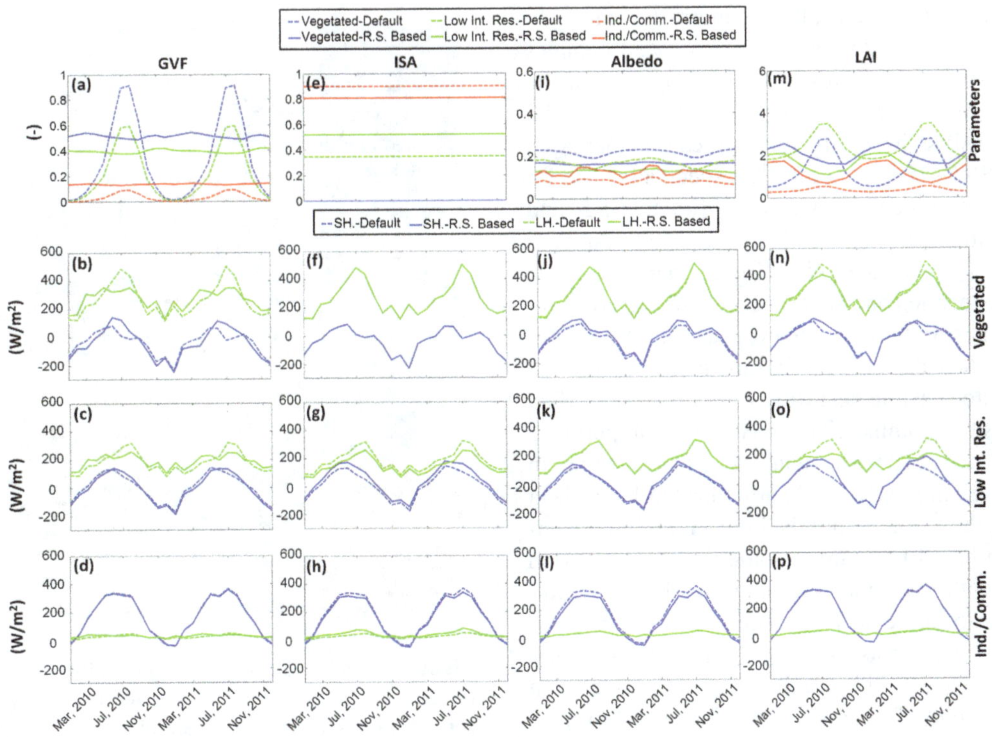

Figure 2. Monthly time series of default Noah UCM compared with remote-sensing-based GVF, ISA, albedo, and LAI (row 1), as well as modeled cumulative monthly sensible and latent heat fluxes ($\mathrm{MJ\,m^{-2}}$) over fully vegetated (row 2), low-intensity residential (row 3), and industrial/commercial areas (row 4) using the default and newly estimated parameters: **(b–d)** GVF, **(f–h)** ISA, **(j–l)** albedo, and **(n–p)** LAI.

of urban fraction in partitioning of the energy fluxes. Over the low-intensity residential areas, higher ISA values minimize the effects of urban vegetation, which leads to latent heat fluxes decreases (up to $62.6\,\mathrm{MJ\,m^{-2}\,month^{-1}}$) and sensible heat fluxes increases (up to $52.4\,\mathrm{MJ\,m^{-2}\,month^{-1}}$), throughout the year, when remotely sensed data replace default urban fractions (Fig. 2g). These changes are reversed and less significant over the industrial and commercial pixels (maximum latent and sensible heat flux changes of 30.0 and $26.5\,\mathrm{MJ\,m^{-2}\,month^{-1}}$, respectively; Fig. 2h). ISA has no influence on the fluxes from fully vegetated pixels which do not include impervious areas (Fig. 2f).

Considerable changes in the monthly albedo averages are detected when incorporating remote sensing data in the parameterization process (Fig. 2i). With the use of fused Landsat and MODIS products, a reduction of averaged albedo values is observed over the fully vegetated and residential areas (up to 48 and 39 %, respectively; Fig. 2i). Moreover, the default seasonal variations are hardly noticeable in the remote-sensing-based albedo values, which is due to the consistent greenness in the study area from irrigation throughout the year. On the other hand, considerable albedo increases (up to 39 %) are detectable over the industrial/commercial pixels (Fig. 2i), which are caused by bright and highly reflective materials seen mainly over the rooftops of industrial/commercial buildings. Albedo affects the radiative energy budget and consequently available energy for the turbulent fluxes. In the current study, decreased albedo values over the fully vegetated and low-intensity residential areas result in reduced loss of solar and longwave radiation, respectively, and, in turn, increases the sensible heat flux (up to 33.8 and $21.5\,\mathrm{MJ\,m^{-2}\,month^{-1}}$; Fig. 2j and k). Albedo-induced sensible heat deceases over industrial/commercial pixels are also noticeable (up to $33.9\,\mathrm{MJ\,m^{-2}\,month^{-1}}$; Fig. 2l).

Distinct seasonal fluctuations of LAI are observed in the remotely sensed data and the default parameter tables (Fig. 2m). This reflects the fact that landscape plantings are quite different from agricultural crops due to their being composed of collections of vegetation species and affected by complex irrigation patterns which are not taken into account in the vegetation parameter tables in the Noah LSM (CDWR, 2000; Vahmani and Hogue, 2013, 2014). Over the heavily vegetated pixels, the default pattern is reversed in the measured parameter sets, with less seasonal variations and peaks in the wintertime, due to the fact that most of the precipitation occurs in the winter months, over the current study domain (Fig. 2m). The industrial and commercial pixels illustrate higher LAI values in the remotely sensed parameter maps, year-round, when compared to the default values (Fig. 2m). LAI is a critical parameter in the Noah LSM, which is involved in the parameterization of the canopy resistance, controlling canopy ET rates. In the presented results (Fig. 2n and o), LAI-induced changes in the simulated turbulent fluxes are more apparent in the summer months and over fully vegetated and residential pixels, where sensible heat flux is significantly increased (up to 57.2 and 86.5 $\mathrm{MJ\,m^{-2}\,month^{-1}}$, respectively) and latent heat flux is significantly decreased (up to 65.5 and 97.9 $\mathrm{MJ\,m^{-2}\,month^{-1}}$, respectively). This is due to the considerable deceases in the LAI values in summertime which lead to elevations of the canopy resistance and therefore reductions of the transpiration from the vegetation, causing decreases in latent heat fluxes. This in turn partitions the net radiation more into sensible heat fluxes. LAI does not affect fluxes from industrial/commercial pixels with small pervious fractions (Fig. 2p). It is worth mentioning that changes in the turbulent fluxes time series, in particular the latent heat flux decreases in the summer months induced by implementation of satellite-based LAI, are to some extent captured in the simulations with the remote-sensing-based GVF (compare Fig. 2b with n and c with o). This reflects our previous point that GVF controls assigned LAI values to vegetated pixels in the Noah LSM and that realistic presentation of GVF in the modeling framework can enhance LAI inputs in the model when LAI measurements are not available.

Remotely sensed emissivity maps are also utilized to replace the default values in the Noah-UCM simulations, which results in changes in the emissivity values (up to 5.1 %). However, the new surface parameterization leads to insignificant changes in turbulent fluxes (results now shown). The largest emissivity induced alterations in sensible heat fluxes are seen over industrial and commercial pixels (up to 31.2 $\mathrm{MJ\,m^{-2}\,month^{-1}}$). Latent heat fluxes are changed most significantly over fully vegetated areas (up to 2.56 $\mathrm{MJ\,m^{-2}\,month^{-1}}$).

6.2 Spatial evaluation

The spatial distributions of newly assigned GVF, ISA, albedo, and LAI are next compared with those based on the Noah-UCM lookup tables. Different urban surface parameterizations, along with their impacts on the simulated maps of turbulent sensible and latent heat fluxes, are presented (Fig. 3; valid at 11:00 LST on 14 April 2011). As expected, during the spring period (April), GVF values are significantly higher when remote sensing products are utilized, due to the irrigation effects which are ignored in the default parameters (Fig. 3a and b). Over fully vegetated and low-intensity residential pixels, where a significant portion of the energy goes into evaporation and transpiration, latent heat flux increases (about 300 and 230 $\mathrm{W\,m^{-2}}$, respectively) and sensible heat fluxes decreases (about 160 and 120 $\mathrm{W\,m^{-2}}$, respectively) are found (Fig. 3c and d) when utilizing the remote sensing GVF.

The spatial distributions of ISA, or urban fraction, between the remote sensing and default values show similar patterns

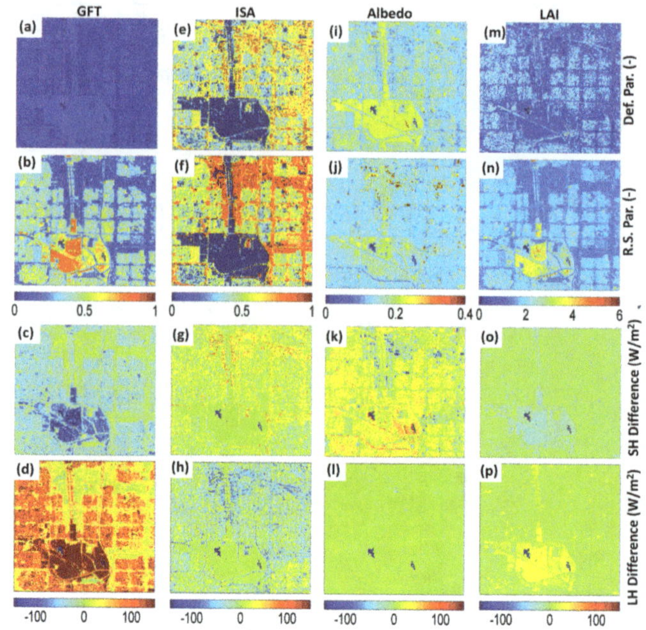

Figure 3. Spatial distributions of GVF, ISA, albedo, and LAI based on Noah-UCM lookup tables (row 1) compared with remotely sensed values (row 2) and simulated difference maps of sensible (row 3) and latent (row 4) heat fluxes using default and remotely sensed urban surface parameters: (**c** and **d**) GVF, (**g** and **h**) ISA, (**k** and **l**) albedo, and (**o** and **p**) LAI. Valid at 11:00 LST on 14 April 2011.

(Fig. 3e and f). However, industrial/commercial and high-intensity residential areas are assigned noticeably higher urban fraction values in the remote-sensing-based maps (compare Fig. 3e with f), which leads to lower latent heat fluxes (bias of up to about 130 $\mathrm{W\,m^{-2}}$) and higher sensible (bias of up to about 100 $\mathrm{W\,m^{-2}}$) in these pixels (Fig. 3g and h).

The Noah-UCM parameters, based on lookup tables, underestimate surface albedo values over highly urbanized pixels when compared with remote sensing data (Fig. 3i and j). In particular, the industrial/commercial buildings with highly reflective rooftops are completely ignored. Over the highly vegetated areas, however, albedo values are slightly overestimated in lookup tables. With the energy budget altered, the newly developed albedo data sets lead to lower Noah-UCM-simulated sensible heat fluxes over intensely developed pixels (Fig. 3k). The sensible heat flux differences are only significant over industrial/commercial pixels, which include buildings with bright roofs (up to $\sim$ 300 $\mathrm{W\,m^{-2}}$). The changes in absolute surface albedos do not affect simulated latent heat fluxes as these reflective roofs are located in industrial/commercial areas with negligible pervious surfaces and simulated latent heat flux (Fig. 3l).

The remote sensing data detect higher LAI values over all pixel types, particularly over fully vegetated areas, where new LAI values are significantly higher (Fig. 3m and n). By influencing the canopy resistance, these changes redefine the spatial distribution of turbulent fluxes (Fig. 3o and p). Over the densely vegetated areas, increases in latent heat flux (up to $50\,\mathrm{W\,m^{-2}}$) and decreases in sensible heat flux (up to $35\,\mathrm{W\,m^{-2}}$) are found (Fig. 3o and p). It is noteworthy that, as illustrated before (Fig. 3n and o), the most significant influences of LAI alterations are detected in the summer months. Thus, it is not surprising that the turbulent fluxes do not show significant sensitivity to the LAI changes in April.

Remotely sensed emissivity maps, implemented in the Noah-UCM simulations, show minimal effect on the output turbulent fluxes maps (results not shown). Our results (Figs. 2 and 3) agree with previous sensitivity studies performed with the Noah UCM, which indicated high sensitivity of the model to GVF, ISA, albedo, and LAI, and less model sensitivity to emissivity (Loridan et al., 2010; Wang et al., 2011). Loridan et al. (2010) highlighted the critical role of ISA and LAI in the simulations of latent heat flux and albedo role in the sensible heat flux simulations. Investigating the peaks of diurnal turbulent fluxes, Wang et al. (2011) reported that latent heat flux is the most sensitive to the GVF. They also found that emissivity has minimal effects on the model outputs.

7 Evaluation of Noah-UCM performance

After initial sensitivity tests, the model performance in reproducing ET and LST fields is evaluated using remotely sensed (independent from derived parameters) and in situ measurements. The comparisons of observed and simulated ET and LST, using different urban surface parameterizations (scenarios 1, 7, 8, and 9 in Table 1), are presented in Figs. 4, 5, and 6.

7.1 ET simulations

The temporal variations of ET, simulated by the Noah-UCM model and averaged over fully vegetated pixels, are evaluated against CIMIS-based ET measurements, spanning 2010 and 2011 (Fig. 4). The presented observations are averages over fully vegetated pixels in the study domain, calculated using ET maps based on ET_0 measurements from 10 CIMIS stations, landscape coefficients, and inverse-distance weighting (second power; see Sect. 5.2). The model reproduces similar ET behaviors when the default parameters and the second group of remotely sensed parameters (albedo, LAI, and emissivity) are implemented (Fig. 4a and c). The ET differences between observations and the default simulation are minimal in the winter and fall months, due to the limited energy available for ET in those months. Over the warmer months, the observed and modeled ETs show distinct behaviors. CIMIS stations report two peaks: one in the spring and one in the summertime. Simulated ETs, however, illustrate one peak in July. The Noah UCM, using these parameterizations, underestimates ET rates for the most of winter and spring months and overestimates them in the summertime (Fig. 4a and c). Including remotely sensed albedo, LAI, and emissivity does

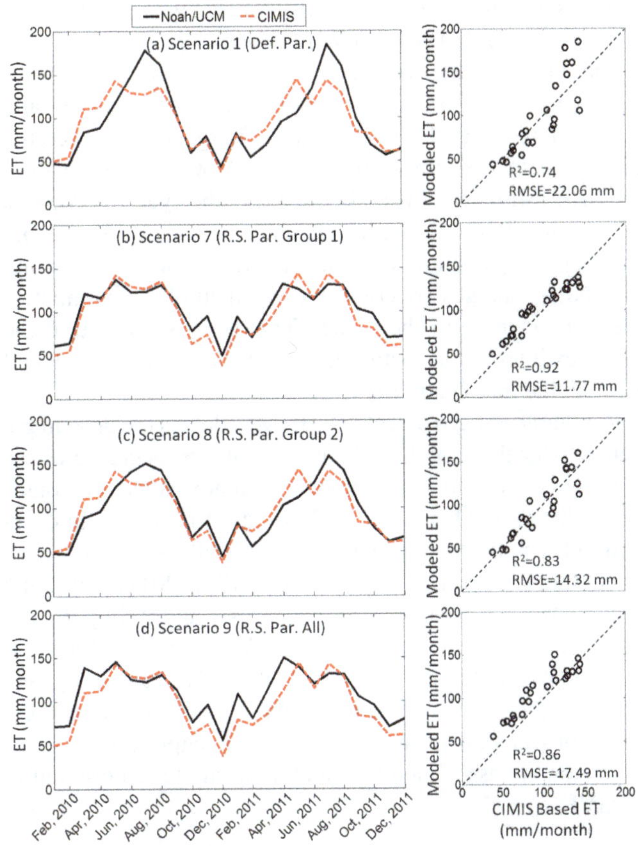

Figure 4. Noah-UCM-simulated cumulative monthly ET, averaged over fully vegetated pixels using different urban surface parameterizations: scenarios **(a)** 1, **(b)** 7, **(c)** 8, and **(d)** 9 in the Table 1 and their comparisons with CIMIS-based ET measurements spanning 2010 and 2011. Scatter plots of these comparisons are also included (right).

not change the general seasonal pattern deviations of ET (Fig. 4a and c), but it reduces the biases considerably (with $R^2 = 0.83$ and RMSE $= 14.32\,\mathrm{mm\,month^{-1}}$). We note that model improvement is mostly associated with inclusion of remotely sensed LAI maps in the model since albedo and emissivity have minimal influence on latent heat fluxes from heavily vegetated pixels (see Fig. 2j).

The new GVF and ISA values alter ET seasonal fluctuations significantly in scenario 7 (Fig. 4b). In agreement with CIMIS observations, the model with inclusion of remotely sensed parameters results in significantly higher ET values in the warming months (February–May) and lower ETs in the summertime. With it noted that ISA has minimal effects over the fully vegetated pixels, one explanation for this pattern is that higher green vegetation fraction detected by Landsat in late winter and early spring increases transpiration rates as soon as the required energy is available and lower measured GVFs in the summertime suppress the transpiration rates, resulting in the lower ET values. These changes enhance the model performance significantly (with $R^2 = 0.92$ and RMSE = 11.77 mm month^{-1}).

Including all the measured parameter sets (Fig. 4d) reduces the behavioral disagreements between observed and modeled monthly ET ($R^2 = 0.86$). Large biases over the summer months are also reduced. However, ET values are overestimated over the rest of the year (RMSE = 17.49 mm month^{-1}). Although each newly developed parameter group enhances the model performance in predicting ET, the advantages are countered when all of the parameters are implemented in the model. This is possibly due to the complex interactions between the parameters (e.g., GVF and LAI) in the model structure.

A notable pattern detected by CIMIS data is the drop in ET values over the month of June. The sudden decrease in ET corresponds to the "June gloom" weather pattern in southern California, when onshore flows result in persistent overcast skies with cool temperatures, as well as fog and drizzle in late spring and early summer (NWS, 2011). The June gloom effects are captured in scenarios 7 and 9 (Fig. 4b and d) and not seen in scenarios 1 and 8 (Fig. 4a and c). Since ISA has minimal influence on ET from the fully vegetated pixels and the second group fails to simulate June gloom influence, the improvements in scenarios 7 and 9 in capturing this phenomenon are associated with a more accurate representation of GVF.

7.2 LST simulations

In order to further evaluate model performance and examine the impacts of different remote-sensing-based parameter sets, Landsat-based LST measurements are utilized (Figs. 5 and 6). Statistics (R^2 and RMSE) are also included to quantify the model performance using different urban surface parameterizations (Fig. 5). The observed LSTs over fully vegetated pixels are estimated with fair accuracy by the default model ($R^2 = 0.86$ and RMSE = 3.21 °C; Fig. 5a). The model performance has almost the same level of accuracy over low-intensity residential areas and is slightly worse (< 1 °C) over industrial/commercial pixels (Fig. 5e). Using remote sensing data over fully vegetated and low-intensity residential pixels weakly improves the biases (with < 1 °C improvement; Fig. 5b–d and f–h). Over industrial/commercial areas, a systematic underestimation of the observed LST is identified (RMSE = 3.96 – 4.59 °C; Fig. 5i–l) which seems to be persistent after using different remotely sensed parameter sets. We speculate that this underestimation of LST over highly developed areas is due to lack of representation of anthropogenic heating in the current study.

A comparison of LST at 11:00 LST on 14 April 2011 with four simulation cases is also presented (Fig. 6). Alterations due to use of remote sensing products are more noticeable in this spatial examination of the results. Using all the default parameters (scenario 1), observed LST is overestimated over

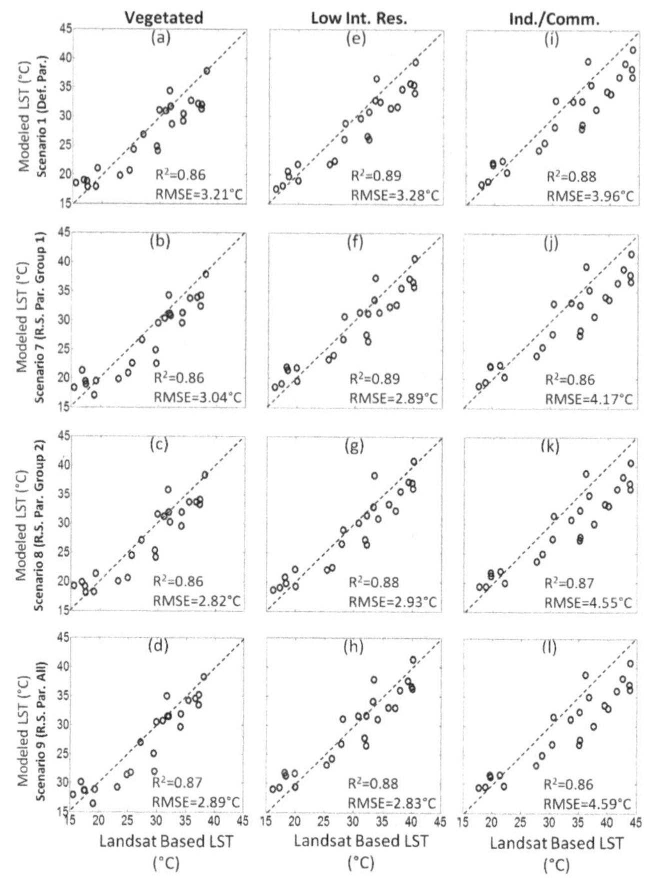

Figure 5. Scatter plots of observed (Landsat-based) versus simulated LSTs averaged over different land cover types using different urban surface parameterizations, including scenarios 1 (row 1), 7 (row 2), 8 (row 3), and 9 (row 4) in Table 1.

the heavily vegetated areas and underestimated over highly developed pixels (Fig. 6a and b).

Remotely sensed GVF and ISA (in scenario 7) significantly decrease LSTs over fully vegetated and low-intensity residential pixels and increase temperatures over industrial and commercial areas, resulting in a better match with the observed LST map. The decreased simulated surface temperatures over heavily vegetated areas is due to higher GVF and consequently higher ET rates, which in turn lead to lower sensible heat flux and LSTs (see Fig. 3b). The increased LSTs over highly developed areas is likely due to lower GVF and higher ISA values detected in Landsat imagery compared with the default values, which partition net radiation more into sensible heat flux (see Fig. 3b and f). The noticed changes in LST maps are small, when using remotely sensed albedo, LAI, and emissivity in scenario 8 (compare Fig. 6a with e). Although simulated LSTs over fully vegetated areas are decreased, the observed temperatures are still overestimated (Fig. 6f). The LST decreases in scenario 8 may be explained by evaporative cooling effect of the higher LAI values over heavily vegetated areas (see Fig. 3n). Similar to

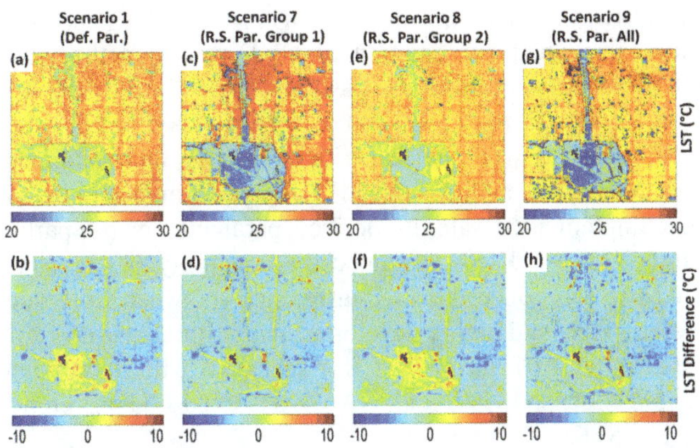

Figure 6. Noah-UCM-simulated LST maps using different urban surface parameterizations: scenarios 1, 7, 8, and 9 from Table 1 (top row) as well as differences between simulated and observed land surface temperature at 1100 LST on 14 April 2011 (bottom row).

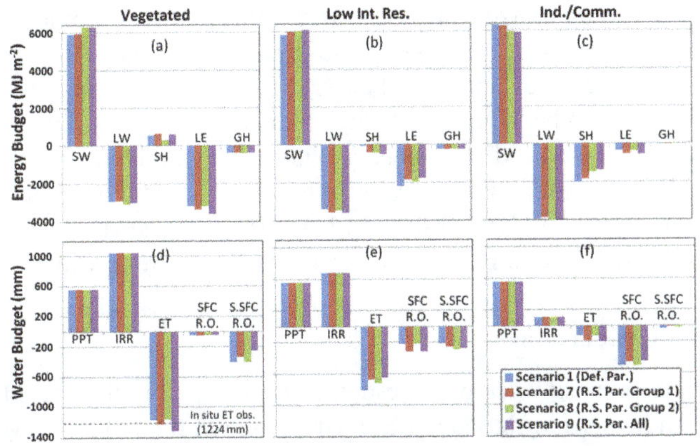

Figure 7. Differences in simulated energy (top) and water (bottom) budgets for WY 2011, using different urban surface parameterization and averaged over different land cover types. Energy budget terms include shortwave radiation (SW) and longwave radiation (LW), as well as sensible (SH), latent (LH), and ground (GH) heat fluxes. Water budget terms include precipitation (PPT), irrigation water (IRR), evapotranspiration (ET), surface runoff (SFC R.O.), and subsurface runoff (S. SFC R.O.).

scenario 7, considerable GVF-induced LST reductions over fully vegetated areas improve the observed LST estimations in scenario 9 (Fig. 6h). Our assessment indicates that implemented satellite-derived parameter maps, particularly GVF and ISA used in scenarios 7 and 9, enhance the Noah-UCM capability to reproduce the LST differences between fully vegetated pixels and highly developed areas (simulated LST differences of 3.07, 6.78, 3.48, and 7.30 °C for scenarios 1, 7, 8, and 9 vs. observed LST difference of 11.25 °C).

7.3 Energy and water budget evaluation

Differences in the simulated energy and water budgets, with different surface parameterizations (scenarios 1, 7, 8, and 9 in the Table 1) are summarized for WY 2011 (Fig. 7). The emissivity induced changes to the energy and water budgets are insignificant and not included. The illustrated radiative and turbulent heat fluxes show that, unlike the longwave radiative fluxes, the simulated available solar radiations are altered considerably using different urban parameter sets (up to 6 %), particularly over fully vegetated (Fig. 7a) and industrial/commercial pixels (Fig. 7c). These changes are induced by new surface albedo values utilized in scenarios 8 and 9. It is also observed that most of the incoming radiative energy is dissipated through latent heat fluxes, over heavily vegetated pixels (Fig. 7a and b), and sensible heat fluxes over industrial/commercial areas (Fig. 7c). These turbulent fluxes are also altered when different surface parameterizations are incorporated. Implementing all the remotely sensed parameters (scenario 9), the annual latent heat flux is increased (12 %) over fully vegetated pixels (Fig. 7a), and the annual sensible heat flux is decreased (32 %) over industrial/commercial pixels (Fig. 7c). Ground heat fluxes, however, are insignificant and unchanged.

Water budget terms also show variable behavior using different parameter sets over different land cover types (Fig. 7d–f). Annual irrigation amounts exceed received precipitations over the pixels with significant vegetation fractions (Fig. 7d and e). This pattern is not rare in semiarid regions (CDWR, 1975; Mini et al., 2014). In these areas, most of incoming water is lost through ET (Fig. 7d and e). Areas with high coverage of impervious surfaces, however, dissipate most of the incoming moisture through surface runoff (Fig. 7f). The alterations in the annual ET rates are, for the most part, due to the changes in the GVF parameterizations (scenarios 7 and 9; Fig. 7d–f). Sub-surface runoff annual rates, on the other hand, are altered using new ISA values (scenarios 7 and 9; Fig. 7e and f). Changes in the annual ET values are as large as 145, 156, and 79.4 mm over fully vegetated, low-intensity residential, and industrial/commercial pixels, respectively (Fig. 7d–f).

To further verify the capability of Noah UCM to reproduce observed ET quantities, additional evaluation of the model is conducted utilizing ground-based chamber ET measurements in the greater Los Angeles area (Moering, 2011). Instantaneous ET measurements over an irrigated park in the study domain during WY 2011 are converted to daily and then annual ET estimates (1224 mm) and compared with the simulated ET values over the parks (Fig. 7d). As expected, the observed ET is best reproduced by scenario 7 (bias of 1.47 mm) due to more accurate representation of GVF in the model. scenarios 1 (with the default parameters) and 8 underestimate, with biases of 58.65 and 65.32 mm, respectively. Scenario 9, with all the remotely sensed parameters, overestimates the measured ET (with bias of 86.24 mm). These shortcomings are likely due to (1) a lack of accurate representation of GVF in the default parameter sets, used in scenarios 1 and 8; (2) the uncertainties associated with the estimated

LAI values utilized in scenarios 8 and 9; and (3) complex interactions between GVF and LAI noted in scenario 9.

The presented analysis of energy balance (Fig. 7) suggests that GVF, albedo, and LAI play an important role in regulating simulated radiative energy budget and turbulent fluxes, mainly by affecting the available net radiation and transpiration quantities. GVF, ISA, and LAI also alter the study area transpiration and ET values, as well as surface runoff rates.

8 Conclusions

In the current work we investigate the utility of a select set of remote-sensing-based surface parameters in the Noah-UCM modeling framework over a highly developed urban area. It was found that remote sensing data show significantly different magnitudes and seasonal patterns of GVF when compared with the default values. The reason for this mismatch is the significant and year-round irrigation in the Los Angeles area which is not accounted for in the default parameter tables. Irrigated landscapes maintain constant greenness rather than a seasonal behavior of greening in the winter/spring and browning in the summer. The noticed differences between the monthly LAI values from default tables and remotely sensed data are also due to complex irrigation patterns. Another factor that contributes to this mismatch is the fact that landscape plantings are quite different from agricultural crops due to their being composed of collections of vegetation species, which is not taken into account in the vegetation parameter tables in the Noah LSM (CDWR, 2000; Vahmani and Hogue, 2013, 2014). There are also considerable deviations between the lookup-table-based ISA, albedo and emissivity maps, and the remotely sensed values. The results of our analysis agree with previous studies which show high sensitivity of the Noah UCM to GVF, ISA, albedo, and LAI, and minimal model sensitivity to emissivity (Loridan et al., 2010; Wang et al., 2011). Our results show that GVF, ISA, and LAI are critical in the simulations of latent and sensible heat flux, and that albedo plays a key role in the sensible heat flux simulations.

Our assessment of the Noah-UCM ET estimation shows that using the default parameters leads to significant errors in the model predictions of monthly ET fields (RMSE = 22.06 mm month^{-1}) over the study domain in Los Angeles. Results show that accurate representation of GVF is critical to reproduce observed ET patterns over vegetated areas in the urban domains. LAI also plays an important role in ET simulations. However, simulations incorporating the remotely sensed GVF values outperform (RMSE = 11.77 mm month^{-1}) simulations with the new LAI estimates (RMSE = 14.32 mm month^{-1}). This could be due to several reasons. First, there are uncertainties associated with the remote-sensing-based LAI retrieval, including nonlinearity of LAI–vegetation index (RSR) relationships (Latifi and Galos, 2010), which do not apply to NDVI-based GVF. Second, more accurate representation of GVF values in the Noah UCM not only improves the assigned LAI values to the vegetated pixels in the model but also enhances other parameters' inputs as well (i.e., albedo, emissivity, and roughness length). Further analysis of the model performance indicates that implemented satellite-derived parameter maps, particularly GVF and ISA, enhance the capability of the Noah UCM to reproduce the LST differences between fully vegetated pixels and highly developed areas (simulated LST differences of 3.07 and 6.78 °C for scenarios with default and remotely sensed GVF and ISA vs. observed LST difference of 11.25 °C).

Our analysis of energy balance suggests that GVF, albedo, and LAI play an important role in regulating simulated radiative energy budget and turbulent fluxes, mainly by affecting the available net radiation and ET quantities. With regard to urban water balance, GVF, ISA, and LAI play a key role in surface hydrologic fluxes, including ET and surface runoff. When compared with in situ observations, Noah UCM shows the capacity to reproduce ET fields with relatively high accuracy (Bias of 1.47 mm) when GVF maps are updated using remote sensing data.

In summary, the current study highlights the significant deviations between the spatial distributions and seasonal fluctuations of the default and remotely sensed parameter sets in the Noah UCM. We illustrate that replacing default parameters with the measured values reduces significant biases in model predictions of the surface fluxes within irrigated urban areas. This ultimately has key implications in feedback processes to the atmosphere when the Noah UCM is coupled with the widely used WRF model, which has been increasingly applied over urban areas to examine the exchange of heat, moisture, momentum, or pollutants. Semiarid urban cities, in particular, are receiving much attention in the literature, given their accelerated growth and increasing dependence on external water sources. More accurate representation of both water and energy fluxes in commonly used modeling frameworks is critical for regional resource management as well as predictions of urban processes under future climate conditions. Although this study focuses on the widely used single-layer UCM, we speculate that implementation of the more accurate remote-sensing-based parameters (particularly GVF and ISA) may also enhance performance of the Noah BEP (building effect parameterization; Martilli et al., 2002) which is currently the most sophisticated urban scheme in WRF. In this multilayer UCM a similar approach to the single-layer UCM is used based on an urban fraction (or ISA) parameter that couples the Noah outputs over pervious portion of pixels and UCM outputs over developed surfaces.

Acknowledgements. Funding for this research was supported by an NSF Hydrologic Sciences Program CAREER grant (no. EAR0846662), a 2012 NASA Earth and Space Science Fellowship (no. NNX12AN63H), and an NSF Water Sustainability and Climate (WSC) grant (no. EAR12040235).

Edited by: P. Gentine

References

Arnfield, A. J.: Two decades of urban climate research: A review of turbulence, exchanges of energy and water, and the urban heat island, Int. J. Climatol., 23, 1–26, doi:10.1002/joc.859, 2003.

Artis, D. A. and Carnahan, W. H.: Survey of emissivity variability in thermography of urban areas, Remote Sens. Environ., 12, 313–329, 1982.

Bauer, M. E., Loeffelholz, B., and Wilson, B.: Estimating and mapping impervious surface area by regression analysis of Landsat imagery, Remote Sensing of Impervious Surfaces, 3–20, Boca Raton, Florida: CRC Press, 2007.

Bornstein, R.: Urban climate models: Nature, limitations, and applications, Meteorol. Atmos. Phys., 38, 185–194, 1987.

Burian, S. J., Stetson, S. W., Han, W., Ching, J., and Byun, D.: High resolution dataset of urban canopy parameters for Houston, Texas, Preprint proceedings, Fifth Symposium on the Urban Environment, Vancouver, BC, Canada, 23–26 August, Am. Meteorol. Soc.: Boston, MA, 2004.

Burian, S. J., Brown, M., McPherson, T. N., Hartman, J., Han, W., Jeyachandran, I., and Rush, J.: Emerging urban databases for meteorological and dispersion, Sixth Symposium on the Urban Environment, Atlanta, GA, 28 January–2 February, Am. Meteorol. Soc.: Boston, MA, Paper 5.2, 2006.

Burian, S. J., Brown, M. J., and Augustus, N.: Development and assessment of the second generation National Building Statistics database, Seventh Symposium on the Urban Environment, San Diego, CA, 10–13 September, Am. Meteorol. Soc.: Boston, MA, Paper 5.4, 2007.

California Department of Water Resources (CDWR): California's ground water, Bulletin 118, Department of Water resources, State of California, 1975.

California Department of Water Resources (CDWR): A guide to estimating irrigation water needs of landscape plantings in California: The landscape coefficient method and WUCOLS III, Department of Water Resources, State of California, 1–150, 2000.

Changnon, S. A. and Huff, F. A.: The urban-related nocturnal rainfall anomaly at St. Louis, J. Climate Appl. Meteorol., 25, 1985–1995, doi:10.1175/1520-0450(1986)025<1985:TURNRA>2.0.CO;2, 1986.

Changnon, S. A.: Inadvertent weather modification in urban area: Lessons for global climate change, B. Am. Meteorol. Soc., 73, 619–627, doi:10.1175/1520-0477(1992)073<0619:IWMIUA>2.0.CO;2, 1992.

Chen, F. and Dudhia, J.: Coupling an advanced land-surface/hydrology model with the Penn State/NCAR MM5 modeling system. Part I: model implementation and sensitivity, Mon. Weather Rev., 129, 569–585, 2001.

Chen, F., Mitchell, K., Schaake, J., Xue, Y., Pan, H.-L., Koren, V., Duan, Q. Y., Ek, M., and Betts, A.: Modeling of land-surface evaporation by four schemes and comparison with FIFE observations, J. Geophys. Res., 101, 7251–7268, doi:10.1029/95JD02165, 1996.

Chen, F., Janjic, Z., and Mitchell, K.: Impact of atmospheric surface layer parameterization in the new land-surface scheme of the NCEP mesoscale Eta model, Boundary-Lay. Meteorol., 85, 391–421, doi:10.1023/A:1000531001463, 1997.

Chen, F., Kusaka, H., Tewari, M., Bao, J. W., and Harakuchi, H.: Utilizing the coupled WRF/LSM/urban modeling system with detailed urban classification to simulate the urban heat island phenomena over the Greater Houston area, Paper 9.11, Am. Meteorol. Soc., Fifth Symposium on the Urban Environment, Vancouver, BC, Canada, 2004.

Chen, F., Kusaka, H., Bornstein, R., Ching, J., Grimmond, C. S. B., Grossman-Clarke, S., Loridan, T., Manning, K. W., Martilli, A., Miao, S., Sailor, D., Salamanca, F. P., Taha, H., Tewari, M., Wang, X., Wyszogrodzki, A. A., and Zhang, C.: The integrated WRF/urban modelling system: development, evaluation, and applications to urban environmental problems, Int. J. Climatol., 31, 273–288, doi:10.1002/joc.2158, 2011.

Cheng, F. Y. and Byun, D. W.: Application of High Resolution Land Use and Land Cover Data for Atmospheric Modeling in the Houston-Galveston Metropolitan Area: Part I, Meteorological Simulation Results, Atmos. Environ., 42, 7795–7811, doi:10.1016/j.atmosenv.2008.04.055, 2008.

Cheng, F. Y., Hsu, Y. C., Lin, P. L., and Lin, T. H.: Investigation of the Effects of Different Land Use and Land Cover Patterns on Mesoscale Meteorological Simulations in the Taiwan Area, J. Appl. Meteorol. Climatol., 52, 570–587, doi:10.1175/JAMC-D-12-0109.1, 2013.

Ching, J., Brown, M., McPherson, T., Burian, S., Chen, F., Cionco, R., Hanna, A., Hultgren, T., Sailor, D., Taha, H., and Williams, D.: National Urban Database and Access Portal Tool, NUDAPT, B. Am. Meteorol. Soc., 90, 1157–1168, doi:10.1175/2009BAMS2675.1, 2009.

Cotton, W. R. and Pielke, R. A.: Human impacts on weather and climate, Cambridge: Cambridge University Press, 1995.

De Ridder, K., Bertrand, C., Casanova, G., and Lefebvre, W.: Exploring a new method for the retrieval of urban thermophysical properties using thermal infrared remote sensing and deterministic modeling, J. Geophys. Res., 117, D17108, doi:10.1029/2011JD017194, 2012.

Ek, M. B., Mitchell, K. E., Lin, Y., Rogers, E., Grunmann, P., Koren, V., Gayno, G., and Tarpley, J. D.: Implementation of Noah land surface model advances in the National Center for Environmental Prediction operational mesoscale Eta model, J. Geophys. Res., 108, 8851, doi:10.1029/2002JD003296, 2003.

Giannaros, T. M., Melas, D., Daglis, I. A., Keramitsoglou, I., and Kourtidis, K.: Numerical study of the urban heat island over Athens (Greece) with the WRF model, Atmos. Environ., 73, 103–111, doi:10.1016/j.atmosenv.2013.02.055, 2013.

Gutman, G. and Ignatov, A.: Derivation of green vegetation fraction from NOAA/AVHRR for use in numerical weather prediction models, Int. J. Remote Sens., 19, 1533–1543, 1998.

Hanasaki, N., Kanae, S., Oki, T., Masuda, K., Motoya, K., Shirakawa, N., Shen, Y., and Tanaka, K.: An integrated model for the assessment of global water resources – Part 1: Model description and input meteorological forcing, Hydrol. Earth Syst. Sci., 12, 1007–1025, doi:10.5194/hess-12-1007-2008, 2008a.

Hanasaki, N., Kanae, S., Oki, T., Masuda, K., Motoya, K., Shirakawa, N., Shen, Y., and Tanaka, K.: An integrated model for the assessment of global water resources – Part 2: Applications and assessments, Hydrol. Earth Syst. Sci., 12, 1027–1037, doi:10.5194/hess-12-1027-2008, 2008b.

Jiang, X., Wiedinmyer, C., Chen, F., Yang, Z. L., and Lo, J. C. F.: Predicted impacts of climate and land use change on surface ozone in the Houston, Texas, area, J. Geophys. Res., 113, D20312, doi:10.1029/2008JD009820, 2008.

Jin, M. and Shepherd J. M.: Inclusion of urban landscape in a climate model – How can satellite data help?, B. Am. Meteorol. Soc., 86, 681–689, doi:10.1175/BAMS-86-5-681, 2005.

Johnson, T. D. and Belitz K.: A remote sensing approach for estimating 1 the location and rate of urban irrigation in semi-arid climates, J. Hydrology, 414–415, 86–98, 2012.

Justice, C. O., Townshend, J. R. G., Vermote, E. F., Masuoka, E., Wolfe, R. E., Saleous, N., Roy, D. P., and Morisette, J. T.: An overview of MODIS land data processing and product status, Remote Sens. Environ., 83, 3–15, 2002.

Kalnay, E. and Cai, M.: Impact of urbanization and land-use change on climate, Nature, 423, 528–531, 2003.

Kusaka, H. and Kimura, F.: Thermal effects of urban canyon structure on the nocturnal heat island:Numerical experiment using a mesoscale model coupled with an urban canopy model, J. Appl. Meteorol., 43, 1899–1910, 2004.

Kusaka, H., Kondo, H., Kikegawa, Y., and Kimura, F.: A simple single-layer urban canopy model for atmospheric models: Comparison with multi-layer and slab models, Boundary-Lay. Meteorol., 101, 329–358, 2001.

Landsberg, H. E.: The urban climate, New York: Academic Press, 1981.

Latifi, H. and Galos, B.: Remote sensing-supported vegetation parameters for regional climate models: a brief review, iForest, 3, 98–101, doi:10.3832/ifor0543-003, 2010.

Li, D. and Bou-Zeid, E.: Quality and Sensitivity of High-Resolution Numerical Simulation of Urban Heat Islands, Environ. Res. Lett., 9, 055001, doi:10.1088/1748-9326/9/5/055001, 2014.

Liang, S.: Narrowband to broadband conversions of land surface albedo: I Algorithms, Remote Sens. Environ., 76, 213–238, 2000.

Loridan, T., Grimmond, C. S. B., Grossman-Clarke, S., Chen, F., Tewari, M., Manning, K., Martilli, A., Kusaka, H., and Best, M.: Trade-offs and responsiveness of the single-layer urban parameterization in WRF: an offline evaluation using the MOSCEM optimization algorithm and field observations, Q. J. R. Meteorol. Soc., 136, 997–1019, doi:10.1002/qj.614, 2010.

Lowry, W.: Urban effects on precipitation amount, Prog. Phys. Geog., 22, 477–520, doi:10.1177/030913339802200403, 1998.

Marshall, C. H., Pielke, R. A., Steyaert, L., and Willard, D.: The impact of anthropogenic land-cover change on the Florida peninsula sea breezes and warm season sensible weather, Mon. Weather Rev., 132, 28–52, doi:10.1175/1520-0493(2004)132<0028:TIOALC>2.0.CO;2, 2004.

Martilli, A., Clappier, A., and Rotach, M. W.: An urban surface exchange parameterization for mesoscale models, Boundary-Lay. Meteorol., 104, 261–304, 2002.

McPherson, E. G., Simpson, J. R., Xiao, Q., and Wu, C.: Los Angeles 1-million tree canopy cover assessment, Gen. Tech. Rep. PSW-GTR-207, Albany, CA: US Department of Agriculture, Forest Service, Pacific Southwest Research Station, 52 p., 2008.

Miao, S. and Chen, F.: Formation of horizontal convective rolls in urban areas, Atmos. Res., 89, 298–304, doi:10.1016/j.atmosres.2008.02.013, 2008.

Miao, S., Chen F., LeMone, M. A., Tewari M., Li, Q., and Wang, Y.: An observational and modeling study of characteristics of urban heat island and boundary layer structures in Beijing, J. Appl. Meteorol. Climatol., 48, 484–501, doi:10.1175/2008JAMC1909.1, 2009.

Mini, C., Hogue, T. S., and Pincetl, S.: Estimation of Residential Outdoor Water Use in Los Angeles, California, Landscape Urban Plan., 127, 124–135, doi:10.1016/j.landurbplan.2014.04.007, 2014.

Moering, D. C.: A comparative study of evapotranspiration rates between irrigated and non-irrigated parks in Los Angeles, M.S. thesis, Dep. of Civil and Env. Eng., University of California Los Angeles, Los Angeles, California, 2011.

National Oceanographic and Atmospheric Administration-Coastal Services Center (NOAA-CSC): Southern California 2000-Era Land Cover/Land Use, LANDSAT-TM, 10m, NOAA-CSC, Charleston, SC, 2003.

National Weather Service (NWS): Jet Stream – The Marine Layer, NOAA National Weather Service, available at: http://www.srh.noaa.gov/jetstream/ocean/marine.htm (last access: 24 February 2013), 2011.

Niyogi, D., Holt, T., Zhong, S., Pyle, P. C., and Basara, J.: Urban and land surface effects on the 30 July 2003 mesoscale convective system event observed in the Southern Great Plains, J. Geophys. Res., 111, D19107, doi:10.1029/2005JD006746, 2006.

Pielke, R. A. Sr., Marland, G., Betts, R. A., Chase, T. N., Eastman, J. L., Niles, J. O., Niyogi, D., and Running, S.: The influence of land-use change and landscape dynamics on the climate system: Relevance to climate change policy beyond the radiative effect of greenhouse gases, Phil. Trans. R. Soc. London A, 360, 1705–1719, 2002.

Pokhrel, Y., Hanasaki, N., Koirala, S., Cho, J., Kim, H., Yeh, P. J.-F., Kanae, S., and Oki, T.: Incorporating anthropogenic water regulation modules into a land surface model, J. Hydrometeorol., 13, 255–269, doi:10.1175/JHM-D-11-013.1, 2012.

Schaaf, C. B., Gao, F., Strahler, A. H., Lucht, W., Li, X. W., Tsang, T., Strugnell, N. C., Zhang, X. Y., Jin, Y. F., Muller, J. P., Lewis, P., Barnsley, M., Hobson, P., Disney, M., Roberts, G., Dunderdale, M., Doll, C., d'Entremont, R. P., Hu, B. X., Liang, S. L., Privette, J. L., and Roy, D.: First operational BRDF, albedo nadir reflectance products from MODIS, Remote Sens. Environ., 83, 135–148, 2002.

Shuai, Y., Schaaf, C. B., Strahler, A. H., Liu, J., and Jiao, Z.: Quality assessment of BRDF/albedo retrievals in MODIS operational system, Geophys. Res. Lett., 35, L05407, doi:10.1029/2007GL032568, 2008.

Shuai, Y., Masek, J. G., Gao, F., and Schaaf, C. B.: An algorithm for the retrieval of 30-m snow-free albedo from Landsat surface reflectance and MODIS BRDF, Remote Sens. Environ., 115, 2204–2216, doi:10.1016/j.rse.2011.04.019, 2011.

Sobrino, J. A. and Raissouni, N.: Toward remote sensing methods for land cover dynamic monitoring: Application to Morocco, Int. J. Remote Sens., 21, 353–366, doi:10.1080/014311600210876, 2000.

Sobrino, J. A., Raissouni, N., and Li, Z. L.: A comparative study of land surface emissivity retrieval from NOAA data, Remote Sens. Environ., 75, 256–266, 2001.

Song, M. and Civco, D. L.: A knowledge-based approach for reducing cloud and shadow, Proceedings of the American Society of Photogrammetry and Remote Sensing annual convention, Washington, DC: Am. Soc. Photogramm. Remote Sens., 7 p., 2002.

State of California Department of Water Resources (SCDWR): California irrigation management information system, Sacramento, CA: State of California Department of Water Resources, available at: http://www.water.ca.gov/ (last access: December 2013), 2009.

Stathopoulou, M. and Cartalis, C.: Daytime urban heat islands from Landsat ETM+ and Corine land cover data: An application to major cities in Greece, Sol. Energy, 81, 358–368, doi:10.1016/j.solener.2006.06.014, 2007.

Stathopoulou, M., Cartalis, C., and Petrakis, M.: Integrating Corine Land Cover data and Landsat TM for surface emissivity definition: application to the urban area of Athens, Greece, Int. J. Remote Sens., 28, 3291–3304, doi:10.1080/01431160600993421, 2007.

Stenberg, P., Rautiainen, M., Manninen, T., Voipio, P., and Smolander, H.: Reduced simple ratio better than NDVI for estimating LAI in Finnish pine and spruce stands, Silva Fennica, 38, 3–14, 2004.

Taha, H.: Meso-urban meteorological and photochemical modeling of heat island mitigation, Atmos. Environ., 42, 8795–8809, doi:10.1016/j.atmosenv.2008.06.036, 2008.

Taha, H., Douglas, S., and Haney, J.: Mesoscale meteorological and air quality impacts of increased urban albedo and vegetation, Energ. Buildings, 25, 169–177, doi:10.1016/S0378-7788(96)01006-7, 1997.

Tan, M. and Li, X.: Integrated assessment of the cool island intensity of green spaces in the mega city of Beijing, Int. J. Remote Sens., 34, 3028–3043, doi:10.1080/01431161.2012.757377, 2013.

Tewari, M., Kusaka, H., Chen, F., Coirier, W. J., Kim, S., Wyszogrodzki, A., and Warner, T. T.: Impact of coupling a microscale computational fluid dynamics model with a mesoscale model on urban scale contaminant transport and dispersion, Atmos. Res., 96, 656–664, doi:10.1016/j.atmosres.2010.01.006, 2010.

US Census: US Census Bureau Releases Data on Population Distribution and Change in the US Based on Analysis of 2010 Census Results, US Census Bureau, 24 March 2011.

Vahmani, P. and Hogue, T. S.: Modelling and analysis of the impact of urban irrigation on land surface fluxes in the Los Angeles metropolitan area, Climate and Land Surface Changes in Hydrology Proceedings of H01, IAHS-IAPSO-IASPEI Assembly, Gothenburg, Sweden, July 2013, IAHS Publ. 359, 2013.

Vahmani, P. and Hogue, T. S.: Incorporating an Urban Irrigation Module into the Noah Land Surface Model Coupled with an Urban Canopy Model, J. Hydrometeorol., 15, 1440–1456, doi:10.1175/JHM-D-13-0121.1, 2014.

Van Wevenberg, K., De Ridder, K., and Van Rompaey, A.: Modeling the contribution of the Brussels heat island to a long temperature time series, J. Appl. Meteorol. Climatol., 47, 976E990, doi:10.1175/2007JAMC1482.1, 2008.

Wang, X. M., Chen, F., Wu, Z. Y., Zhang, M. G., Tewari, M., Guenther, A., and Wiedinmyer, C.: Impacts of weather conditions modified by urban expansion on surface ozone: comparison between the Pearl River Delta and Yangtze River Delta regions, Adv. Atmos. Sci., 26, 962–972, doi:10.1007/s00376-009-8001-2, 2009.

Wang, Z. H., Bou-Zeid, E., Au, S. K., and Smith, J. A.: Analyzing the sensitivity of WRF's single-layer urban canopy model to parameter uncertainty using advanced Monte Carlo simulation, J. Appl. Meteor. Climatol., 50, 1795–1814, doi:10.1175/2011JAMC2685.1, 2011.

Wei-guang, M., Yan-xia, Z., Jiang-nan, L., Wen-shi, L., Guang-feng, D., and Hao-rui, L.: Application of WRF/UCM in the simulation of a heat wave event and urban heat island around Guangzhou, J. Trop. Meteorol., 3, 257–267, doi:10.3969/j.issn.1006-8775.2011.03.007, 2011.

12

Geostatistical assessment of warm-season precipitation observations in Korea based on the composite precipitation and satellite water vapor data

Sojung Park[1,3,4], **Seon Ki Park**[1,2,3,4], **Jeung Whan Lee**[5], **and Yunho Park**[5]

[1]Department of Climate and Energy Systems Engineering, Ewha Womans University, Seoul, Republic of Korea
[2]Department of Environmental Science and Engineering, Ewha Womans University, Seoul, Republic of Korea
[3]Severe Storm Research Center, Ewha Womans University, Seoul, Republic of Korea
[4]Center for Climate/Environment Change Prediction Research, Ewha Womans University, Seoul, Republic of Korea
[5]Korea Meteorological Administration, Seoul, Republic of Korea

Correspondence: Seon Ki Park (spark@ewha.ac.kr)

Abstract. Among the meteorological disasters, heavy rainfalls cause the second largest damage in Korea, following typhoons. To manage the potential disasters due to heavy rainfalls, understanding the observational characteristics of precipitation is of utmost importance. In this study, we investigate the spatial and temporal characteristics of warm-season precipitation in Korea, according to the precipitation types, by conducting the geostatistical analyses such as an autocorrelogram, Moran's I and general G on the composite (radar + station) precipitation data. The e-folding distance of precipitation ranges from 15 to 35 km, depending on the spatial distribution, rather than intensity, of precipitation, whereas the e-folding time ranges from 1 to 2 h. The directional analyses revealed that the warm-season precipitation systems in Korea, especially those with a high precipitation amount, have high spatial autocorrelations in the southwest–northeast and west–east directions, in association with the frontal rainfalls, convection bands, etc. Furthermore, the cluster versus dispersion patterns and the hot versus cold spots are analyzed through Moran's I and general G, respectively. Water vapor, represented by the brightness temperature, from three Himawari-8 water vapor bands also shows similar characteristics with precipitation but with strong spatial correlation over a much longer distance ($\sim$ 100 km), possibly due to the continuity of water vapor. We found that, under the e-folding-based standard, the current observation network of Korea is sufficient to capture the characteristics of most precipitation systems; however, under a strict standard (e.g., autocorrelation of 0.6), a higher-resolution observation network is essentially required – especially in local areas with frequent heavy rainfalls – depending on the directional features of precipitation systems. Establishing such an observation network based on the characteristics of precipitation enables us to improve monitoring, tracking, and prediction skills of high-impact weather phenomena as well as to enhance the utilization of numerical weather prediction.

1 Introduction

Heavy rainfall causes one of the most serious meteorological disasters in Korea. Most precipitation systems in Korea are highly influenced by the East Asian summer monsoon, called *Changma* in Korean (Riyu et al., 2001; Kim et al., 2002; Lee et al., 2017b), and hence have high seasonal variation; they also show high spatial variation due to complex topographic features (e.g., Ko et al., 2005; Chang and Kwon, 2007; Park and Lee, 2007; In et al., 2014). Since heavy rainfalls including typhoon cause serious damages to both human life and property, improving the their forecast accuracy is of utmost importance. Many studies have been done to improve the quantitative precipitation forecast using numerical weather prediction models and satellite or radar data (e.g., Fritsch et al., 1998; Lee et al., 2006; Kim and Oh, 2010; Feng and Kitzmiller, 2006; Yu et al., 2013). In order to improve the forecast accuracy, it is essential to understand the spatial and temporal characteristics as well as the occurrence and development mechanisms of the precipitation systems.

The ground-based rainfall observation data, in Korea, are collected from the automated synoptic observing systems (ASOS), the automatic weather stations (AWS), and the automated agriculture observing system (AAOS). The observation density is about 67 km for ASOS and approximately 13 km by including AWS. The agrometeorological observation network consists of 11 AAOS stations (Choi et al., 2015). Based on these precipitation observations, the mesoscale structures of precipitation as well as the hydrologic budgets in Korea have been extensively studied (e.g., Kim and Lee, 2006; Cassardo et al., 2009; Jeong et al., 2012, 2016; Jung and Lee, 2013; Lee et al., 2017a). Capturing the spatiotemporal features of precipitation systems from the observation networks is essential to runoff forecast, especially at the catchment scale and for the flooding cases (Volkmann et al., 2010). Given the limited resources, it is desirable to optimally design the observation networks, which should be efficient in using least amount of available instruments but sufficient in capturing the precipitation characteristics – both amount and spatiotemporal variability. In other words, understanding temporal and spatial characteristics of precipitation is fundamental and essentially required for the establishment of the optimal observation networks (see Ciach and Krajewski, 2006; Volkmann et al., 2010). Moreover, the appropriateness of such observation networks, in analyzing and/or forecasting the precipitation systems, has seldom been assessed.

In this study, we investigate the spatial and temporal characteristics of precipitation in Korea using the geostatistical analysis methods – autocorrelogram, Moran's I, and general G (e.g., Ciach and Krajewski, 2006; Schiemann et al., 2011; Emmanuel et al., 2012; Fu et al., 2014). Ciach and Krajewski (2006) estimated the spatial correlation functions in small-scale rainfall, based on rain gauges covering an area of about 3 km by 3 km in central Oklahoma. They showed a dependence of the correlogram parameters on the averaging timescale, large differences of the correlograms in the individual storms, and the ambiguities in correlation estimates based on rainfall intensities. Bacchi and Kottegoda (1995) addressed the effects of weather systems, topography, and temporal and spatial scale of observation networks on the spatial distribution of rainfall. Emmanuel et al. (2012) analyzed the characteristics of spatiotemporal variability by rainfall types using a variogram.

In Korea, the spatial and temporal characteristics of precipitation have been reported in several previous studies (e.g., Chang and Kwon, 2007; Hong et al., 2006; Ha et al., 2007; In et al., 2014). Ha et al. (2007) found that the e-folding distance and time in correlation coefficients of AWS precipitation ranged from 50 to 110 km and 1 to 2 h, respectively. In summer, precipitation showed localized features with relatively more precipitation mainly in the western part and southern coastal areas of Korea (Kim et al., 2005; Hong et al., 2006; In et al., 2014). In et al. (2014) also analyzed the degree of localization in summer precipitation and noted a high locality of rainfalls, i.e., more frequent occurrence of short-lived local precipitation, in July and August. In addition to these studies, we provide further analyses on the variation of spatial correlation associated with the distribution and intensity of precipitation.

Water vapor is also strongly linked to the precipitation system because not only does it become the absolute source of precipitation (Eltahir and Bras, 1996; Park, 1999; Bretherton et al., 2004; Stohl et al., 2008; Gimeno et al., 2012) but also its transport in the lower troposphere provides diabatic heating and hence promotes development of convective precipitation systems (Tompkins, 2001; Trenberth and Stepaniak, 2003a, b; Smith and Yuter, 2010). Therefore, the amount and behavior of water vapor are the crucial factors in the precipitation system. For instance, a large amount of water vapor originated from the adjacent oceans is transported to the East Asian monsoon region through the large-scale monsoon circulations. The origins of water vapor supply can also distinguish the anomalous rainfall patterns (e.g., Zhou and Yu, 2005). Furthermore, about 70 % of precipitation in extratropical cyclones is generated by water vapor already present when the storm formed in the atmosphere (Trenberth, 1999). Hence, we also conduct the spatial and temporal analysis of water vapor to enhance the understanding of precipitation characteristics over Korea and adjacent areas.

This study aims at classifying the precipitation types statistically based on the spatial distribution of precipitation and at identifying the spatial and temporal characteristics of warm-season precipitation in Korea according to the precipitation types via the geostatistical analysis methods, including autocorrelogram, Moran's I, and general G. Furthermore, we investigate the characteristics of water vapor over the Korean Peninsula associated with the precipitation types, using satellite data. Section 2 describes the data set used for the analyses, and Sect. 3 briefly introduces the geostatistical methods employed in this study. Results and discussions are provided in Sect. 4. Section 5 is devoted to conclusions.

2 Data description

We use the precipitation data from weather stations (see Fig. 1) to categorize the precipitation systems. We classify four different precipitation types statistically, based on two criteria: the portion of weather stations with precipitation (C1; in %) and the station average precipitation rate (C2; in mm h^{-1}). In order to determine the threshold values for classifying the precipitation types, we have conducted a preliminary statistical analysis on precipitation events in the period of 2011–2015 (not shown). As the precipitation events occur in a given time period and space interval, our precipitation data are assumed to follow the Poisson distribution, which represents a probability situation of a large number of obser-

Table 1. Classification of precipitation types with the number of events (in bold).

		The portion of weather stations with precipitation (C1)	
		$< 20\,\%$	$\geq 20\,\%$
The station average precipitation rate (C2)	$< 3\,\mathrm{mm\,h^{-1}}$	Low precipitation at a few points (LPFP) **5858**	Low precipitation at many points (LPMP) **1157**
	$\geq 3\,\mathrm{mm\,h^{-1}}$	High precipitation at a few points (HPFP) **980**	High precipitation at many points (HPMP) **594**

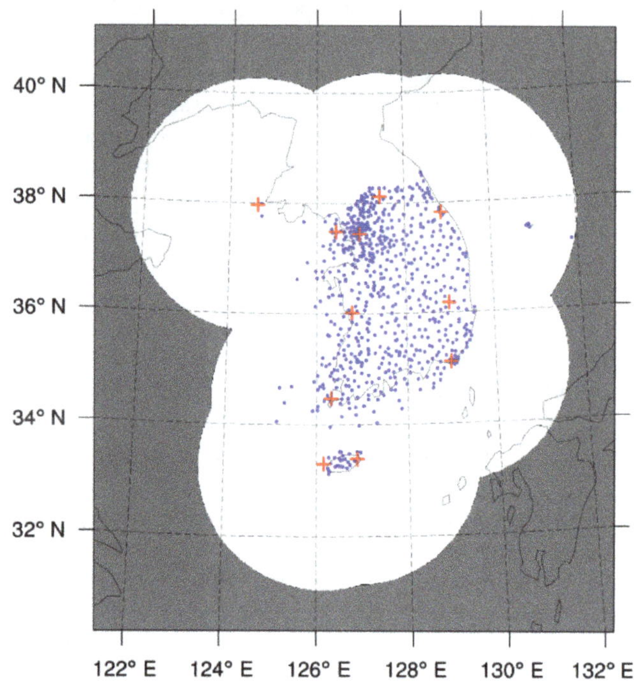

Figure 1. The weather station locations (blue dots) and radar locations (red "+" symbols) and coverages (white area) in Korea.

vations with a small probability of occurrence. Many studies developed the Poisson distribution models to estimate rainfall and to cluster the rainfall systems (e.g., Rodriguez-Iturbe et al., 1987; Lee et al., 2014; Barton et al., 2016; Ritschel et al., 2017). We have chosen the threshold values when the cumulative percentage of precipitation events for each criterion (i.e., C1 and C2) reached approximately 80 %, thus obtaining the threshold values of 20 % for C1 and $3\,\mathrm{mm\,h^{-1}}$ for C2, respectively. Our preliminary statistical analysis showed that, in general, most precipitation events occur over small areas and precipitation events with high intensity rarely occur over large areas. The locality of precipitation appeared higher as the precipitation intensity was higher, in accordance with Nam et al. (2014). In particular, precipitation systems with the highest intensity ($\geq 10\,\mathrm{mm\,h^{-1}}$) were mostly confined to a small area with the number of stations less than 10 % of total weather stations. This implies that the locality feature of precipitation systems may depend on the threshold value in precipitation intensity.

Based on these criteria, we define four different precipitation types, as shown in Table 1: (1) low precipitation at a few points (LPFP) for $C1 < 20\,\%$ and $C2 < 3\,\mathrm{mm\,h^{-1}}$, (2) low precipitation at many points (LPMP) for $C1 \geq 20\,\%$ and $C2 < 3\,\mathrm{mm\,h^{-1}}$, (3) high precipitation at a few points (HPFP) for $C1 < 20\,\%$ and $C2 \geq 3\,\mathrm{mm\,h^{-1}}$, and (4) high precipitation at many points (HPMP) for $C1 \geq 20\,\%$ and $C2 \geq 3\,\mathrm{mm\,h^{-1}}$. We practically exclude the LPFP type in our analyses, i.e., the case with $C1 < 20\,\%$ and $C2 < 3\,\mathrm{mm\,h^{-1}}$, since it may be less effective.

The Korea Meteorological Administration (KMA) has produced the composite precipitation data over Korea using the data from radars, weather stations, and satellites through the following steps (see Hwang et al., 2015): (1) remove non-precipitation echoes from the radar data using the satellite cloud-type data, (2) calculate the difference between the station precipitation and the radar-estimated precipitation, (3) perform the objective analysis on the precipitation difference field and on the station precipitation data, (4) correct the bias using the objectively analyzed difference field, and (5) combine the corrected radar-estimated precipitation data and the objectively analyzed station precipitation data to produce the composite precipitation data (in $\mathrm{mm\,h^{-1}}$). In order to analyze the precipitation systems with the evenly distributed high-resolution data, we used these composite precipitation data, which cover 1153 km × 1441 km over the Korean Peninsula, with a grid size of 1 km and a time resolution of 1 h. Geostatistical analyses are conducted using the composite precipitation data sets from April to October in the period of 2013–2015 to investigate the spatial and temporal characteristics of warm-season rainfall.

We also used the Himawari-8 water vapor bands to investigate the distribution feature of water vapor in the lower to upper atmosphere in association with precipitation. The water vapor bands are strongly linked with the properties of moisture, thermodynamics, and dynamics of the troposphere. As they are sensitive to the moisture and temperature profiles in the radiation path, they can provide information for a wide range of atmospheric processes (Georgiev et al., 2016), including atmospheric motions such as the upper-level lows, jet stream, and blocking. Himawari-8, operated by the Japan

Meteorological Agency since 7 July 2015, carries a new unit called the Advanced Himawari Imager (AHI) with higher radiometric, spectral, and spatial resolution in the geostationary orbit (Bessho et al., 2016). The AHI has 16 observation bands – 3 visible (0.47, 0.51, 0.64 µm), 3 near-infrared (0.86, 1.6, 2.3 µm), and 10 infrared (3.9, 6.2, 6.9, 7.3, 8.6, 9.6, 10.4, 11.2, 12.4, 13.3 µm). Among them the water vapor absorption bands – 8 (6.2 µm), 9 (6.9 µm), and 10 (7.3 µm) represent upper-, mid-, and mid- to lower-level water vapor, respectively. The calibration of the Himawari-8 water vapor bands is accurate to within 0.2 K through a validation approach developed under the Global Space-based Inter-Calibration System (GSICS) project with hyper-spectral infrared sounders (e.g., Okuyama et al., 2015; Bessho et al., 2016). In this study, we used the scan area for the full disk by the AHI that covers the East Asia region: the full disk images are taken every 10 min with a 2 km spatial resolution. The analyses are conducted on the data of the hour over the domain of 120.132–134.243°E and 30.436–44.068° N (600 × 770 grids), similar to the area of the precipitation data.

3 Geostatistical methods

In this study, we use several methods to analyze the spatial characteristics – the spatial autocorrelation, Moran's I, and general G. The spatial autocorrelation shows how the precipitation observations at a certain distance are correlated with each other. Thus, it is possible to determine the degree of spatial scale, or the effective influence range, of a single precipitation system or associated precipitation systems. Through Moran's I, we can figure out the property of spatial correlations, especially distinguishing between the cluster and dispersion type in the patterns and/or areas of precipitation. In other words, Moran's I identifies local cluster areas of strong precipitation, large areas of precipitation, or precipitation boundaries. General G indicates whether the cluster region has high or low precipitation.

3.1 Spatial autocorrelation

Autocorrelation in spatial analysis implies that a variable is correlated with itself at different points. Simply speaking, pairs of precipitation data close to each other are likely to have more similar values than those far apart from each other. When the precipitation data are spatially autocorrelated, precipitation at one place is related to that at another place, and they are not independent of each other. The autocorrelation coefficient can be obtained once the autocovariance has been calculated. The autocovariance indicates the degree of similarity of a variable itself at a certain distance. Generally, the autocovariance is greater as the separation distance decreases, becomes smaller as the distance increases, and does not show any tendency (e.g., converge) after a certain distance. For the one-dimensional data, the autocovariance (cov) in terms of the grid distance (or lag; k) between the data pairs, $X=(x_1, x_2, \cdots, x_{n-k})$ and $X'=(x_{1+k}, x_{2+k}, \cdots, x_n)$, is defined as

$$\begin{aligned}\mathrm{cov}(X,X') &= E\left[(X-\mu_X)\left(X'-\mu_{X'}\right)\right]\\ &= E(XX')-E(X)E(X')\\ &= \mu_{XX'}-\mu_X\mu_{X'}. \end{aligned} \tag{1}$$

Here, $E(XX')$, $E(X)$, and $E(X')$ are the means of sub-data sets (i.e., XX', X, and X', respectively) and are calculated as follows:

$$\begin{aligned} E(XX') &= \frac{1}{n-k}\sum_{i=1}^{n-k} x_i x_{i+k} = \mu_{XX'},\\ E(X) &= \frac{1}{n-k}\sum_{i=1}^{n-k} x_i = \mu_X, \text{ and}\\ E(X') &= \frac{1}{n-k}\sum_{i=1}^{n-k} x_{i+k} = \mu_{X'}, \end{aligned} \tag{2}$$

where n is the number of total points and i is the grid point index. The lag index, k, depicts the grid distance between two point values, x_i and x_{i+k}. The autocorrelation coefficient (ρ) is defined as

$$\rho(X,X') = \frac{\mathrm{cov}(X,X')}{\sqrt{\mathrm{var}(X)\mathrm{var}(X')}} = \frac{\mu_{XX'}-\mu_X\mu_{X'}}{\sqrt{\mathrm{var}(X)\mathrm{var}(X')}}, \tag{3}$$

where var(X) means the variance of X and can be regarded as the covariance of X with itself, i.e., var(X) = cov(X, X). At the zero separation distance, the autocovariance has the same value as the variance and the autocorrelation coefficient becomes unity. For the two-dimensional data, the autocorrelation coefficient of all data pairs within the separation distance is obtained by using the relaxed separation distance and by following the same calculation principle as in the one-dimensional data. In order to compute the relaxed separation distance, the distance between the data pairs is rounded.

3.2 Moran's I

Moran's I is a statistical index for measuring the similarity of neighboring data (Moran, 1948). It is obtained by comparing the values of the target and neighboring feature pairs with the average of the entire targets:

$$I = \frac{n\sum_i \sum j w_{ij}\left(x_i-\overline{x}\right)\left(x_j-\overline{x}\right)}{\sum_i\sum_j w_{ij}\sum_i\left(x_i-\overline{x}\right)^2}, \tag{4}$$

where n is the number of the entire target observations, x_i is the variable value at the ith location and $\overline{x}$ is the mean of the entire targets. Here, w_{ij} is the spatial weight of the link between i and j, which is defined by the inverse distance weight, i.e., $w_{ij}=1/d_{ij}$, with d_{ij} representing the distance between grids i and j. Moran's I ranges from -1 to 1 with the following interpretation: (1) the positive value means that the variable of interest and its spatial lags are positively autocorrelated, defining a *cluster* pattern; (2) the negative value indicates spatially negative autocorrelation, defining a *dispersion* pattern; and (3) the zero value represents a *random* spatial pattern. Equation (4) denotes the global Moran's I that identifies the pattern – either cluster or dispersion – but with no information on the magnitude of variables. In addition, the local Moran's I (I_i), showing the location of the cluster and dispersion, is computed as the following (Anselin, 1995):

$$I_i=\frac{(x_i-\overline{x})}{\sum\limits_{k=1}^{n}(x_k-\overline{x})^2/(n-1)}\sum_{j=1}^{n}w_{ij}\left(x_j-\overline{x}\right). \tag{5}$$

I_i is interpreted the same as I but with no specific range. The Z score (Z) can be analyzed together to confirm the significance of I_i and is calculated as

$$Z\left(I_i\right)=\frac{I_i-E\left(I_i\right)}{\mathrm{SD}\left(I_i\right)}, \tag{6}$$

where $E(I_i)=-\sum\limits_j w_{ij}/(n-1)$ is the expected value of I_i if random and SD(I_i) is the standard deviation of I_i. The Z score suggests whether we can reject the null hypothesis or not. In this case, the null hypothesis states "there is no spatial clustering". To determine if the Z score is statistically significant, we compare it to the range of values for a particular confidence level. For example, at the significance level of 0.1, 0.05, 0.01, the absolute value of the Z score would have to be greater than 1.65, 1.96, 2.57, respectively, to be statistically significant, and the clusters are not created by chance.

3.3 General *G*

We can identify the cluster patterns through Moran's I; however, it does not indicate whether the clustered variable values are low or high. In addition to Moran's I, it is necessary to determine whether the properties of the cluster are hot spots (i.e., high values) or cold spots (i.e., low values) by using the Getis–Ord general G (Getis and Ord, 1992; hereafter referred to as general G). The global general G, like the global Moran's I, shows the cluster characteristics for the entire study area as a statistical index. Therefore, we use the local general G which can represent locations where features with either high or low values are spatially clustered. The following equation calculates the local general G ($G_i(d)$):

$$G_i(d)=\frac{\sum\limits_j w_{ij}(d)x_j}{\sum\limits_j x_j},\ j\neq i, \tag{7}$$

where d is the distance between the target feature and the neighboring feature, and w_{ij} is the same spatial weight (i.e., the inverse distance weight) used for calculating Moran's I as in Eqs. (4) and (5). To be a statistically significant hot spot or cold spot, the Z score of the general G can be calculated. Ord and Getis (1995) redefined $G_i(d)$, which can be considered as the Z score ($Z(G_i)$), as

$$Z\left(G_i\right)=\frac{G_i-E\left(G_i\right)}{\mathrm{SD}\left(G_i\right)},\ j\neq i, \tag{8}$$

where $E(G_i)=\sum\limits_j w_{ij}(d)/(n-1)$ is the expected value if random and SD(G_i) is the standard deviation of G_i. High positive values indicate the possibility of a local cluster of high values, whereas negative values indicate a similar cluster of low values. An absolute value of the Z score would have to be greater than 1.65, 1.96, 2.57 to be statistically significant at a significance level of 0.1, 0.05, 0.01, respectively.

4 Results and discussions

4.1 Spatial autocorrelation

4.1.1 Precipitation

The general characteristics of the spatial autocorrelation structure in precipitation for each type are shown in Fig. 2. We have calculated the correlation coefficients for all points in the range from 0 to 100 km. In order to compare the characteristics between different precipitation types, we defined the *e*-folding value as the threshold. The *e*-folding distance means the separation distance when autocorrelation becomes e^{-1} (i.e., 0.3674). In other words, it refers to the spatial scale of precipitation that autocorrelation decreases to $\rho=e^{-1}$ from the value at the zero separation distance (i.e., $\rho=1$). The *e*-folding value, under the assumption that atmospheric variables decrease exponentially with time and distance, has often been used to estimate both the temporal and spatial structure of precipitation (e.g., Skøien et al., 2003; Ha et al., 2007; Kursinski and Mullen, 2008; Zeweldi and Gebremichael, 2009; In et al., 2014). As classified in Table 1, the LPMP and HPMP cases (i.e., C1 is high, regardless of C2) show the *e*-folding distance at $\sim$35 km. On the other hand, the HPFP cases (i.e., C1 is low) depict the *e*-folding distance at $\sim$15 km.

This result can be expected from the classification of precipitation types: precipitation with smaller spatial scale brings about more rapid decrease in the spatial correlation, thus showing smaller correlation at the *e*-folding distance. Ha

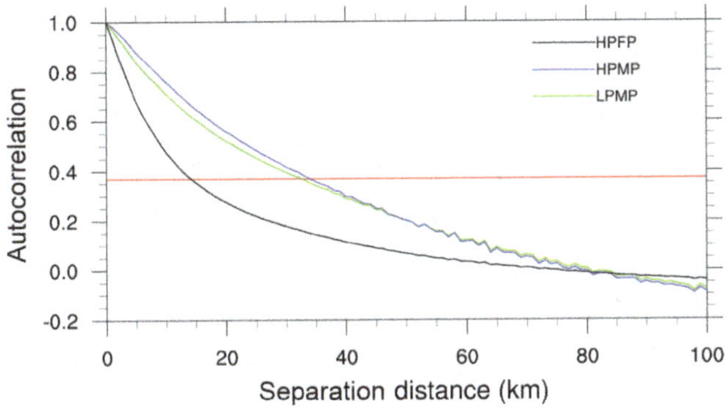

Figure 2. The averaged spatial autocorrelations of precipitation for each precipitation type. The red line indicates the *e*-folding value.

et al. (2007) estimated the *e*-folding distance of warm-season precipitation as 50–110 km, and In et al. (2014) pointed out that the *e*-folding distance varied depending on months and regions with the minimum being 40 km. Our results show shorter *e*-folding distances than those from the previous studies: it is ascribed to exclusion of the LPFP cases (see Table 1). Nam et al. (2014) examined the relationship between precipitation intensity and spatial autocorrelation and noticed that the latter became smaller as the former appeared stronger, thus depicting a larger locality in precipitation. They found that the *e*-folding distance was 14 km for the precipitation intensity in excess of 30 mm h^{-1}, whereas it was about 40 km for the intensity greater than 5 mm h^{-1}. However, our results indicate that the spatial autocorrelations of HPMP and LPMP are in similar shape. Generally, the spatial locality is proportional to precipitation intensity; however, strictly speaking, the number of points with precipitation has a greater effect on spatial characteristics than the precipitation intensity.

In Fig. 3, in order to identify the spatial characteristics of precipitation for each case, we show a histogram depicting the number of cases according to the *e*-folding distance. We found that the *e*-folding distance at the mode (the most frequent value) was 15 km for HPFP, 30 km for LPMP, and ∼ 34.7 km for HPMP, respectively. It is evident that, as shown in Fig. 2, the *e*-folding distance depends on the precipitation spatial scale rather than the precipitation intensity.

We additionally analyzed the directional features of precipitation distribution over the Korean Peninsula in order to know the spatial structures of precipitation systems (see Figs. 4 and 5). We classified the direction range from 0 to 180° with an interval of 45°, considering the symmetry. Here, the direction is a measure of the angle from the origin–east axis, whose magnitude increases counterclockwise (see the radar chart in Fig. 5). Figure 4 shows the averaged directional autocorrelation for different precipitation types. The autocorrelation curves show similar shapes and properties in HPMP (Fig. 4a) and HPFP (Fig. 4c), which depict two distinct directional characteristics: the precipitation structures have similar spatial features in directional pairs of 45 and 0°

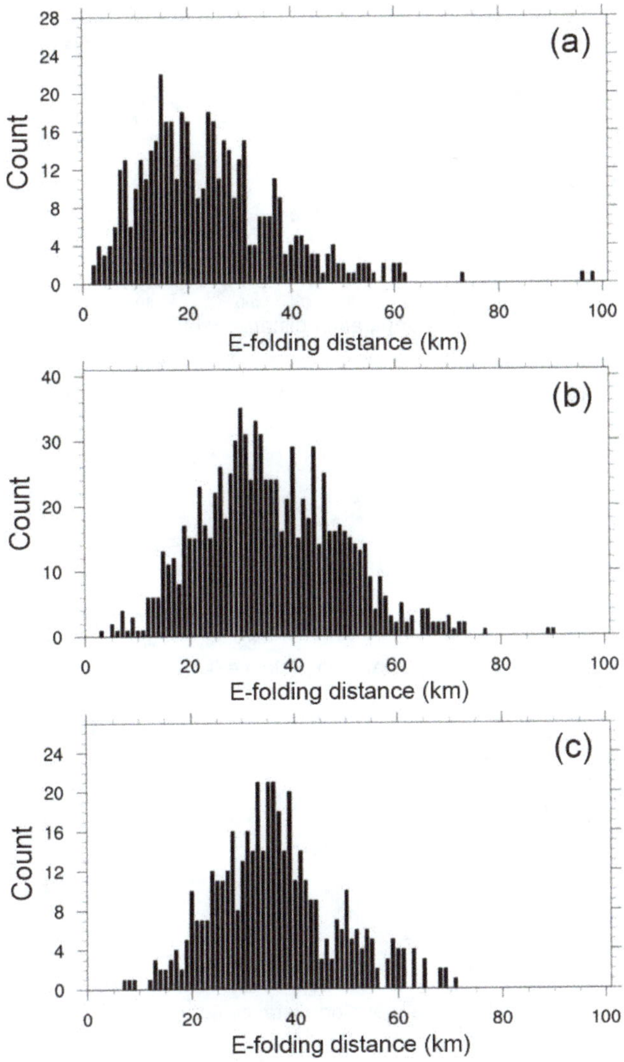

Figure 3. A histogram depicting the number of cases according to the *e*-folding distance for **(a)** HPFP, **(b)** LPMP, and **(c)** HPMP.

and of 135 and 90°. For HPFP where the precipitation spatial scale is small, the autocorrelations of all directions decrease more rapidly than those for LPMP and HPMP, resulting in a smaller gap between the two distinct curves compared to HPMP. For LPMP, i.e., low precipitation rate with a large spatial scale, the spatial correlation is highest with the precipitation axis of 45° (i.e., southwest–northeast), whereas it is lowest with that of 135° (northwest–southeast).

Figure 5 shows a radar chart representing the directional *e*-folding distance at the mode in the directional histogram of case numbers (not shown). In general, the directional *e*-folding distances (and hence the spatial correlations) are larger for the cases with large spatial scale (i.e., LPMP and HPMP) than the cases with small spatial scale (i.e., HPFP). It is noteworthy that the *e*-folding distances in LPMP have no characteristic directionality (i.e., isotropic), indicating that the precipitation systems in LPMP are mostly related to both

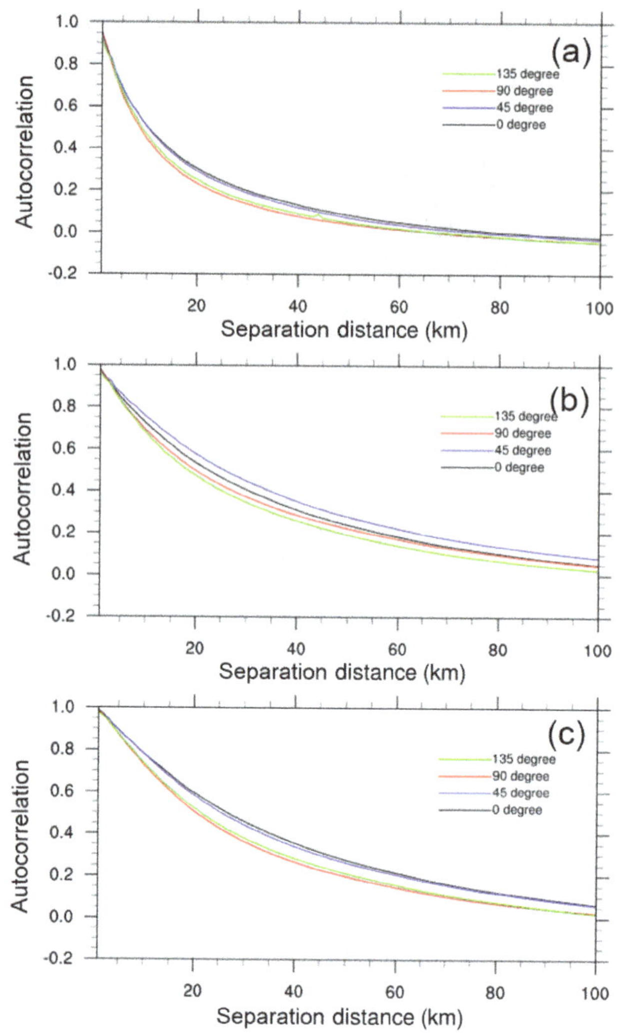

Figure 4. The averaged directional spatial autocorrelation of precipitation for (a) HPFP, (b) LPMP, and (c) HPMP.

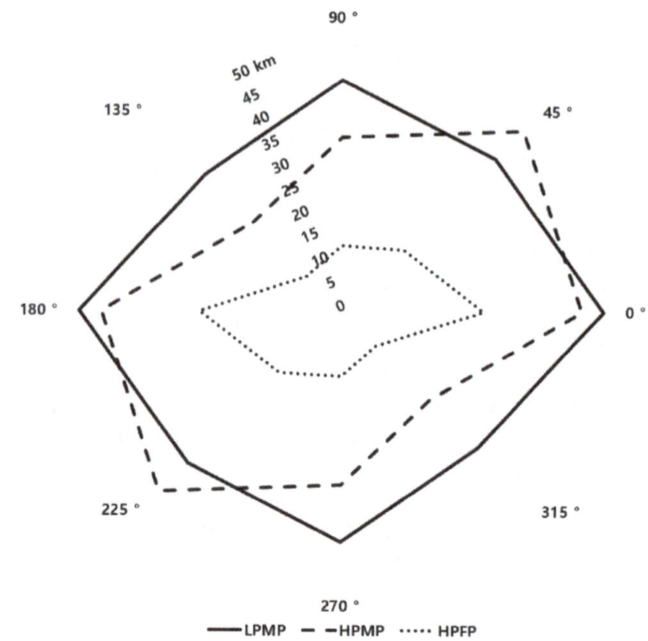

Figure 5. A radar chart representing the directional *e*-folding distance (in km) for each precipitation type at the mode in the directional histogram (not shown) depicting the number of cases according to the *e*-folding distance in different directions.

the migratory cyclones that move into Korea from various directions and the cloud clusters where moderate convective systems in large area are closely connected to each other. On the other hand, the cases with high precipitation intensity, i.e., HPFP and HPMP, both show strong directionality (i.e., anisotropic) with large *e*-folding distances along the axes of 45 and 0° (i.e., the southwest–northeast and west–east directions, respectively). They are considered to be strongly linked to the frontal rainfalls, convection bands, etc. (e.g., Sun and Lee, 2002; Kim and Lee, 2006; Lee et al., 2008; Jeong et al., 2012, 2016; Jung and Lee, 2013).

As we have used the 3-year data, the spatial autocorrelations here are considered to include the temporal features implicitly. All the weather systems evolve in time (i.e., develop, mature, and decay) and move in space during their life cycles. The fact that precipitation systems have a high correlation along a specific direction implies that those systems in that direction, even in the far distances, have similar/common structures or are originated from the same weather system. For instance, the meteorological systems with strong directionality include a squall line or a frontal system in which several thunderstorms banded together and a multicell cluster that comprises a series of individual storm at a different stage of life cycle with the same movement direction. For the multicell cluster, new cells form along the upwind edge of the cluster, and decaying cells are found along the downwind side with mature cells located at the center; thus, it includes evolutions in both space and time.

In terms of the summertime heavy precipitation systems over the Korean Peninsula, Lee and Kim (2007) classified them into four major types such as isolated thunderstorms, convection bands, squall lines and cloud clusters, whereas Song and Sohn (2015) classified them into two types: (i) the cold type characterized by an eastward-moving cloud system with an oval shape and (ii) the warm type with a comparatively wide spatial distribution over an area extending from the southwest to northeast. In our case, the precipitation events in HPFP may correspond to the isolated thunderstorms or the cold-type heavy rainfalls while those in HPMP correspond to convection bands, squall lines, cloud clusters, or the warm-type heavy rainfalls. On the other hand, Ha et al. (2007) investigated the monthly characteristics in the directional features of rainfall from May to September and reported that the cell type was dominant in May and September while the southwest–northeast tilted band type was ruling in

Table 2. Monthly occurrences of precipitation systems according to the precipitation types from April to October in the period of 2013–2015.

	LPMP	HPMP	HPFP
April	277	31	21
May	106	54	37
June	130	50	186
July	191	213	250
August	193	158	326
September	138	56	116
October	122	32	44

June–August. They addressed that the band type is mainly due to frontal rainfall during the rainy season. In our results, most cases of HPMP and HPFP occurred in July and August (see Table 2); thus, indicating that the warm-season precipitation systems in Ha et al. (2007) are mostly related to the long-lived (HPMP) and short-lived (HPFP) band-type heavy rainfalls along the monsoon frontal system.

4.1.2 Himawari-8 water vapor bands

Water vapor is the core element and driver of the precipitation development through dynamical processes (e.g., advection and convection) and physical processes (e.g., evaporation and condensation). For example, the East Asian monsoon starts when a huge amount of water vapor from the adjacent ocean is transported to the monsoon region by the large-scale atmospheric circulation. Thus, the spatial analysis of water vapor will contribute to further understanding of the spatial patterns of precipitation. Many studies have been carried out on the relationship between satellite water vapor and extratropical or tropical cyclones and storms (e.g., Velden, 1987; Milford and Dugdale, 1990; Krennert and Zwatz-Meise, 2003; De Haan et al., 2004; Rabin et al., 2004; Mukhopadhyay et al., 2005; Chosh et al., 2008).

By analyzing the water vapor imagery, we can detect not only the horizontal distribution of tropospheric water vapor but also the dynamical behavior of atmospheric flow such as the middle and upper troughs, vortexes, and jet streams, even in the absence of clouds. Therefore, analyzing the spatial characteristics of tropospheric water vapor is essential to improve the understanding of precipitation systems. In particular, we are focusing on the extent and direction of spatial correlation of water vapor flowing into the Korean Peninsula in association with the precipitation types.

In this study, we analyze the brightness temperatures from the Himawari-8 water vapor bands to characterize the lower to upper atmosphere related to the precipitation systems. A humid atmosphere absorbs more longwave radiation from the Earth, resulting in a lower brightness temperature. On the other hand, a dry atmosphere absorbs less longwave radiation, bringing about a higher brightness temperature. Although we cannot directly quantify the amount of water vapor through the water vapor imager, we can sufficiently recognize the spatial distribution of water vapor. Moreover, using two water vapor bands (i.e., 6.2 and 7.3 µm), we can clearly identify the moisture boundaries at the zone between the warm/moist and cold/dry side of the jet/wind maximums at two different levels in the troposphere (Georgiev et al., 2016). The spatial analyses were performed with the mixed images of clouds and water vapor because it was hard to distinguish between clouds and water vapor without a cloud detection algorithm. Since both water vapor and clouds are strongly linked to precipitation as its sources, analysis of the mixed variables from the satellite data would not make a serious problem in understanding and relating to the precipitation systems. As we focus on the spatial distribution of water vapor when precipitation occurs, we analyze water vapor for each precipitation type.

The general characteristics of the spatial autocorrelation structure in water vapor bands for each precipitation type are shown in Fig. 6. The patterns of spatial autocorrelations of water vapor bands, for three precipitation types, are generally similar to those of precipitation; however, the separation distances extend much longer (cf. Fig. 2). Even at a separation distance of 100 km, the spatial correlation is higher than 0.6. This strong spatial correlation is considered to be due to the continuity of water vapor. The largest spatial autocorrelation appears in LPMP, followed by HPMP and HPFP. The LPMP depicts similar correlation coefficients regardless of bands and hence heights of atmospheric layers. This implies that the spatial structure of water vapor bands is similar in the vertical. On the other hand, the spatial correlations in HPMP and HPFP become smaller along the downward path, from band 8 to band 10.

In Fig. 7, the brightness temperature of water vapor bands characterizes directionality in the spatial correlations. In general, for all precipitation types and water vapor bands, the highest spatial correlations appear in the axis of 45° (southwest–northeast) while the lowest in the axis of 135° (northwest–southeast), as in precipitation (cf. Fig. 4). In LPMP, the spatial correlations in the axes of 0 and 90° coincide even at longer distances. We also notice that, for a given precipitation type and separation distance, the spatial autocorrelation of water vapor becomes larger along the upward path, from band 10 to band 8. Figure 8 shows the directional separation distances at the mode in the directional histogram of case numbers (not shown) for each precipitation type with the autocorrelation coefficients of 0.7 (blue), 0.8 (black), and 0.9 (grey), respectively. The separation distances generally show high values mainly in the axis of 45°, similar to the directional features of precipitation (cf. Fig. 5). Note that, in LPMP, the brightness temperature shows a large separation distance (i.e., high spatial correlation) in the axis of 45° while precipitation depicts no directionality (cf. Fig. 5). This is possible because water vapor can form as a band type, e.g., over a stationary front, but precipitation can occur in cell type

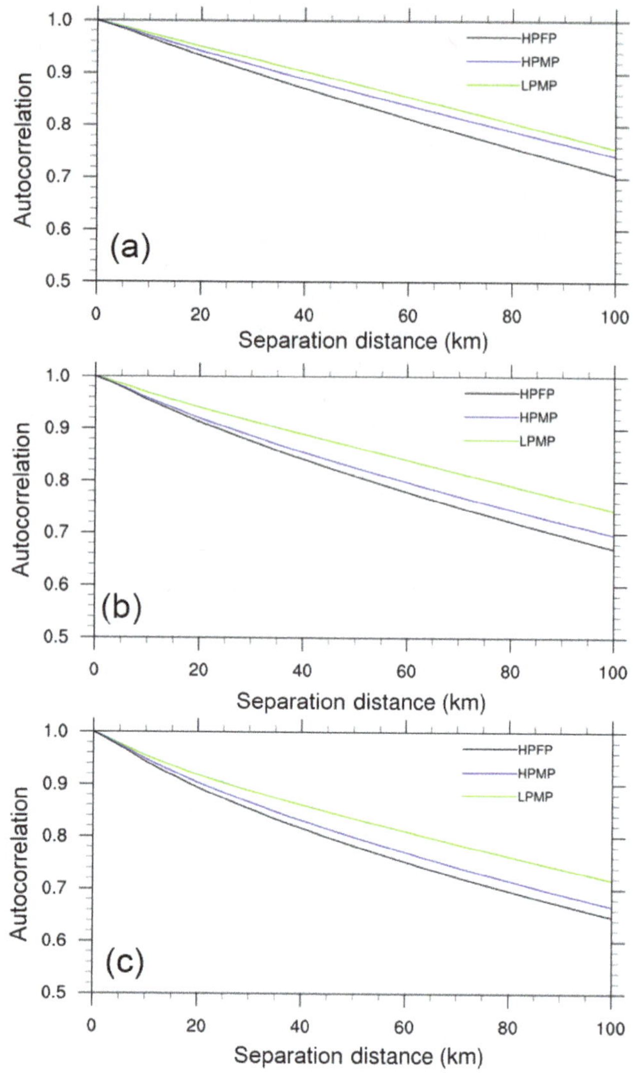

Figure 6. The averaged spatial autocorrelation of brightness temperature of the AHI water vapor band **(a)** 8, **(b)** 9, and **(c)** 10 for each precipitation type.

due to local features of convection over the same front. In addition, as the autocorrelation coefficient decreases, the directionality tends to disappear even though the autocorrelation coefficient is as large as 0.7. As in Fig. 7, we note that for a given precipitation type and spatial autocorrelation, the separation distance of water vapor becomes larger along the upward path. This is related to the stronger wind at upper atmosphere that carries the characteristic feature of water vapor further downstream.

Through the satellite water vapor analyses, we found both similarity and dissimilarity in spatial correlations between water vapor and precipitation. Similar to the results of precipitation analyses, the spatial autocorrelations of water vapor for HPFP decreased more rapidly than those for HPMP and LPMP (cf. Figs. 2 and 6). Water vapor and precipitation showed additional similarity in terms of characteristic directionality in spatial autocorrelations (cf. Figs. 4 and 7, and Figs. 5 and 8). However, both fields showed significant dissimilarity in the separation distances, i.e., the spatial scales (cf. Figs. 2 and 6). Note that water vapor makes phase changes and the conversion from water vapor to precipitation includes a bunch of nonlinear processes; thus, it is not surprising to see such dissimilarity. Further studies on the relationship between precipitation systems and satellite water vapor in Korea are essentially required.

4.2 Moran's *I* and general *G*

Moran's I represents the spatial distribution pattern, i.e., cluster or dispersion: the cluster (dispersion) patterns appear in the areas with positive (negative) spatial autocorrelation. General G determines whether the cluster area is a hot or cold spot: a hot (cold) spot means an area with high (low) values and hence high (low) precipitation amount. For all precipitation cases in this study, the values of the global Moran's I showed positive values, meaning that clusters were detected, in the domain-average sense. Through the local Moran's I (I_i), we can define the similarity between the target feature and its neighbors and check the spatial outliers as well (Lalor and Zhang, 2001; McGrath and Zhang, 2003). For example, with a negative I_i, the target feature may have high precipitation while its surroundings have low precipitation, and vice versa. Conversely, with a positive I_i, the target feature has a similar amount of precipitation as its neighbors. Thus, the local Moran's I can detect strong precipitation over local clusters, precipitation boundaries, or precipitation over large areas. As Moran's I cannot distinguish between low and high precipitation, we employed the local general G to identify the hot/cold spots (i.e., high/low precipitation areas, respectively). In order to calculate the indices, we selected a sub-domain for each precipitation case where the precipitation values were extracted at an interval of 5 km. An inverse distance weighting function is used as the spatial weight.

We present the analysis results for three cases – one from each precipitation type (see Figs. 9–11). For the selected cases, the values of the global Moran's I were all positive: 0.1207, 0.1711, and 0.2357 for the cases of LPMP, HPMP, and HPFP, respectively, with a higher value representing a stronger clustering. This indicates that all three cases show the cluster patterns. Figure 9 shows an LPMP case, which occurred at 05:00 KST (Korean Standard Time) on 25 August 2015, in association with a tropical cyclone. Typhoon Goni (2015) approached from Japan and brought rainfall all over Korea (Fig. 9a). In particular, heavy precipitation systems, with more than 30 mm h^{-1}, developed around the mountain range along the east-central coast of the Korean Peninsula. In Fig. 9b, the southeastern region is identified as hot spots (areas of high precipitation) while the southwestern region, over both land and sea, as cold spots (areas of low precipitation). Figure 9c depicts several precipitation

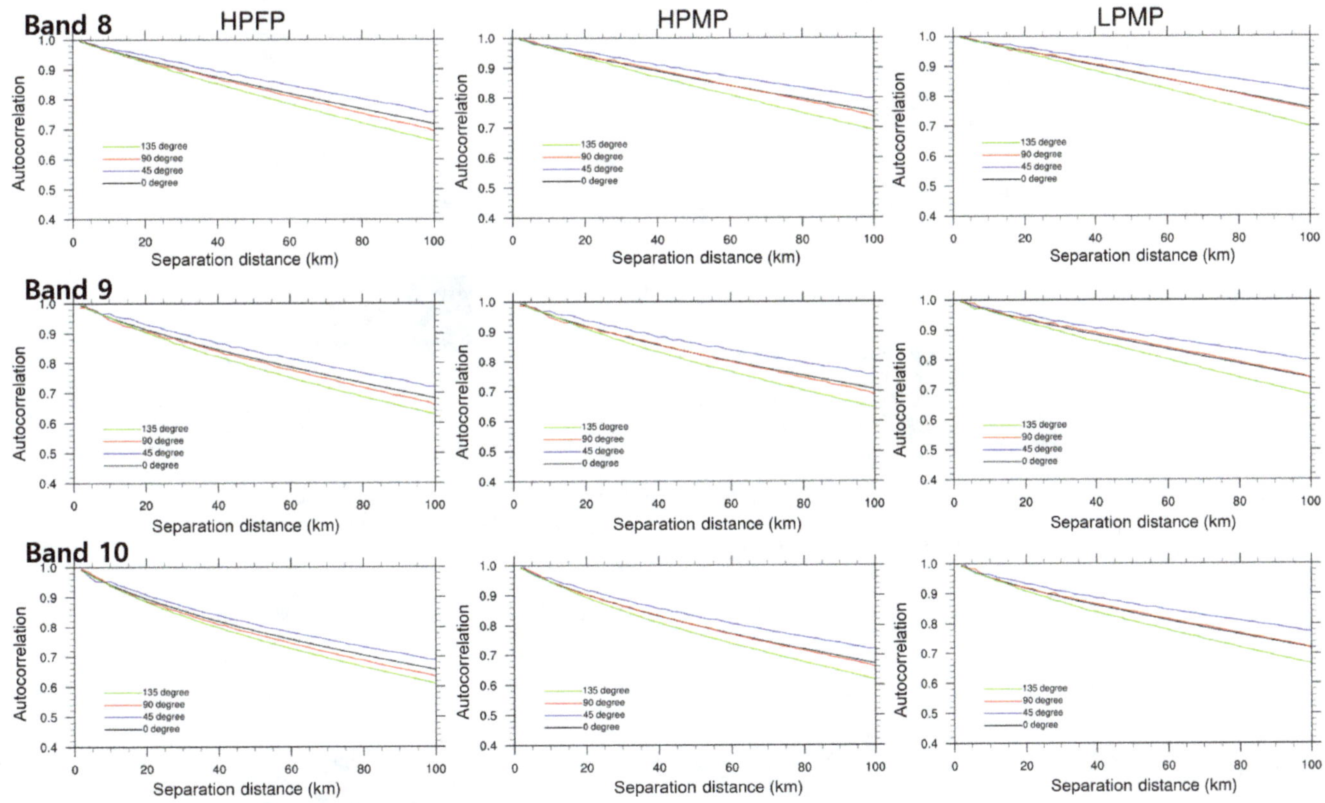

Figure 7. The averaged directional spatial autocorrelation of the brightness temperature of the AHI water vapor band 8, 9, and 10 by each precipitation type (i.e., HPFP, HPMP, and LPMP) for directions of 0° (black), 45° (blue), 90° (red), and 135° (green). The direction (angle) is measured counterclockwise from the origin–east axis (i.e., 0°).

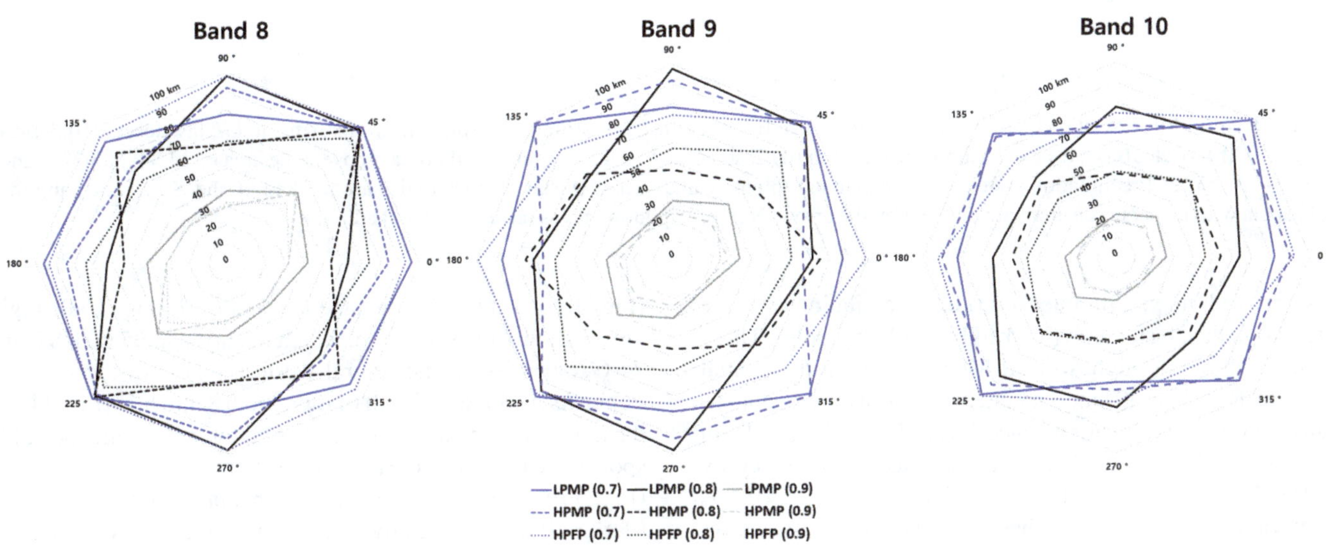

Figure 8. A radar chart representing the directional separation distance (in km) of brightness temperature of the AHI water vapor band 8, 9, and 10 at the mode in the histogram of case numbers for LPMP (solid), HPMP (dashed), and HPFP (dotted). The colors indicate autocorrelation coefficients of 0.7 (blue), 0.8 (black), and 0.9 (grey), respectively.

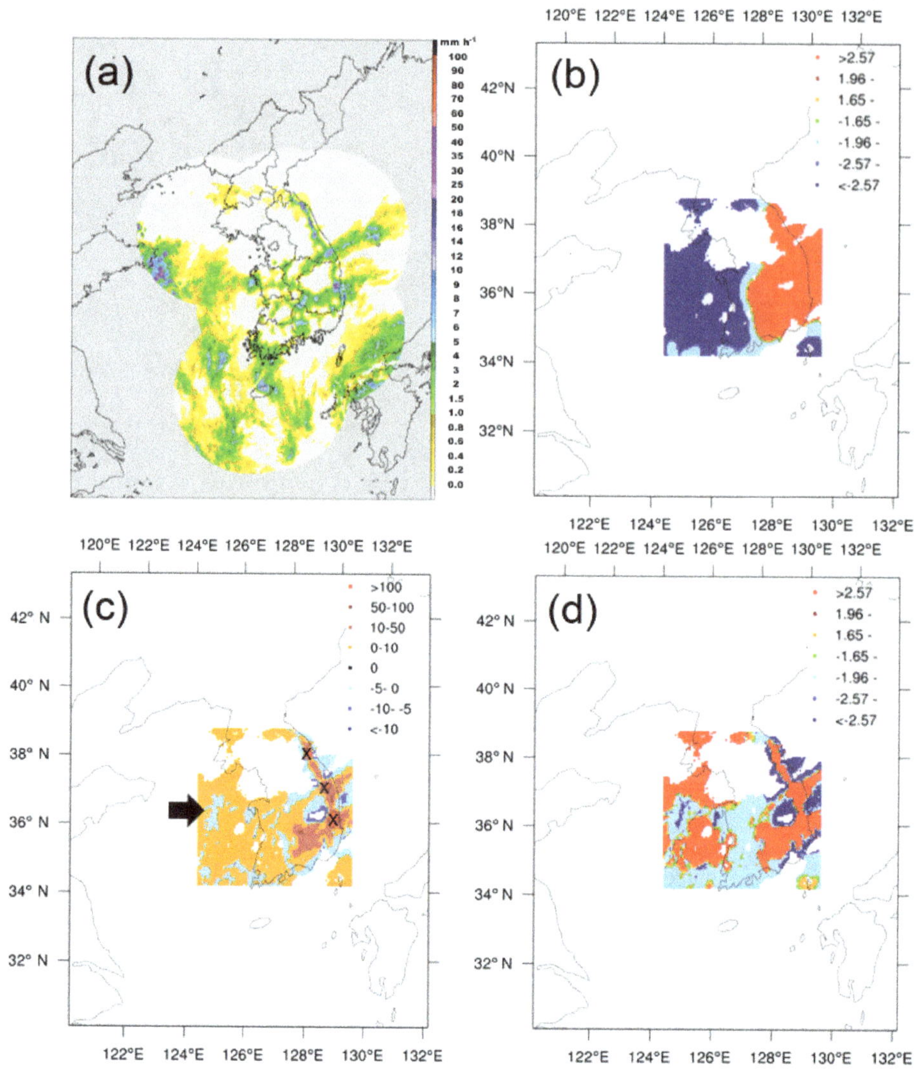

Figure 9. An LPMP case at 05:00 KST on 25 August 2015: **(a)** precipitation distribution (from https://afso.kma.go.kr/, last access: 10 February 2018), **(b)** local $Z(G_i)$, **(c)** local Moran's I (I_i), and **(d)** Z score of I_i. The computational domain covers the area of 34.34–38.97° N and 124.25–130.05° E. Precipitation systems with maximum intensity and strong cluster characteristics are marked by the X symbols, and the cold spots with dispersion pattern are denoted by the arrow. Non-precipitating areas have no color shading.

systems with maximum intensity (i.e., precipitation rate; see Fig. 9a) in the cluster area (marked by the X symbols). These systems show the highest local Moran's I with the spatial scale of less than 30 km. The cluster patterns were statistically significant at a significance level of 0.01 (Fig. 9d). In Fig. 10, we illustrate a heavy rainfall case in the category of HPMP, which occurred at 17:00 KST on 27 May 2013 as a migratory cyclone approached from the west, passed over the Korean Peninsula slowly, and brought rainfall throughout the country, mainly in the western and southern regions (Fig. 10a). Such precipitation areas matched well with the hot spots in Fig. 10b. The map of local Moran's I shows that a precipitation system with a strong cluster feature has developed, over the southwestern sea of the Korean Peninsula, along with the successive cluster patterns in a line type (Fig. 10c), with a significance level of 0.01 (Fig. 10d). Figure 11 shows an HPFP case, which occurred at 05:00 KST on 24 October 2015. The central-western region of the Korean Peninsula received rainfall of more than 8 mm h^{-1} (Fig. 11a): the hot spots appeared in the precipitation area while the cold spots were located at the region of relatively low precipitation (Fig. 11b). Figure 11c illustrates an intense cluster pattern (dark red) over the relatively high rainfall area and a weaker cluster pattern over the cold spot, with a significance level of about 0.01 (Fig. 11d).

Overall, the spatial structures of the selected precipitation systems are well presented by the analyses of Moran's I and general G. The overall spatial pattern (cluster vs. dispersion) of a precipitation system is characterized by the global Moran's I, whereas the spatial distributions of clus-

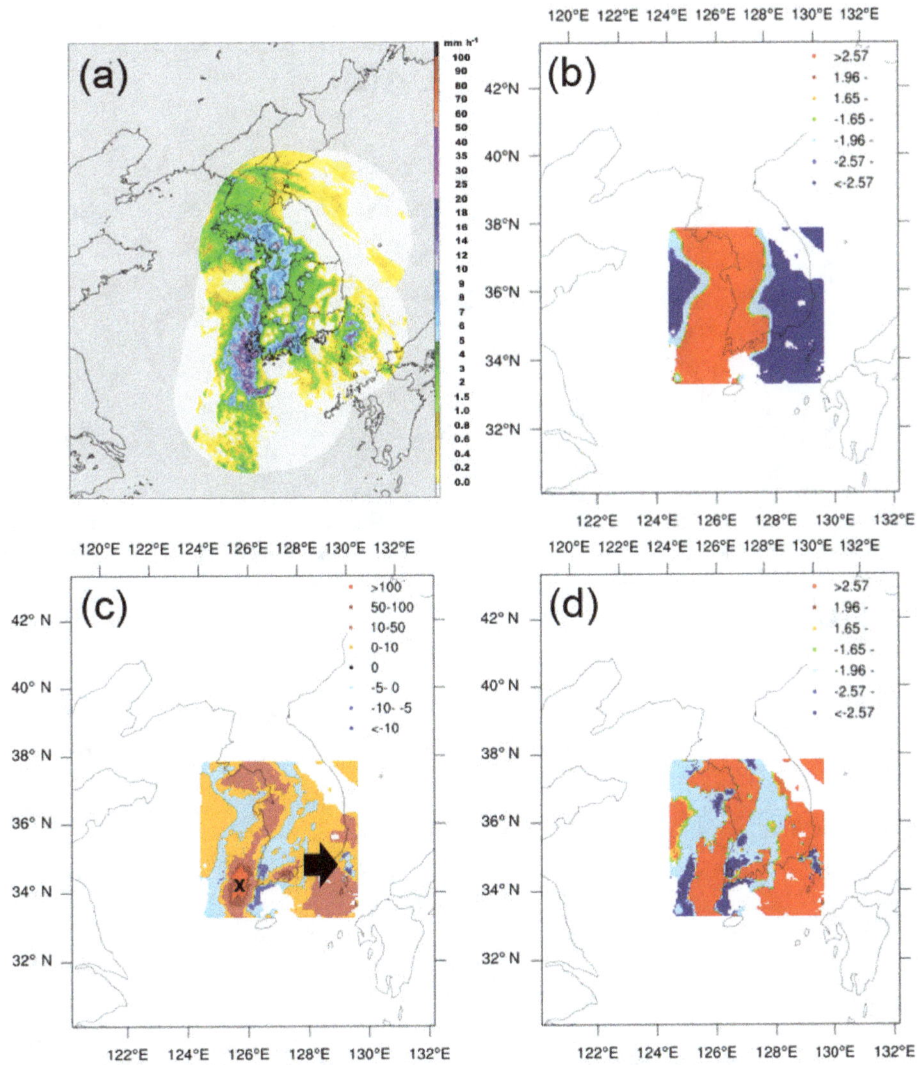

Figure 10. Same as in Fig. 9 but for an HPMP case at 17:00 KST on 27 May 2013 and the computational domain of 33.43–38.05° N and 124.25–130.04° E.

tered areas are represented by the local Moran's I. The overall regional features of precipitation amount (i.e., hot vs. cold spot) are captured by the local general G. For example, precipitation in the HPFP case, relevant to convective clusters, appeared over a small local area (Fig. 11a) but the value of the global Moran's I was much higher (0.2357) than the others (0.1207 and 0.1711), implying a stronger cluster pattern; however, considering that the global Moran's I is a domain-averaged value, this high value may be due to fewer dispersion (negative correlation) areas. In fact, the HPFP case shows a relatively weaker dispersion area (Fig. 11c), compared to the LPMP and HPMP cases (Figs. 9c and 10c, respectively). It is notable that the cluster pattern shows different spatial characteristics depending on the spatial scale and distribution feature of precipitation systems: several strong local clusters develop, especially over areas of high precipitation, when precipitation occurs over a large region (Figs. 9 and 10), whereas weaker clusters develop when precipitation occurs over a small region (Fig. 11). Through the integrated interpretation of these indices, we discovered some important spatial features of the precipitation systems: (1) the area where the cluster pattern overlaps with the hot spots is characterized by heavy rainfall (e.g., the X-marked or dark red places in Figs. 9c and 10c); (2) the area where the dispersion pattern concurs with the cold spots is distinguished by a relatively high precipitation surrounded by low precipitation (e.g., the rain cells denoted by arrows in Figs. 9c and 10c); (3) the precipitation gets weaker as it moves away from the strong cluster area – that is, the cluster pattern appears at the center and the dispersion pattern appears around; and (4) the secondary cluster pattern in a precipitation system over a wide area indicates clusters of low-intensity precipitation.

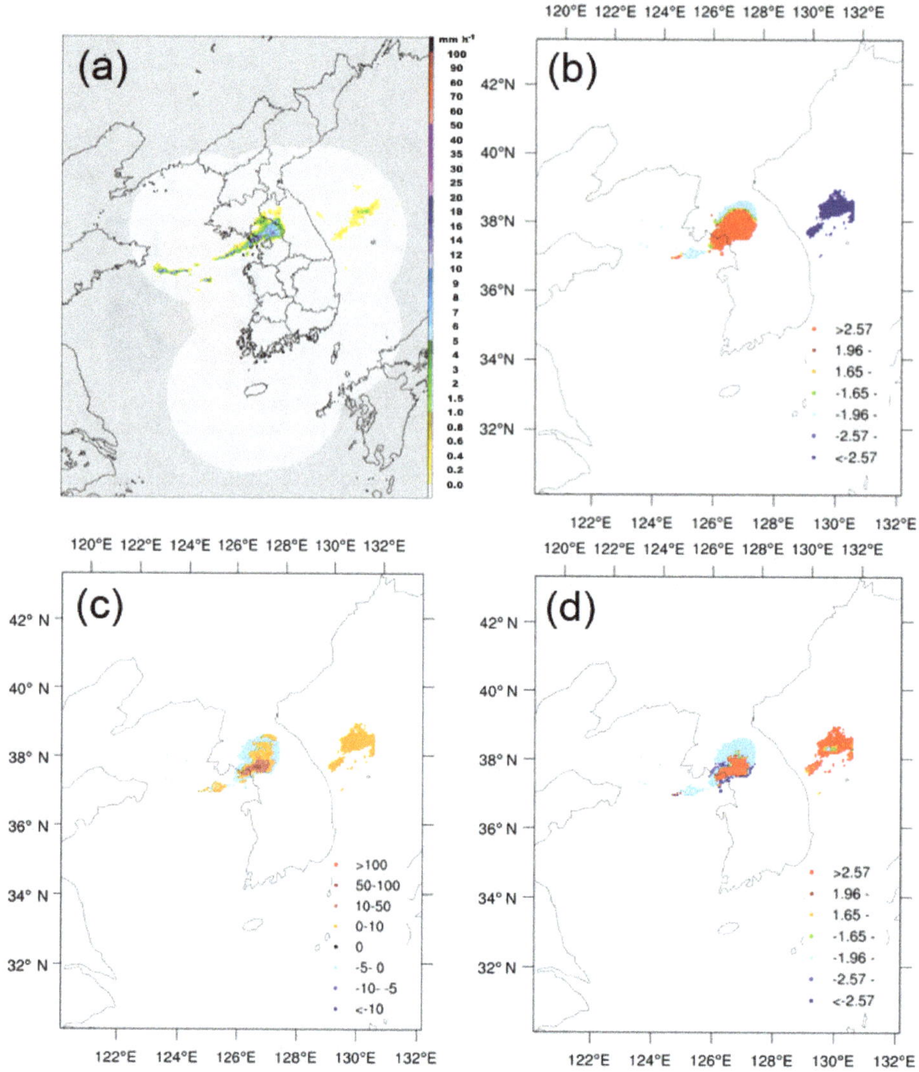

Figure 11. Same as in Fig. 9 but for an HPFP case at 05:00 KST on 24 October 2015 and the computational domain of 37.14–39.06° N and 123.32–131.21° E.

4.3 Temporal autocorrelation

We additionally calculated the temporal autocorrelation coefficients of precipitation and water vapor bands for each precipitation type. For precipitation, HPFP represents the shortest *e*-folding time, followed by LPMP and HPMP (Fig. 12). This order in the *e*-folding time is the same as in the *e*-folding distance: this implies that, for the precipitation types classified here, the temporal correlation as well as the spatial correlation appears greater in the precipitation systems over a larger area than that over a smaller area. The range of the *e*-folding time for all precipitation types is 1–2 h. The averaged temporal autocorrelations of water vapor bands show similar features to those of precipitation for different precipitation types, though the correlations decrease slowly over time (Fig. 13). In addition, compared with the spatial correlations, the temporal correlations do not show significant dependence on the height of atmospheric layer.

Through this temporal correlation analyses, we noticed that the *e*-folding time is in a short range (1–2 h), and its difference among different precipitation types is only about 30 min. This short *e*-folding timescale for all precipitation types implies that the precipitation systems affecting Korea in the warm season are mostly characterized by convective-type precipitation, from either isolated storm cells or clustered bands, at least for the given analysis period. The *e*-folding timescale suggests an adequate time interval for data collection and analysis for capturing the detailed structure of and better forecasting of precipitation systems. It also implies a proper time interval for incorporating observations in the operational data assimilation system, for more accurate numerical forecasting of the precipitation systems.

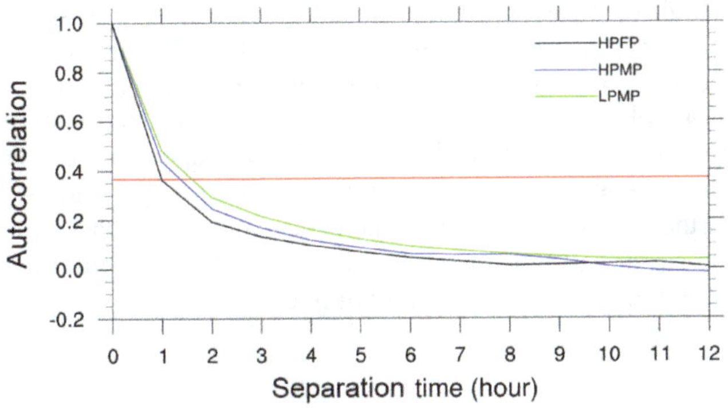

Figure 12. Same as in Fig. 2 but for the averaged temporal autocorrelation.

Moreover, Ha et al. (2007) reported that the *e*-folding time of precipitation in Korea is 1–2 h regardless of months (from May to September) and that the monthly difference of the *e*-folding time is approximately 30–40 min. Therefore, we can conclude that the typical *e*-folding time of the warm-season precipitation systems in Korea is 1–2 h, regardless of the precipitation types and months. This conclusion is based on the analyses of the hourly precipitation data: we may find different temporal characteristics for different precipitation types using a data set of shorter interval (e.g., 10 min). In terms of the satellite water vapor data, a further study is also necessary to investigate the relationship between water vapor transport and precipitation, regarding the temporal scale as well as the spatial scale, with more detailed analyses.

5 Conclusions

Heavy rainfall causes many casualties as well as property damage, and thus its accurate forecast is very important. To improve the forecast accuracy, it is necessary to understand the characteristics of the precipitation systems through analyses of data from observation and/or numerical modeling. Furthermore, a well-designed observation network is essentially required to capture the characteristic features of precipitation systems.

In this study, we identify the characteristics of warm-season precipitation systems in Korea, via the geostatistical analyses on the composite precipitation and satellite water vapor data. We have classified the precipitation cases into four types, based on the average rainfall amount per point and the ratio of the points with precipitation (i.e., spatial scale): (1) low precipitation at a few points (LPFP); (2) low precipitation at many points (LPMP); (3) high precipitation at a few points (HPFP); and (4) high precipitation at many points (HPMP), among which the LPFP cases are excluded from the analysis.

We have conducted analyses of the spatial and temporal autocorrelations of precipitation and water vapor for the precipitation cases of each classification type and found that the

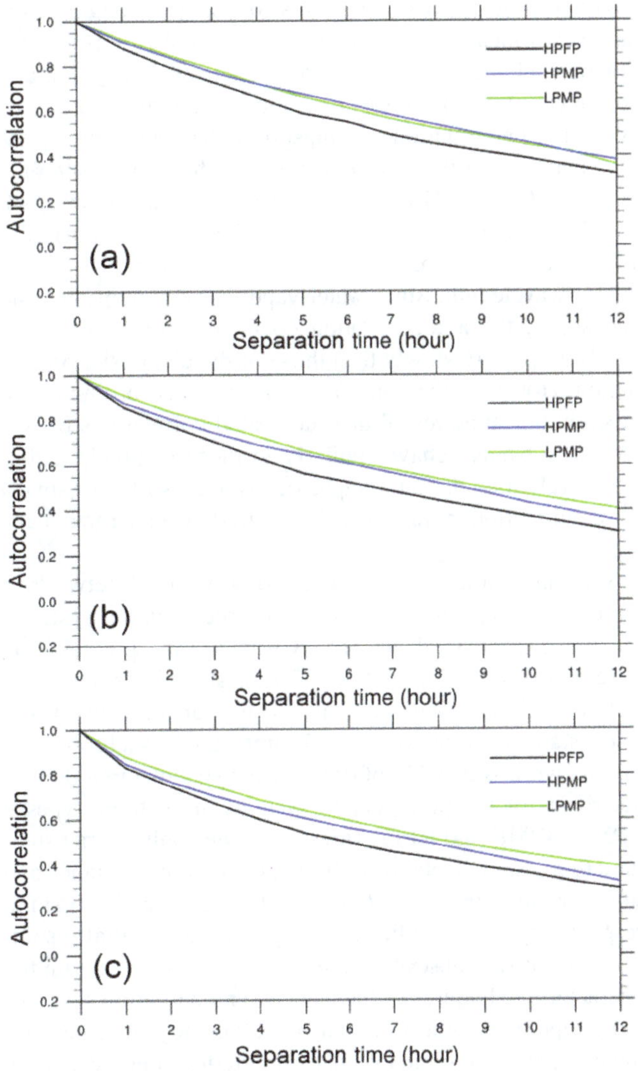

Figure 13. Same as in Fig. 6 but for the averaged temporal autocorrelation.

e-folding distance of precipitation ranges between 15 and 35 km while the *e*-folding time ranges between 1 and 2 h. The *e*-folding distance is as short as 15 km for an HPFP case (e.g., a local shower), but as large as ~ 35 km for the cases of both LPMP and HPMP, implying that the spatial correlation becomes larger as precipitation occurs over a wider area and is less affected by the precipitation amount. We also noted that the spatial autocorrelations have the following characteristic directionality: (1) precipitation systems with high precipitation amount (i.e., HPFP and HPMP) have high spatial autocorrelations in the southwest–northeast and west–east directions, mainly associated with frontal rainfalls during the monsoon season; and (2) those with low precipitation amount (i.e., LPMP) have no directionality, pertaining to both the migratory cyclones moving in various directions and the moderate convective systems in the cloud clusters. Other spatial characteristics were identified as well, including cluster versus dispersion patterns, hot versus cold spots, strong precipitation over local clusters, precipitation boundaries, precipitation over a large area, and so forth. In general, heavy rainfalls appear over the hot spots with the cluster pattern while an isolated high-precipitation system, surrounded by low precipitation, occurs over the cold spots with the dispersion pattern.

The Advanced Himawari Imager, onboard Himawari-8, has three water vapor bands – 8 (6.2 µm), 9 (6.9 µm), and 10 (7.3 µm) – representing the lower atmospheric layer with higher wavelength. Since water vapor has continuity in time and space, the autocorrelation coefficients of water vapor bands drop more slowly than those of the composite precipitation. However, the comparative analyses according to the precipitation type reveal that the spatial correlations of water vapor bands behave similarly to those of precipitation: the correlations of water vapor bands are also large (small) for precipitation systems which occurred over a large (small) area.

We can also assess the current observational network in Korea, focusing on the warm-season precipitation systems, via the geostatistical analyses. Currently, the precipitation observation network in Korea has a spatial resolution of ~ 13 km, distributing the analysis over an interval of 1 h. For a standard based on the e-folding values, such observation network is capable of deriving the spatiotemporal characteristics of the precipitation systems in all three types – LPMP, HPMP, and HPFP. However, if we apply a strict standard (e.g., autocorrelation of 0.6), the separation distance and time for the HPFP cases fall down to ~ 6 km and ~ 35 min, respectively. This implies that, regarding the spatial resolution, the current observation network can hardly capture the characteristic features of the localized precipitation systems; the temporal resolution might be sufficiently high because the composite precipitation data are produced every 10 min. Therefore, for more accurate analyses of the precipitation systems with a higher correlation, a denser observation network should be considered in local areas where a heavy rainfall often occurs. This study has enabled us not only to explore the characteristics of precipitation systems in Korea but also to suggest the criteria for evaluating the observation network.

We note that the spatial distribution and movement features of the precipitation systems are strongly affected by terrain, but the orographic effect is not fully assessed in this study. Bacchi and Kottegoda (1995) demonstrated the orographic effect of the Alps mountain range on the spatial distribution of the rainfalls and addressed the directional characteristics of the spatial correlations. Other studies also confirmed the direct and induced effect of mountains on the spatial distribution of rainfall (e.g., Mass, 1981; Carruthers and Choularton, 1983; Barros and Kuligowski, 1997; Park and Lee, 2007). As the Korean Peninsula is represented by a highly complex terrain with many mountains, the spatial properties of precipitation systems will be distinctively affected. It is essential to conduct further research about the orographic effect on the spatiotemporal characteristics of precipitation, specifically over different slope sides of various mountain ranges in Korea. We additionally suggest a further study to analyze the autocorrelations by considering the storm cell movement and the occurrences of initial precipitation and maximum rainfall intensity.

Competing interests. The authors declare that they have no conflict of interest.

Acknowledgements. This work is supported by the Korea Meteorological Administration (KMA) Research and Development Program under Grant KMIPA, mainly through KMIPA2016-1010 and partly through KMI2018-06710. We acknowledge KMA for supplying the composite precipitation and the Himawari-8 water vapor bands data.

Edited by: Luis Samaniego

References

Anselin, L.: Local Indicators of Spatial Association – LISA, Geogr. Anal., 27, 93–115, https://doi.org/10.1111/j.1538-4632.1995.tb00338.x, 1995.

Bacchi, B. and Kottegoda, N. T.: Identification and calibration of spatial correlation patterns of rainfall, J. Hydrol., 165, 311–348, 1995.

Barros, A. P. and Kuligowski, R. J.: Orographic effects during a severe wintertime rainstorm in the Appalachian mountain, Mon. Weather Rev., 126, 2648–2672, 1997.

Barton, Y., Giannakaki, P., Von Waldow, H., Chevalier, C., Pfahl, S., and Martius, O.: Clustering of regional-scale extreme precipitation events in southern Switzerland, Mon. Weather Rev., 144, 347–369, 2016.

Bessho, K., Date, K., Hayashi, M., Ikeda, A., Imai, T., Inoue, H., Kumagai, Y., Miyakawa, T., Murata, H., Ohno, T., Okuyama, A., Oyama, R., Sasaki, Y., Shimazu, Y., Shimoji, K., Sumida, Y., Suzuki, M., Taniguchi, H., Tsuchiyama, H., Uesawa, D., Yokota, H., and Yoshida, R.: An introduction to Himawari-8/9 – Japan's new-generation geostationary meteorological satellites, J. Meteorol. Soc. Jpn. Ser. II, 94, 151–183, https://doi.org/10.2151/jmsj.2016-009, 2016.

Bretherton, C. S., Peters, M. E., and Back, L. E.: Relationships between water vapor path and precipitation over the tropical oceans, J. Climate, 17, 1517–1528, https://doi.org/10.1175/1520-0442(2004)017<1517:RBWVPA>2.0.CO;2, 2004.

Carruthers, D. J. and Choularton, T. W.: A model of the feeder-seeder mechanism of orographic rain including stratification and wind-drift effects, Q. J. Roy. Meteorol. Soc., 109, 575–588, 1983.

Cassardo, C., Park, S. K., Thakuri, B. M., Priolo, D., and Zhang, Y.: Soil surface energy and water budgets during a monsoon season in Korea, J. Hydrometeorol., 10, 1379–1396, 2009.

Chang, H. and Kwon, W.-T.: Spatial variations of summer precipitation trends in South Korea, 1973–2005, Environ. Res. Lett., 2, 045012, https://doi.org/10.1088/1748-9326/2/4/045012, 2007.

Choi, S.-W., Lee, S.-J., Kim, J. Lee, B.-L., Kim, K.-R., and Choi, B.-C.: Agrometeorological observation environment and periodic report of Korea Meteorological Administration: Current status and suggestions, Korean J. Agr. Forest Meteorol., 17, 144–155, https://doi.org/10.5532/KJAFM.2015.17.2.144, 2015.

Chosh, A., Lohar, D., and Das, J.: Initiation of Nor'wester in relation to mid-upper and low-level water vapor patterns on METEOSAT-5 images, Atmos. Res., 87, 116–135, 2008.

Ciach, G. J. and Krajewski, W. F.: Analysis and modeling of spatial correlation structure in small-scale rainfall in Central Oklahoma, Adv. Water Resour., 29, 1450–1463, 2006.

De Haan, S., Barlag, S., Baltink, H. K., Debie, F., and Van der Marel, H.: Synergetic use of GPS water vapor and Meteosat images for synoptic weather forecasting, J. Appl. Meteorol., 43, 514–518, 2004.

Eltahir, E. A. B. and Bras, R. L.: Precipitation recycling, Rev. Geophys., 34, 367–378, https://doi.org/10.1029/96RG01927, 1996.

Emmanuel, I., Andrieu, H., Leblois, E., and Flahaut, B.: Temporal and spatial variability of rainfall at the urban hydrological scale, J. Hydrol., 430, 162–172, 2012.

Feng, Y. and Kitzmiller, D. H.: A short-range quantitative precipitation forecast algorithm using back-propagation neural network approach, Adv. Atmos. Sci., 23, 405–414, 2006.

Fritsch, J. M., Houze Jr., R. A., Adler, R., Bluestein, H., Bosart, L., Brown, J., Carr, F., Davis, C., Johnson, R. H., Junker, N., Kuo, Y.-H., Rutledge, S., Smith, J., Toth, Z., Wilson, J. W., Zipser, E., and Zrnic, D.: Quantitative precipitation forecasting: Report of the eighth prospectus development team, U.S. Weather Research Program, B. Am. Meteorol. Soc., 79, 285–299, 1998.

Fu, W. J., Jiang, P. K., Zhou, G. M., and Zhao, K. L.: Using Moran's *I* and GIS to study the spatial pattern of forest litter carbon density in a subtropical region of southeastern China, Biogeosciences, 11, 2401–2409, https://doi.org/10.5194/bg-11-2401-2014, 2014.

Georgiev, C. G., Santurette, P., and Maynard K.: Weather Analysis and Forecasting: Applying Satellite Water Vapor Imagery and Potential Vorticity Analysis, 2nd Edn., Academic Press, Cambridge, USA, 2016.

Getis, A. and Ord, J. K.: The analysis of spatial association by use of distance statistics, Geogr. Anal., 24, 189–206, 1992.

Gimeno, L., Stohl, A., Trigo, R. M., Dominguez, F., Yoshimura, K., Yu, L., Drumond, A., Durán-Quesada, A. M., and Nieto, R.: Oceanic and terrestrial sources of continental precipitation, Rev. Geophys., 50, RG4003, https://doi.org/10.1029/2012RG000389, 2012.

Ha, K.-J., Jeon, E.-H., and Oh, H.-M.: Spatial and temporal characteristics of precipitation using an extensive network of ground gauge in the Korean Peninsula, Atmos. Res., 86, 330–339, 2007.

Hong, K.-O., Shu, M.-S., and Rha, D.-K.: Temporal and spatial variations of precipitation in South Korea for recent 30 Years (1976–2005) and geographic environments, J. Korean Earth Sci. Soc., 27, 433–449, 2006.

Hwang, D.-I., Lee, J. W., Seo, D.-I., Na, H., and Seo, Y.-K.: Prediction system using extrapolation-based nowcasting techniques for very short-range forecasting and improvement of 1 hr accumulated precipitation forecasts, in: Proceedings of the Autumn Meeting of Korean Meteorological Society, 12–14 October 2015, Jeju, Korea, 260–261, http://203.247.66.28/, 2015.

In, S.-R., Han, S.-O., Im, E.-S., Kim, K.-H., and Shim, J.: Study on temporal and spatial characteristics of summertime precipitation over Korean Peninsula, Atmos., 24, 159–171, 2014 (in Korean with English abstract).

Jeong, J.-H., Lee, D.-I., Wang, C.-C., Jang, S.-M., You, C.-H., and Jang, M.: Environment and morphology of mesoscale convective systems associated with the Changma front during 9–10 July 2007, Ann. Geophys., 30, 1235–1248, https://doi.org/10.5194/angeo-30-1235-2012, 2012.

Jeong, J.-H., Lee, D.-I., Wang, C.-C., and Han, I.-S.: Characteristics of mesoscale-convective-system-produced extreme rainfall over southeastern South Korea: 7 July 2009, Nat. Hazards Earth Syst. Sci., 16, 927–939, https://doi.org/10.5194/nhess-16-927-2016, 2016.

Jung, W. and Lee, T.-Y.: Formation and evolution of mesoscale convective systems that brought the heavy rainfall over Seoul on September 21, 2010, Asia-Pacific J. Atmos. Sci., 49, 635–647, https://doi.org/10.1007/s13143-013-0056-4, 2013.

Kim, B. J., Kripalani, R., Oh, J. H., and Moon, S.-E.: Summer monsoon rainfall patterns over South Korea and associated circulation features, Theor. Appl. Climatol. 72, 65–74, https://doi.org/10.1007/s007040200013, 2002.

Kim, E.-H., Kim, M.-K., and Lee, W.-S.: The regional characteristics of daily precipitation intensity in Korea for recent 30 years, J. Korean Earth Sci. Soc., 26, 404–416, 2005.

Kim, H. W. and Lee, D. K.: An observational study of mesoscale convective systems with heavy rainfall over the Korean Peninsula, Weather Forecast., 21, 125–148, https://doi.org/10.1175/WAF912.1, 2006.

Kim, O. Y. and Oh, J. H.: Verification of the Performance of the High Resolution QPF Model for Heavy Rainfall Event over the Korean Peninsula, Asia-Pacific J. Atmos. Sci., 46, 119–133, 2010.

Ko, J.-W., Baek, H.-J., and Kwon, W.-T.: The characteristics of precipitation and regionalization during rainy season in Korea, J. Korean Meteorol. Soc., 41, 101–114, 2005.

Krennert, T. and Zwatz-Meise, V.: Initiation of convective cells in relation to water vapour boundaries in satellite images, Atmos. Res., 67–68, 353–366, 2003.

Kursinski, A. L. and Mullen, S. L.: Spatiotemporal variability of hourly precipitation over the eastern contiguous United States from stage IV multisensor analyses, J. Hydrometeorol., 9, 3–21, 2008.

Lalor, G. and Zhang, C. S.: Multivariate outlier detection and remediation in geochemical databases, Sci. Total Environ., 281, 99–109, 2001.

Lee, D.-K., Park, J.-G., and Kim, J.-W.: Heavy Rainfall Events Lasting 18 Days from July 31 to August 17, 1998, over Korea, J. Meteorol. Soc. Jpn. Ser. II, 86, 313–333, https://doi.org/10.2151/jmsj.86.313, 2008.

Lee, J., Yoon, J., and Jun, H. D.: Evaluation for the correction of radar rainfall due to the spatial distribution of raingauge network, J. Korea Soc. Hazard Mitig., 14, 337–345, https://doi.org/10.9798/KOSHAM.2014.14.6.337, 2014.

Lee, J.-Y., Kim, W., and Lee, T.-Y.: Physical and dynamic factors that drove the heavy rainfall event over the middle Korean Peninsula on 26–27 July 2011, Asia-Pacific J. Atmos. Sci., 53, 101–120, https://doi.org/10.1007/s13143-017-0009-4, 2017a.

Lee, J.-Y., Kwon, M., Yun, K.-S., Min, S.-K., Park, I.-H., Ham, Y.-G., Jin, E. K., Kim, J.-H., Seo, K.-H., Kim, W., Yim, S.-Y., and Yoon, J.-H.: The long-term variability of Changma in the East Asian summer monsoon system: A review and revisit, Asia-Pacific J. Atmos. Sci., 53, 257–272, https://doi.org/10.1007/s13143-017-0032-5, 2017b.

Lee, T.-Y. and Kim, Y.-H.: Heavy precipitation systems over the Korean Peninsula and their classification, J. Korean Meteorol. Soc., 43, 367–396, 2007.

Lee, Y. H., Park, S. K., and Chang, D.-E.: Parameter estimation using the genetic algorithm and its impact on quantitative precipitation forecast, Ann. Geophys., 24, 3185–3189, https://doi.org/10.5194/angeo-24-3185-2006, 2006.

Mass, C.: Topographically forced convergence in western Washington State, Mon. Weather Rev., 109, 1335–1347, 1981.

McGrath, D. and Zhang, C.: Spatial distribution of soil organic carbon concentrations in grassland of Ireland, Appl. Geochem., 18, 1629–1639, 2003.

Milford, J. R. and Dugdale, G.: Estimation of rainfall using geostationary satellite data, in: Applications of Remote Sensing in Agriculture, edited by: Steven, M. D. and Clark, J. A., Butterworth, London, 97–110, 1990.

Moran, P. A. P.: The interpretation of statistical maps, J. Roy. Stat. Soc. B, 10, 243–251, 1948.

Mukhopadhyay, P., Singh, H. A. K., and Singh, S. S.: Two severe Nor'westers in April 2003 over Kolkata using Doppler Radar observations and satellite imagery, Weather, 60, 343–353, 2005.

Nam, J.-E., Lee, Y. H., Ha, J.-C., and Cho, C.-H.: A study on the *e*-folding distance of summer precipitation using precipitation reanalysis data, in: Proc. the Autumn Meeting of Korean Meteorological Society, Korean Meteorol. Soc., 13–15 October 2014, Jeju, Korea, 657–658, 2014.

Okuyama, A., Andou, A., Date, K., Hosaka, K., Mori, N., Murata, H., Tabata, T., Takahashi, M., Yoshino, R. and Bessho, K.: Preliminary validation of Himawari-8/AHI navigation and calibration. Proc. SPIE, 9607, 96072E, https://doi.org/10.1117/12.2188978, 2015.

Ord, J. K. and Getis, A.: Local spatial autocorrelation statistics: distributional issues and an application, Geogr. Anal., 27, 286–306, 1995.

Park, S. K.: Nonlinearity and predictability of convective rainfall associated with water vapor perturbations in a numerically simulated storm, J. Geophys. Res., 104, 31575–31587, https://doi.org/10.1029/1999JD900446, 1999.

Park, S. K. and Lee, E.: Synoptic features of orographically enhanced heavy rainfall on the east coast of Korea associated with Typhoon Rusa (2002), Geophys. Res. Lett., 34, L02803, https://doi.org/10.1029/2006GL028592, 2007.

Rabin, R. M., Corfidi, S. F., Brunner, J. C., and Hane, C. E.: Detecting winds aloft from water vapor satellite imagery in the vicinity of storms, Weather, 59, 251–257, https://doi.org/10.1256/wea.182.03, 2004.

Ritschel, C., Ulbrich, U., Névir, P., and Rust, H. W.: Precipitation extremes on multiple timescales – Bartlett–Lewis rectangular pulse model and intensity-duration-frequency curves, Hydrol. Earth Syst. Sci., 21, 6501–6517, https://doi.org/10.5194/hess-21-6501-2017, 2017.

Riyu, L., Oh, J. H., Kim, B. J., Baek, H.-J., and Ronghui, H.: Associations with the interannual variations of onset and withdrawal of the Changma, Adv. Atmos. Sci., 18, 1066–1080, https://doi.org/10.1007/s00376-001-0023-3, 2001.

Rodriguez-Iturbe, I., Cox, D. R., and Isham, V.: Some models for rainfall based on stochastic point processes, P. Roy. Soc. Lond. A, 410, 269–288, 1987.

Schiemann, R., Erdin, R., Willi, M., Frei, C., Berenguer, M., and Sempere-Torres, D.: Geostatistical radar-raingauge combination with nonparametric correlograms: methodological considerations and application in Switzerland, Hydrol. Earth Syst. Sci., 15, 1515–1536, https://doi.org/10.5194/hess-15-1515-2011, 2011.

Skøien, J. O., Blöschl, G., and Western, A. W.: Characteristic space scales and timescales in hydrology, Water Resour. Res., 39, 1304, https://doi.org/10.1029/2002WR001736, 2003.

Smith, B. L. and Yuter, S. E.: Water vapor fluxes and orographic precipitation over northern California associated with a landfalling atmospheric river, Mon. Weather Rev., 138, 74–100, https://doi.org/10.1175/2009MWR2939.1, 2010.

Song, H.-J. and Sohn, B.-J.: Two heavy rainfall types over the Korean Peninsula in the humid East Asian summer environment: A satellite observation study, Mon. Weather Rev., 143, 363–382, https://doi.org/10.1175/MWR-D-14-00184.1, 2015.

Stohl, A., Forster, C., and Sodemann, H.: Remote sources of water vapor forming precipitation on the Norwegian west coast at 60° N – a tale of hurricanes and an atmospheric river, J. Geophys. Res., 113, D05102, https://doi.org/10.1029/2007JD009006, 2008.

Sun, J. and Lee, T.-Y.: A Numerical Study of an Intense Quasi-stationary Convection Band over the Korean Peninsula, J. Meteorol. Soc. Jpn. Ser. II, 80, 1221–1245, https://doi.org/10.2151/jmsj.80.1221, 2002.

Tompkins, A. M.: Organization of tropical convection in low vertical wind shears: The role of water vapor, J. Atmos. Sci., 58, 529–545, https://doi.org/10.1175/1520-0469(2001)058<0529:OOTCIL>2.0.CO;2, 2001.

Trenberth, K. E.: Atmospheric moisture recycling: Role of advection and local evaporation, J. Climate, 12, 1368–1381, 1999.

Trenberth, K. E. and Stepaniak, D. P.: Co-variability of components of poleward atmospheric energy transports on seasonal and interannual timescales, J. Climate, 16, 3690–3704, 2003a.

Trenberth, K. E. and Stepaniak, D. P.: Seamless poleward atmospheric energy transports and implications for the Hadley circulation, J. Climate, 16, 3705–3721, 2003b.

Velden, C. S.: Satellite observations of Hurricane Elena (1985) using the VAS 6.7 μm "watervapor" channel, B. Am. Meteorol. Soc., 68, 210–215, 1987.

Volkmann, T. H. M., Lyon, S. W., Gupta, H. V., and Troch, P. A.: Multicriteria design of rain gauge networks for flash flood prediction in semiarid catchments with complex terrain, Water Resour. Res., 46, W11554, https://doi.org/10.1029/2010WR009145, 2010.

Yu, X., Park, S. K., Lee, Y. H., and Choi, Y. S.: Quantitative precipitation forecast of a tropical cyclone through optimal parameter estimation in a convective parameterization, Sci. Online Lett. Atmos., 9, 36–39, 2013.

Zeweldi, D. A. and Gebremichael, M.: Evaluation of CMORPH precipitation products at fine space-time scales, J. Hydrometeorol., 10, 300–307, 2009.

Zhou, T.-J. and Yu, R.-C.: Atmospheric water vapor transport associated with typical anomalous summer rainfall patterns in China, J. Geophys. Res., 110, D08104, https://doi.org/10.1029/2004JD005413, 2005.

13

Assessment of the Weather Research and Forecasting (WRF) model for simulation of extreme rainfall events in the upper Ganga Basin

Ila Chawla[1,5], **Krishna K. Osuri**[3,4], **Pradeep P. Mujumdar**[1,2], **and Dev Niyogi**[4,5]

[1]Department of Civil Engineering, Indian Institute of Science, Bangalore, 560012, India
[2]Divecha Centre for Climate Change, Indian Institute of Science, Bangalore, 560012, India
[3]Department of Earth and Atmospheric Sciences, NIT Rourkela, Odisha, 769008, India
[4]Department of Agronomy- Crops, Soils, Water Sciences, Purdue University, West Lafayette, IN 47907, USA
[5]Department of Earth, Atmospheric and Planetary Sciences, Purdue University, West Lafayette, IN 47907, USA

Correspondence: Pradeep P. Mujumdar (pradeep@iisc.ac.in)

Abstract. Reliable estimates of extreme rainfall events are necessary for an accurate prediction of floods. Most of the global rainfall products are available at a coarse resolution, rendering them less desirable for extreme rainfall analysis. Therefore, regional mesoscale models such as the advanced research version of the Weather Research and Forecasting (WRF) model are often used to provide rainfall estimates at fine grid spacing. Modelling heavy rainfall events is an enduring challenge, as such events depend on multi-scale interactions, and the model configurations such as grid spacing, physical parameterization and initialization. With this background, the WRF model is implemented in this study to investigate the impact of different processes on extreme rainfall simulation, by considering a representative event that occurred during 15–18 June 2013 over the Ganga Basin in India, which is located at the foothills of the Himalayas. This event is simulated with ensembles involving four different microphysics (MP), two cumulus (CU) parameterizations, two planetary boundary layers (PBLs) and two land surface physics options, as well as different resolutions (grid spacing) within the WRF model. The simulated rainfall is evaluated against the observations from 18 rain gauges and the Tropical Rainfall Measuring Mission Multi-Satellite Precipitation Analysis (TMPA) 3B42RT version 7 data. From the analysis, it should be noted that the choice of MP scheme influences the spatial pattern of rainfall, while the choice of PBL and CU parameterizations influences the magnitude of rainfall in the model simulations. Further, the WRF run with Goddard MP, Mellor–Yamada–Janjic PBL and Betts–Miller–Janjic CU scheme is found to perform "best" in simulating this heavy rain event. The selected configuration is evaluated for several heavy to extremely heavy rainfall events that occurred across different months of the monsoon season in the region. The model performance improved through incorporation of detailed land surface processes involving prognostic soil moisture evolution in Noah scheme compared to the simple Slab model. To analyse the effect of model grid spacing, two sets of downscaling ratios – (i) 1 : 3, global to regional (G2R) scale and (ii) 1 : 9, global to convection-permitting scale (G2C) – are employed. Results indicate that a higher downscaling ratio (G2C) causes higher variability and consequently large errors in the simulations. Therefore, G2R is adopted as a suitable choice for simulating heavy rainfall event in the present case study. Further, the WRF-simulated rainfall is found to exhibit less bias when compared with the NCEP FiNaL (FNL) reanalysis data.

1 Introduction

Indian summer monsoon rainfall (ISMR) is often associated with very heavy (124.5 to 244.4 mm day^{-1}) to extremely heavy (more than 244.5 mm day^{-1}) rainfall (Indian Meteorological Department, Terminologies and Glossary; http://imd.gov.in/section/nhac/termglossary.pdf), particularly during June to September months. The extremely heavy rainfall events usually occur due to the presence of organized meso-convective systems (MCSs) embedded in large-scale monsoonal features such as offshore troughs and vortices, depressions over the Bay of Bengal and Arabian Sea, and mid-tropospheric cyclones (Sikka and Gadgil, 1980; Webster et al., 1998; Fasullo and Webster, 2003).

Extremely heavy rainfall on shorter timescales are particularly difficult to predict in mountainous terrains and continue to be a challenge to operational and research communities (Das et al., 2008; Li et al., 2017). Global models have been employed in several studies to understand the large-scale circulation pattern and for quantitative analysis of the monsoon rainfall, but due to their coarse resolution, they are unable to represent the local to regional characteristics of monsoon rainfall. Regional models, however, can explicitly simulate the interactions between the large-scale weather phenomenon and regional topography, making the climate simulations reliable (Gadgil and Sajani, 1998; Ratna et al., 2011; Srinivas et al., 2013). Furthermore, regional models have a better representation of convection, thus offsetting one of the major sources of errors and uncertainties in the global models. Therefore, regional models become a preferred choice to study seasonal monsoon rainfall.

The advanced research version of the Weather Research and Forecasting model (hereafter referred to as the WRF model) is a regional popular community model that is widely used for both studying as well as forecasting a variety of high-impact meteorological events, such as rainfall (Vaidya and Kulkarni, 2007; Deb et al., 2008; Kumar et al., 2008; Chang et al., 2009; Routray et al., 2010; Mohanty et al., 2012), tropical cyclones (Raju et al., 2011; Routray et al., 2016; Osuri et al., 2017b) and thunderstorms (Madala et al., 2014; Osuri et al., 2017a). Several works are reported in the literature which have considered the WRF model over the Himalayan region. Kumar et al. (2012) used the WRF model to simulate the cloudburst event of 2010 in the Leh area over the north-western Himalayan belt, while Kumar et al. (2014) and Thayyen et al. (2013) used the WRF model to gain insight into the atmospheric processes and the MCSs that led to the 2010 Leh event. Similarly, Chevuturi et al. (2015) simulated the heavy precipitation event of September 2012 in the central Himalayas using the WRF model. Medina et al. (2010) used the WRF model to understand how topography and land surface conditions affect the extreme convection in western and eastern Himalayas. Particularly for the 2013 heavy rainfall episode in the Uttarakhand region, the WRF model is used in several studies, including those by Kotal et al. (2014), Vellore et al. (2016) and Hazra et al. (2017), to understand the physical processes leading to the event. Shekhar et al. (2015), Chevuturi and Dimri, (2016), and Dimri et al. (2016) performed in-depth synoptic analysis of the June 2013 heavy rainfall event using the WRF model. Rajesh et al. (2016) presented the role of land surface conditions in simulating the heavy rainfall event. Therefore, from the existing literature, it can be established that the regional model performs considerably well over the region. However, finding the optimal set of physics parameterization schemes (along with the selection of appropriate model grid spacing and resolution) to simulate heavy to extremely heavy rainfall events, and understanding the effect of the combination of different parameterization schemes on rainfall estimates over the Indian monsoon region is still an active area of research.

Earlier studies such as those by Krishnamurthy et al. (2009), Misenis and Zhang, (2010), Rauscher et al. (2010), Mohanty et al. (2012), and Chevuturi et al. (2015) indicated that heavy rainfall predictions can be improved through ensemble model techniques and fine grid resolution. However, the influence of the interaction between model parameterization schemes on mesoscale rainfall simulations over India is still an understudied issue. In particular, heavy rain simulation studies have reviewed the impact of individual parameterization options such as the microphysics (MP) scheme (Rajeevan et al., 2010; Raju et al., 2011; Kumar et al., 2012), cumulus (CU) parameterization scheme (Deb et al., 2008; Mukhopadhyay et al., 2010; Srinivas et al., 2013; Madala et al., 2014), planetary boundary layer (PBL) scheme (Li and Pu, 2008; Hu et al., 2010; Hariprasad et al., 2014) and land surface model (LSM) options (Chang et al., 2009). However, the ensemble analysis that reviews the relative impact of different configurations and the associated variability (uncertainty) is lacking. It is important to study the impact of different parameterizations in an ensemble mode because it is often likely that the performance of one scheme depends on other model configurations considered. For example, the conclusions regarding which CU scheme performs best would be intimately tied to the choice of the MP or land surface options considered in conducting the numerical experiments. With this perspective, this paper seeks to assess the sensitivity of the WRF model in predicting heavy to extremely heavy rainfall episodes over the Ganga Basin in the foothills of the Himalayas. Specific tasks undertaken in this work are as follows: (i) quantitative verification of the WRF model to simulate an extremely heavy rainfall event; (ii) assessment of the sensitivity of the model-simulated rainfall to different parameterization options, downscaling ratios and land surface models; (iii) validation of the selected configuration for other rainfall events over the region; and (iv) comparison of the WRF-simulated rainfall with the global reanalysis data to investigate the impact of local versus global factors on rainfall simulations. A related objective is to provide suitable recommendations on a possible optimal choice for model configuration to simulate such heavy rainfall events in the region.

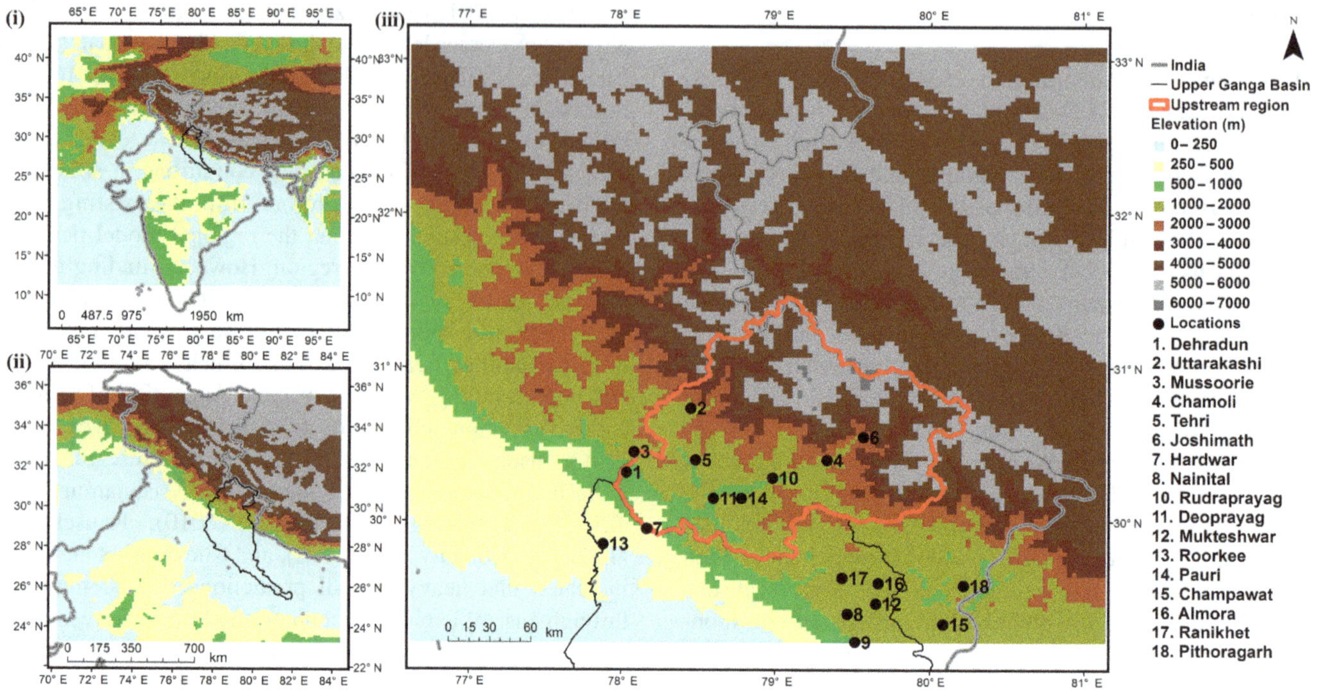

Figure 1. Topography of the study region (shown with black outline) as represented in the WRF model for **(i)** Domain 1–27 km grid spacing, **(ii)** Domain 2a–9 km grid spacing (downscaling ratio–1 : 3) and **(iii)** Domain 2b–3 km grid spacing (downscaling ratio–1 : 9). Locations of the rain gauge stations within the UGB are presented as black dots in **(iii)**.

Description of the heavy rainfall event

The 2013 summer monsoon had a normal onset but the trough advanced rapidly, covering the whole of India by mid-June, instead of mid-July (Ray et al., 2014). This large-scale setting is thought to have created a platform for interaction between two synoptic-scale events – a north-west moving depression from the Bay of Bengal and a pre-existing westerly trough in the mid-troposphere. Meteorological studies conducted over the region (Kotal et al., 2014; Ray et al., 2014; Chevuturi and Dimri, 2016; Rajesh et al., 2016) established that there was a monsoon low-pressure system during this period. The longitudinal time section for 850 hPa geopotential height along with anomaly averaged over 20–26° N showed a high negative anomaly on 14 June, which migrated to the west, moving over 75° E by 17 June. The meridional wind anomaly within the belt of 35–45° N showed a westerly wave, moving from 10° E on 12 June to 70° E on 17 June. These two anomalies are found to be in phase, consequently causing interaction between the eastward-moving trough in the mid- to upper troposphere and westward-moving monsoon low in the lower troposphere. The monsoon low provided the moisture feed and the upper-level westerly trough provided the divergence to lift the moisture. This whole system eventually led to an unanticipated heavy rainfall during 15–18 June 2013 in the Kedarnath valley and adjoining areas in the state of Uttarakhand, India (Kotal et al., 2014; Ray et al., 2014; Chevuturi and Dimri, 2016; Rajesh et al., 2016). The region received rainfall greater than 370 mm in 1 day (16–17 June 2013), which is 375 % above the daily normal rainfall (65.9 mm) during the monsoon season (Ray et al., 2014). Consequently, heavy floods occurred in the region, causing unprecedented damage to life and property.

The synthesis of the synoptic setting of the event has been carried out in a number of studies such as Dube et al. (2014), Kotal et al. (2014), Ray et al. (2014), Shekhar et al. (2015), Chevuturi and Dimri, (2016), and Rajesh et al. (2016), but the mesoscale assessment pertaining to the simulation of this rainfall event is still lacking. Therefore, the present study emphasizes quantitatively evaluating and conducting the sensitivity analysis of the WRF model in predicting extreme rainfall. The ability of the WRF model to simulate heavy rainfall events is further verified by considering additional episodes (apart from the June 2013 event, details of which are presented in Sect. 2.1) that occurred within this region across different monsoon months.

A study region comprising of the upstream part of the Ganga Basin in India, referred as Upper Ganga Basin (UGB) hereafter, is selected for the analysis in this paper. Figure 1 presents the topography of the UGB as described for the three domains of the WRF model (Domain 1, Domain 2a and Domain 2b, with 27, 9 and 3 km grid resolution respectively) along with the 18 rain gauge stations located within the region. The region is of social, cultural and economic importance to India, further making this study necessary.

2 Data and experimental setup

2.1 Observed data

Figure 2 presents daily and cumulative rainfall data from 15 to 18 June 2013 obtained from the Indian Meteorological Department (IMD) and the literature (Ray et al., 2014) for the 18 official rain gauges located within the UGB.

It is noticed that the north-west part of the region received higher rainfall compared to the north-east, with stations such as Tehri and Dehradun showing 327 and 210 % (respectively) more rainfall than their historic means. A few stations like Chamoli in the north-east region received 250 mm of cumulative rainfall over the 4-day period, which is 144 % higher than the historic mean. In general, most of the stations in the southern part of the basin, which are located at a lower elevation, recorded relatively less rainfall with a cumulative range of 445 mm, in comparison to the northern part (at higher altitude), which had a rainfall range of 515 mm. Additionally, three stations in the south-east region, i.e. Mukteshwar, Haldwani and Nainital received extremely heavy rainfall with a cumulative average of 498 mm. From the above analysis, it is evident that the system moved from east to west with two distinct regions in the UGB – south-east and north-west – receiving extremely heavy rainfall.

The region has complex topography and a limited number of rain gauges because of the difficulty in operating a network in this region. To further capture the spatial variability in rainfall, the Tropical Rainfall Measuring Mission (TRMM) Multi-Satellite Precipitation Analysis (TMPA) 3B42RT (version 7) product, which is available at 0.25° resolution on a daily scale, is analysed (Fig. 3). It is to be noted that, since the focus area for the analyses is the upstream region of the UGB (Figs. 1iii and 2), results are presented with respect to the geographical extent of Domain 2b throughout this paper. From Fig. 3, it can be noted that the TMPA data are able to capture the spatial variability in the rainfall – with distinct clusters corresponding to heavy rainfall in the north-west and south-east regions of the study area. However, the rainfall amount is significantly underestimated by the TMPA product, with the maximum value of 265 mm against the recorded 650 mm. This under-reporting for gridded satellite products versus rain gauges in the ISMR region is a well-known feature (Rahman et al., 2009; Mishra and Srinivasan, 2013; Kneis et al., 2014; Bharti et al., 2016). The TMPA estimates are verified against the IMD station observations for a baseline quality check. Mean absolute error (MAE), root mean square error (RMSE) and bias (β) are computed using the nearest neighbourhood mapping approach and are presented in Table 1.

TMPA data are observed to behave differently for different ranges of rainfall values over the study domain. TMPA overestimated the rainfall at stations with cumulative rainfall less than 200 mm (e.g. Pauri and Almora). In contrast, rainfall at stations receiving more than 250 mm of cumulative rainfall is underestimated. Stations that received rainfall of 200–250 mm are well represented in the TMPA data (e.g. Mussorie, Pithoragarh and Champawat). From the analysis, it could be inferred that in the TMPA data rainfall values are clustered towards the mean value. Errors noticed in the TMPA data could be attributed to two factors: first, large spatial coverage and coarse resolution of the TMPA data, and second, for comparison with the observed data a simple approach of selecting the nearest grid point is implemented.

Table 1. Comparison of TMPA data with station data.

Station	Mean absolute error (mm)	Root mean square error (mm)	Bias* (%)
Uttarakashi	64	97	−40
Tehri	63	83	−44
Mussorie	75	94	−6
Dehradun	126	191	−70
Joshimath	51	55	−21
Chamoli	44	55	−27
Rudraprayag	48	53	−22
Pauri	34	44	37
Deoprayag	95	98	−56
Hardwar	83	114	−76
Roorkee	64	82	−39
Ranikhet	71	77	27
Almora	26	30	63
Mukteshwar	90	118	−60
Nainital	86	111	−67
Haldwani	115	157	−80
Pithoragarh	56	69	−6
Champawat	100	126	1

* Bias (%) = $\frac{\text{(Station data} - \text{TMPA data)}}{\text{Station data}} \times 100$.

In addition to June 2013 case, five additional heavy to extremely heavy rainfall events are also considered in the present study for the analysis, details of which are presented in Table 2. Rainfall from the IMD gridded data at 0.25° resolution (Pai et al., 2014) is considered as the observed data for these events.

It is to be noted that on 13–14 September 2012, a cloudburst event was reported in the region and the total amount of rainfall on 14 September was recorded as approximately 210 mm (Chevuturi et al., 2015). This event is significantly underestimated in the IMD gridded data, indicating that caution must be exercised while using the data for applications involving heavy rainfall events, such as flood modelling and validating the rainfall simulations from the mesoscale models. Figure 4 presents the spatially averaged daily and cumulative rainfall received during different events (as specified in Table 2).

2.2 Model configuration and experimental setup

The simulation experiments in this paper are conducted using the WRF model, version 3.8. WRF is a numerical weather

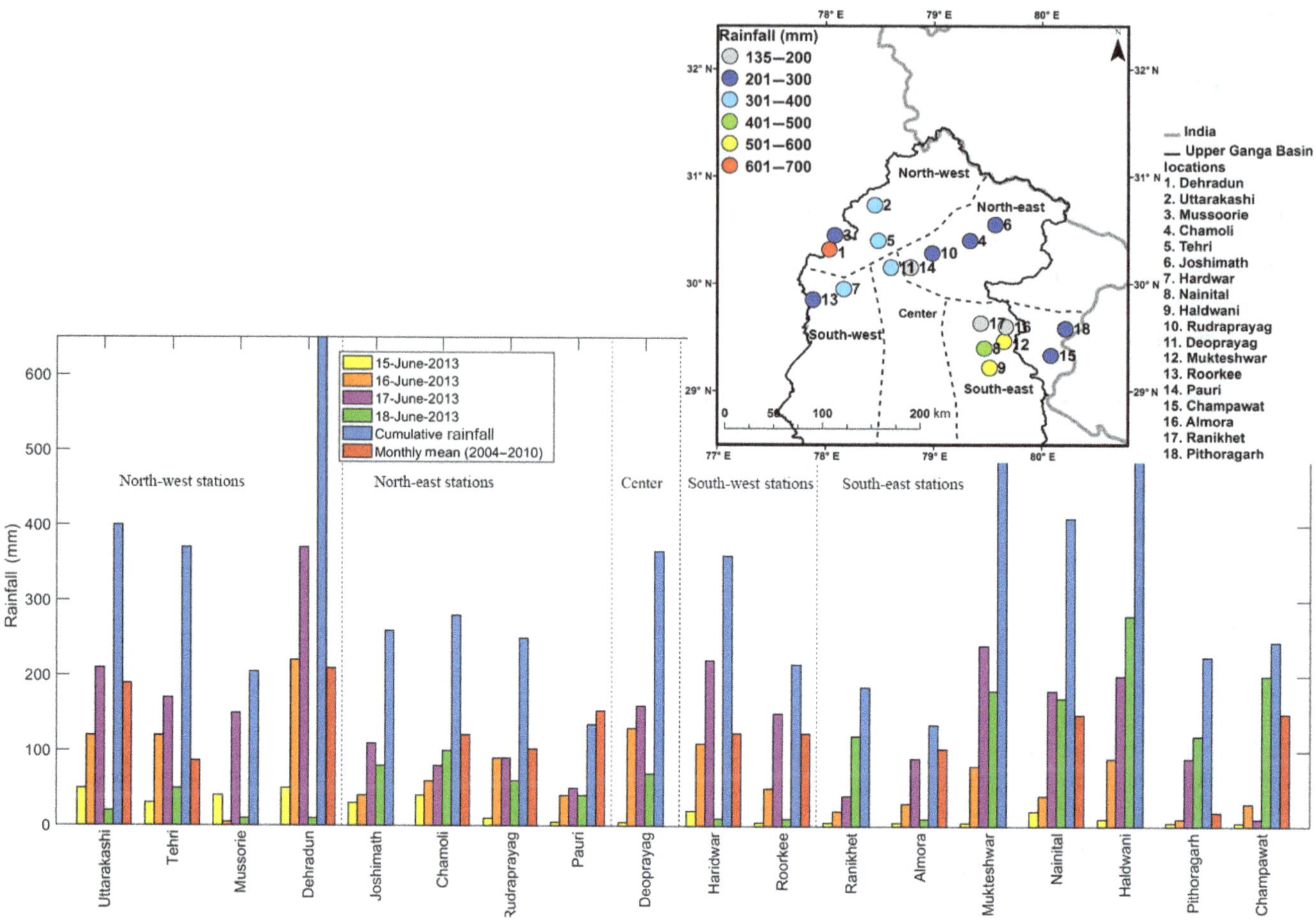

Figure 2. Observed daily and cumulative rainfall along with historic monthly mean (2004–2010) values at the 18 rain gauges in the upstream region of the UGB.

Table 2. Heavy to extremely heavy rainfall events recorded in the UGB region.

Event no.	Time period	Maximum rainfall day	Maximum rainfall amount (mm)
1	18–22 June 2008	20 June	126
2	29 July–2 August 2010	31 July	271
3	15–19 August 2011	16 August	234
4	17–21 September 2010	19 September	218
5	11–15 September 2012	14 September	38

prediction (NWP), non-hydrostatic, mesoscale model, available with several advanced physics and numerical schemes, designed for better prediction of atmospheric processes. The model description and updates can be found from Skamarock et al. (2005) and the WRF user web page (http://www2.mmm.ucar.edu/wrf/users/).

The WRF model utilizes large-scale atmospheric forcing as input for initialization and lateral boundary condition. These large-scale conditions are regridded by the model domain, considering the grid spacing and local topographical as well as other terrain conditions. As is common for most WRF studies over the Indian region, the National Centers for Environmental Prediction (NCEP) global FiNaL (FNL) analysis dataset, based on Global Data Assimilation System (GDAS) with Global Forecast System (GFS), is considered. The FNL data are available at a coarse resolution of $1^\circ \times 1^\circ$, at 6 h intervals – 00:00, 06:00, 12:00 and 18:00 UTC (Coordinated Universal Time) – and is used to provide initial and boundary conditions to the model. The lateral boundary conditions in the WRF model are updated at 6 h inter-

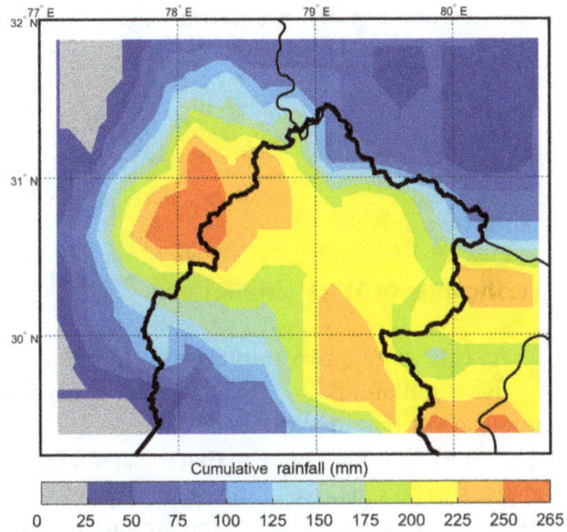

Figure 3. Cumulative rainfall in the upstream region of the UGB obtained from the Tropical Rainfall Measuring Mission (TRMM) Multi-Satellite Precipitation Analysis (TMPA) 3B42RTV7 product at 0.25° resolution. The domain shown is similar to the insert shown in Fig. 2.

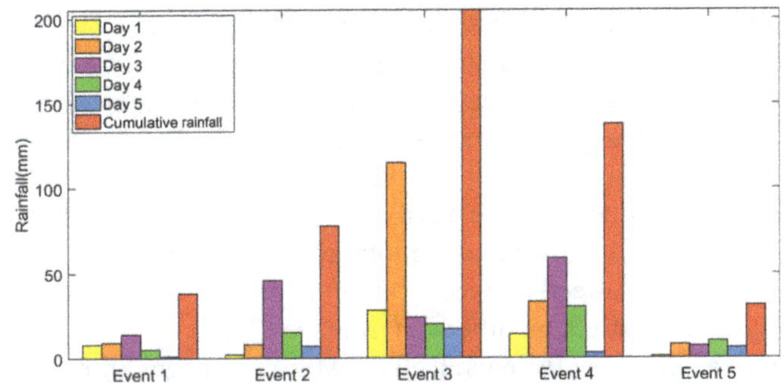

Figure 4. Spatially averaged daily and cumulative rainfall for Event 1 (18–22 June 2008), Event 2 (29 July–2 August 2010), Event 3 (15–19 August 2011), Event 4 (17–21 September 2010) and Event 5 (11–15 September 2012) in the upstream region of the UGB.

vals. Considering the short duration of the run, the model was forced with fixed sea surface temperature (SST) throughout the integration, and no regional data assimilation is carried out. The land surface boundary conditions are taken from the Moderate Resolution Imaging Spectroradiometer (MODIS) International Geosphere–Biosphere Programme (IGBP) 21-category land use and land cover fields that are available with a horizontal grid spacing of 10 min. Three telescopically nested domains are used in this study – the parent domain (Domain 1) is fixed between 60 and 100° E with grid spacing of 27 km; the first nested domain (Domain 2a) covers 70–85° E, 22–37° N with 9 km grid spacing and is indicative of "global to regional scale" (G2R) downscaling, and the second nested domain (Domain 2b) covers 76–81.5° E and 28.5–34° N at 3 km grid spacing (Fig. 1), for global to convective scale (G2C) downscaling (Trapp et al., 2007). The parent domain provides lateral boundary conditions to the inner domains, resulting in the downscaling ratios for simulations as 1 : 3 and 1 : 9. The three domains use 30 vertical pressure levels, with the top fixed at 50 hPa. The model time steps were a function of grid spacing: 135, 45 and 15 s respectively for the three domains.

The model configuration used default parameterization options following Osuri et al. (2012). For example, shortwave radiation is based on Dudhia (Dudhia, 1989) and longwave radiation is based on the Rapid Radiative Transfer Model (RRTM; Mlawer et al., 1997) scheme. Other physical parameterization options such as microphysics, cumulus parameterization schemes, planetary boundary layer and land surface models were selected as outlined ahead. There is currently no known unique configuration that can best simulate an extremely heavy rainfall event. Therefore, based on the literature (e.g. Kumar et al., 2008; Hong and Lee, 2009; Misenis and Zhang, 2010; Mukhopadhyay et al., 2010; Argüeso et al., 2011; Cardoso et al., 2013; Efstathiou et al., 2013), four MP schemes, two CU schemes, two PBL schemes and two LSMs are considered to obtain an ensemble of rainfall simulations. The two PBL schemes considered are the Yonsei University (YSU) scheme (Hong et al., 2006) and the Mellor–Yamada–Janjic (MYJ) scheme (Janić, 2001). YSU is a non-local scheme, wherein fluxes are calculated at a certain height in the PBL considering the profile of the entire domain. The MYJ scheme, however, is a local scheme in which fluxes are calculated at various heights within the PBL and are related to vertical gradient in the atmospheric variables at the same height. Further details regarding the difference between the YSU and the MYJ schemes can be obtained from Misenis and Zhang (2010) and Efstathiou et al. (2013). The two CU schemes considered are the Kain–Fritsch (KF) scheme (Kain, 2004) and the Betts–Miller–Janjic (BMJ) scheme (Janjić, 1994, 2000). The KF scheme is both a shallow and deep convection scheme, wherein shallow convection is allowed for updrafts that do not reach minimum precipitating cloud depth. This scheme is based on entrainment and detrainment plume model with updrafts and downdrafts of mass flux. Potential energy is removed in the convective timescale within this scheme. Furthermore, it includes cloud, rain, snow and ice detrainment at cloud tops. BMJ also considers convection at both shallow and deep levels. However, there is no updraft and downdraft of mass flux and no cloud detrainment. Domain 2b is configured without any CU scheme, assuming MP can explicitly solve the convection at the finer resolution (Sikder and Hossain, 2016). The four MP schemes considered are the Purdue Lin (PLin) scheme (Lin et al., 1983; Chen and Sun, 2002), the Eta Ferrier (Eta) scheme (NOAA, 2001), the WRF Single-Moment 6-class (WSM6) scheme (Hong and Lim, 2006) and the Goddard scheme (Tao et al., 1989). Both the PLin scheme and the WSM6 scheme are based on the parameterization from Rutledge and Hobbs (1984) and have 6-class microphysics, which includes the mixing ratios of water vapour, cloud water, cloud ice, snow, rain and graupel. The main difference between these two schemes is related to the treatment of ice-phase microphysical processes. Details of the PLin and the WSM6 schemes are available in Hong et al. (2009). The Eta scheme was designed primarily for computational efficiency in NWP models, wherein the total condensate and the water vapours are directly advected into the model. The Goddard scheme is a slight modification from the PLin scheme for ice-water saturation. In general, all the MP schemes are known to influence the rainfall simulations at fine grid resolution by influencing the water phase component (Li et al., 2017). Since each physics scheme is associated with a distinct feature, it is important to examine the effect of their interactions on the rainfall simulations.

The sensitivity of various WRF configurations in simulating heavy rainfall events is assessed using the Noah LSM (Chen and Dudhia, 2001; Ek et al., 2003; Tewari et al., 2004). The Noah LSM is a community model that is included in the WRF suite with the prime aim of providing reliable boundary conditions to the atmospheric model. As a result, Noah LSM is a moderately detailed model, which includes a single canopy layer with the canopy resistance scheme of Noilhan and Planton (1989) and four soil layers (at 0.1, 0.3, 0.6 and 1.0 m), with a total soil depth of 2 m. The last soil layer of 1 m acts as a reservoir for drainage of water under gravity and the above three layers serve as root-zone depths. There is a provision in the model to allow default root-zone depths to be replaced with the actual values from the field, subject to data availability. In the Noah LSM, surface (skin) temperature is obtained using a single linearized surface energy balance equation, which effectively considers the ground and vegetation surface. Frozen soil parameterization based on Koren et al. (1999) and the surface runoff scheme of Schaake et al. (1996) are also included in this model. Soil moisture, soil temperature, water intercepted by the canopy and snow stored on the ground are also included as the prognostic variables in the model. More detailed information on the Noah LSM can be obtained from Ek et al. (2003).

To assess the effect of the land surface scheme on simulations, the Noah LSM is replaced with the simple five-layer soil model (Slab; Dudhia, 1996). In contrast to the relatively sophisticated Noah LSM, Slab is based on simple thermal diffusion in the soil layers that has constant soil moisture availability but a prognostic soil temperature term (Deardorff, 1978). Further differences between the two LSMs are presented in Sect. 3.1.3.

Table 3 provides the summary of the WRF physics schemes considered to simulate the extremely heavy rainfall events.

The ability of the WRF model configuration to simulate an extreme rainfall event is evaluated by comparing the simulated rainfall with the observations through indices such as scale error (SE), which is the ratio of standard deviation of model simulations to the observed standard deviation, and coefficient of variation (CV) in addition to MAE, RMSE and β.

3 Results and discussion

3.1 Sensitivity analysis

3.1.1 Verification of WRF simulations

Figure 5 presents cumulative rainfall for 15–18 June 2013 from 16 WRF simulations (4 MP, 2 CU and 2 PBL) corresponding to each of the three domains.

From Fig. 5i to iii, it may be seen that the spatial pattern of rainfall appears to be sensitive to the microphysics, i.e. PLin, Eta and WSM6 MP schemes, while the amount of rainfall is more dependent on the PBL and CU scheme options. There is a considerable difference in the rainfall amount simulated with the Goddard MP scheme option compared to other MP schemes. Further, most of the model runs are able to reproduce the spatial gradient in the rainfall amount, which is perhaps primarily due to the topographical variation in the region. For locales below 1000 m, observations show distinctly lower rainfall compared to the high elevation regions (> 1000 m). Further, distinct clusters corresponding to heavy rainfall event are observed in the north-east and north-west areas of the study region. These clusters are found to be consistent with the TMPA data; however, due to lack of surface rain gauge observations, the amount of rainfall in these regions could not be verified at this stage. Incidentally, the observed heavy rainfall event in the south-east part of the region is seen in a few WRF configurations, such as configuration (b) and (c). In general, WRF-simulated rainfall fields show a similar spatial pattern as that of the TMPA rainfall product. However, the magnitude of WRF rainfall is significantly high compared to TMPA and is attributed to the negative bias in TMPA for heavy rains. Figure 6 summarizes the comparison of WRF rainfall with rain gauge observations, accumulated over the 4-day period (15–18 June 2013) for the three domains. For comparison, grid points from the WRF domains closest to the gauge location are considered.

Figure 6 indicates that Domain 1 captures rainfall within the range of 150 to 400 mm for most of the WRF configurations. For Domain 2a and Domain 2b, an increase in the predicted rainfall amount is noted, particularly for small rainfall thresholds. Further, the WRF runs still under-predict extremely heavy rainfall and each of the configuration considered (across all the three domains) underestimated the rainfall amount more than 400 mm. However, the underestimation of rainfall is less in Domain 2b (G2C scale) compared to others, indicating the necessity of finer grid spacing as the first-order requirement for simulating the magnitudes of the extremely heavy rainfall events. The bias in the WRF simulations is typically due to a number of interactive factors: (i) scale feedback between mesoscale convection and

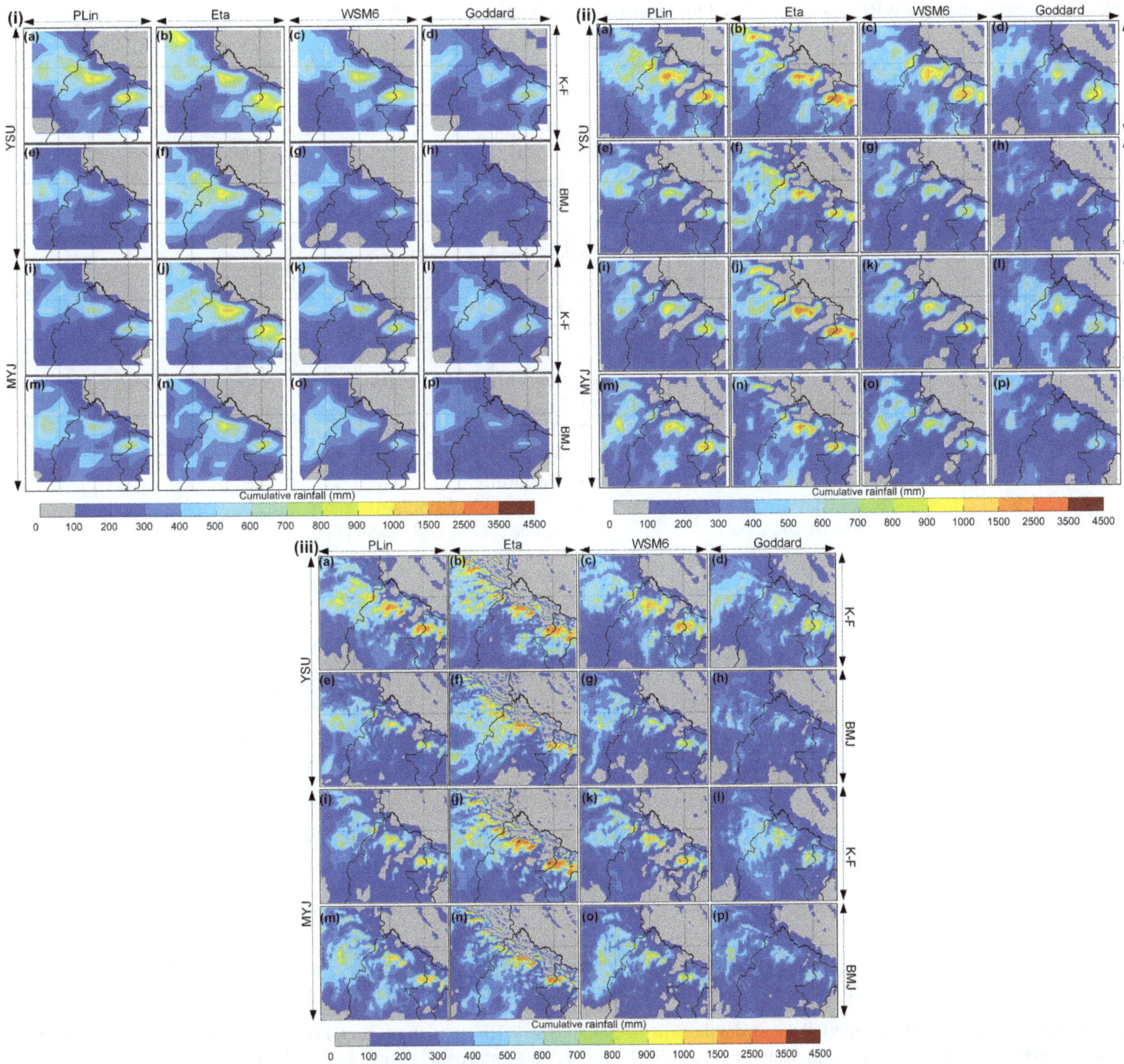

Figure 5. Spatial plots showing rainfall estimates obtained for **(i)** Domain 1, **(ii)** Domain 2a and **(iii)** Domain 2b. Arrows on the left indicate the PBL scheme, arrows on the right represent the CU scheme and the top arrows represent the MP scheme considered for the simulation runs. Subpanels (a) to (p) are the WRF configurations; for instance, (a)–(i) represent the WRF configuration with the YSU PBL scheme, KF CU scheme and PLin MP scheme for Domain 1. Refer to Appendix A (Table A1) for further details of WRF configurations (a) to (p).

large-scale processes within the model (Bohra et al., 2006), (ii) lack of local observations that can add mesoscale features (Osuri et al., 2012; Osuri et al., 2015), (iii) lack of proper land surface processes (Niyogi et al., 2006; Chang et al., 2009; Osuri et al., 2017a), and (iv) the inability of the model to fully resolve the complex topography (Argüeso et al., 2011; Cardoso et al., 2013; Chevuturi et al., 2015). To assess the performance of the WRF simulations, quantitative scores (MAE and RMSE) with respect to the observed data are computed for daily rainfall data, which is then averaged over the 4-day period. The results are shown in Fig. 7. The last column in the figure presents the spatially averaged values obtained for different model configurations.

Figure 7 indicates that there is more error at the stations Dehradun and Haldwani, which received higher rainfall. The highest rainfall obtained in different WRF configurations for these stations was less than 500 mm and this underestimation is highlighted in the error statistics. The model results show

Table 3. Configuration of the WRF model considered for simulation of rainfall.

Model options	Dataset or value
Domains	3
Grid resolution (spacing)	27; 9; 3 km
Downscaling ratio	1 : 3; 1 : 9
Projection system	Mercator
Land surface boundary condition	21-class MODIS
Initial conditions	NCEP FNL
Shortwave radiation scheme	MM5 shortwave or Dudhia
Longwave radiation scheme	Rapid radiative transfer model (RRTM)
PBL Schemes	1. Yonsei University (YSU)
	2. Mellor–Yamada–Janjic (MYJ)
Cumulus schemes	1. Kain–Fritsch (KF)
	2. Betts–Miller–Janjic (BMJ)
Microphysics schemes	1. Lin (Purdue)
	2. Eta (Ferrier)
	3. WSM6
	4. Goddard
Surface layer option	Monin–Obukhov similarity theory
Land surface models	1. Simple 5-layer soil model (Slab)
	2. Noah

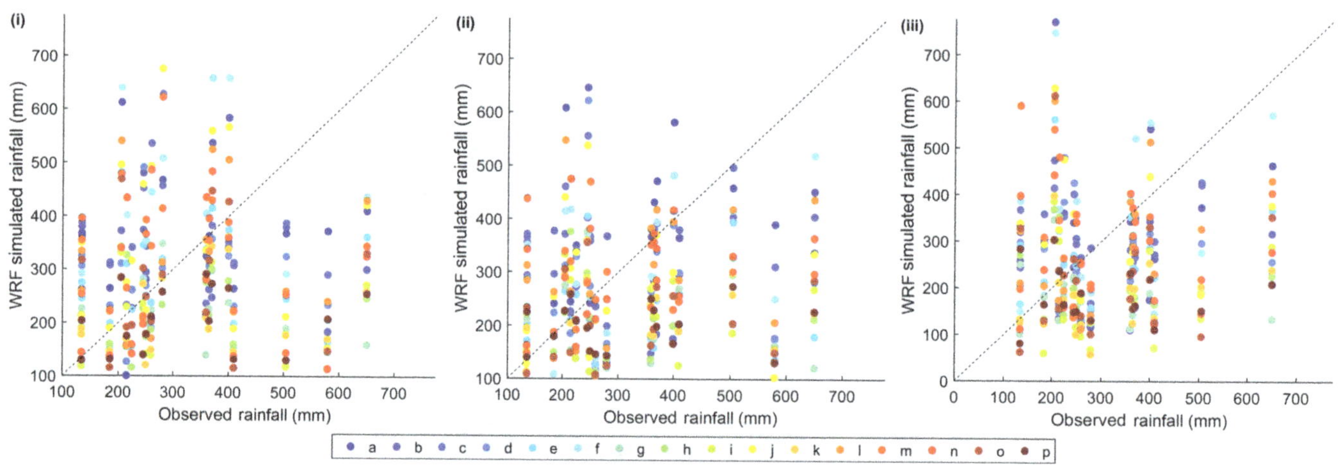

Figure 6. Scatter plots between the rainfall data from the rain gauges and the WRF simulations for **(i)** Domain 1, **(ii)** Domain 2a and **(iii)** Domain 2b for WRF configurations (a) to (p). Refer to Appendix A (Table A1) for the detailed list of the WRF configurations.

higher error and variability in the simulations for the northern part of the domain compared to the southern part. This is likely due to the complex terrain in the northern part of the domain.

To identify the "best" and the "worst" performing configurations, temporal errors (wherein MAEs across the 18 locations are summed up) and spatial errors (wherein MAEs obtained over the entire region for the 4 days are summed up) for all the three domains are obtained (Appendix B), which indicate that the configuration (b), with YSU PBL, KF CU and Eta MP, produces maximum error, whereas configuration (p), with MYJ PBL, BMJ CU and Goddard MP, gives minimum error.

To further assess the sensitivity of configuration (p) and configuration (b) in capturing the extreme rainfall events in the region, additional simulations pertaining to other heavy to extremely heavy rainfall events (as mentioned in Table 2) are conducted. Spatial plots showing the cumulative rainfall estimates obtained for the three domains in comparison to the observed IMD gridded data and the TMPA data are presented in Appendix C. To summarize the performance of configuration (p) and configuration (b) against the observations (IMD gridded data), spatio-temporal MAE values are computed, which are presented in Table 4.

From the analysis conducted over the additional rainfall events, it is noted that configuration (p) gives less error in

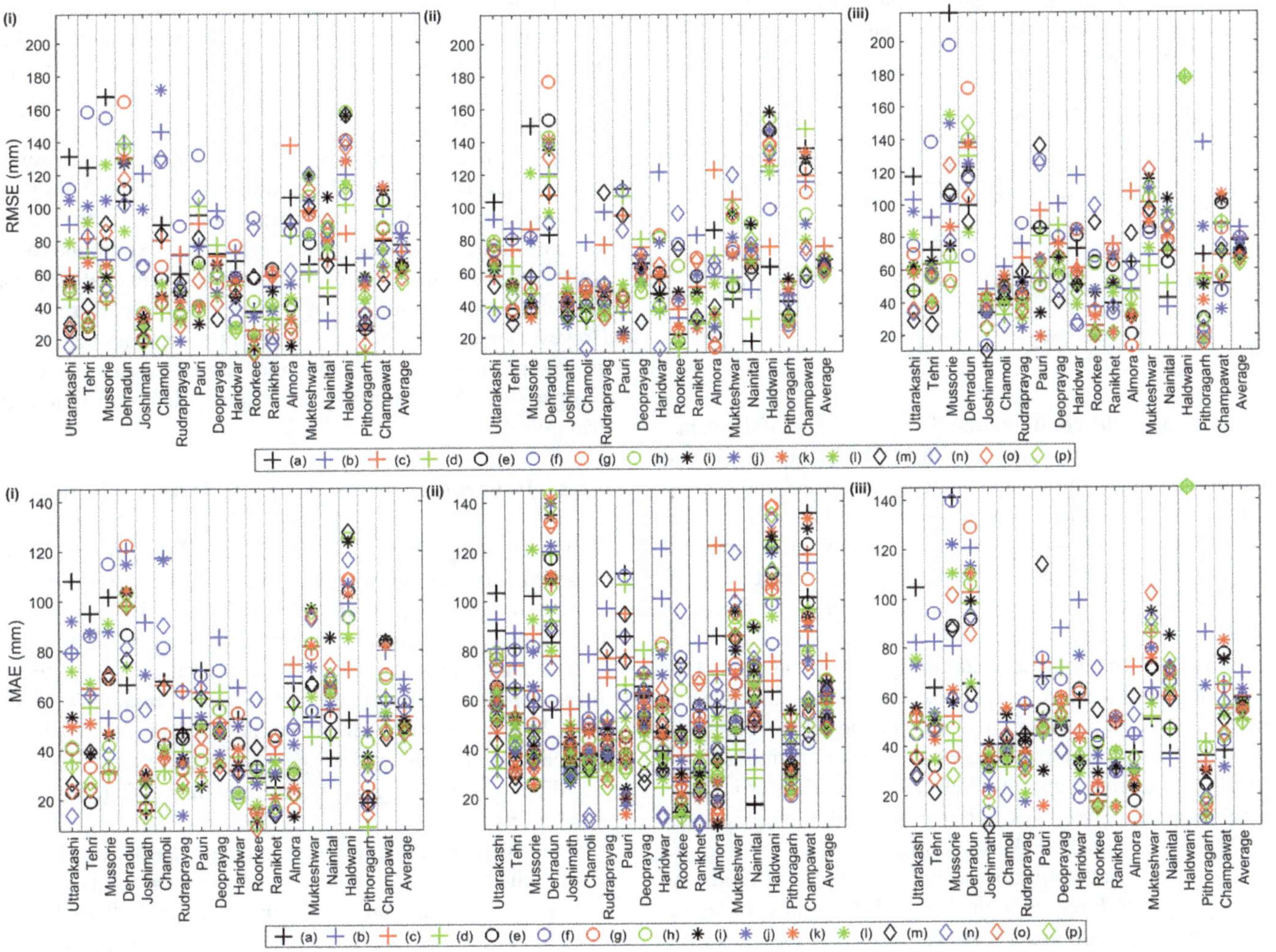

Figure 7. Root mean square error (top panels) and mean absolute error (bottom panels) computed temporally for **(i)** Domain 1, **(ii)** Domain 2a and **(iii)** Domain 2b for WRF configurations (a) to (p). Refer to Appendix A (Table A1) for the detailed list of the WRF configurations.

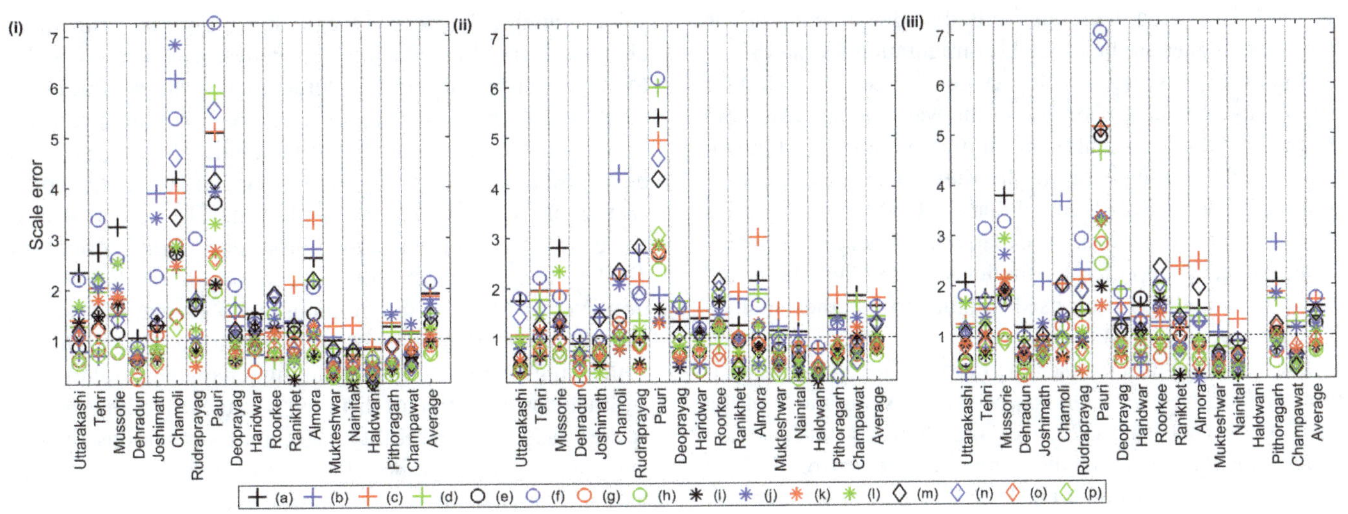

Figure 8. Scale error (SE) in WRF configurations (a) to (p). for 18 locations in the UGB for **(i)** Domain 1, **(ii)** Domain 2a and **(iii)** Domain 2b. Refer to Appendix A (Table A1) for the detailed list of the WRF configurations.

Table 4. Spatio-temporal mean absolute error (MAE) values (in mm) corresponding to WRF configuration (p) and configuration (b) for the three domains.

Event no.	Domain 1		Domain 2a		Domain 2b	
	(p)	(b)	(p)	(b)	(p)	(b)
1	10	13	10	14	11	14
2	18	23	18	23	21	22
3	39	45	38	44	40	46
4	23	28	23	28	24	29
5	12	12	9	13	12	11

comparison to the configuration (b) for all the rainfall events. This makes configuration (p) with MYJ PBL, BMJ CU and Goddard MP the "best" at simulating the spatial and temporal variability of the extremely heavy rainfall over the upstream region of the UGB. Why this combination emerged as the best performing is an intriguing but difficult question to address at this stage. Note that the rainfall prediction is the combination of many nonlinear, interactive factors including the behaviour of each configuration and cannot be realistically studied with the sparse rainfall data and absence of vertical sounding observations. Some possible factors that could contribute would be that local boundary formulation in MYJ may be more appropriately capturing the vertical environment in the complex terrain compared to the non-local YSU scheme which seeks to simulate vertical mixing and boundary layer evolution using averaged and grid representative fields. With regard to the BMJ CU emerging in the top configuration, there are a number of studies for the ISMR where it has emerged as performing "overall best" (Vaidya and Singh, 2000; Ratnam and Kumar, 2005; Vaidya, 2006; Rao et al., 2007; Kumar et al., 2010; Mukhopadhyay et al., 2010; Srinivas et al., 2013; Sikder and Hossain, 2016). As for the MP scheme, there are limited studies in comparison to those that have studied the CU configuration for the ISMR. Further, the MP scheme performance has been evaluated for tropical cyclone cases because of the warm versus cold pool processes that are critical in the simulation of the cyclone intensity. Of those available in the literature, studies such as Sing and Mandal (2014) found that the Goddard scheme has a "slightly better" performance than other schemes. This conclusion is also supported by studies such as Choudhury and Das (2017) and has been used in hailstorm studies such as Chevuturi et al. (2014).

The impact of the downscaling ratio on the rainfall simulations is addressed next. On comparing the simulations of the June 2013 event from G2R and G2C domains with the rain gauge data, it is noted that the former gives less error for most of the locations (Appendix D). The G2C scale has a large resolution (grid spacing) gap from the outer to the inner domain in comparison to G2R, which could result in less accurate initial and lateral boundary conditions and, consequently, more simulation errors in G2C. Another possibility is that the metric being used, which is the rainfall observation from in situ data, is itself more conservative with regards to the grid in which rainfall occurs in the coarser domain and may slightly favour the G2R. However, on reviewing the overall structure of rainfall fields and the amounts across the domain, results suggest that the G2R scale with a moderate downscaling ratio may be better suited for simulation of the extreme rainfall event as in the present case study. The results are found to be consistent with other studies, such as that of Liu et al. (2012), wherein the moderate ratio of 1 : 3 is found to perform best. However, it is to be noted that errors corresponding to the grid point nearest to the rain gauge are considered here for comparison. The result may vary upon selection of another grid point.

3.1.2 Impact of different parameterization schemes

Although configuration (p), with MYJ PBL, BMJ CU and Goddard MP, appears to be the "best" physics configuration for the study region, significant variability exists among the simulations pertaining to different configurations of the WRF model. This variability causes significant uncertainty across different runs, which is quantified through computation of SE and CV (Figs. 8 and 9) for the June 2013 event. Deviation in model simulations with respect to observed data provides the SE; however, CV gives variation within different model simulations.

Most of the model configurations have an SE value clustered around 1 (Fig. 8), indicating that the variability in simulated rainfall is similar to the observed rainfall. However, variability in the north-eastern part of the domain is observed to be high compared to others. The same is reflected in the CV plot (Fig. 9), wherein grid points around the Chamoli station (on the north-eastern side) have CV between 41 and 51 %, whereas stations closer to Uttarakashi and Tehri have values ranging between 11 and 30 %. Further, grid points closer to Dehradun have low CV values, which could be due to the models consistently underestimating the rainfall in this subdomain. The southern part of the region, which received low rainfall, also exhibited high variability. In general, it can be inferred that uncertainty in rainfall is greater in the north-eastern part compared to the north-west. The regions that received very high or very low rainfall during this period also displayed higher uncertainty. Uncertainty in rainfall simulations varies between the domains, with Domain 2b having maximum uncertainty. This could be attributed to high variability in the simulated values at higher spatial resolution.

Since consideration of different parameterization schemes is the reason for variability in rainfall simulations, it is of interest to understand how the former influences the model output. For this, the average cumulative rainfall over the region and across different configurations is considered. The differences between various configurations are evaluated to assess the influence of PBL, CU and MP parameterization

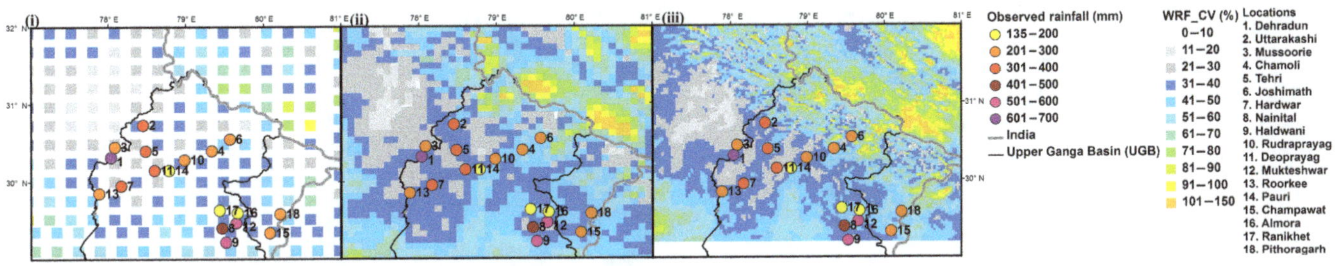

Figure 9. Coefficient of variation (CV) value across different WRF configurations in the UGB for **(i)** Domain 1, **(ii)** Domain 2a and **(iii)** Domain 2b.

schemes on the rainfall simulations. Results for the same are presented in Fig. 10.

It is noticeable that, in general, the WRF configurations with the KF convective scheme produce rainfall of higher magnitudes. This result is consistent with previously conducted studies (Gallus, 1999; Fonseca et al., 2015; Pieri et al., 2015). For PLin MP, it is noted that considering YSU PBL along with KF CU scheme has a synergistic effect, leading to the maximum amount of rainfall over the region. This additive effect could be attributed to the YSU being a non-local scheme, making it suitable for convective, unstable PBL conditions (Bright and Mullen, 2002). Upon changing the PBL scheme (from YSU to MYJ), and maintaining the convective scheme as KF, a notable difference in the fields is simulated (as shown by red circles in Fig. 10). This difference obtained for changing the PBL (with PLin MP and KF CU) is found to be equivalent to the case when only the CU is changed (from KF to BMJ) under the YSU PBL scheme (shown by blue plus sign in Fig. 10). Similarly, the difference in rainfall obtained for two cases – changing the PBL (from YSU to MYJ) with PLin MP and BMJ CU; and changing the CU (from KF to BMJ) with PLin MP and MYJ PBL is also found to be approximately equal. This indicates that the average cumulative rainfall values obtained under two configurations – PLin MP, BMJ CU and YSU PBL and PLin MP, KF CU and MYJ PBL are almost equal. Further, BMJ CU, irrespective of the PBL scheme, results in less simulated rainfall across the region. For WSM6 MP, within the YSU PBL scheme, changing the CU (from KF to BMJ) parameterization produces significant variability (displayed by blue plus sign) in rainfall than changing the PBL scheme itself (from YSU to MYJ PBL with KF CU). However, with MYJ PBL, the effect of changing the CU scheme is insignificant (yellow star in Fig. 10). Furthermore, with BMJ the difference in rainfall produced due to changing the PBL is minimal. With Eta and Goddard MP, changing the PBL (irrespective of the CU scheme) produces a negligible difference in the rainfall simulations, as represented by red circles and purple diamonds in Fig. 10. However, changing the CU schemes (irrespective of the PBL condition) is seen to have a significant influence on rainfall. It can be concluded from this section that the relationship between PBL and CU is interlinked, wherein YSU and MYJ PBL complement (contradict) the effect of KF (BMJ) and BMJ (KF) CUs with regard to the quantity of rainfall. Overall, the choice of CU appears to have a significant impact on the simulation of rainfall over the region. This conclusion is consistent with the earlier studies such as those by Sikder and Hossain (2016), where they found ISMR to be more sensitive to CU than to MP.

3.1.3 Impact of land surface boundary condition

It is well established that the soil moisture plays a significant role in weather predictions (Chen and Brutsaert, 1995; Betts et al., 1996, 1997; Entekhabi et al., 1996). Therefore, Slab and Noah LSMs, which differ significantly in two factors – (i) the soil depths along with the inclusion of land surface processes and (ii) the temporal evolution of soil moisture – are selected in the present study to assess the impact of land surface conditions on simulation of heavy rainfall events. The Slab is a relatively simple LSM with five soil layers (at 1, 2, 4, 8 and 16 cm depths) and uses a thermal diffusion equation to compute surface fluxes based on a surface temperature and drag coefficient formulations. The Noah LSM is modestly detailed (compared to Slab) with four soil layers (at 10, 30, 60 and 100 cm depths) and explicit representation of land surface parameters, which includes the effect of soil moisture changes, snow cover, evapotranspiration and hydrologic processes such as runoff and drainage (to subsurface layers). Further, in the Noah LSM soil moisture and temperature are prognostically computed for each of the four soil layers, whereas in the Slab LSM only soil temperature is prognostic and moisture is considered as a constant value based on the land use. Slab lacks the feature of predicting the snow cover and does not capture the evaporation and runoff processes over the region.

To understand the influence of each LSM on the rainfall estimation, simulations using the Slab LSM are conducted for the "best" and the "worst" performing configurations from the Noah LSM case (configuration (p) and (b), respectively). Results comparing the two land-model-based runs are presented in Fig. 11. The top panel in Fig. 11i–iii presents the scatter plot of the cumulative rainfall obtained with Noah and Slab LSM runs versus the observed cumulative rainfall. Bottom panel (Fig. 11i–iii) presents the mean

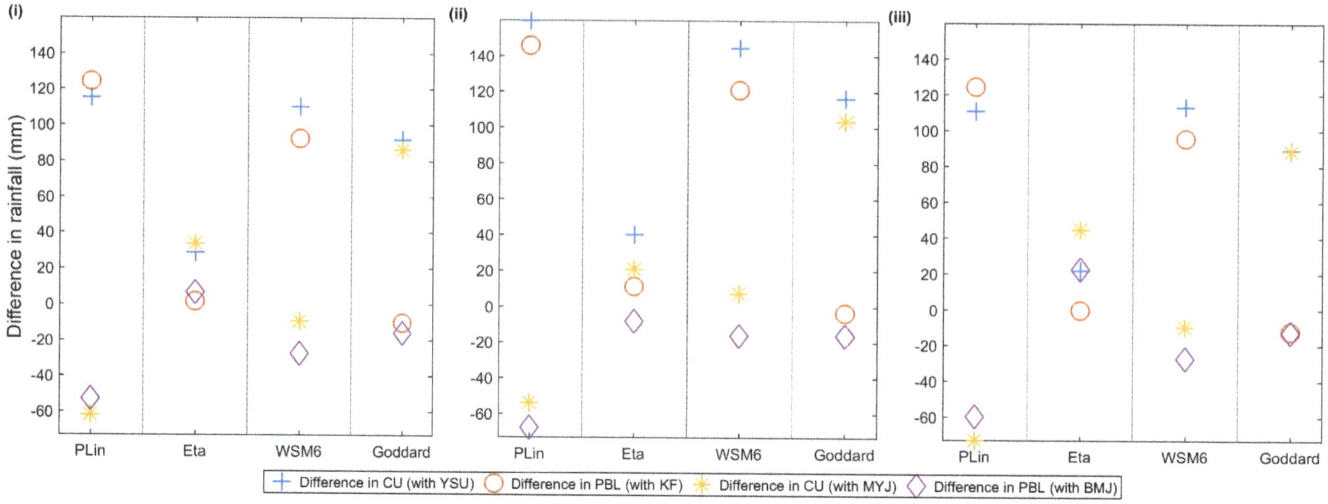

Figure 10. Difference in simulated rainfall due to PBL, CU and MP parameterization schemes corresponding to **(i)** Domain 1, **(ii)** Domain 2a and **(iii)** Domain 2b over the UGB region.

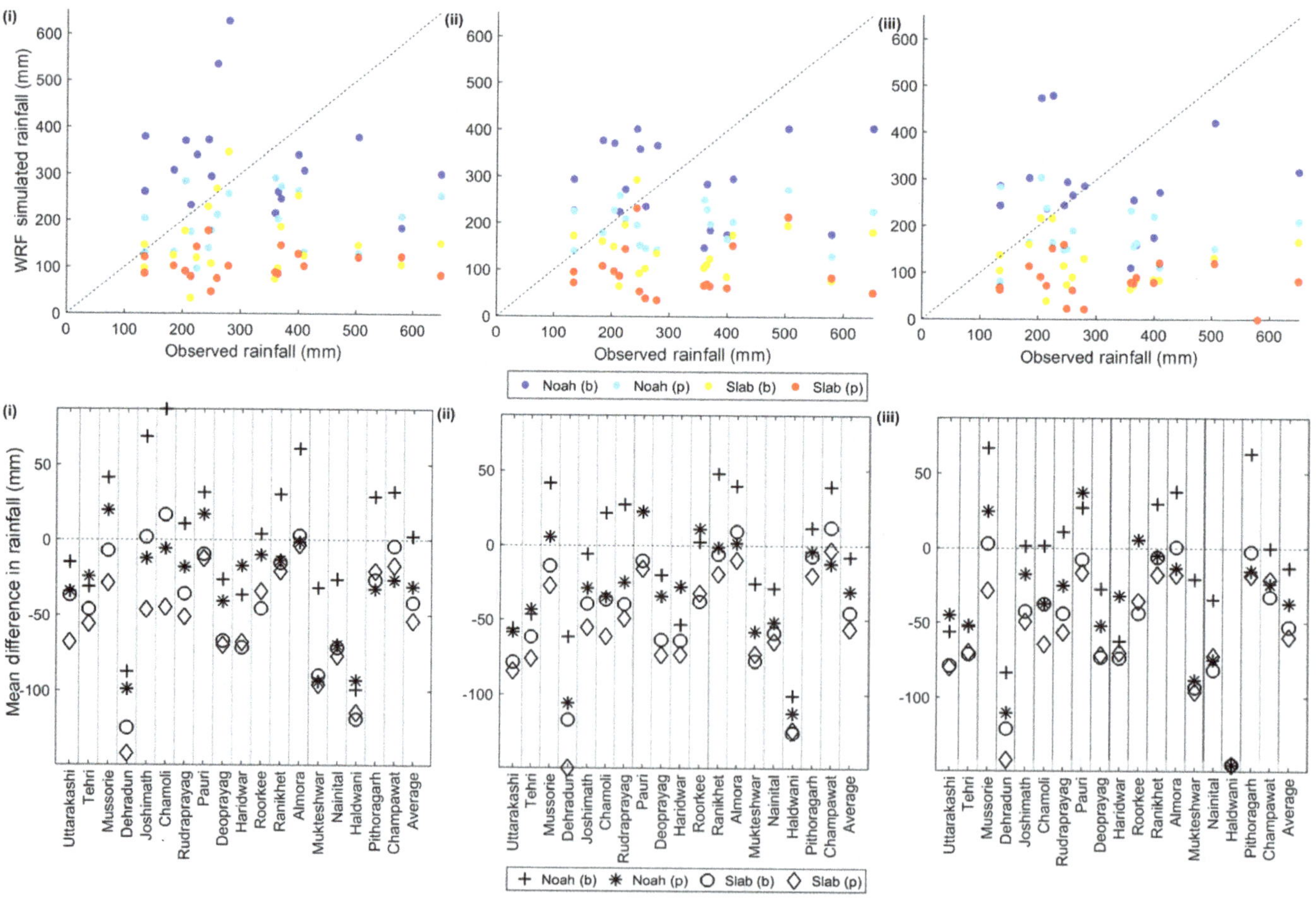

Figure 11. Scatter plot (top panels) and mean difference in rainfall (bottom panels) for the observed rainfall data and the WRF simulations (for (b) and (p) configurations) pertaining to Noah and Slab LSMs corresponding to **(i)** Domain 1, **(ii)** Domain 2a and **(iii)** Domain 2b.

difference in the simulated and the observed rainfall values over the 4-day period for 18 locations along with the spatially averaged values (last columns in Fig. 11i–iii) within the region.

The Slab-LSM-based run significantly underestimates the rainfall in comparison to the Noah LSM. For example, the locations which recorded rainfall greater than 400 mm have the Slab-LSM-based simulated values in the range of 100–150 mm. As stated earlier, although the Noah LSM also underestimated the rainfall for such stations, the bias with the Noah LSM is significantly less than the Slab LSM (−26 % with the Noah LSM, in contrast to −64 % with the Slab LSM for Domain 2a and the (p) configuration). Further, the mean difference in rainfall obtained with Slab LSM is found to be higher in comparison to the Noah LSM. This is essentially due to significant underestimation of rainfall during 16 and 17 June 2013 by the Slab LSM.

In a number of studies, the differences in the surface energy fluxes simulated by the choice of different LSMs (i.e. Slab versus Noah) has been discussed (see Niyogi et al., 2016 for a review). The main reason being that the surface processes affect the boundary layer feedbacks which in turn create zones of mesoscale convergence that can affect the location and intensity of convection. These convective systems then contribute to the simulated rainfall. The results obtained in this study emphasize this feature with differences in the rain amounts and locations in response to the change in LSM. The better performance of using the Noah model could be attributed to the temporal evolution of soil moisture fields. Analysing the soil moisture in Slab and Noah models, the soil is noted to be relatively dry in Slab (soil moisture less than 0.05 $m^3\,m^{-3}$) and the value is constant throughout the model run (since in the Slab model there is no prognostic soil moisture term). In the case of Noah, soil moisture varies in response to the rainfall and is found to vary between 0.25 and 0.45 $m^3\,m^{-3}$. Higher surface moisture conditions improve mass flux, convective updrafts and diabatic heating in the boundary layer that contributes to low-level positive potential vorticity or convective potential, which leads to enhanced rainfall potential (Osuri et al., 2017a). The importance of representing soil moisture variability over India for extreme weather conditions is also highlighted through this work.

3.2 Comparison between rainfall from the WRF and the FNL dataset

Simulated rainfall from the WRF model runs is assessed with respect to the NCEP FNL reanalysis dataset. To achieve this, it is necessary to bring both the datasets to a common spatial resolution. Therefore, the WRF-simulated rainfall is upscaled, through averaging of the grids, to match the resolution of NCEP FNL (1° × 1°). For the analysis, only simulations pertaining to the "best" performing configuration (p), are considered. Bias (β) in rainfall simulations from the two datasets corresponding to 18 rain gauge locations is obtained, results for which are presented in Fig. 12.

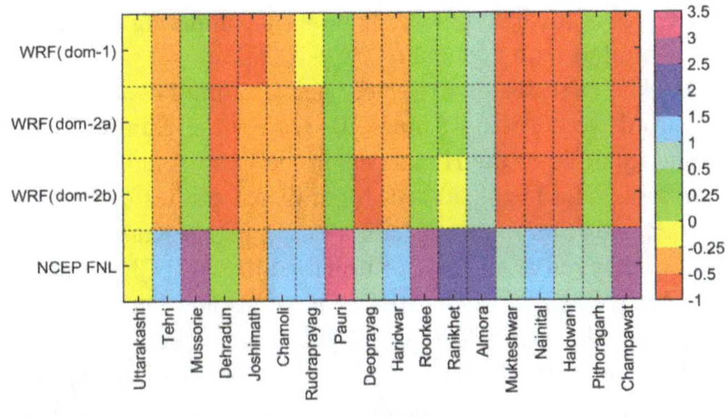

Figure 12. Bias (β) in rainfall simulations obtained from the NCEP FNL and WRF (upscaled to 1° × 1°) data.

From Fig. 12 it can be observed that the NCEP FNL data overestimate the rainfall for most of the locations. Upon dynamic downscaling of the FNL data through the WRF model, rainfall simulations improved over the UGB region. Locations such as Mussorie, Pauri and Roorkee, which have shown β between 2.5 and 3.5 in the FNL data, reduced to 0–0.25 in the WRF simulations. Uttarakashi and Pithoragarh locations, which have a small bias in the FNL data, show similar small bias in the WRF simulations. Dehradun along with three stations from the south-eastern region, (Mukteshwar, Haldwani and Nainital), which recorded heavy rainfall (Sect. 2.1), are observed to have a small bias in the FNL data, and the rainfall at this location is underestimated by the WRF model. Overall, rainfall simulations from the WRF model (for all the three domains) have less β compared to the FNL data even after upscaling to the resolution of 1° × 1°. As expected, upon upscaling, the spatial variability between the domains is reduced due to averaging across several grid points.

From the above analysis, it is evident that the WRF model can simulate extreme precipitation better than the reanalysis data. This can be attributed to increase in spatial resolution, and better representation of surface and meteorological features, with respect to the lateral boundary conditions as suggested in some of the previous works such as those by Argüeso et al. (2011), Mishra et al. (2014), Giorgi and Gutowski, (2015), and Singh et al. (2017).

4 Summary and conclusions

The main focus of this paper is to provide a general guideline for setting up the WRF model configuration to simulate heavy rainfall events. In this regard, sensitivity of the WRF model to predict heavy to extremely heavy rainfall events is examined through (a) quantitative verification of the rainfall simulated by the WRF model, (b) investigating sensitivity of the simulated rainfall to different parameterization schemes, downscaling ratios and land surface models, (c) testing the selected scheme for other rainfall events, and (d) assessing the effect of local and global factors by comparing the simulated rainfall with a global reanalysis dataset.

For the analysis, an extremely heavy rainfall event, which occurred from 15 to 18 June 2013, over the Ganga Basin, in the foothills of the Himalayas in the Uttarakhand State of northern India is considered. Most of the studies conducted earlier over this region (Medina et al., 2010; Kumar et al., 2012; Thayyen et al., 2013; Kumar et al., 2014; Chevuturi et al., 2015; Shekhar et al., 2015; Chevuturi and Dimri, 2016; Rajesh et al., 2016; Hazra et al., 2017) are based on the general or default WRF configuration with WSM6 microphysics, a Kain–Fritsch cumulus parameterization scheme and planetary boundary layer of the Yonsei University scheme. In this paper, ensemble experiments are conducted using the WRF model with different grid spacing, four microphysics schemes, two cumulus parameterization schemes, two planetary boundary layer schemes and two land surface model conditions. The rainfall simulations are evaluated against the observed rain gauge data and the Tropical Rainfall Measuring Mission Multi-Satellite Precipitation Analysis precipitation data. The WRF configuration with Goddard microphysics, a Mellor–Yamada–Janjic planetary boundary layer condition and a Betts–Miller–Janjic cumulus parameterization scheme is found to perform "best" in simulating an extremely heavy rain event of June 2013. The selected configuration is then verified by simulating several other heavy to extremely heavy rainfall events that occurred across different months in the monsoon season over the upstream region of the UGB. The results for the additional events indicate that the selected configuration (Goddard microphysics, Mellor–Yamada–Janjic planetary boundary layer condition and Betts–Miller–Janjic cumulus parameterization scheme) is indeed the "best" in simulating the spatial and temporal variability of the extremely heavy rainfall over the region. Therefore, through the exhaustive analysis conducted in this paper, the recommended WRF configuration for extreme rainfall simulations in the Himalayan region is Goddard microphysics, Mellor–Yamada–Janjic planetary boundary layer condition and Betts–Miller–Janjic cumulus parameterization scheme.

Although complex interactions are observed between different physics options, microphysics schemes can be seen to influence the spatial pattern of the rainfall, while the choice of cumulus scheme is found to modulate the magnitude of the simulated rainfall. Upon analysing the impact of downscaling ratios on rainfall simulations, it is concluded that downscaling from global to regional scale with a moderate downscaling ratio may give the least model errors and, thus, be considered as suitable for reproducing the extreme rainfall event. In addition to this, the effect of land surface models on rainfall simulations is also assessed in this paper. The Slab LSM significantly underestimates the rainfall values, and incorporating Noah helped improve the performance. The underperformance of the Slab model is attributed to dry soil conditions in the region for this LSM.

In addition to the sensitivity experiments, the WRF-simulated rainfall is also compared with the NCEP FNL reanalysis data. The NCEP FNL data are found to overestimate the rainfall, whereas the WRF-simulated rainfall exhibited less bias. The comparison results indicated that care must be taken while employing global datasets for regional analysis. Through this, it can be established that the rainfall values obtained from the high-resolution mesoscale model can be effectively used in hydrologic models for realistic streamflow estimates.

The analyses presented in this paper are subject to a few limitations: first, results are limited to the physics parameterization schemes considered in this paper, and may vary upon inclusion of other schemes; second, only two sets of downscaling ratios, i.e. 1 : 9 and 1 : 3, are tested in the current work. The sensitivity of simulations pertaining to other downscaling ratios should be tested in future; and third, only G2R and G2C sensitivity are assessed in this work.

Data availability. The meteorological data were purchased from the India Meteorological Department using the requisition form given at the following link: http://www.imd.gov.in/advertisements/20170320_advt_34.pdf (Pai et al., 2014). The station data for June 2013 event can be obtained from Ray et al. (2013). The following are the links for procuring the publicly accessible data: TMPA data – https://pmm.nasa.gov/data-access/downloads/trmm (Huffman, 2016); FNL data – https://doi.org/10.5065/D6M043C6 (NCEP, 2017). The source code for the WRF model, used in the study, can be obtained from http://www2.mmm.ucar.edu/wrf/users/downloads.html (Skamarock et al., 2005). The rainfall simulation obtained from the WRF model are available at https://github.com/ilachawla/Chawla_2018_HESS_Data (Chawla, 2018).

Appendix A

Table A1. List of different WRF configuration.

WRF configuration	Microphysics scheme (MP)	Cumulus scheme (CU)	Planetary boundary layer scheme (PBL)
a	PLin	KF	YSU
b	Eta	KF	YSU
c	WSM6	KF	YSU
d	Goddard	KF	YSU
e	PLin	BMJ	YSU
f	Eta	BMJ	YSU
g	WSM6	BMJ	YSU
h	Goddard	BMJ	YSU
i	PLin	KF	MYJ
j	Eta	KF	MYJ
k	WSM6	KF	MYJ
l	Goddard	KF	MYJ
m	PLin	BMJ	MYJ
n	Eta	BMJ	MYJ
o	WSM6	BMJ	MYJ
p	Goddard	BMJ	MYJ

Appendix B

Table B1. Mean absolute error (MAE) values corresponding to different WRF configuration for the three domains.

WRF configuration	Domain 1		Domain 2a		Domain 2b	
	Temporal	Spatial	Temporal	Spatial	Temporal	Spatial
a	1028	228	943	210	1066	237
b	1227	273	1097	244	1249	278
c	960	213	1048	233	1077	239
d	825	183	852	189	908	202
e	877	195	885	197	995	221
f	1068	237	886	197	1055	234
g	887	197	950	211	979	218
h	885	197	976	217	1010	225
i	932	207	939	209	1024	228
j	1161	258	970	215	1131	251
k	894	199	888	197	987	219
l	883	196	855	190	918	204
m	890	198	913	203	974	217
n	1012	225	870	193	976	217
o	837	186	863	192	976	217
p	740	164	843	187	886	197

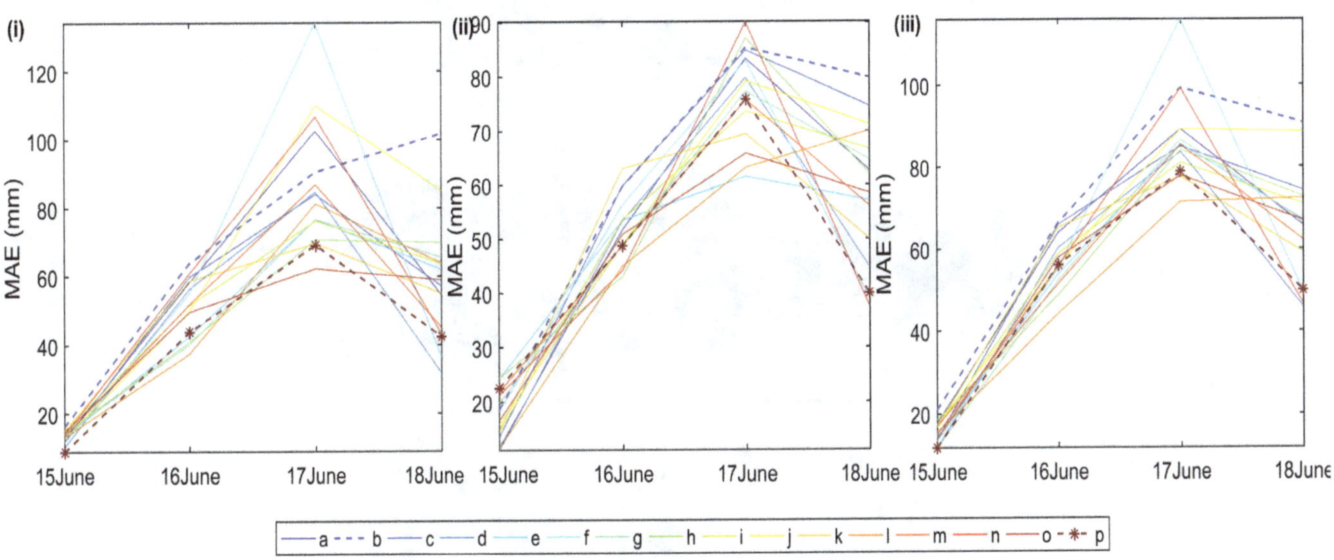

Figure B1. Mean absolute errors in space corresponding to different WRF configurations for **(i)** Domain 1, **(ii)** Domain 2a and **(iii)** Domain 2b. Blue dotted lines present the "worst" performing configuration, i.e. configuration (b), and red dotted lines show the "best" performing configuration, i.e. configuration (p).

Appendix C

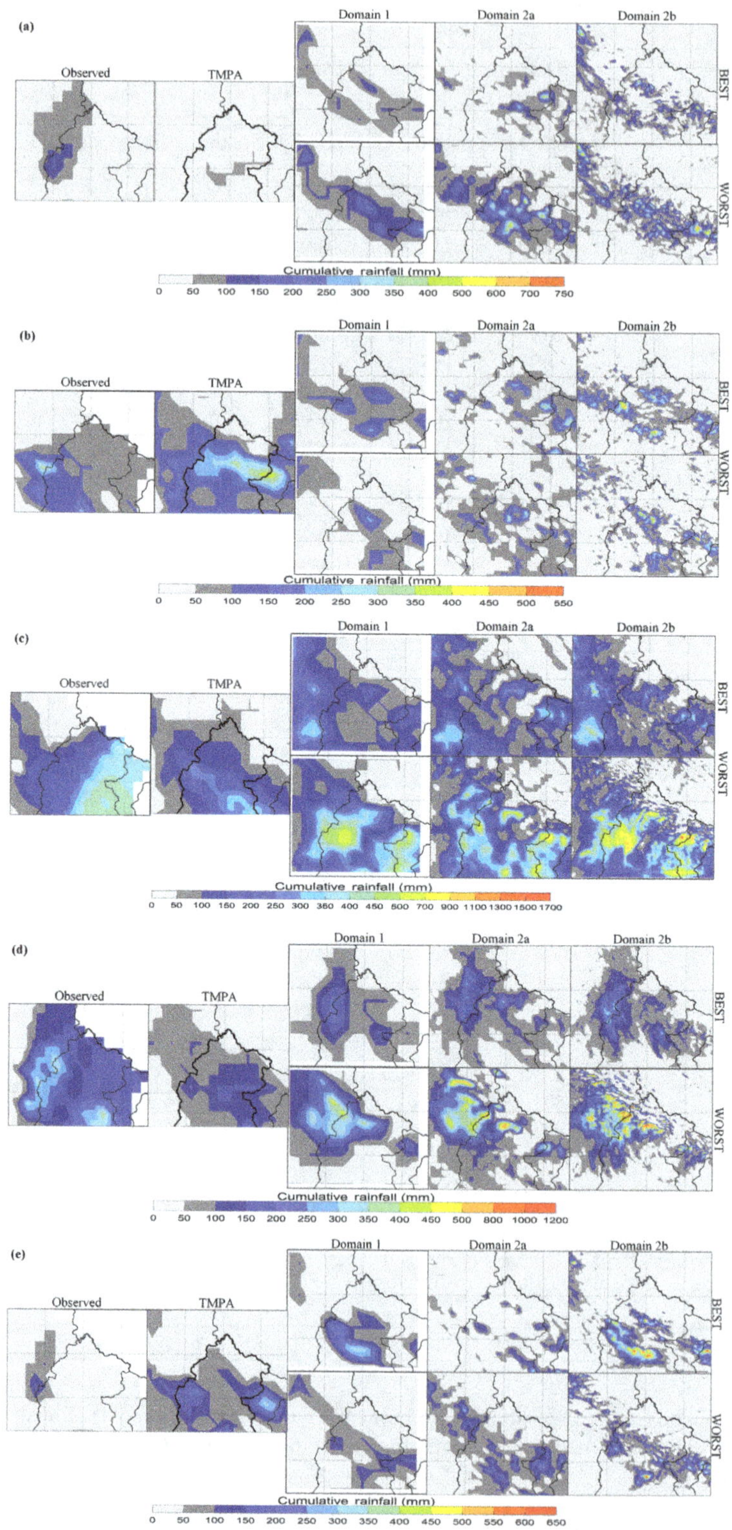

Figure C1. Spatial plots presenting the rainfall simulations obtained across the three domains for the best and the worst configuration for heavy to extremely heavy rainfall events during **(a)** Event 1 (18–22 June 2008), **(b)** Event 2 (29 July–2 August 2010), **(c)** Event 3 (15–19 August 2011), **(d)** Event 4 (17–21 September 2010) and **(e)** Event 5 (11–15 September 2012).

Appendix D

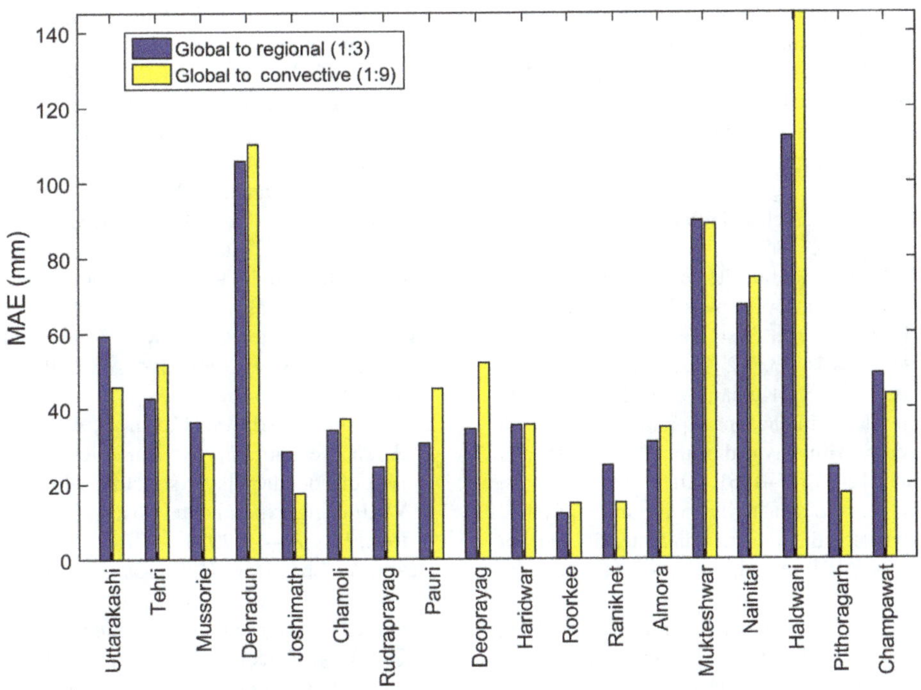

Figure D1. Bar plot representing the mean absolute errors in simulating rainfall across the 18 rain gauge locations on global to regional (G2R) scale (1 : 3) and global to convection-permitting (G2C) scale (1 : 9).

Competing interests. The authors declare that they have no conflict of interest.

Special issue statement. This article is part of the special issue "The changing water cycle of the Indo-Gangetic Plain". It is not associated with a conference.

Acknowledgements. The work is part of the study supported by the US National Science Foundation (CAREER AGS0847472) and the Government of India/MOES National Monsoon Mission (grant no./project no.: MM/SERP/CNRS/2013/INT-10/002) at Purdue University. The second author gratefully acknowledges the financial support of ESSO, Ministry of Earth Sciences, Government of India and Science and Engineering Research Board (SERB), Department of Science and Technology, Government of India.

Edited by: Ana Mijic

References

Argüeso, D., Hidalgo-Muñoz, J. M., Gámiz-Fortis, S. R., Esteban-Parra, M. J., Dudhia, J., and Castro-Díez, Y.: Evaluation of WRF parameterizations for climate studies over Southern Spain using a multistep regionalization, J. Climate, 24, 5633–5651, 2011.

Betts, A. K., Ball, J. H., Beljaars, A., Miller, M. J., and Viterbo, P. A.: The land surface-atmosphere interaction: A review based on observational and global modeling perspectives, J. Geophys. Res.-Atmos., 101, 7209–7225, 1996.

Betts, A. K., Chen, F., Mitchell, K. E., and Janjić, Z. I.: Assessment of the land surface and boundary layer models in two operational versions of the NCEP Eta model using FIFE data, Mon. Weather Rev., 125, 2896–2916, 1997.

Bharti, V., Singh, C., Ettema, J., and Turkington, T.: Spatiotemporal characteristics of extreme rainfall events over the Northwest Himalaya using satellite data, Int. J. Climatol., 36, 3949–3962, 2016.

Bohra, A., Basu, S., Rajagopal, E., Iyengar, G., Gupta, M. D., Ashrit, R., and Athiyaman, B.: Heavy rainfall episode over Mumbai on 26 July 2005: Assessment of NWP guidance, Curr. Sci., 90, 1188–1194, 2006.

Bright, D. R. and Mullen, S. L.: The sensitivity of the numerical simulation of the southwest monsoon boundary layer to the choice of PBL turbulence parameterization in MM5, Weather Forecast., 17, 99–114, 2002.

Cardoso, R., Soares, P., Miranda, P., and Belo-Pereira, M.: WRF high resolution simulation of Iberian mean and extreme precipitation climate, Int. J. Climatol., 33, 2591–2608, 2013.

Chang, H.-I., Kumar, A., Niyogi, D., Mohanty, U., Chen, F., and Dudhia, J.: The role of land surface processes on the mesoscale simulation of the July 26, 2005 heavy rain event over Mumbai, India, Global Planet. Change, 67, 87–103, 2009.

Chawla, I.: Chawla_2018_HESS_Data, GitHub repository, available at: https://github.com/ilachawla/Chawla_2018_HESS_Data, last access: 5 February 2018.

Chen, D. and Brutsaert, W.: Diagnostics of land surface spatial variability and water vapor flux, J. Geophys. Res.-Atmos., 100, 25595–25606, 1995.

Chen, F. and Dudhia, J.: Coupling an advanced land surface–hydrology model with the Penn State–NCAR MM5 modeling system. Part I: Model implementation and sensitivity, Mon. Weather Rev., 129, 569–585, 2001.

Chen, S.-H. and Sun, W.-Y.: A one-dimensional time dependent cloud model, J. Meteorol. Soc. Jpn., 80, 99–118, 2002.

Chevuturi, A. and Dimri, A.: Investigation of Uttarakhand (India) disaster-2013 using weather research and forecasting model, Nat. Hazards, 82, 1703–1726, 2016.

Chevuturi, A., Dimri, A. P., and Gunturu, U. B.: Numerical simulation of a rare winter hailstorm event over Delhi, India on 17 January 2013, Nat. Hazards Earth Syst. Sci., 14, 3331–3344, https://doi.org/10.5194/nhess-14-3331-2014, 2014.

Chevuturi, A., Dimri, A., Das, S., Kumar, A., and Niyogi, D.: Numerical simulation of an intense precipitation event over Rudraprayag in the central Himalayas during 13–14 September 2012, J. Earth Syst. Sci., 124, 1545–1561, 2015.

Choudhury, D. and Das, S.: The sensitivity to the microphysical schemes on the skill of forecasting the track and intensity of tropical cyclones using WRF-ARW model, J. Earth Syst. Sci., 126, 1–10, 2017.

Das, S., Ashrit, R., Iyengar, G. R., Mohandas, S., Gupta, M. D., George, J. P., Rajagopal, E., and Dutta, S. K.: Skills of different mesoscale models over Indian region during monsoon season: Forecast errors, J. Earth Syst. Sci., 117, 603–620, 2008.

Deardorff, J.: Efficient prediction of ground surface temperature and moisture, with inclusion of a layer of vegetation, J. Geophys. Res.-Ocean, 83, 1889–1903, 1978.

Deb, S., Srivastava, T., and Kishtawal, C.: The WRF model performance for the simulation of heavy precipitating events over Ahmedabad during August 2006, J. Earth Syst. Sci., 117, 589–602, 2008.

Dimri, A., Thayyen, R., Kibler, K., Stanton, A., Jain, S., Tullos, D., and Singh, V.: A review of atmospheric and land surface processes with emphasis on flood generation in the Southern Himalayan rivers, Sci. Total Environ., 556, 98–115, 2016.

Dube, A., Ashrit, R., Ashish, A., Sharma, K., Iyengar, G., Rajagopal, E., and Basu, S.: Forecasting the heavy rainfall during Himalayan flooding – June 2013, Weather Clim. Extr., 4, 22–34, 2014.

Dudhia, J.: Numerical study of convection observed during the winter monsoon experiment using a mesoscale two-dimensional model, J. Atmos. Sci., 46, 3077–3107, 1989.

Dudhia, J.: A multi-layer soil temperature model for MM5, Preprints, The Sixth PSU/NCAR mesoscale model users' workshop, 22–24, 1996.

Efstathiou, G., Zoumakis, N., Melas, D., Lolis, C., and Kassomenos, P.: Sensitivity of WRF to boundary layer parameterizations in simulating a heavy rainfall event using different microphysical schemes. Effect on large-scale processes, Atmos. Res., 132, 125–143, 2013.

Ek, M., Mitchell, K., Lin, Y., Rogers, E., Grunmann, P., Koren, V., Gayno, G., and Tarpley, J.: Implementation of Noah land surface model advances in the National Centers for Environmental Prediction operational mesoscale Eta model, J. Geophys. Res.-Atmos., 108, 12-1–12-16, 2003.

Entekhabi, D., Rodriguez-Iturbe, I., and Castelli, F.: Mutual interaction of soil moisture state and atmospheric processes, J. Hydrol., 184, 3–17, 1996.

Fasullo, J. and Webster, P.: A hydrological definition of Indian monsoon onset and withdrawal, J. Climate, 16, 3200–3211, 2003.

Fonseca, R. M., Zhang, T., and Yong, K.-T.: Improved simulation of precipitation in the tropics using a modified BMJ scheme in the WRF model, Geosci. Model Dev., 8, 2915–2928, https://doi.org/10.5194/gmd-8-2915-2015, 2015.

Gadgil, S. and Sajani, S.: Monsoon precipitation in the AMIP runs, Clim. Dynam., 14, 659–689, 1998.

Gallus Jr., W. A.: Eta simulations of three extreme precipitation events: Sensitivity to resolution and convective parameterization, Weather Forecast., 14, 405–426, 1999.

Giorgi, F. and Gutowski Jr., W. J.: Regional dynamical downscaling and the CORDEX initiative, Annu. Rev. Env. Resour., 40, 467–490, 2015.

Hariprasad, K., Srinivas, C., Singh, A. B., Rao, S. V. B., Baskaran, R., and Venkatraman, B.: Numerical simulation and intercomparison of boundary layer structure with different PBL schemes in WRF using experimental observations at a tropical site, Atmos. Res., 145, 27–44, 2014.

Hazra, A., Chaudhari, H. S., Ranalkar, M., and Chen, J. P.: Role of interactions between cloud microphysics, dynamics and aerosol in the heavy rainfall event of June 2013 over Uttarakhand, India, Q. J. Roy. Meteorol. Soc., 143, 986–998, 2017.

Hong, S.-Y. and Lee, J.-W.: Assessment of the WRF model in reproducing a flash-flood heavy rainfall event over Korea, Atmos. Res., 93, 818–831, 2009.

Hong, S.-Y. and Lim, J.-O. J.: The WRF single-moment 6-class microphysics scheme (WSM6), J. Korean Meteor. Soc, 42, 129–151, 2006.

Hong, S.-Y., Noh, Y., and Dudhia, J.: A new vertical diffusion package with an explicit treatment of entrainment processes, Mon. Weather Rev., 134, 2318–2341, 2006.

Hong, S.-Y., Sunny Lim, K.-S., Kim, J.-H., Jade Lim, J.-O., and Dudhia, J.: Sensitivity study of cloud-resolving convective simulations with WRF using two bulk microphysical parameterizations: ice-phase microphysics versus sedimentation effects, J. Appl. Meteorol. Clim., 48, 61–76, 2009.

Hu, X.-M., Nielsen-Gammon, J. W., and Zhang, F.: Evaluation of three planetary boundary layer schemes in the WRF model, J. Appl. Meteorol. Clim., 49, 1831–1844, 2010.

Huffman, G.: TRMM (TMPA-RT) Near Real-Time Precipitation L3 3 hour 0.25 degree × 0.25 degree V7, Greenbelt, MD, Goddard Earth Sciences Data and Information Services Center (GES DISC), available at: https://disc.gsfc.nasa.gov/datacollection/TRMM_3B42RT_7.html (last access: 2 November 2017), 2016.

Janjić, Z. I.: The step-mountain eta coordinate model: Further developments of the convection, viscous sublayer, and turbulence closure schemes, Mon. Weather Rev., 122, 927–945, 1994.

Janjić, Z. I.: Comments on "Development and evaluation of a convection scheme for use in climate models", J. Atmos. Sci., 57, 3686–3686, 2000.

Janić, Z. I.: Nonsingular implementation of the Mellor-Yamada level 2.5 scheme in the NCEP Meso model, US Department of Commerce, National Oceanic and Atmospheric Administration, National Weather Service, National Centers for Environmental Prediction, 2001.

Kain, J. S.: The Kain–Fritsch convective parameterization: an update, J. Appl. Meteorol., 43, 170–181, 2004.

Kneis, D., Chatterjee, C., and Singh, R.: Evaluation of TRMM rainfall estimates over a large Indian river basin (Mahanadi), Hydrol. Earth Syst. Sci., 18, 2493–2502, https://doi.org/10.5194/hess-18-2493-2014, 2014.

Koren, V., Schaake, J., Mitchell, K., Duan, Q. Y., Chen, F., and Baker, J.: A parameterization of snowpack and frozen ground intended for NCEP weather and climate models, J. Geophys. Res.-Atmos., 104, 19569–19585, 1999.

Kotal, S., Roy, S. S., and Roy Bhowmik, S.: Catastrophic heavy rainfall episode over Uttarakhand during 16–18 June 2013 – observational aspects, Curr. Sci, 107, 234–245, 2014.

Krishnamurthy, C. K. B., Lall, U., and Kwon, H.-H.: Changing frequency and intensity of rainfall extremes over India from 1951 to 2003, J. Climate, 22, 4737–4746, 2009.

Kumar, A., Dudhia, J., Rotunno, R., Niyogi, D., and Mohanty, U.: Analysis of the 26 July 2005 heavy rain event over Mumbai, India using the Weather Research and Forecasting (WRF) model, Q. J. Roy. Meteorol. Soc., 134, 1897–1910, 2008.

Kumar, A., Houze Jr., R. A., Rasmussen, K. L., and Peters-Lidard, C.: Simulation of a flash flooding storm at the steep edge of the Himalayas, J. Hydrometeorol., 15, 212–228, 2014.

Kumar, M. S., Shekhar, M., Krishna, S. R., Bhutiyani, M., and Ganju, A.: Numerical simulation of cloud burst event on August 05, 2010, over Leh using WRF mesoscale model, Nat. Hazards, 62, 1261–1271, 2012.

Kumar, R. A., Dudhia, J., and Roy Bhowmik, S.: Evaluation of Physics options of the Weather Research and Forecasting (WRF) Model to simulate high impact heavy rainfall events over Indian Monsoon region, Geofizika, 27, 101–125, 2010.

Li, L., Gochis, D. J., Sobolowski, S., and Mesquita, M. D.: Evaluating the present annual water budget of a Himalayan headwater river basin using a high-resolution atmosphere-hydrology model, J. Geophys. Res.-Atmos., 122, 4786–4807, 2017.

Li, X. and Pu, Z.: Sensitivity of numerical simulation of early rapid intensification of Hurricane Emily (2005) to cloud microphysical and planetary boundary layer parameterizations, Mon. Weather Rev., 136, 4819–4838, 2008.

Lin, Y.-L., Farley, R. D., and Orville, H. D.: Bulk parameterization of the snow field in a cloud model, J. Climate Appl. Meteorol., 22, 1065–1092, 1983.

Liu, J., Bray, M., and Han, D.: Sensitivity of the Weather Research and Forecasting (WRF) model to downscaling ratios and storm types in rainfall simulation, Hydrol. Processes, 26, 3012–3031, 2012.

Madala, S., Satyanarayana, A., and Rao, T. N.: Performance evaluation of PBL and cumulus parameterization schemes of WRF ARW model in simulating severe thunderstorm events over Gadanki MST radar facility – case study, Atmos. Res., 139, 1–17, 2014.

Medina, S., Houze, R. A., Kumar, A., and Niyogi, D.: Summer monsoon convection in the Himalayan region: Terrain and land cover effects, Q. J. Roy. Meteorol. Soc., 136, 593–616, 2010.

Misenis, C. and Zhang, Y.: An examination of sensitivity of WRF/Chem predictions to physical parameterizations, horizontal grid spacing, and nesting options, Atmos. Res., 97, 315–334, 2010.

Mishra, A. and Srinivasan, J.: Did a cloud burst occur in Kedarnath during 16 and 17 June 2013?, Curr. Sci., 105, 1351–1352, 2013.

Mishra, V., Kumar, D., Ganguly, A. R., Sanjay, J., Mujumdar, M., Krishnan, R., and Shah, R. D.: Reliability of regional and global climate models to simulate precipitation extremes over India, J. Geophys. Res.-Atmos., 119, 9301–9323, 2014.

Mlawer, E. J., Taubman, S. J., Brown, P. D., Iacono, M. J., and Clough, S. A.: Radiative transfer for inhomogeneous atmospheres: RRTM, a validated correlated-k model for the longwave, J. Geophys. Res.-Atmos., 102, 16663–16682, 1997.

Mohanty, U., Routray, A., Osuri, K. K., and Prasad, S. K.: A study on simulation of heavy rainfall events over Indian region with ARW-3DVAR modeling system, Pure Appl. Geophys., 169, 381–399, 2012.

Mukhopadhyay, P., Taraphdar, S., Goswami, B., and Krishnakumar, K.: Indian summer monsoon precipitation climatology in a high-resolution regional climate model: Impacts of convective parameterization on systematic biases, Weather Forecast., 25, 369–387, 2010.

NCEP (National Centers for Environmental Prediction): National Weather Service/NOAA/U.S. Department of Commerce, 2000, updated daily, NCEP FNL Operational Model Global Tropospheric Analyses, continuing from July 1999, Research Data Archive at the National Center for Atmospheric Research, Computational and Information Systems Laboratory, available at: https://doi.org/10.5065/D6M043C6, last access: 2 November 2017.

Niyogi, D., Holt, T., Zhong, S., Pyle, P. C., and Basara, J.: Urban and land surface effects on the 30 July 2003 mesoscale convective system event observed in the southern Great Plains, J. Geophys. Res.-Atmos., 111, 1–20, D19107, 2006.

Niyogi, D., Subramanian, S., and Osuri, K. K.: The Role of Land Surface Processes on Tropical Cyclones: Introduction to Land Surface Models, in: Advanced Numerical Modeling and Data Assimilation Techniques for Tropical Cyclone Prediction, Springer, 221–246, 2016.

NOAA: National Oceanic and Atmospheric Administration Changes to the NCEP Meso Eta Analysis and Forecast System: Increase in resolution, new cloud microphysics, modified precipitation assimilation, modified 3DVAR analysis, 2001.

Noilhan, J. and Planton, S.: A simple parameterization of land surface processes for meteorological models, Mon. Weather Rev., 117, 536–549, 1989.

Osuri, K., Nadimpalli, R., Mohanty, U., Chen, F., Rajeevan, M., and Niyogi, D.: Improved prediction of severe thunderstorms over the Indian Monsoon region using high-resolution soil moisture and temperature initialization, Scientific Reports, 7, 1–12, 41377, 2017a.

Osuri, K. K., Nadimpalli, R., Mohanty, U. C., and Niyogi, D.: Prediction of rapid intensification of tropical cyclone Phailin over the Bay of Bengal using the HWRF modelling system, Q. J. Roy. Meteorol. Soc., 143, 678–690, 2017b.

Osuri, K. K., Mohanty, U., Routray, A., Kulkarni, M. A., and Mohapatra, M.: Customization of WRF-ARW model with physical parameterization schemes for the simulation of tropical cyclones over North Indian Ocean, Nat. Hazards, 63, 1337–1359, 2012.

Osuri, K. K., Mohanty, U., Routray, A., and Niyogi, D.: Improved prediction of Bay of Bengal Tropical cyclones through assimilation of doppler weather radar observations, Mon. Weather Rev., 143, 4533–4560, 2015.

Pai, D., Sridhar, L., Rajeevan, M., Sreejith, O., Satbhai, N., and Mukhopadhyay, B.: Development of a new high spatial resolution (0.25 × 0.25) long period (1901–2010) daily gridded rainfall data set over India and its comparison with existing data sets over the region, available at: http://www.imd.gov.in/advertisements/20170320_advt_34.pdf, Mausam, 65, 1–18, 2014.

Pieri, A. B., von Hardenberg, J., Parodi, A., and Provenzale, A.: Sensitivity of precipitation statistics to resolution, microphysics, and convective parameterization: A case study with the high-resolution WRF climate model over Europe, J. Hydrometeorol., 16, 1857–1872, 2015.

Rahman, S., Sengupta, D., and Ravichandran, M.: Variability of Indian summer monsoon rainfall in daily data from gauge and satellite, J. Geophys. Res.-Atmos., 114, 1–14, D17113, 2009.

Rajeevan, M., Kesarkar, A., Thampi, S. B., Rao, T. N., Radhakrishna, B., and Rajasekhar, M.: Sensitivity of WRF cloud microphysics to simulations of a severe thunderstorm event over Southeast India, Ann. Geophys., 28, 603–619, https://doi.org/10.5194/angeo-28-603-2010, 2010.

Rajesh, P., Pattnaik, S., Rai, D., Osuri, K., Mohanty, U., and Tripathy, S.: Role of land state in a high resolution mesoscale model for simulating the Uttarakhand heavy rainfall event over India, J. Earth Syst. Sci., 125, 475–498, 2016.

Raju, P., Potty, J., and Mohanty, U.: Sensitivity of physical parameterizations on prediction of tropical cyclone Nargis over the Bay of Bengal using WRF model, Meteorol. Atmos. Phys., 113, 125–137, 2011.

Rao, Y. R., Hatwar, H., Salah, A. K., and Sudhakar, Y.: An experiment using the high resolution Eta and WRF models to forecast heavy precipitation over India, Pure Appl. Geophys., 164, 1593–1615, 2007.

Ratna, S. B., Sikka, D., Dalvi, M., and Venkata Ratnam, J.: Dynamical simulation of Indian summer monsoon circulation, rainfall and its interannual variability using a high resolution atmospheric general circulation model, Int. J. Climatol., 31, 1927–1942, 2011.

Ratnam, J. V. and Kumar, K. K.: Sensitivity of the simulated monsoons of 1987 and 1988 to convective parameterization schemes in MM5, J. Climate, 18, 2724–2743, 2005.

Rauscher, S. A., Coppola, E., Piani, C., and Giorgi, F.: Resolution effects on regional climate model simulations of seasonal precipitation over Europe, Clim. Dynam., 35, 685–711, 2010.

Ray, K., Bhan, S., and Sunitha Devi, S.: A Meteorological Analysis Of Very Heavy Rainfall Event Over Uttarakhand During 14–17 June 2013, Monsoon 2013 a report, IMD Met. Monograph: Synoptic Meteorology No.: ESSO/IMD/SYNOPTIC MET/01-2014/15, available at: http://www.tropmet.res.in/~kolli/MOL/Monsoon/year2013/Monsoon-2013-NEW.pdf (last access: 7 February 2018), 37–54, 2014.

Routray, A., Mohanty, U., Niyogi, D., Rizvi, S., and Osuri, K. K.: Simulation of heavy rainfall events over Indian monsoon region using WRF-3DVAR data assimilation system, Meteorol. Atmos. Phys., 106, 107–125, 2010.

Routray, A., Mohanty, U., Osuri, K. K., Kar, S., and Niyogi, D.: Impact of satellite radiance data on simulations of Bay of Bengal tropical cyclones using the WRF-3DVAR modeling system, IEEE Trans. Geosci. Remote Sens., 54, 2285–2303, 2016.

Rutledge, S. A. and Hobbs, P. V.: The mesoscale and microscale structure and organization of clouds and precipitation in midlatitude cyclones. XII: A diagnostic modeling study of precipitation development in narrow cold-frontal rainbands, J. Atmos. Sci., 41, 2949–2972, 1984.

Schaake, J. C., Koren, V. I., Duan, Q. Y., Mitchell, K., and Chen, F.: Simple water balance model for estimating runoff at different spatial and temporal scales, J. Geophys. Res.-Atmos., 101, 7461–7475, 1996.

Shekhar, M., Pattanayak, S., Mohanty, U., Paul, S., and Kumar, M. S.: A study on the heavy rainfall event around Kedarnath area (Uttarakhand) on 16 June 2013, J. Earth Syst. Sci., 124, 1531–1544, 2015.

Sikder, S. and Hossain, F.: Assessment of the weather research and forecasting model generalized parameterization schemes for advancement of precipitation forecasting in monsoon-driven river basins, J. Adv. Model. Earth Sy., 8, 1210–1228, 2016.

Sikka, D. and Gadgil, S.: On the maximum cloud zone and the ITCZ over Indian, longitudes during the southwest monsoon, Mon. Weather Rev., 108, 1840–1853, 1980.

Sing, K. S. and Mandal, M.: Sensitivity of Mesoscale Simulation of Aila Cyclone to the Parameterization of Physical Processes Using WRF Model, in: Monitoring and Prediction of Tropical Cyclones in the Indian Ocean and Climate Change, Springer, 300–308, 2014.

Singh, S., Ghosh, S., Sahana, A., Vittal, H., and Karmakar, S.: Do dynamic regional models add value to the global model projections of Indian monsoon?, Clim. Dynam., 48, 1375–1397, 2017.

Skamarock, W. C., Klemp, J. B., Dudhia, J., Gill, D. O., Barker, D. M., Wang, W., and Powers, J. G.: A description of the advanced research WRF version 2, National Center For Atmospheric Research Boulder Co Mesoscale and Microscale Meteorology Div, available at: http://www2.mmm.ucar.edu/wrf/users/downloads.html (last access: 2 June 2017), 2005.

Srinivas, C., Hariprasad, D., Bhaskar Rao, D., Anjaneyulu, Y., Baskaran, R., and Venkatraman, B.: Simulation of the Indian summer monsoon regional climate using advanced research WRF model, Int. J. Climatol., 33, 1195–1210, 2013.

Tao, W.-K., Simpson, J., and McCumber, M.: An ice-water saturation adjustment, Mon. Weather Rev., 117, 231–235, 1989.

Tewari, M., Chen, F., Wang, W., Dudhia, J., LeMone, M., Mitchell, K., Ek, M., Gayno, G., Wegiel, J., and Cuenca, R.: Implementation and verification of the unified NOAH land surface model in the WRF model, 20th conference on weather analysis and forecasting/16th conference on numerical weather prediction, 2004,

Thayyen, R. J., Dimri, A., Kumar, P., and Agnihotri, G.: Study of cloudburst and flash floods around Leh, India, during August 4–6, 2010, Nat. Hazards, 65, 2175–2204, 2013.

Trapp, R. J., Halvorson, B. A., and Diffenbaugh, N. S.: Telescoping, multimodel approaches to evaluate extreme convective weather under future climates, J. Geophys. Res.-Atmos., 112, 1–13, D20109, 2007.

Vaidya, S.: The performance of two convective parameterization schemes in a mesoscale model over the Indian region, Meteorol. Atmos. Phys., 92, 175–190, 2006.

Vaidya, S. and Kulkarni, J.: Simulation of heavy precipitation over Santacruz, Mumbai on 26 July 2005, using mesoscale model, Meteorol. Atmos. Phys., 98, 55–66, 2007.

Vaidya, S. and Singh, S.: Applying the Betts–Miller–Janjic scheme of convection in prediction of the Indian monsoon, Weather Forecast., 15, 349–356, 2000.

Vellore, R. K., Kaplan, M. L., Krishnan, R., Lewis, J. M., Sabade, S., Deshpande, N., Singh, B. B., Madhura, R., and Rao, M. R.: Monsoon-extratropical circulation interactions in Himalayan extreme rainfall, Clim. Dynam., 46, 3517–3546, 2016.

Webster, P. J., Magana, V. O., Palmer, T., Shukla, J., Tomas, R., Yanai, M., and Yasunari, T.: Monsoons: Processes, predictability, and the prospects for prediction, J. Geophys. Res.-Ocean, 103, 14451–14510, 1998.

Permissions

All chapters in this book were first published in HESS, by Copernicus Publications; hereby published with permission under the Creative Commons Attribution License or equivalent. Every chapter published in this book has been scrutinized by our experts. Their significance has been extensively debated. The topics covered herein carry significant findings which will fuel the growth of the discipline. They may even be implemented as practical applications or may be referred to as a beginning point for another development.

The contributors of this book come from diverse backgrounds, making this book a truly international effort. This book will bring forth new frontiers with its revolutionizing research information and detailed analysis of the nascent developments around the world.

We would like to thank all the contributing authors for lending their expertise to make the book truly unique. They have played a crucial role in the development of this book. Without their invaluable contributions this book wouldn't have been possible. They have made vital efforts to compile up to date information on the varied aspects of this subject to make this book a valuable addition to the collection of many professionals and students.

This book was conceptualized with the vision of imparting up-to-date information and advanced data in this field. To ensure the same, a matchless editorial board was set up. Every individual on the board went through rigorous rounds of assessment to prove their worth. After which they invested a large part of their time researching and compiling the most relevant data for our readers.

The editorial board has been involved in producing this book since its inception. They have spent rigorous hours researching and exploring the diverse topics which have resulted in the successful publishing of this book. They have passed on their knowledge of decades through this book. To expedite this challenging task, the publisher supported the team at every step. A small team of assistant editors was also appointed to further simplify the editing procedure and attain best results for the readers.

Apart from the editorial board, the designing team has also invested a significant amount of their time in understanding the subject and creating the most relevant covers. They scrutinized every image to scout for the most suitable representation of the subject and create an appropriate cover for the book.

The publishing team has been an ardent support to the editorial, designing and production team. Their endless efforts to recruit the best for this project, has resulted in the accomplishment of this book. They are a veteran in the field of academics and their pool of knowledge is as vast as their experience in printing. Their expertise and guidance has proved useful at every step. Their uncompromising quality standards have made this book an exceptional effort. Their encouragement from time to time has been an inspiration for everyone.

The publisher and the editorial board hope that this book will prove to be a valuable piece of knowledge for researchers, students, practitioners and scholars across the globe.

List of Contributors

Y. Wang
Key Laboratory of Agricultural Water Resources, Hebei Key Laboratory of Agricultural Water-Saving, Center for Agricultural Resources Research, Institute of Genetics and Developmental Biology, Chinese Academy of Sciences, Shijiazhuang, 050021, China
University of Chinese Academy of Sciences, Beijing, 10049, China
Research Center for Hebei Ecological Environmental Construction, Hebei Academy of Social Sciences, Shijiazhuang, 050051, China
Research School of Biology, The Australian National University, Canberra, 0200, Australia

M. L. Roderick
Research School of Biology, The Australian National University, Canberra, 0200, Australia
Research School of Earth Science, The Australian National University, Canberra, 0200, Australia
Australian Research Council Centre of Excellence for Climate System Science, Canberra, Australia

Y. Shen
Key Laboratory of Agricultural Water Resources, Hebei Key Laboratory of Agricultural Water-Saving, Center for Agricultural Resources Research, Institute of Genetics and Developmental Biology, Chinese Academy of Sciences, Shijiazhuang, 050021, China

F. Sun
Research School of Biology, The Australian National University, Canberra, 0200, Australia
Australian Research Council Centre of Excellence for Climate System Science, Canberra, Australia

Mathieu Vrac
Laboratoire des Sciences du Climat et de l'Environnement (LSCE-IPSL, CNRS), Centre d'Etudes de Saclay, Orme des Merisiers, 91190 Gif-sur-Yvette, France

Xingguo Mo, Xuejuan Chen and Suxia Liu
Key Laboratory of Water Cycle & Related Land Surface Processes, Institute of Geographic Sciences and Natural Resources Research, Chinese Academy of Sciences, Beijing 100101, China
College of Resources and Environment, University of Chinese Academy of Sciences, Beijing, 100190, China

Shi Hu and Jun Xia
Key Laboratory of Water Cycle & Related Land Surface Processes, Institute of Geographic Sciences and Natural Resources Research, Chinese Academy of Sciences, Beijing 100101, China

Kristian Förster
Leibniz Universität Hannover, Institute of Hydrology and Water Resources Management, Hanover, Germany
Centre for Climate Change Adaptation, Innsbruck, Austria
Institute of Geography, University of Innsbruck, Innsbruck, Austria

Florian Hanzer
Institute of Geography, University of Innsbruck, Innsbruck, Austria
Wegener Center for Climate and Global Change, University of Graz, Graz, Austria

Elena Stoll and Matthias Huttenlau
Centre for Climate Change Adaptation, Innsbruck, Austria

Ulrich Strasser
Institute of Geography, University of Innsbruck, Innsbruck, Austria

Adam A. Scaife
Met Office Hadley Centre, Exeter, Devon, UK
College of Engineering, Mathematics and Physical Sciences, University of Exeter, Exeter, UK

Craig MacLachlan
Met Office Hadley Centre, Exeter, Devon, UK

Johannes Schöber
TIWAG, Tiroler Wasserkraft AG, Innsbruck, Austria

Stefan Achleitner
Unit of Hydraulic Engineering, Institute of Infrastructure, University of Innsbruck, Innsbruck, Austria

Jonas Olsson
Research and Development (hydrology), Swedish Meteorological and Hydrological Institute, 601 76 Norrköping, Sweden

Cintia Bertacchi Uvo
Department of Water Resources Engineering, Lund University, 221 00 Lund, Sweden

Kean Foster
Research and Development (hydrology), Swedish Meteorological and Hydrological Institute, 601 76 Norrköping, Sweden
Department of Water Resources Engineering, Lund University, 221 00 Lund, Sweden

D. Michel, M. Hirschi and S. I. Seneviratne
Institute for Atmospheric and Climate Science, ETH Zürich, Zürich, Switzerland

C. Jiménez
Estellus, Paris, France
LERMA, Paris Observatory, Paris, France

D. G. Miralles
Department of Earth Sciences, VU University Amsterdam, Amsterdam, the Netherlands
Laboratory of Hydrology and Water Management, Ghent University, Ghent, Belgium

B. Martens
Laboratory of Hydrology and Water Management, Ghent University, Ghent, Belgium

M. Jung
Max Planck Institute for Biogeochemistry, Jena, Germany

A. Ershadi and M. F. McCabe
Division of Biological and Environmental Sciences and Engineering, King Abdullah University of Science and Technology, Thuwal, Saudi Arabia

J. B. Fisher
Jet Propulsion Laboratory, California Institute of Technology, Pasadena, California, USA

Q. Mu
Department of Ecosystem and Conservation Sciences, University of Montana, Missoula, Montana, USA

E. F. Wood
Department of Civil and Environmental Engineering, Princeton University, Princeton, New Jersey, USA

D. Fernández-Prieto
ESRIN, European Space Agency, Frascati, Italy

Maoya Bassiouni, Chad W. Higgins and Stephen P. Good
Department of Biological and Ecological Engineering, Oregon State University, Corvallis, OR97331, USA

Christopher J. Still
College of Forestry, Oregon State University, Corvallis, OR97331, USA

Cassandra Normandin, Bertrand Lubac, Vincent Marieu, Arthur Robinet and Léa Guiastrennec-Faugas
EPOC, UMR 5805, Université de Bordeaux, Allée Geoffroy Saint-Hilaire, 33615 Pessac, France

Frédéric Frappart
GET-GRGS, UMR 5563, CNRS/IRD/UPS, Observatoire Midi-Pyrénées, 31400 Toulouse, France
LEGOS-GRGS, UMR 5566, CNRS/IRD/UPS, Observatoire Midi-Pyrénées, 31400 Toulouse, France

Fabien Blarel
LEGOS-GRGS, UMR 5566, CNRS/IRD/UPS, Observatoire Midi-Pyrénées, 31400 Toulouse, France

Simon Bélanger
Dép. Biologie, Chimie et Géographie, groupe BOREAS and Québec-Océan, Université du Québec à Rimouski, 300 allée des ursulines, Rimouski, Qc, G5L 3A1, Canada

Zongxue Xu
College of Water Sciences, Beijing Normal University, Beijing, 100875, China
Beijing Key Laboratory of Urban Hydrological Cycle and Sponge City, Beijing, 100875, China

Yiheng Chen and Dawei Han
Department of Civil Engineering, University of Bristol, Bristol, BS8 1TR, UK

Qi Chu
College of Water Sciences, Beijing Normal University, Beijing, 100875, China
Beijing Key Laboratory of Urban Hydrological Cycle and Sponge City, Beijing, 100875, China
Department of Civil Engineering, University of Bristol, Bristol, BS8 1TR, UK

Sibo Zhang
CNRM – UMR3589 (Météo-France, CNRS), Toulouse, France
Fondation STAE, Toulouse, France

Jean-Christophe Calvet
CNRM – UMR3589 (Météo-France, CNRS), Toulouse, France

Nicolas Roussel, Minh Cuong Ha and José Darrozes
GET – UMR5563, CNRS/Université Paul Sabatier, UR254 IRD, Toulouse, France

Karen Boniface
Fondation STAE, Toulouse, France
GET - UMR5563, CNRS/Université Paul Sabatier, UR254 IRD, Toulouse, France
CESBIO, Université de Toulouse, CNES/CNRS/IRD/UPS, Toulouse, France
Joint Research Centre/European Commission, Ispra, Italy

Frédéric Frappart
GET - UMR5563, CNRS/Université Paul Sabatier, UR254 IRD, Toulouse, France
LEGOS - UMR566 (CNES, CNRS, IRD, UPS), Toulouse, France

Frédéric Baup
CESBIO, Université de Toulouse, CNES/CNRS/IRD/UPS, Toulouse, France

P. Vahmani
University of California, Los Angeles, CA, USA

T. S. Hogue
University of California, Los Angeles, CA, USA
Colorado School of Mines, Golden, CO, USA

Sojung Park
Department of Climate and Energy Systems Engineering, Ewha Womans University, Seoul, Republic of Korea
Severe Storm Research Center, Ewha Womans University, Seoul, Republic of Korea
Center for Climate/Environment Change Prediction Research, Ewha Womans University, Seoul, Republic of Korea

Jeung Whan Lee and Yunho Park
Korea Meteorological Administration, Seoul, Republic of Korea

Seon Ki Park
Department of Climate and Energy Systems Engineering, Ewha Womans University, Seoul, Republic of Korea
Department of Environmental Science and Engineering, Ewha Womans University, Seoul, Republic of Korea
Severe Storm Research Center, Ewha Womans University, Seoul, Republic of Korea
Center for Climate/Environment Change Prediction Research, Ewha Womans University, Seoul, Republic of Korea

Ila Chawla
Department of Civil Engineering, Indian Institute of Science, Bangalore, 560012, India
Department of Earth, Atmospheric and Planetary Sciences, Purdue University, West Lafayette, IN 47907, USA

Krishna K. Osuri
Department of Earth and Atmospheric Sciences, NIT Rourkela, Odisha, 769008, India
Department of Agronomy- Crops, Soils, Water Sciences, Purdue University, West Lafayette, IN 47907, USA

Pradeep P. Mujumdar
Department of Civil Engineering, Indian Institute of Science, Bangalore, 560012, India
Divecha Centre for Climate Change, Indian Institute of Science, Bangalore, 560012, India

Dev Niyogi
Department of Agronomy- Crops, Soils, Water Sciences, Purdue University, West Lafayette, IN 47907, USA
Department of Earth, Atmospheric and Planetary Sciences, Purdue University, West Lafayette, IN 47907, USA

Index

A

Absorbed Dry Season Dynamics, 103, 111

Agro-ecosystems, 33

Albedo, 44, 63, 87, 101, 172-173, 175, 177-184, 186-187

Alpine Water Balance and Runoff Estimation Model, 49, 52

Anthropogenic, 2, 30-31, 46, 49-50, 84, 100, 118, 176, 178, 182, 186

Anthropogenic Greenhouse Gas Emissions, 49-50

Atmospheric Modeling Systems, 173

B

Bayesian Inference Framework, 103

Biogeochemical Cycles, 1, 118

C

California Irrigation Management Information System, 172-173, 187

Carbon Cycles, 33-34

Climate Models, 9, 27-28, 31, 49, 52-53, 57, 59, 81, 101, 185-186, 224-225

D

Domain Configurations, 137

E

Earth Observation Satellites, 154

Ecohydrological Parameters, 103-105, 108, 110-115

Ecohydrological Processes, 103-105, 115

Emergency Events Database, 137

Evapotranspiration, 33-34, 41, 46-48, 53, 83-84, 92, 95, 97, 99-102, 104, 107-109, 111, 114-117, 172-173, 183, 186, 217

Evapotranspiration Algorithms, 83

G

Global Energy and Water Cycle Exchanges Project, 49

Global Hydrologic Cycles, 103

Global Land Evaporation Amsterdam Model, 83-84

Global Navigation Satellite System, 154

Green Vegetation Fraction, 172-174, 181, 185

Gross Primary Productivity, 33, 46

H

Haloxylon Ammodendron, 2

Heihe River Basin, 1-3, 8-11

Hydrologic Response Times, 137, 150

Hydrological Model Storage, 49, 60

Hydrological Seasonal Forecast System, 65-66, 80

Hyper-arid Region, 1-2, 5

I

Impervious Surface Area, 172-173, 175, 178, 185

Individual Modelling Chains, 66-67

Inn Headwaters, 49, 51, 53

L

Land Surface Modeling, 172

Land Surface Temperature, 85, 87, 102, 172-173, 176, 183

Leaf Area Index, 9, 35-37, 41, 46, 86, 168, 172-173, 175

M

Mackenzie Delta, 118-123, 125, 127-136

Mesoscale Circulation Systems, 137

Moderate Resolution Imaging Spectroradiometer, 1, 4, 83-84, 101-102, 173, 211

Multi-model Ensemble, 66, 69, 71, 74, 77

Multi-model System, 66

Multivariate Bias Adjustment, 12, 28

N

Normalized Difference Vegetation Index, 1, 119, 174

Numerical Weather Prediction, 50, 138, 153, 185, 188, 226

O

Ocean Surface Topography, 119

P

Planetary Boundary Layer, 84, 94, 141, 173, 207, 211, 220, 224-225

Plate Boundary Observatory, 155, 170

Populus Euphratica, 2

Priestley-taylor Jet Propulsion Laboratory Model, 83-84

Probability Density Functions, 103, 106

R

Rank Resampling, 12, 14-15, 25, 28

Regional Climate Model, 12, 25, 30-31, 82, 152, 225-226

Remote Sensing Parameters, 172, 177

Reservoir Management, 65

River Discharge, 2-3, 118, 130, 132-133, 135

S

Satellite Multi-mission Data, 118

Satellite Water Vapor Data, 188, 201

Satellite-observed Vegetation, 1

Signal to Noise Ratio, 154

Snow Water Equivalent, 49-51, 62-63, 89, 100

Soil Moisture Active Passive, 155

Soil Moisture and Ocean Salinity, 155, 170
Soil Salinization Dynamics, 104
Spatial Correlation Coefficient, 137, 144-145
Spatiotemporal Scales, 33
Spring Flood Volume, 65-66, 69
Sub-daily Extreme Rainfall, 137
Surface Energy Balance System, 83-84, 100-101
Surface Water Volume, 118, 123, 130-131, 133

T

Terrestrial Hydrological, 33
Terrestrial Vegetation, 1, 10
Tower-scale Evaluation, 83
Tropical Rainfall Measuring Mission Multi-satellite Precipitation Analysis, 206, 220

U

Urban Canopy Model, 172, 186-187

W

Warm-season Precipitation, 188-189, 193, 195, 201-202
Water Cycle Multi-mission Observation Strategy, 83-84
Weather Prediction Models, 185, 188
Weather Research and Forecasting Model, 137, 151-152, 207, 224, 226
World Climate Research Programme, 49

www.ingramcontent.com/pod-product-compliance
Lightning Source LLC
LaVergne TN
LVHW060541210926
848485LV00005B/1238